MODERN
MAPS & ATLASES

....

AN OUTLINE GUIDE TO
TWENTIETH CENTURY PRODUCTION

Frontispiece: the Wild A7 in use, demonstrating how the pencil moves on the plotting table; in this photograph, the lines have been inked over, so that they show up more clearly.

MODERN
MAPS AND ATLASES

....

AN OUTLINE GUIDE TO
TWENTIETH CENTURY PRODUCTION

C B MURIEL LOCK
BA PhD ALA FRGS

CLIVE BINGLEY  LONDON

FIRST PUBLISHED 1969 BY CLIVE BINGLEY LTD
16 PEMBRIDGE ROAD LONDON W11
SET IN 10 ON 11 POINT TIMES AND PRINTED BY
THE CENTRAL PRESS (ABERDEEN) LTD
85157 072 0

' CARTOGRAPHY
IS A LANGUAGE—
THE PRECISE AND VERSATILE LANGUAGE
OF GEOGRAPHY. IT IS CONCERNED
WITH CONVERTING ALL KINDS OF FACTS INTO
MAP OR CHART FORM.'

From the brochure prepared for the 'Britain Makes Maps' exhibition of contemporary and new techniques in British map making, held at The Geological Museum, London, on the occasion of the 20th International Geographical Congress, 1964.

CONTENTS

INTRODUCTION

A VAST AND erudite literature has accumulated on the subject of historical cartography, whereas—certainly not in the English language—little co-ordinated work has been produced on recent and current world, national and thematic mapping; the available information is scattered. Professor Dr Eduard Imhof calculated recently that about thirty thousand newly prepared or revised maps appeared annually, and this figure takes no account of all the original cartographic material issued as part of organised or individual research papers or monographs. It is obvious, therefore, that an outline only of the total production can be offered in a comparatively small work, with representative examples cited to illustrate the variety of publications available. It is hoped that a second edition may be possible in due time, which may also include hydrographic charts and air charts, which are not touched upon in this work except by passing reference; neither are celestial or space charts included. Relevant bibliographical citations are quoted for further reference. Many national current bibliographies include maps, but much of the current output, including the British, is not recorded even in this manner.

The dates of initial scientific national or thematic mapping vary greatly from one country to another; in each case, a period of experimental work has inevitably preceded the establishment of a regular cartographic body or service, in the way, for example, that the mapping of disease was attempted long before any co-ordinated work was considered. In each section of this work it should be obvious from the text how the transition to the ' modern ' period has been interpreted. Some recently produced historical atlases and maps have been included, for the approach to their planning and the cartographical techniques involved are relevant to this study.

The first section attempts to formulate some of the ideas and factual knowledge essential to an effective study of these specialised documents, and the final section brings together some aspects of map librarianship, which has so far been strangely neglected in the published literature.

The text has attempted to emphasise examples of internationally organised research and publications, and it is suggested that the reader should observe throughout the degree to which international co-operation and the interchange of ideas, information and standards

7

have become operative during the period under review, tending towards a completely revised attitude to mapping and the documentation of mapping throughout the world, which will become increasingly apparent. Three reasons can be distinguished for the remaining anomalies. The isolation or backwardness of a country may indeed have resulted in very recent establishment of a central cartographical agency, and the coverage of such a country by aerial photographic surveys, followed by the production of medium-scale and large-scale maps may be in its infancy, but improving all the time; secondly, political problems have disorganised such services in some areas of the world; thirdly, in a few cases, although it is known that much activity is in progress, it may be difficult to learn precise details. For example, it is extremely difficult to obtain copies of the Survey of India sheet maps and, while such fine atlases as the *Morskoi atlas* and the *Fiziko-geografičeskij atlas mira* are freely available in the west, official topographical map sheets of the USSR are not.

It is my pleasure to acknowledge the primary information and instruction received from several organisations specialising in new developments in cartography, notably from the Ordnance Survey, the Cartographic Department of the Clarendon Press, Oxford, the Esselte Corporation, Stockholm, the Georg Westermann Publishing House, Brunswick, the Hunting Organisation, Fairey Surveys, the VEB Carl Zeiss, Jena and the Monotype Corporation Limited.

CBML
November 1968

1

THE TECHNIQUES OF MODERN CARTOGRAPHY

' I daylie see many that delight to look on mappes . . .
but yet for want of skill in Geography, they knowe not
with what manner of lines they are traced, nor what those
lines do signify nor yet the true use of Mappes.' THOMAS
BLUNDEVILLE *A brief description of universal Mappes and
Cardes and their use* (London, 1589).

THE FRAMEWORK OF MODERN CARTOGRAPHY : The definition of cartography, as pronounced by the experts at the meeting of the United Nations Secretariat and representatives in 1949, runs as follows: ' the science of preparing all types of maps and charts, and includes every operation from original surveys to final printing of copies. The types of maps and charts included are 1) topographic maps; 2) geologic maps, hydrographic charts, and aeronautical charts, all of which are prepared upon a topographic map base; and 3) office-compiled maps showing the location, extent and characters of physical, economic and social phenomena.' Topographic mapping was considered to be a social service and therefore a function of government, whether it was actually prepared by governmental agencies or under contract, for all purposes of development and administration. The need for accurate map coverage is the same in all parts of the world, but the kinds of maps needed vary with the terrain and stage of development. Stress is now laid on the essential advantages of surveying and mapping developing areas before any projects are planned.

The Council of the British Cartographic Society agreed the following definition on 24 April 1964 : ' Cartography is the art, science and technology of making maps, together with their study as scientific documents and works of art. In this context maps may be regarded as including all types of maps, plans, charts and sections, three-dimensional models and globes, representing the earth or any heavenly body at any scale. In particular, cartography is concerned with all stages of evaluation, compilation, design and draughting required to produce a new or revised map document from all forms

1*

of basic data; It also includes all stages in the reproduction of maps. It encompasses the study of maps, their historical evolution, methods of cartographic presentation and map use.' Compare also Professor Imhof's preface to the second *International yearbook of cartography* —' Maps are inexhaustible mines of information for geographers and scientists, indispensable elements of regional and country planning, comparisons and guides for all tourists and globe-trotters. More than ever we can nowadays state the truth of Napoleon's words: " When preparing maps, one should only produce good ones " '.

The American Congress on Survey and Mapping, on the other hand, in 1952 (*Survey and mapping*, XII, 1, 1952), decided on a definition restricted both at the beginning and the end of the entire process—'. . . the compilation of essentially new maps in general through final drafting and photo-processing stages preparatory to reproduction ', omitting the specialised knowledge of surveyor and map printer.

During the past twenty years or so, a number of national and international glossaries have been undertaken. Since 1943, the Geographical Section, General Staff, of the War Office, has produced a series of *Short glossaries for use on foreign maps* in thirty volumes; these include all terms commonly found on maps, in thirty languages. In 1944, Alain Bargilliat compiled a *Vocabulaire pratique Anglais-Français et Français-Anglais des termes concernant la cartographie,* for the Institut Géographique National; this includes terms used in geodesy, topography and all processes involved in the drawing and printing of maps. The *Lexique Anglais-Français des termes appartenent aux techniques en usage à l'Institut Géographique National* was published in three volumes, 1956-58. Two practical works appeared in Washington in 1948: *Definitions of terms used in geodetic and other surveys* was published by the United States Coast and Geodetic Survey (Special publication no 242) and *Glossary of cartographic* terms was issued by the United States Army Map Service. The United States Geological Survey, Topographical Division, compiled a *Glossary of names for topographic forms,* in 1954, and in the same year, the American Society of Civil Engineers, New York, published *Definitions of surveying, mapping and related terms.* A very careful work, *English-Russian dictionary on cartography, geodesy and aerial phototopography,* compiled by G L Galperin and edited by G V Gospodinov, was produced by the State Publishing Office for Physical and Mathematical Literature, Moscow, 1958; and a *Multilingual dictionary,* compiled by a Special Committee of the International Federation of Surveyors, was published by the Federation in 1963. More than 5,500 entries, including definitions, are in French, English/American and German.

The discussions involved in all these projects revealed a good deal of discrepancy in the meaning attached to terms, and much consultation is still required before agreement is reached. Agreement on terminology is of central importance, not only for the complete understanding of research and textbooks published in different countries, but also looking ahead to co-ordinated indexing projects, programming for computer use, etc. The Group of Experts on Geographical Names was appointed by the Secretary-General of the United Nations, following a resolution of the Economic and Social Council at the end of June 1960; the *Report* of the Group (UN Economic and Social Council, New York, 1961) stressed the need for standardisation of geographical names. The problems involved were briefly set out in twenty two sections intended for the guidance of national names authorities. Particularly stressed were the publication of gazetteers of standardised names after due research, and close liaison between national cartographical agencies and national names authorities. Considerable progress has already been made in international agreement and in the United Nations cartographical bulletin, *World cartography,* Volume VI there is a bibliography of gazetteers.

In 1964, therefore, the International Cartographic Association appointed a Commission to establish some degree of standardisation in the use of terms by map-makers and users throughout the world; the first working meeting of the ICA Commission on the Definition, Classification and Standardisation of Cartographic Terms was held in September 1965, at which it was agreed that the principal task should be the compilation of cartographic terms published in a *Multilingual dictionary of cartographic terms,* contributed by the National Committees. The chief contributors are the Cartography Sub-Committee for Geography, the Comité de Cartographie, the Deutsche Gesellschaft für Kartographie, the Hungarian Geodetic and Cartographic Society, the Japan Cartographic Association, the Seminario de Estudios Cartograficos and the National Council of the Cartographers of the USSR. The United Kingdom contribution was published separately in 1966, with the title *Glossary of cartographic terms.* Four hundred terms are included; three appendices relate to 'Map projections', 'Isograms' and 'Paper sizes' and the work is introduced by W D C Wiggins, Director of the Overseas Surveys. The standard form of entry agreed by the Special Commission includes full definition, with explanatory notes in English, French, German and Russian, possibly in Spanish also. Single word equivalents will be listed in a number of other languages, but not defined. All such glossaries must be frequently revised in this age of constant development and change. The Deutsche Forschungsgemeinschaft and the Federal Government have granted facilities as an exchange centre for evaluation of the lists and the aim is to complete the whole work in three or four years.

The currently accepted definition for an atlas, given in the *Glossary of technical terms in cartography* is 'A collection of maps designed to be kept (bound or loose) in a volume'. In a paper presented at the ICA Symposium at Edinburgh in 1964, O M Miller, Secretary of the American Geographical Society, discussing the theme 'Concepts and purposes of a serial atlas', likened such an atlas to a journal appearing at regular or irregular intervals. He was referring in particular to the *Antarctic map folio series* and the *Serial atlas of the marine environment*. The loose-leaf format is the new element in atlas production, of which several examples will be mentioned later; the analogy with the journal form can be carried too far in that the individual sheets are expected to be used in conjunction with one another. Production in this form allows, in the first place, for the research first completed to be made available and, secondly, revision of individual sheets is made possible.

The Cartographic Office of the United Nations has as its aims the sponsorship of maps, charts and other geographic aids and information, assistance in the co-ordination of the activities of the United Nations and the specialised agencies in the field of cartography, interpretation of cartography in all aspects and the encouragement of projects and conferences. Promotion of continuous effort and uniformly high standards are the main ideals. United Nations meetings on surveying and mapping have been of central importance, especially the 'Third United Nations Cartographic Conference for Asia and the Far East', held in 1961 at Bangkok. 120 participants from twenty five countries took part, including representatives from FAO, ICAO and UNESCO, the European Organisation for Experimental Photogrammetry Research, the International Geographical Union, the International Hydrographic Bureau and the International Union of Geodesy and Geophysics. Technical Committees were established on Geodesy; Topography and photogrammetry, Photo-interpretation and topical maps; the IMW World Aeronautical Charts, and Geographical Names; Magnetic surveys and world magnetic charts, hydrography and oceanography. The establishment of a regional intergovernmental centre for surveying and mapping was discussed, also new techniques and developments, electronic distance measuring instruments, the materials of scribing, use of computers and the co-ordination of aerial photographic surveys with the production and maintenance of nautical charts. The

12

Proceedings were issued in two volumes, the 'Report of the Conference', by G R C Rimington, and 'Proceedings of the conference and technical papers' (Canberra: Department of National Development, 1962).

These conferences continue every two or three years. The fifth conference was outstanding in its results, held in March 1967 at The Academy of Science in Canberra. 149 delegates represented thirty one countries. Cartographic progress throughout the world was reported and discussed, five technical committees were established, on geodesy and control surveys, on topographical cartography, on topical cartography, on general cartography and on hydrography and oceanography, the latter significantly reflecting the growing economic interest in the oceans and the ocean floors. Another obvious trend was the stress on thematic and rapid topical mapmaking, shown by such resolutions as the following: the use of airborne electronic measuring systems, such as Aerodist, for establishing control surveys in difficult country; the use of satellite geodesy for establishing or improving international geodetic connections; the repetition of accurate horizontal and vertical surveys in seismic areas to determine the topographical effect of earthquakes; and the use of colour and infra-red photography in the preparation of topical maps.

The International Union of Geodesy and Geophysics and its predecessors have undertaken important projects in international latitude services, of a bureau for isostatic reductions and the adoption of standards of accuracy for geodetic operations. An international bibliography on geodesy has been published, and *Chronique* (1957-), which superseded the original *Bulletin d'information de l'UGGI*, is issued monthly in English, German, French and Russian, with bibliographical supplements on geodesy, hydrology and oceanography from 1958.

The International Cartographic Association had its origin in a meeting of cartographers from both national and commercial mapping organisations near Stockholm in 1956, under the sponsorship of the Swedish Esselte Corporation. A small committee of representatives from France, Germany, Great Britain, Sweden, Switzerland and the United States was formed to look into the potentialities of an internationally founded body and, in 1958, a second international conference was held, with hospitality from the house of Rand McNally and Company, Chicago. At this conference, national delegates were advised to see what could be done in their own countries to co-ordinate cartographic work, and from this time on many new national cartographic organisations were formed. In Britain, the Sub-Committee for Cartography of the National Committee for Geography, under the auspices of the Royal Society, meets at quarterly intervals. These meetings, together with the discussions at four well-attended symposia, have done much to co-ordinate cartographic ideas. A two-day cartographic symposium organised by the Ordnance Survey at Southampton in November 1961 gave an opportunity for contact between government representatives, commercial mapmakers and academic geographers. Topics such as the representation of relief, the potentialities in the use of plastic materials and the whole concept of 3-D maps were discussed and papers were given on the relative methods of drawing and scribing, on the preparation of six-inch maps for mountain and moorland areas, particularly in Scotland and Ireland and on production methods for small-scale maps. The use of machines in the various processes of mapmaking was naturally very much to the fore throughout the discussions. The symposium was held under the aegis of the Sub-Committee of the Royal Society and the *Report,* to which an explanatory preface was added by M A Shaw, of the Cartographic Department of the Clarendon Press, was published by the OUP, 1964. Then there were the two symposia held in 1962 and 1963 by the Universities of Glasgow and Edinburgh and of Leicester respectively, which led directly to the formation of the British Cartographic Society in 1963 to promote the development of cartography. The fourth symposium was held at Oxford in 1963 by the Cartographic Department of the Clarendon Press on the subject ' Experimental cartography '. The first object of the discussions was to look closely at the use of maps as research tools, in the mapping of data in industry, geology, demographic distributions, climate, transport, vegetation, flora and fauna, history and archaeology, hydrography and oceanography. The second object was to consider whether the cartographic analysis of different subjects has anything to contribute to topographic mapping. The discussions at the Ordnance Survey symposium proved so valuable that two more such meetings were arranged; at the third, called the Map Users' Conference, held in London in 1964, it was decided to continue regular November meetings. Discussion of the use of machines in the various processes of mapmaking has been increasingly in evidence.

Meanwhile, later in 1958, the German Cartographic Society organised a conference at Mainz, at which the name ' International Cartographic Association ' was adopted and an Executive Committee appointed. At the International Geographical Congress at Stockholm, 1960, a Special Commission on Cartography was set up to discuss the affiliation of the ICA to the International Geographical Union. The first General Assembly of the Association was held in Paris in 1961, the second in London in 1964, arranged to coincide with the Twentieth International Geographical Congress, with the hope of facilitating liaison between the two organisations. The objects of the ICA are the advancement of the study of cartography and its problems, the encouragement and co-ordination of cartographic research, the exchange of ideas and documents, the furtherance of training in cartography and the dissemination of cartographic knowledge. *The cartographic journal*, December 1964, includes an outline of the progress and objects of the ICA, and the *International yearbook of cartography*, 1965, contains a selection of papers and discussions from the 1964 technical conference. Member nations of the ICA have each appointed a national committee to represent them at ICA meetings.

The First General Assembly of the ICA, held in Paris in May 1961, was organised by the Comité Française de Techniques Cartographiques; twenty five member nations were represented and Professor Dr Carl Troll represented the International Geographical Union. A *Bulletin* began in 1962. The ICA holds three types of meetings: the general assembly every four years, technical conferences held independently, and a technical conference held in conjunction with the International Geographical Congress. Special Commissions are set up as required, of which details are given in *Orbis geographicus* and in the issues of the *Newsletter* of the ICA. The Third General Assembly of the ICA was particularly important, held in New Delhi in 1968, co-ordinating with the Twenty-first International Geographical Congress. At this congress, a combined session of the IGU and the ICA was held to demonstrate the affiliation of the two organisations and to encourage the collaboration between geographers and cartographers. The main themes of the technical conference highlight the questions currently under debate among both geographers and cartographers, namely, the education and training of cartographers, the standardisation of technical terms, automation in cartography, the collection and recording of information for map revision, maps of the future, considering any new styles and types required, the problems of map production in small quantities and frequent editions, maps libraries and their problems and the mapping of developing countries. The programme of work set in progress by the United Nations and the International Cartographic Association has done much to stimulate the development of separate national cartographic bodies.

The British Cartographic Society, formally founded in 1963, was the first individual organisation in Britain to be concerned entirely with cartography. Annual meetings and separate symposia, a library housed in the Edinburgh Public Library and *The cartographic journal*, issued twice a year since 1964 have been among the society's achievements so far. The journal has quickly won world-wide acclaim and has in this short time established itself as a central source of information on technical developments; the inclusion of map supplements has been a recent feature. A particularly useful symposium of the society was held at the University College of Swansea, in September 1965, of which some of the papers were printed in the June 1969 issue of the *Journal,* on cartographical research carried out in the University of Swansea, on map libraries and on cartographic training. In 1965-66, the Map Curators' Group was formed. The Natural Environmental Research Council has been set up under the auspices of the Royal Society to establish a National Research Centre for Cartographic Sciences.

The American Congress on Surveying and Mapping has been held since 1941, having its headquarters in Washington DC; sub-divisions of the Congress are devoted to Cartography, Control surveys, Property surveys, Topography, Instruments, and Education. Among the other organisations which have made outstanding contributions to the subject, and which will be constantly mentioned below, are the International Association of Geodesy, the International Society for Photogrammetry, the Centre de Documentation Cartographique et Géographique, the Institute of Geodesy, Photogrammetry and Cartography of Ohio State University and the Ontario Institute of Chartered Cartographers. The research of such organisations as the Cartographic Department of the Clarendon Press, Oxford; of the Hunting Surveys and Consultants Limited; Fairey Surveys Limited; the Generalstabens Litografiska Anstalt, Esselte AB, Stockholm; the Bertelsmann Cartographic Institute, Gütersloh and the International Society for Photogrammetry has made outstanding contributions, also that of the specialist instrument and equipment makers, without whose constant innovations and improvements the fine existing work could not be achieved.

The Commission on Interpretation of Aerial Photographs was established at the Geographical Congress in London, 1964, continuing the work of the previous Sub-Commission on Air Photo Interpretation of the Special Commission on Cartography, set up in 1961. Regional reports have been circulated in mimeographed form and some have been published in *Photogrammetria.* Miscellaneous information on congress, conferences, research projects, practical surveys etc has been disseminated by the commission in general circulars, February 1965, November 1965, January 1967 and February 1968. The commission has first concentrated on aerial photography as a means of collecting analytical data; later, the scope of the commission widened to include the study of aerial photography as a unique source of information for integrated and

16

synthetical geographical studies. Separate working parties have been investigating the photo-interpretation of glacial geomorphological features and of rural land use. Especially on the latter subject, a vast amount of information collected from many countries has been processed and compiled into an index, ' Geographical interpretation of aerial photographs', which comprises two parts, a list of research or practical mapping projects with details concerning type of photographs used, methods employed, maps produced etc, and secondly, a bibliography printed out from punched cards, published as part one of a new serial publication of the Bundesanstalt für Landeskunde und Raumsforschung, Bad Godesberg. A number of regional reports on photo-interpretation and mapping of rural land use have been published as *Collection of papers on photo interpretation and mapping of rural land use*: papers written for the Commission on Interpretation of Aerial Photographs, reprinted from *Photogrammetria*. An *Air photo atlas of rural land use* has been begun, with the aim of bringing together annotated aerial and, possibly, space photographs showing various types of rural land uses in regions of the world, at suitable scales.

A conference at Toulouse in 1964 on the subject 'Aerial surveys and integrated studies', organised jointly by UNESCO, the Centre National de la Recherche Scientifique and the University of Toulouse, was attended by 190 scientists from forty-five countries. Papers covered the uses of aerial photography in the study of geology, geography, vegetation, hydrology, ecology, geomorphology, soils etc, and the application of these methods to the integrated surveys being carried out all over the world, particularly in Australia, Africa and Latin America. Selected papers were published in English and French in the Natural Resources Research Series VI, 1968.

In Delft, two famous departments in adjacent buildings on Kanaalweg, the International Training Centre for Aerial Survey and the Survey Department of the Rijkswaterstaat have produced the major research and publications from the Netherlands in photogrammetry, survey and topographical cartography in postwar years. The latest developments include large-scale plotting automatically by computer. Of central importance is the *International bibliography of photogrammetry,* prepared by the ICT, to which centres of bibliographical work in photogrammetry throughout the world contribute.

On the last occasion of checking, the following organisations co-operated in the bibliography, representing a roll-call of important organisations in this field: Clair A Hill and Associates, Civil Engineers, Redding, California; Yale University, Forestry Library; Air Force Missile Test Center Library, Florida; Photogrammetry Inc, Rockville; MIT, Department of Civil and Sanitary Engineering,

Photogrammetry Laboratories, Lexington, Mass; University of Missouri, Forestry Department; Dartmouth College, Hanover, New Hampshire, Baker Library Division of Acquisitions and Preparations; Eastman Kodak, Research Library, Rochester, New York; Lockwood, Kesslen and Bartlett Inc, New York; Fairchild Camera and Instrument Corporation, Long Island, New York; Syracuse University, Department of Civil Engineering; The Thomas Engineering and Surveying Company, Columbus, Ohio; Ohio State University; Aero Service Corporation, Philadelphia; Stephen F Austin State College, Nacogdoches, Texas; American Society of Photogrammetry, Washington; Coast and Geodetic Survey, Washington; Library of Congress, Exchange and Gift Division; US Geological Survey, Washington; National Bureau of Standards, Washington; the Australian Institute of Cartographers; University of Queensland Library, Brisbane; Lands and Surveys Department, Tasmania; Ontario Department of Mines and Technical Surveys, Ottawa; University of New Brunswick, Surveying Engineering, Fredericton, NB; Department of Forestry Library, Ottawa; National Research Council, Division of Applied Physics, Ottawa; Université Laval Library, Quebec; Survey of India, Dehra Dun; The Royal College, Nairobi, Kenya; Surveyor General, Northern Nigeria Surveying; Ministry of Land and Survey, Kaduna; Survey General, Salisbury, Rhodesia; Director of Conservation and Extension and Department of Trigonometrical and Topographical Surveys, Salisbury, Rhodesia; University of Cape Town, Department of Land Surveying, Rondebosch, Cape Province; University of Natal, Department of Land Surveying, Durban; Ferdinand Postma Biblioteek, Potchefstroom University, Potchefstroom; Australian Institute of Cartographers; the Library of the University of Queensland; the Lands and Surveys Department of Tasmania; the Royal Institute of Chartered Surveyors; the Ordnance Survey and the Directorate of Overseas Surveys, and several other relevant interested departments in British universities; the Geodetic Library, no 1, Survey Production Centre, Feltham, Middlesex; Road Research Laboratory, Harmondsworth, West Drayton, Middlesex; Department of Surveying and Geodesy, University of Oxford; BR Section, Ordnance Survey; University of Leeds, Department of Civil Engineering; Queen's University, Belfast, David Keir Library; and the Department of Geography, Glasgow University. University College, London, includes a Department of Photogrammetry and Surveying, also the headquarters of The Photogrammetric Society, which publishes the *Photogrammetric record*.

The Premier Congrès de Géographie Aérienne, organised by the Union Syndicale des Industries Aéronautiques, took place at the

18

end of 1938. Emmanuel de Martonne summarised the development of the subject at that time in his *Géographie aérienne* (Paris: Michel, 1948); a summary nearly twenty years later is given in Volume XVI, 1967, of the *Archives internationales de photogrammetrie,* which contains the seventy three papers in English, French and German, presented at the International Symposium on Photo-interpretation at the Sorbonne, 1966, illustrated with photographs, maps, diagrams and bibliographies.

The International Society for Photogrammetry, founded in 1910, has proved an important forum for the discussion of the latest developments in both techniques and theory. There are seven auto-nomous technical commissions and every four years the International Congress of Photogrammetry is organised under the direction of the Committee of the Society. No library is maintained at present. Publications are normally in English, French, German and Italian; *Photogrammetria* is the official journal of the Society and *Archives* are published after each Congress, usually in several volumes, consisting of proceedings, national reports, reports of the technical commissions and invited papers.

GROUND AND AERIAL SURVEYING FOR MAPMAKING

All maps, whether topographic or thematic, are based on a survey of one kind or another. In the case of topographic maps, which depict the earth's surface, the surveyor measures the land with precise instruments to fix the positions of reference points by triangulation and traversing for horizontal control. In difficult terrain, progress in topographical mapping was slow or impossible until the development of aerial photographic techniques, which are still best combined with ground survey if possible. Most areas of the world have now been mapped at small scales or at least have photographic coverage, but improved mapping, revision or special purpose mapping goes on continuously. Marine surveying is a never-ending task, especially in coastal waters, where the siting of navigational aids must be plotted on charts and the checking and amendment of charts are vital, especially when shifting seabeds are concerned. The Thames Estuary Surveys, for example, are regularly checked twice a year, as are other particularly busy shipping lanes and some less stable offshore areas may need much more frequent checking. Much greater attention has been given to the precise mapping of coastal waters, of recent years. The various methods used in portraying coastal features should be noticed, as, for example, on the new maps of British coasts in the *Atlas of Britain and Northern Ireland.* The polder coastal areas are finely portrayed in the *Atlas van Nederland,* and the Russian meticulous mapping

19

of estuaries and deltaic areas is most striking in both the *Atlas mira* and the new *Fisiko-geografičeskij atlas mira.*

The chief problem affecting the plotting of air charts is the ever-increasing speed at which aircraft pass over the ground mapped and charting has had to adapt to a form suitable for projection from a 35mm film on to a display screen operated by push-button controls. As stated in the introduction, details of hydrographic and air charts are not included in this volume.

The purpose of the map to be drawn must first be firmly established; with this aim constantly in mind, the selection and collection of relevant data are the next steps. The distinction is usually made between geodetic cartography, which supplies the basic control data and topographic cartography, which fills in the details required, using the results of more localised research.

The use of satellites for geodetic exploration now develops rapidly through two American organisations, the National Aeronautics and Space Administration (NASA), and the Environmental Science Services Administration (ESSA). The first international symposium on the use of artificial satellites for geodesy was held in Washington in April 1962, the *Proceedings,* edited by G Veis, being issued in 1963. NASA's first explorer spacecraft, GEOS-A, was launched from Cape Kennedy in November 1965, the objective being to compare different techniques of measurement. The next, PAGEOS, satellite aimed to establish a world-wide geodetic satellite network with over forty stations. The sun-illuminated satellite, at an altitude of over 2,500 miles, will serve as a survey beacon for ground stations located approximately 2,500 miles apart. Meanwhile, a more limited series of observations has been conducted with ESSA by the United States Coast and Geodetic Survey, in co-operation with the Canadian, British, Danish, Norwegian and Icelandic Governments. By photographing satellites Echo 1 and Echo 2 simultaneously from two or more positions on the earth's surface, ground distances can be determined more exactly than by any other method. The immediate aim is to ' tie in ' the geodetic networks of the European and North American continents, by using accurate inter-continental measurements.

The origins of photography in mapmaking began in the balloon-photography experiments carried out by Gaspard Felix Tournachon (known as Nadar) in 1856. Four years later, aerial photographs were taken over Boston from a balloon and by 1864 Henri Negretti and James Glaisher took photographs from free balloons in Britain. During the following few years, many experiments were carried out on both sides of the Atlantic, cameras being carried by various types of balloons, rockets and kites. Early in 1909, the first aerial photographs were taken from an aircraft. Before this time, however, aircraft had been used for photo-mapping in many countries, including the United Kingdom, Canada and Switzerland, using the photo-theodolite for ground surveys. Between the wars, much progress was made; for example, the United States Coast and Geodetic Survey had mapped the Mississippi delta from aerial photographs and the Hydrographic Office had surveyed Cuba in 1928, using the same methods. During the second world war, the need for rapid surveys of difficult terrain led to vastly improved techniques and, by the end of the war, the aerial camera was established as a powerful new mapping tool.

Aerial pictures, used in conjunction with stereoplotting machines and co-ordinatographs, are of use not only to cartographers, but to mining geologists, agriculturists, road and railway engineers; changes in vegetation, and therefore of underlying soil and rock, are clearly brought out. A network of ground control points or co-ordinates which can be identified on the photographs must be available before plotting can begin, and some ground survey is necessary to fill in the local details. But hundreds of square miles of inaccessible territory in jungles, swamps, icefields and mountain ranges have been mapped by aerial photography, which would otherwise have remained unknown. Successful results, however, obviously depend very much on suitable weather conditions.

Aerial photographs are either vertical or oblique and they may be taken and used as pinpoints, strips or block coverage. Vertical photographs can be used directly for mapping with exact contours plotted by viewing stereoscopic pairs. Each print must overlap adjacent prints sufficiently to give a three-dimensional view. Hydrographic features stand out particularly well with this method. Oblique photographs show ground features as they would appear from the windows of an aircraft. Needless to say, great skill must be used in interpreting the photographs and improvements in instruments are being made all the time. A device has recently been perfected, for example, by which cloud shadow on a negative can be prevented from obscuring ground detail on a subsequent print.

In areas difficult of access, both vertical and oblique air photography are used; and in the Himalaya, for example, and other high mountain ranges, even these methods are impossible. In such areas, use is made of the terrestrial phototheodolite and, more recently, increasing use has been made of modern distance-measuring equipment using radio or light waves. Ultra-high-frequency radio can be used for sounding the depths of ice sheets to about three thousand metres. A Royal Society symposium was held at

Oxford in September 1965, at which geodesists and physicists from twenty five countries presented thirty seven papers on such subjects as these; and in the same year, electromagnetic distance measurement was demonstrated at a symposium of the International Association of Geodesy. The Stanford Research Institute of California has developed a method for converting radar signals to map symbols, by which weather data can be automatically plotted on a map and this data may also be transmitted over telephone or teletype wires to any distance.

Land capability studies must be carried out by a team of workers, as so many factors are involved. Maps have been prepared as part of routine government survey programmes and for oil or mining companies in planning their prospective work; the geological surveys range from rapid small scale mapping of many thousands of square miles of territory to detailed surveys of complex areas involving specialised studies in petrology, micropalaeontology and applied geology. The new methods are now commonly used in urban mapping for town planning, resettlement, surveys for re-development of railways and roads, and the special requirements needed for irrigation planning, reconnaissance surveys, hydro-electric schemes and mineral exploitation.

Two of the most influential private surveying organisations are the Hunting Group of Companies and Fairey Surveys, which operate throughout the world, using these latest developments in instruments and equipment, photographic and printing facilities. The Hunting Group, founded in 1874, with headquarters in London, comprises many specialised companies. The most international of the group's activities are those of the Hunting Survey Group. In Britain, there are four operating companies under the overall management of Hunting Surveys and Consultants Limited; their work includes mapping, ranging from topographical maps to large-scale plans for civil engineering works, precision topographical and architectural model-making. Hunting Technical Services Group is a group of scientists providing a land-use and agricultural consultancy service. Teams of experts work in all parts of the world, especially in South America, Africa, the Middle East and Asia, including the Lower Indus Project in Pakistan, the largest land rehabilitation programme ever mounted. The firm specialises in applied geological sciences, including airborne, ground and marine geophysical surveys, hydrogeological and engineering geology studies. The Aerofilms Library has specialised in oblique aerial photography since 1919 and the aerial photographic library is believed to be the largest in the world, fully indexed.

Fairey Surveys, founded in India in 1923, also works in all parts of the world; its Research and Engineering Division has planned and patented the Fairey Plotterscope, a viewing aid for stereoplotting machines with separate co-ordinatograph tables. The Plotterscope provides an accurately magnified image of both pencil and plot, enabling the operator to plot control points, spot heights, contours and other detail without leaving his seat to examine the trace. The Survey has had long experience in the field of aerial photography, including a ten-year period of continuous operation as flight trials contractors to the British Ministry of Aviation, and it now covers the whole range of surveying, mapping and photography. Among its most interesting projects has been the topographical survey for the Kariba Dam on the Zambesi.

Aircraft can now be fitted with a precise radar altimeter, the Airborne Profile Recorder, which was developed within the Hunting Group. It records a continuous profile of the ground along the line of flight and supplies data from which ground heights can be determined, therefore providing a rapid means of covering an area with a dense network of height control. For precision work the Doppler Navigator is used—a radar aid which measures continuously throughout flight the ground speed and distance travelled by the aircraft to an accuracy of 1 : 1 per cent. Hunting Surveys Limited was among the first organisations to carry out aerial surveys in colour.

Photomosaics consist of an assembly of aerial photographs, with the detail matched in order to provide a comprehensive picture of the area, and they are of great value in the first reconnaissance stage of planning a new road or route-finding through poorly mapped country, as the basis for compiling information on land use, selection of sites for industrial development, selection of routes for power or pipe lines, records of tide levels and location of sand and mud banks, preliminary studies in irrigation and drainage projects, etc. Photo-maps are produced from standard aerial photomosaics, using photomechanical meshes to separate the main types of terrain into different colours. The United States Army Service published ' Provisional specifications for the pictomaps ' in 1966.

The ' Preliminary Plot ' method developed by the Directorate of Overseas Surveys, which produces an outline base map mainly for use in plotting subsequent field observations, resulted in an economy of time and has enabled the directorate to complete more than a thousand square miles of 1 : 50,000 scale mapping since its creation in 1946. In some cases, the maps thus produced show little more than the drainage pattern, vegetation types and an index to the aerial photography used in compilation. The photographic cover of the world is considerably greater than the existing coverage by topographical maps, for adverse weather factors are still more than offset by the time taken in traditional ground survey, draughtsmanship and cartographic reproduction.

Pre-eminent in the vast periodical literature on photogrammetry are the *Proceedings* of the various Congresses on Aerial Photography and Photogrammetry: the *Monthly abstract bulletin,* issued by the Eastman Kodak Company, Rochester, New York, is a most useful source. In addition to those periodicals mentioned above, the quarterly *Photogrammetric engineering,* 1934- by the American Society of Photogrammetry, includes articles, statistics, notes and reviews concerning photographic procedures, instruments and air photo-interpretation; from the Brussels Société Belge de Photogrammétrie comes a scholarly *Bulletin.* Mention should also be made of the *Österreichische zeitschrift für vermessungswesen* of the Austrian Society of Photogrammetry at Baden bei Wien, and of the *Revista Brasileira de fotogrametria,* issued by the Sociedade Brasileira de Fotogrametria at Rio de Janeiro. *Allgemeine vermessungs-nachrichten* is the quarterly journal of the German Society of Photogrammetry, Berlin; and a new series of the quarterly *Revue Française de photogrammétrie* began in 1950, issued by Section Láussedat de la Société Française de Photographie et de Cinématographie, Société Française de Photogrammétrie.

The Transactions of the Moscow Institute of Geodetic, Aerial Photographic and Cartographic Engineers and of the Central Scientific Research Institute of Geodesy, Aerial Photography and Cartography are the respective bulletins of Moscow and Leningrad. Papers on techniques and equipment, photogrammetric mapping, photointerpretation techniques, photogeology, on use in forestry and soil surveys, aeromagnetic surveying, etc form the record of the United Nations Seminar on Aerial Survey Methods and Equipment, Bangkok, 1960, published under the title *Proceedings of the United Nations Seminar on Aerial Survey Methods and Equipment. Photo interprétation,* produced in Paris in six issues a year since 1963, constitutes a well-conceived attempt to present a wide range of problems of air-photo interpretation in an accessible and useful format for teaching purposes. Each issue contains eight stereoscopic pairs of photographs showing areas dissimilar in location and character. For each pair, there is a transparent overlay illustrating the principal features shown on the photograph, a short article explaining the interpretation and a punched card classifying the photograph according to its location and principle features of interest, thus providing a growing collection of useful examples for teaching the elements of air photo-interpretation, geographical features and relationships. The Oxford symposium on experimental cartography, mentioned above, led to the publication of *Experimental cartography*: report on the Oxford Symposium October 1963 (OUP, 1964), which contains the papers and discussions; pertinent bibliographical references are also appended.

Organisation Européenne d'études photogrammétriques expérimentales was the official publication of the Frankfurt am Main Institut für Angewandte Geodäsie, 1966. Thirty four maps and diagrams enclosed inside the back cover illustrate the report of the Commission of the Organisation on the preparation of a photogrammetric map at 1 : 100,000 for developing countries, where speed and straightforward methods are important. The report is in French and English, summarising the tests with small scale aerial photographs. 'Air photography and geography' was the theme of another symposium of significance, of which the papers were reported in *Drumlin*, II, 2, 1966, with a sketch-map, diagrams and bibliography. In March 1967, with financial support from UNESCO, a two-day meeting was held in Ottawa in conjunction with the Second Canadian Symposium on Photo Interpretation, organised by the Canadian Institute of Surveying and the Inter-Departmental Committee on Aerial Surveys. A report is being prepared for UNESCO, which will contain the papers presented, the discussions and the relevant literature references.

Most of the great cartographical departments of the world have prepared technical manuals, such as *A guide to the compilation and revision of maps* published by the United States Department of the Army, Washington, 1955; the *Technical manual of the War Department*, Corps of Engineers, Washington and the *Topographic instructions* of the United States Geological Survey, with their supplementary pamphlets and technical memoranda. The Surveys and Mapping Branch, Department of Mines and Technical Surveys, Ottawa, published a *Cartographic manual*, 1954, and a *Surveying manual*, by L C Ripa, 1964, which includes diagrams, a bibliography and glossary. A fourth revised edition, by Major-General G Cheetham, of the *Textbook of topographical surveying*, was published by the Ministry of Defence in 1964 (HMSO, 1965), illustrated by maps, diagrams and a bibliography. Some other relevant references are :

A L Allan *et al*: *Practical field surveying and computations.* (Heinemann, 1968).

American Society of Photogrammetry: *Manual of photographic interpretation* (Washington: The Society, 1960), edited by R N Colwell, with illustrations, bibliographies and glossaries. Development and interpretation in various applications, including geography,

25

geology, agriculture, soils, forestry, hydrology, urban area analysis. Third edition, 1965.

E Arnberger: *Lehrbuch der kartographie* (Vienna: Deuticke, 1965).

T E Avery: *Interpretation of aerial photographs: an introductory college textbook and self-instruction manual* (Minneapolis: Burgess Publishing Company, 1962, repr 1965).

H F von Bandat: *Aerogeology* (Brill, 1962). The practical use of air photography: the methods used in constructing photogeologic maps for use in the field.

A Bannister and S Raymond: *Surveying* (Pitman, second edition 1965).

C E Bardsley and E W Carlton: *Surveyor's field-note forms* (Scranton, third edition 1952).

L Barth: *Bild und karte im erdkundeunterricht* (Berlin: Volk und Wissen, 1963).

Jacqueline Beaujeu-Garnier: *Practical work in geography* (Arnold; New York, St Martin's Press, 1963).

Hallert Bertil: *Photogrammetry* (McGraw-Hill, 1960). Includes many photographs of instruments for air survey.

T W Birch: *Map and photo reading* (Arnold, new edition 1960). Contains samples of ten Ordnance Survey maps on various scales, with matching photographs.

Guy Bomford: *Geodesy* (Oxford: Clarendon Press, second edition 1962).

F Bonasera: *Fondamenti di cartografia* (Palermo, Univ degli Studi di Palermo, 1965).

W Bormann: *Allgemeine kartenkunde* (Lahr, Schwarzwald, Astra; Bailey and Swinfen, 1954).

H Bosse: *Kartengestaltung und Kartenentwurf* (Philadelphia, Chilton, 1962).

R E Bowyer and G A German: *A guide to map projections* (Murray, 1958).

C B Breed and A J Bone: *Surveying* (Wiley, second edition revised 1957).

G C Brock: *Physical aspects of air photography* (Longmans, 1952).

L A Brown: *Mapmaking: the art that became a science* (Boston, Little Brown, 1960).

R Chevallier, *editor: Photographie aérienne: panorama inter-technique* (Paris, Gauthier-Villers, 1965). Twenty two contributions by specialist authors.

J J Cirrito: 'Radar photogrammetry: a cartographic process', (*Professional geographer*, January 1963).

James Clendinning: *The principles and use of surveying instruments* (Blackie, second edition 1959).

James Clendinning: *The principles of surveying* (Blackie, second edition 1960).

C A Close and H S L Winterbotham: *Textbook of topographical and geographical surveying* (HMSO, third edition 1925). Still a standard and an indispensable reference book.

R S Coggins and R K Hefford: *Practical geographer* (Melbourne, Longmans, 1957).

C H Cotter: *The astronomical and mathematical foundations of geography* (Hollis and Carter, 1966).

D R Crone: *Elementary photogrammetry* (Arnold, 1963).

W G Curtin and R F Lane: *Concise practical surveying* (EUP, 1955).

Ryer Daniel: *La photogrammétrie appliquée à la topographie* (Paris, Eyrolles, 1952).

R E Davis: *Elementary plane surveying* (McGraw-Hill, third edition 1955).

R E Davis and F S Foote: *Surveying: theory and practice* (McGraw-Hill, fourth edition 1953).

Frank Debenham: *Map making: surveying for the amateur* (Blackie, third edition 1955).

C H Deetz and O S Adams: *Elements of map projections: with applications to map and chart construction* (Washington DC: US Coast and Geodetic Survey, fifth edition 1945).

G H Dury: *Map interpretation* (Pitman, third edition 1968).

G H Dury and J A Morris: *The land from the air: a photographic geography* (Harrap, third impression 1958).

A G F Elwood: *Essentials of map-reading* (Harrap, fourth edition 1964).

C C Fagg and G E Hutchings: *An introduction to regional surveying* (CUP, 1930).

F Fiala: *Mathematische kartographie* (Berlin, VEB Verlag Technik, 1957).

Richard Finsterwalder: *Photogrammetrie* (Berlin, de Gruyter, 1952).

W Flichmer et al.: *Route-mapping and position locating in unexplored regions* (Basle, Birkhauser, 1957).

Nicola Franchi: *Elementi di cartografia* (Firenze, Istituto Geografico Militare, 1950).

L S Garaevskia: *Kartografiva* (Moscow, Geodezizdat, 1955).

David Greenwood: *Mapping* (University of Chicago Press, revised edition 1964).

E A Gutkind: *Our world from the air: an international survey of man and his environment* (New York, Doubleday, 1952).

J A Gwyer and V G Waldon: *Photo-interpretation techniques: a bibliography* (United States Library of Congress: Technical Information Division, 1956).

Exhaustive, annotated bibliography of literature published 1935-1953.

Bertil Hallert: *Photogrammetry: basic principles and general survey* (McGraw-Hill, 1960).

R H Hammond: *Air survey in economic development* (Muller, 1967). Air survey methods, equipment and examples of air surveys.

C A Hart: *Air survey* (Royal Geographical Society pamphlet).

C A Hart: 'Air survey: the modern aspect' (*The geographical journal*, April 1947).

Viktor Heissler: *Kartographie* (Berlin, de Gruyter, 1962).

A L Higgins: *Elementary surveying* (Longmans, second edition enlarged by L A Beaufoy 1965).

A R Hinks: *Map projections* (CUP, 1921). A standard work to its date.

A R Hinks: *Maps and survey* (CUP, fifth edition 1947).

C B Hitchcock and O M Miller: 'Concepts and purposes of a serial atlas' (ICA Technical Symposium, Edinburgh, 1964).

H L Hitchins and W E May: *From lodestone to gyro-compass* (Hutchinson Scientific and Technical Publications, 1952).

Jean Hurault: *Aerial photography and map-making* (Paris, Institut du Transport Aérien, 1959).

In effect, a summary of the methods employed by the Institut Géographique National, with English translation.

Jean Hurault: *Applications de la photographie aérienne aux recherches de sciences humaines dans les régions tropicales* (Paris, Ecole Pratique des Hautes Etudes, Sorbonne 1963). With folder of photographs.

Jean Hurault: *L'exam stereoscopique des photographies aériennes* (Théorie et pratique) (Paris, Institut Géographique National, two volumes 1960). Also a box each of photographs and of maps.

Jean Hurault: *Manuel de photogrammétrie* (Paris, Institut Géographique National, new edition, two volumes 1956).

Eduard Imhof: *Gelände und karte* (third revised edition 1968).

H W Kaden: *Kartographie: praktischer: leitfaden für kartographen, kartolithographen und landkartenzeichner* (Leipzig, Fachbuchverlag, 1955).

G P Kellaway: *Map projections* (Methuen, second edition 1958). (Methuen's Advanced Geographies series).

28

W K Kilford: *Elementary air survey* (Pitman, 1963). Appendix: List of subscribers to the LTC *International bibliography of photogrammetry*, Delft.

Philip Kissam: *Surveying* (McGraw-Hill, second edition 1956).

Philip Kissam: *Surveying instruments and methods for surveys of limited extent* (McGraw-Hill, second edition 1956).

J J Klawe: 'Photography in the service of cartography', *Graphic arts focal point*, no 6, 1962.

H P Kosack and K H Meine: *Die kartographie, 1943-1954: eine bibliographische übersicht* (Lahr, Schwarzwald, Astra; Bailey and Swinfen, 1956).

David Landen: 'Fotocartas para la planificacion urbana', *Revista cartografía*, XIV, 1965, with English summary.

S Laurila: *Electronic surveying and mapping* (Ohio State UP, 1960).

W T Lee: *The face of the earth as seen from the air: a study in the application of airplane photography to geography* (American Geographical Society of New York, 1922). A pioneer work and still interesting.

D R Leuder: *Aerial photographic interpretation: principles and applications* (McGraw-Hill, 1959).

Sir Clinton Lewis: *The making of a map* (Royal Geographical Society pamphlet).

A Libault: *La cartographie* (Paris, PUF, 1962). ('Que sais-je? series ').

A Libault: *Les mesures sur les cartes et leurs incertitude* (Paris, Editions Géographiques du France, 1961).

J D Lines: 'Spot photography for map revision' (*Cartography*, September 1962).

J W Low: *Plane table mapping* (New York, Harper, 1952).

Karl-Heinz Meine,: 'Aviation cartography', *The cartographic journal*, June 1966.

R E Middleton and O Chadwick: *A treatise on surveying* (Spon, sixth edition, two volumes 1955).

V C Miller: *Photogeology* (McGraw-Hill, 1961).

The application of aerial photography to geology.

G F Morris and F R Flooks: *Background to surveying* (Blackwell 1964).

P G Mott: 'Some aspects of colour aerial photography in practice and its applications' (*The photogrammetric record*, October, 1966).

I I Mueller: *Introduction to satellite geodesy* (Constable, 1964).

W E Powers and C F Kohn: *Aerial photo-interpretation of land-forms and rural cultural features in glaciated and coastal regions* (Northwestern University, Studies in geography no 3, 1959).

N J D Prescott: 'The geodetic satellite—" SECOR " ', *The geographical journal*, March 1966.

H F Rainsford: *Survey adjustments and least squares* (Constable, 1957).

T F Rasmussen: *Kartlaere* (Oslo, Universitetsforlag, 1963; Scandinavian University Books).

W H Rayner: *Advanced surveying* (Van Nostrand, 1950).

W H Rayner and M O Schmidt: *Surveying* (Van Nostrand, 1957).

François Reignier: *Les systèmes de projection et leurs applications à la géographie, à la cartographie, à la navigation, à la topométrie*, etc (Paris, Institut Géographique National, two volumes 1957); text and 48 plates in looseleaf form.

W W Ristow: *Aviation cartography: a historico-bibliographic study of aeronautical charts* (Library of Congress, second edition 1960).

Paul Rossier: *Géographie mathématique* (Paris, Société d'Éditions d'Enseignement Supérieur, 1953).

F Ruellan: *Photogrammétrie et interpretation de photographies stéréoscopiques terrestres et aériennes* (Paris, Masson, 1967).

J K S St Joseph: 'Air photography and archaeology', *The geographical journal*, February 1945.

J K S St Joseph: *The uses of air photography: nature and man in a new perspective* (John Baker for the Cambridge University Committee for Aerial Photography, 1966). Thirteen contributions by experts.

J A Sandover: *Plane surveying* (Arnold, 1961). Contains chapters on photogrammetry by D H Maling.

J A Sandover: *Theodolite practice* (Cleaver-Hume Press, 1960).

K Schwidefsky: *Grundriss der photogrammetrie* (Bielefeld, Verlag für Wissenschaft und Fachbuch GmbH, 1950).

K Schwidefsky: *An outline of photogrammetry,* translated by John Fosberry (Pitman, 1959).

H E Seely: *Air photography and its application to forestry* (Ottawa, Forest Air Survey, Department of Resources and Development, 1949).

H E Seely: *The forestry tri-camera method of air photography* (Ottawa, Forest Air Survey, Department of Resources and Development, 1948).

H O Sharp: *Practical photogrammetry* (New York, Macmillan, 1951).

30

A I Shersen: *Aerial photography* (Jerusalem, Israel Program for Scientific Translations, 1961). Authorised by the Ministry of Higher Education of the USSR as a textbook for aerophotogeodesy.

P T Silley: *Topographical maps and photographic interpretation* (Philip, 1955).

J W B Sisam: *The use of aerial survey in forestry and agriculture* (Oxford, Imperial Forestry Bureau; Aberystwyth, Imperial Bureau of Pastures and Field Crops, 1947).

H T U Smith: *Aerial photographs and their application* (New York, Appleton-Century, 1943). Dated, but still useful, especially for the section on the geographical and geological application of aerial photographs.

' Some recent developments in hill-shading from air photographs in the Directorate of Overseas Surveys', *Survey review,* January 1963.

G A Spear: *Photolithographed forest maps* (Ottawa, Forest Air Survey, Department of Resources and Development, 1951).

S H Spurr: *Photogrammetry and photo-interpretation, with a section on applications to forestry,* second edition of *Aerial photographs in forestry,* 1948 (New York, Ronald Press, 1960). Includes also use of photography in geology, soil and plant life study and analysis of human activities and settlement patterns.

J A Steers: *An introduction to the study of map projections* (ULP, fourteenth edition 1965).

C H Strandberg: *Aerial discovery manual* (Wiley, 1967).

K Thoeme: *Karte und Kompasse: eine praktische Anleitung zum Gebrauch von Karte und Kompass* (Berne, Hallwag, 1965).

P D Thomas: *Conformal projections in geodesy and cartography* (Washington DC, Coast and Geodetic Survey, Special publication no 251, 1952).

W N Thomas: *Surveying* (Arnold, fifth edition 1961).

C Thomson: *Map and compass work* (Arco, 1964).

N J W Thrower: *Original survey and land subdivision* (Chicago, Rand McNally for the Association of American Geographers, 1966).

L G Trorey: *Handbook of aerial mapping and photogrammetry* (CUP, second edition 1952).

United States Army Engineer Corps: *Photogrammetric mapping* (Defense Department, Engineer Manual series, 1963). Looseleaf basic manual on mapping from aerial photographs, well illustrated.

United States Census Bureau, Department of Commerce: *How to read aerial photographs for census work* (1947). Dated, but still useful on this aspect.

G W Usill: *Practical surveying: a textbook for students preparing for examination or for survey work at home and overseas* (Technical Press, fifteenth edition 1960).

Vermessungskunde für kartographen, two volumes: 1, edited by E Lehmann *et al;* 2, edited by E Thum *et al* (Gotha, Haack, 1957-60).

F von Ikier: Kartenkunde: *Handbuch für den gebrauch und die benutzung von karten und luftbildern* (Bonn, Verlags und Vertriebs-Gesellschaft die Reserve, 1964).

Frank Walker: *Geography from the air* (Methuen; New York, Dutton 1953, repr 1956). Air photographs and their interpretation; appendix on photogrammetry.

Paul Werkmeister: *Geodätische instrumente* (Leipzig, Geest and Portig, second edition 1950).

MAP CONSTRUCTION AND DESIGN

Any study of map projections *per se* would be beyond the scope of this work, but a word should be said concerning the effect of choice of projection on map accuracy and design. Choice must depend upon the area to be mapped and the purpose of the map. Distribution maps should be plotted on equal area projections, otherwise relative densities cannot be truly compared. The Bonne Projection, for example, allows freedom from distortion along its central meridian, whereas scale and angular distortion increase rapidly with distance away from the central meridian; it is therefore more suitable for mapping an area with greater north-south extent than east-west. The Albers Projection includes two standard parallels free from angular distortion, so that its greatest usefulness is demonstrated in the mapping of middle latitude areas with greater extent in an east-west direction. Other equal area projections in frequent use include the Lambert Azimuthal Equal-Area Projection, the Sinusoidal, Moll-weide and Eckert. Preservation of the shape of small areas on a map is known as conformality; the angles round any point are correctly represented. Air and sea charts, in particular, require this quality. The most widely used projection for charts has been the Mercator, since its inception in 1569. Most of the Admiralty Charts are drawn on the Mercator, but it has the disadvantage of enlarging areas in the higher latitudes on a map covering a wide area; much mis-conception regarding the size of Greenland, for example, in comparison with the Indian sub-continent, is due to this drawback. Conformality is not so vital on topographic maps; the Ordnance Survey maps are now prepared on the Transverse Mercator, which enables the National Grid reference system to be easily operated.

A third quality in a projection, azimuthality, enables all places to be shown in their true direction from the centre of the map and linear and area scale variations are symmetrical round the central point. The Gnomonic is such a projection, frequently used for navigational charts, because all straight lines drawn from the centre of the map represent parts of great circles, the shortest distances between any two points on the globe. ' Interrupted ' projections have been used intermittently from the sixteenth century. Mathematically constructed, they include the use of multiple axes, according to the area and purpose of the map. The various interrupted projections of Goode were of particular value, not only because of their skilful design, but because they encouraged the acceptance and further development of this useful device. In 1916, he prepared an interrupted version of the sinusoidal, which divided the world into seven segments, splitting Eurasia along the 60 degrees east meridian. His interrupted version of the Mollweide preserved the continuity of Eurasia. In 1923, he produced an interrupted homolosine and an interrupted Werner followed in 1928, under the title of Goode's Polar Equal-Area Projection. These projections appeared at a time when the limitations of the Mercator Projection were becoming increasingly obvious; their use in *Goode's school atlas* and in the widely used wall-map series, familiarised the use of such adjusted interruptions. American map companies quickly adopted other interrupted arrangements for basic world maps. Philip and Sons Limited used the interrupted Mollweide as the basic world projection in the *Chambers of commerce atlas,* 1925, but without inserting meridians; later, the House of Bartholomew devised a ' recentred ' sinusoidal and, in the *Nordisk Världs atlas* produced in Sweden in 1926, the sinusoidal projection in interrupted form was used, the central meridians and interruptions being chosen to make the graticule symmetrical about the equator. The majority of contemporary atlases use some form of un-symmetrical projections for world maps, with the exception of navigational or air route maps, for which the Mercator still best serves the purpose.

Little need be said about scales here, as there are so many excellent books currently including this aspect of mapmaking. Suffice it to draw attention to the fact that in the descriptions of topographical mapping services following later, it will be noticed that four scales occur particularly frequently: the 1:25,000, used by several European surveys, particularly by the German, and recently adopted by the Ordnance Survey; 1:50,000 used by many topographical surveys as a ' standard ' scale for base maps, at which reasonable detail may be shown; 1:62,500, an approximate one-inch to the mile scale, used by the United States Geological Survey and any surveys modelled on their work; and the exact one-inch to the mile scale being used by the Ordnance Survey and by most of the Commonwealth Surveys.

Clear distinction must be made between basic maps directly drawn from a field survey of all features, and derived maps, usually on smaller scales and for different purposes. The trend is now towards transfer to the metric system by those mapping agencies not already using it. Frequent experiments have been made in modern mapping to familiarise users with world maps ' in the round ', to emphasise the actual relationships of countries with one another. These concern particularly the polar regions; the *Readers' Digest world atlas* and R E Harrison's *Fortune atlas,* 'Look at the world' section, presented some unusual orientations. A relevant concise source of information is the article by Richard E Dahlberg: 'Evolution of interrupted map projections ', *International yearbook of cartography,* II, 1962, which adds many reference notes.

The base materials upon which the maps are prepared for reproduction must be carefully chosen for stability and ease of working. Some particularly important atlases have had paper specially made for the purpose. Before the second world war, most cartographic drawing was carried out on cartridge paper, although such agencies as the Ordnance Survey and the Esselte Corporation were using coated glass plates even at this time, because of their high stability. Hand engraving on copper plates remains now in only a few cartographic departments; hydrographic charts, for example, are made on copper, which allows of easier revision than some of the more recent materials. By 1946, zinc plates coated with a white enamel drawing surface were in use and soon after, high stability plastic sheets were devised, which required special inks. The ultimate development in this line was the introduction of plastic or glass sheets coated with visually translucent and actinically opaque lacquer-like substances. A scribing tool replaced the pen in drawing very fine lines more quickly and economically. The points, of sapphire or steel, are prepared in different widths, so that uniformity is easily achieved and the drawing can be used directly, in negative or positive form, by using different methods for the preparation of the printed plate. In drawing the Ordnance Survey maps of Ireland, scribing is done on white coated astrafoil with a blue key, and sometimes on the white astrafoil direct over the basic field document. This is quicker and cheaper than drawing and the standard is higher. The use of plastics involves air conditioning and temperature and humidity control in drawing offices as well as in production stages and storage.

Cartography is now beginning a new era: automatic cartography is causing adjustments in techniques as great as the change from hand drawing to copper-engraving. Mr. D P Bickmore, Head of the Cartographic Department of the Oxford University Press, and Dr R Boyle, a Director of Dobbie McInnes, together worked out the prototype of the system which evolved as the Bickmore Boyle System of Automatic Cartography, a result of the co-operation of the two organisations, with financial assistance from the (then) DSIR. The basic concept of the system was described by Mr Bickmore in a talk given on the BBC Third Programme, reprinted in *The listener* of 30 January 1964, in which he spoke of the system as consisting of 'a table on which a hand-made map compilation is placed; the lines on this draft are then followed or " read " with a pointer. As this pointer moves over the map, streams of x and y co-ordinates are fed to magnetic tape, giving thousands of positions per inch of line, and to an accuracy of three-thousandths of an inch. The information on the magnetic tape is coded into such features as rivers, coastlines, contours, outlines of built-up areas, and so on. A second plotting table is then brought into operation by the tape. On this table is mounted a projector with a fine beam of light; this moves again with great precision and plots lines of varying thickness on to sensitized film. The scale of the map can be altered and variations of map projection can be achieved by feeding the tape through a computer. The system will short-circuit many of the laborious production and checking stages that at present separate the map compilation from the finished negatives or positives from which it is printed.'

The four component parts are the compilation reader, the magnetic tape deck and editor's control panel, the automatic plotting table and the name placement projector. Both the choice of data and the symbols for rivers, coastlines etc are monitored from the editorial control panel. The drawing equipment works to an accuracy of 0·07 mm; lines of any thickness or of varying character may be drawn or coloured areas may be produced as negative masks. The editorial control unit enables the editor to select data from the stored library information and to control the drawing units. He selects the area of the map he requires from the relevant magnetic tape and punched cards; chooses the scale, which is not dependent on the scale at which the compilation was read; selects the detail he requires from the tape or cards, combining, if he wishes, information from one tape with that of another; specifies the symbols, line weight etc. If a change of projection is necessary, the tape and cards are in a form that can be fed through any internationally compatible computer. New or revised information can be fed in at any time, replacing superseded information. This bank of information, which is in precise digital form, is stored in terms of the different features of the map and not in terms of the type of line or symbol by which it will be represented. This means that subsequent choice of symbols is not restricted. Names are also keyed on to the punched cards or magnetic tape, in relation to the appropriate point information. All the various automatic plotting instruments work as co-ordinate principles, so that any geographical information capable of being expressed by co-ordinates is capable of being programmed for automatic plotting. It is fortunate, too, that these technical developments have evolved at this time, for increasingly social, economic and political changes demand quick and accurate revised maps.

One of the best-known automatic co-ordinatographs is the Aristomat, produced by Dennert and Pape of Hamburg, distributed in the United Kingdom by Technical Sales Limited. This machine, with linear numerical control, is a drafting unit for locating and marking points. Grid nets can be drawn parallel to the axes. Punched cards or tape are used for input. A similar machine is the Auto-Tool Model 6000 High Speed Plotter, produced by the Auto-Tool Corporation, Aruada in Colorado and another American plotter is the Model 502 Digital Plotter, from Calcomp, California Computer Products, Inc.

The Army Map Service and the United States Naval Oceanographic Office have greatly influenced the development of specialised cartographic equipment in America. Concord Control Inc, of Boston, has perfected a basic drafting machine, Model E-51 Precision Digital Coordinatograph. Input is from computer-programmed magnetic tape or punched paper tape; continuous lines can be drawn or scribed on paper or coated plastic; or, by using a small, general purpose computer, with special equipment, such as the Photoscribe Head, a complete automatic drafting machine, the Concord Mark 8 Coordinatograph system. The Benson-Lehner Data Reduction System, developed by Benson-Lehner Corporation, Van Nuys, California, is another plotter using magnetic tape, punched tape or punched cards, used by a number of United States government departments, universities and private concerns. ' Read-out ' equipment is usually supplied by the same companies, such as the Aristometer by Dennert and Pape, the Model 3700 Series Digitising System by the Auto-Trol Corporation. More specialised still is the Concord Automatic Type Placement System made by the Concord Control Inc for the United States Army Map Service. The Concord equipment is designed as a comprehensive automatic system.

d-Mac Limited of Scotland and Kongsberg Vapenfabrik of Norway are collaborating in offering complete systems for analysing graphic material for computation and producing drawings from computer-generated data. The Cartographic Digitiser will provide the input to the computer and the Kingmatic drawing machine will draw from the computer output. The Cartographic Digitiser Type CF has been developed as an aid to reducing vast amounts of data, obtained over many years of field survey and cartographic work to

36

a form suitable for computer processing. Providing a rapid means of digitising selected information contained in maps, charts, drawings and photographs, it becomes a key unit in the growing employment of computers for the analysis of problems relating to conservation of natural resources and food production.

The first two units have been supplied to the Canadian Department of Forestry and Rural Development, for the reduction of map records needed to assist in the generation of a computer data bank. Because of the complexity of cartographic data, the new digitiser has been designed with a large range of input modes and outputs. A Cartographic Digitiser has been supplied to the Experimental Cartography Unit at Oxford, sponsored by the Natural Environment Research Council, where it is being used as a compilation reader for a system of automatic map production in conjunction with a computer. Another has been ordered by the technical advisers to the Royal Dutch Shell Group. The Kingmatic Automatic drafting machine Model 1215 is a general purpose drafting machine with an accuracy of 0·004 inch. Both companies have marketing arrangements with the Norwegian computer service bureau Byggedata AS, which has evolved programmes for use in conjunction with their equipment. As a result, they will be in a position to provide both hardware and software for a wide range of data processing requirements in the fields of cartography, surveying, town planning, land inventory, civil engineering and road classification.

The necessary height data for the computation of the analytical shaded picture can be collected not only from contour maps but also directly from stereoscopic models of air photographs. Parallel profiles of the relief at suitable densities are measured and the necessary density of height data obtained either by direct measurement or by interpolation, by the electronic computer. 'Topocart' is the latest of the VEB Carl Zeiss Jena stereocartographic instruments for the plotting of aerial photographs at small and medium scales.

A Research Grant has been awarded to the Royal College of Art by the Natural Environment Research Council for a project to develop an automatic technique for presenting data in a cartographic form by means of a computer. Basically, it consists of an electronic plotting table and 'lock on' time follower, both being developed by AEG in Germany, connected to a PDP-9 computer. Lines traced from maps will be converted into the form of x and y coordinates and stored on magnetic tape. At a later stage the tapes will be read to control the movement of a beam of light thrown on to photosensitive film, thereby re-drawing the original. The work has been carried out at the college and at a special laboratory in Oxford.

When the surveyors have completed their work, it is the task of the cartographer to select from this raw material the information he wishes to include in particular maps, to design the maps on those projections and scales most suited, using the most appropriate symbols, colours etc, and to add such text as may be necessary. Finally, the maps must be reproduced. Although these three processes may be clearly distinguished, they must in practice be considered in conjunction from the outset, if the completed maps are to be efficient and effective. Every part of the total production demands knowledge, skill and experience. At their best, modern maps demonstrate artistic excellence as well.

A well-considered map should present individual factors clearly, so that they can easily be studied simultaneously. Contents may express the variations of a single commodity over an area, or may show the distribution of different elements. Relief representation, colours and symbols all play their parts. Great skill must be applied to the selection of items for inclusion or emphasis; it is increasingly the practice to prepare explanatory text for use with the map, outlining purpose, degree of proven accuracy, the date of information incorporated etc.

An improvement was the introduction of coated plastic or glass, on which the map details are scribed or scratched through the coating with sapphire or steel edged cutting tools. Better lines are produced more easily and quickly with a scribing tool than with a pen, and these new techniques have caused a major revision of practical cartography, though the method has still some disadvantages. The Ordnance Survey reports that it can still call on the services of excellent pen draughtsmen, but that the time may come when this ceases to be the case and the advantages of scribing may then outweigh its drawbacks. Vinyl plastic sheets have appeared under such trade names as Vinylite, Astralon and Astrafoil, available in transparent, translucent or opaque forms. Vinyl plastics are stable in changes of relative humidity, which is an improvement on the drawing paper, but they are affected by extreme temperature variations. In their turn, vinyl plastics were to a great extent replaced by sheets of polyester plastic, also marketed under several trade names, such as Stabilene and Cronaflex. Experiments have long been made with papers for different purposes. One of the recent developments has been the Mellotex range of papers, made by Tullis Russell and Company Limited, by the 'twin-wire' process, which gives an identical printing surface on each side and has a very high degree of the dimensional stability necessary for fine printing. 'Syntosil' is another new material, a product of Switzerland, using man-made fibres to create a printing surface outstand-

38

ing for toughness and high strength, which are extremely resistant to water and chemical attack.

Ilford Supermattex and other similar materials are making films with a polyester base, excellent for scribing and photomechanical application, produced by Ilford Graphics Limited, London. A successful process has been the Fluorographic Process patented by Printing Arts Research Laboratories Inc, California, using Mylar or Melinex transparent sheets. The latest in many experiments in waterproof paper is ' Silbond ', a synthetic paper made by Fraserprodukte GmbH and Company, in West Germany and available in the United Kingdom through Reeve Angel International Limited. This astonishing product, made from polyester fibre, is waterproof, weather-resistant, dirt-repellant and washable; it will not crumple or fray and it will take all the normal printing processes. It is naturally more expensive than ordinary papers. During the past ten years or so, much of the discussion at cartographic conferences and technical congresses has been concerned with the relative qualities of these new plastics. A second major consideration has been the choice of inks and the methods of using them. The development of plastics has resulted in the replacement to a large extent of drawing and glass engraving by negative or positive scribing on pre-coated plastics, and new tools have been developed. Sapphire cutters, though more expensive, have been found much more satisfactory than steel needles, when fine line-work is required. The choice of suitable printing format is also vital in planning maps. The tendency is to use larger printing plates, and this is again an economy, providing that consistency in colours and exact registration are not impaired.

The aesthetic appreciation of modern maps, as compared with their forerunners, is of an entirely different quality; thorough appreciation must now depend on some knowledge of the technical miracles performed in producing an accurate and complex, yet artistically pleasing map face. The engaging dolphin creatures and the gay little ships may be missed, but the representation of relief has turned, in one sense, full circle; the modern three-dimensional relief techniques have the same purpose as the ' sugar-loaf ' hill features, plus, of course, now, scientific accuracy. Small pictorial trees still indicate the presence of individual types within a woodland area. At this time, when not only specialists but the general public are using maps more frequently than ever before, contemporary maps are expected to convey their information as directly and as effectively as possible, without ornamentation, but with full documentation and essential explanatory text; absolute registration is expected up to twelve colours, involving extreme accuracy in all stages of production.

The representation of land relief is fascinating in its variety. In spite of the introduction of contours, about 1850, the Ordnance Survey continued to use some additional method of portraying relief. For reasons of economy, hachures were omitted from the fourth or 'popular' edition published after the 1914-18 war, but the omission did not meet with full acceptance. Publication of layered or shaded maps of selected tourist areas followed, and, in due time, the fifth relief edition, in which hachures on the shadow sides of the hills were overprinted in violet. The second world war put an end to the fifth edition, and after this war, the sixth edition, the 'new popular', was produced as quickly as possible, with contours only. The seventh edition also has contours only, but the policy of producing tourist maps of selected areas with hill-shading has been continued. Hill-shading has also been introduced into the fifth edition of the quarter-inch map series. Much thought has been given to the problem of depicting landforms with greater fidelity, using hill-shading to create an impression of three dimensions. Among the more elaborate systems are those requiring the making of three-dimensional models in exact replica of the landforms. The model is then photographed, as if light were directed from the north-west corner, through a half-tone screen for overprinting on the maps in suitable colours, usually brown or grey. Other techniques begin by shading the ground relief in pencil on grained Astrafoil or by painting it with an air brush on a suitable drawing surface, as if it were lit from a north-westerly direction. These originals are then photographed in readiness for making a printing plate. For the best results of all, hill-shading is blended with layer-colouring, with or without contours.

The use of relief model maps has increased, especially for educational purposes. The most frequent method of achieving such a map is to make a key print on to a sheet of Astrafoil of the map area, indicating rivers and main contour intervals. A felt pad is then placed underneath the Astrafoil and, by using suitable tools and a good deal of skill, the Astrafoil is pushed into the felt and stretched as necessary to take on the shape of the ground relief. Wooden pegs cut to scale are used to push up the peaks and outline the hill slopes. The model is sprayed with white paint and set in front of the camera with specially controlled side lighting to give a shadow effect on the hills. Any individual features desired can be painted on the models, which are especially useful for demonstration purposes or framed as wall maps for teaching.

The representation by colours of relief features, especially height, has caused much discussion through the years. By using a colour scheme related to height alone, deserts have been known to be shown in green! If any approximation to verisimilitude is to be attempted, seasonal and other variations should logically be taken into account as well. Some of the 1 : 100,000 maps of Sweden have been published in summer and winter editions. As colour-photogrammetry is increasingly developed, some more effective means for depicting the natural colours of a landscape may be found, particularly on large-scale maps. Modern reproduction techniques enable much more gradual colour changes to be shown. The 'strip' method has been used to show the presence of several factors within an area, or a 'transition' area. A black and white shaded relief map is frequently used as a base for combination with other factors, by overprinting. Several of the distribution maps in *The Atlas of Britain and Northern Ireland,* for example, use a grey relief base, against which colour symbols show up well.

The Directorate of Overseas Survey has been experimenting for some time with different methods of the representation of ground surfaces, especially in parts of Africa and the Middle East, where there are few or no settlements and even scanty vegetation cover. Hachures have by no means disappeared. Relief on some European maps, notably on the mountain areas of Switzerland, is most successfully shown by hachures, with or without contours; the details of rock faces are frequently exquisitely superimposed in this way. The change from contour lines and hypsometric tints to methods based on plastic relief models for the representation of relief results in a more accurate and graphic effect, but requires more accurate base maps and also demands from the draughtsman not only skilled precision, but also an imaginative feeling for surface relief forms.

An international system of colour measurement has been under discussion for many years. A standardised scheme for the distinction of rocks of various periods was agreed in 1881 at the Geological Congress at Bologna. The Munsell system is the most widely known and has been adopted for the IMW at 1 : 1M. The Ordnance Survey has published the scale of hypsometric tints agreed in Bonn in 1962, for the IMW; in the solid colours used, the colour is indicated in Munsell's terms. There are five principal hue names in the Munsell system; the value notation relates to the degree of lightness or darkness of a colour, ranging from a theoretically pure black, having notation 0/, to a theoretically pure white, with notation 10/. The chroma notation of a colour indicates the strength or degree of saturation of a particular hue from a neutral grey of the same value; the scales here range from /0 for a neutral grey to

41

/10, /12, /14 or more. The use of colour in cartography was one of the topics discussed at the Third International Conference for Cartography at Amsterdam, of which reports are included in the *International yearbook of cartography*, VII, 1967.

The factors affecting the selection and use of colour on maps are complex, not the least important being the differing subjective reactions of map-users. During the years of guiding students' evaluation of maps and atlases, for the Part II, c 203 section, ' Bibliography and librarianship of geography ', for the final examination of the Library Association, interesting results have been obtained by noting individual instinctive reactions to certain distinctively coloured map sheets, prior to more objective analysis; particularly useful in this exercise have been the physical sheets of the *Atlas of Britain,* the sheets of the first Land Utilisation Survey and, in the Clarendon Press *Atlas of Britain and Northern Ireland,* the sheets depicting geology, the double sheet depicting the surrounding seas and the final section of regional maps. In each case, without hesitation, some of the group reacted favourably, others unfavourably; of the latter, after closer examination, reasons for choice of colours might be admitted, but the aesthetic reservation remained. By the use of graduated screens, the combination of the three standard colours in all their variations can be made to produce a great number of colour tints; in addition, hatching and other devices can increase the powers of distinguishing factors on the map face.

Portrayal of coast lines on topographic maps presents special problems, notably on the smaller-scale maps. Care must be taken that measurements all relate to, for example, the high tide level. Agreed symbols should indicate the nature of the cliffs. The map of the coasts in the *Atlas of Britain and Northern Ireland* was an interesting pioneer work; in the *Atlas van Nederland,* as would be expected, particular care has been taken in drawing the polder areas; and, in the 1954 and 1962 *Atlas mira,* the skill with which deltas were shown is particularly impressive. The practice of showing shadow relief as a background to thematic mapping of, for example, commodities or resources, is becoming more common, showing clearly the correlation between elevation and crop or livestock production. The scale of the map must also greatly affect the choice of relief symbols; at small scales, contours and hypsometric tints are of little value. The cost limitations of the work in hand must also affect the method chosen to portray the information to be included. With skill, black and white can be perfectly effective. Only on the largest scales can the key signs accurately represent the position of features on the ground. At smaller scales, a reduction in the amount of information shown must result, to avoid misrepresentation and overcrowding. Statistical information presented in visual form can represent only abstractions, not actual values, unless figures are actually affixed at convenient points. A detailed summary of such problems as these, with an extensive bibliography, is given by Erik Arnberger: 'Das topographische, graphische, bildstatistische und bildhafte Prinzip in der Kartographie', *International yearbook of cartography*, IV, 1964.

One of the most exacting aspects of cartography is the reduction to smaller scales from original surveys. Decisions depend on a number of factors—feeling for terrain, for example, and the purpose of the map are paramount considerations. In using maps in conjunction with geographical interpretation, therefore, it is essential to remember that the cartographers have already made a selection from the total features present in the area; the greater the geographical knowledge of the cartographers, or the greater the co-operation between specialised geographers and cartographers, the more reliable and valuable the resulting maps should be. As quantitative data accumulate throughout the world, so do the problems of representing the variety cartographically, or by diagrammatic maps. The data may be continuously collected, such as weather information, or may be collected at intervals, as in censuses and surveys. Maps rather than tables are normally compiled when the aim is the comparison of particular factors from place to place; or it may be required to compare the distribution of the data with physical variations, climate or soils, etc. Very great care must be taken in choice of scale, colouring and symbols, otherwise a distorted view may be given. Exactly similar data, mapped in different ways, can give quite different visual impressions. Yet, skilfully prepared maps are the most powerful media for expressing comparisons or evoking deductions.

The choice of names for inclusion on a map, their linguistic form, the kind of typeface chosen and the placing of names are all-important to the clarity and legibility of the map. Several specifications have been adopted for the form of names; for example, those for the International Map of the World 1 : 1M and for the ICAO World Aeronautical Chart; also for the hydrographic charts issued by the International Hydrographic Bureau, and the Universal Postal Union world map using the form of place names adopted in the Directory of Post Offices. At the national level, among others, the Permanent Committee on Geographical Names for British Official Use has published rules for dealing with geographical names in various languages, and the United States Board on Geographic Names has worked on similar lines for the United States. All countries have their own points of view. A Draft Programme was circulated for discussion at the Second United Nations Regional Cartographic Conference for Asia and the Far East, at Tokyo, 1958. The establishment by the United Nations of a central international body, an 'International Committee on Geographical Names', was recommended; also that the Secretariat of the United Nations should act as a central clearing house of information concerning national systems of transliteration, etc, and that national gazetteers should be compiled. Full details of the progress of the discussions, recommendations and decisions are given in volumes VI, 1958; VII, 1962 of *World cartography* and in the *Official records of the Economic and Social Council, Twenty-seventh Session.*

A H Robinson in particular has emphasised that very few type faces have been designed especially for cartography, and that a main principle of typography is violated when any type is reduced by photography. For the 1964 International Geographical Congress, an experiment was conducted by The Monotype Corporation Limited, in conjunction with the Cartographical Department of the Clarendon Press, to demonstrate the effect of typography on maps. The results were published in the form of a broadsheet, on which eight maps of France are shown, identical except for the variations in typography. The area chosen was a varied one, giving full rein to the contrasts demonstrated. Serif can be compared with sans serif, condensed with expanded letters, upper case with lower case, roman and italic capitals etc. The chief functional requirements of type faces for cartographic work are legibility in small sizes; legibility in areas particularly complex topographically; easy recognition of individual characters; the use of several readily distinguishable styles of type to isolate individual factors or to emphasise the more important cities etc; legibility against a multi-coloured background and unity and pleasing appearance in the finished map. It must be borne in mind that on small-scale maps, the name of a city, even in the smallest of readable type, may spread over hundreds of miles. On the other hand, names of political areas or of an extended range of mountains must be spaced out. Neither will the printed names always have a horizontal base line, for river-names etc, normally follow their flow.

The Monotype photo-lettering machine is one of the most commonly used. The machine photographs individual characters direct on to film or paper; characters are selected manually by a dialling operation, after which the cycle of operations is controlled automatically by sequential electrical and electro-mechanical components. The Ordnance Survey uses a system of letters and footnotes which enables first and subsequent editions of current maps and

44

plans to be easily identified. The edition is marked by a letter printed in the left-hand bottom margin, in the top right hand margin also in the case of large scale plans. The year of publication to which the letter applies is shown in a footnote in the bottom margin. When a sheet is reprinted, incorporating change, an additional footnote is inserted. In the case of facsimile reprints, the footnote remains unchanged. The Ordnance Survey uses two 'Monophoto' filmsetters and five Monotype keyboards, an installation which provides the lettering for all the series of maps. Both Fairey Air Surveys Limited and the Cartographic Department of the Clarendon Press have installed Monotype photo-lettering machines. John Bartholomew and Son Limited use hand-set type, cast on Monotype machines, from which reproduction proofs are taken for sticking up on to artwork prepared for photo-mechanical reproduction. All kinds of lettering, numbers and symbols are offered for sticking up by 'Format' Modern Multi-Use Acetate Graphic Art Aids, attached with a wax-free adhesive. Interesting commentaries on these problems are made in the *Report of the group of experts on geographical names* in *World cartography,* VII, 1962; and in 'The question of adopting a standard method of writing geographical names on maps in the United Nations', *World cartography,* VII, 1962. Most maps carry lettering of many sizes ranging from about 14 point to $4\frac{1}{4}$ point, the vast majority being in the range of 6 point and smaller. Particular type faces are usually kept consistent for the same features. Sans serif has been establishing itself, especially sans serif capitals. Experiments on lettering and type faces for complex maps continues by cartographic departments, by the Chart Reproduction Committee of the Joint Advisory Board and by the Cartographic Sub-Committee of the Royal Society.

The insertion of grid lines on maps has become an increasingly common practice during recent years, especially with the various editions of national maps in series. These lines must be accurate, so that adjacent map sheets show continuous grid lines when used together, and they are plotted by a co-ordinatograph, a measuring machine which enables them to be plotted in two directions at right angles to one another. They work to an accuracy of 0·002 inches and can draw a line up to fifty one inches. Co-ordinatographs have proved indispensable in precision mapping for geodetic, photogrammetric and cartographic purposes. As well as plotting orthogonal grid systems, they are used for the production of masters for graticules for all purposes. Among the most used are the Carl Zeiss Precision Co-ordinatographs, which map points in systems of orthogonal co-ordinates on the following scales: 1:800, 1:1,000, 1:2,000, 1:2,500, 1:4,000, 1:5,000 and multiples of these scales. Use has expanded since various models of the Precision Co-ordinatograph have been equipped with selsyns; when combined with an electro-mechanical recording and computing unit, it is possible to record the co-ordinates of points fixed on a map or plan with the co-ordimeter. Recording is done either in clear text or on punched tape. The 'Carl Zeiss Precision Co-ordinatograph Instruction Manual' enlarges on all aspects of these machines, with a folder of illustrations tipped into the back cover.

Comparable advances have been made in map reproduction processes, where accuracy must now be perfect, even in small-scale work, involving up to twelve colours. Lithography is the form of printing most usually used for map reproduction. The invention of the offset method of lithographic printing and the discovery that zinc could be used instead of limestone were two major stages in the development of modern map-printing. The offset printing machines used for map printing are usually sheet fed rotary machines having a capacity of 10,000 impressions per hour. The stages of map production, using lithographic processes, are compilation, photographic copying, fair drawing, followed again by photographic copying, preparation of the printing plates and the final printing. In the first photographic copying, a process is used which gives prints in pale blue, so that these lines do not interfere with the improved drawing; one print is made for each of the colours to be used in the map. The fair drawing is done in black and white, to give a sharp image, and each drawing contains only the material to be printed in one colour. The precision of the finished map detail depends upon an accurate drawing and accurate superimposition of transparent sheets carrying the various components, line networks, colours and lettering. Where, for example, many pages of an atlas are to be reproduced in several colours on a large sheet, the correlation of pages on successive colour printing plates requires great skill. Examples that spring to mind in English cartography are the vegetation maps in the *Atlas of Britain and Northern Ireland,* and the maps showing glaciation. Step-and-repeat machines, working to a tolerance of less than 0·001 inches, are invaluable for this kind of work.

A regular revision policy should be planned at the time of preparation of any new series of maps, the time interval varying according to the kind of terrain covered and the content and purpose of the maps. Transitional revisions may usually be seen at most survey offices between published editions, as in the case of the Ordnance Survey.

Automation has already been mentioned in passing. The Working Party of the Cartography Sub-Committee of the British National

Committee for Geography—'Automation in Cartography'—has as its main purpose the review of automation processes and developments in cartography within the United Kingdom and liaison, through their corresponding member, with the International Cartographical Association Special Commission. Commission III of the International Geographic Union, 'Automation in cartography', was created in 1964 to study and evaluate the automated aids of the cartographer. The basic function is to transcribe map compilations into finished reproduction material through the media of magnetic tape and punched cards. The repetitive plotting and re-plotting, drawing and re-drawing of the data are therefore avoided, as is expensive camera work and intermediate charting stages. The system consists of four components: the compilation reader, the magnetic tape deck and editor's control panel, the automatic plotting table and the name placement projector. The cartographic editor marks his selection of data to be mapped and this input is then placed on the glass table of the compilation reader where a pointer is guided manually over the lines. Beneath the glass top, a sensitive instrument records the movements and feeds them as a stream of x and y co-ordinates to magnetic tape, giving thousands of positions per inch of line to an accuracy of 0·003 inch. Data concerning surface features, woods, rivers, etc are fed on to magnetic tape; point information, such as the positions of settlements and names, are usually fed on to punched cards. Both the selection of data from the library of information and the control of the drawing units are monitored from the editorial control panel. The map projection can also be changed, if necessary, by feeding the tape and cards through a computer. Lines of any thickness or character can be drawn and information can be converted into colour areas as negative meshes. Point data from the tape and/or cards can be translated into the required symbols. One of the inventors of the system perfected by the Cartographic Department of the Clarendon Press and Dobbie McInnes has stated that it ' does not offer press-button cartography; indeed, the skill and ingenuity of the cartographer will be extended. But what the system should offer is the elimination of much of today's time-consuming drudgery, without loss either of the precision or of the elegance of maps.'

With the development of mechanical equipment, various comparative experiments were made and, in 1956, the Stanley Precision Planimeter proved satisfactory, except that it lacked an automatic recording device. In 1959, Messrs W F Stanley and the Rank Cintel Organisation together developed a prototype automatic reading planimeter, which, with further improvements, has since been used by the Ordnance Survey. Automatic punched card data processing was developed in the mid 1960s and this is being replaced by a larger, tape-operated installation. VEB Carl Zeiss Jena produce a data processing system, under the name 'Cartimat', which can plot all points or curves graphically, the position of the curve of which is in any form given by cartesian or polar co-ordinates. This equipment can be used for the cartography of points determined by surveying methods on processes of analytical photogrammetry, for the cartography of contour lines and profile sections.

Any kind of social and industrial mapping will benefit from the increased use of computer processing of data, so that maps can be produced quickly and revised editions supplied equally quickly, when changes occur. The vast numbers of items involved in meteorological and oceanographical recording and charting have benefited especially from the installation of automated equipment. All the above-mentioned considerations for map production apply to the making of atlases, with the additional factor than an atlas should be carefully planned in its entirety before detailed preparation begins. The purpose of the publication must determine the scope, contents, overall arrangement and emphasis on individual sections. The sequence of maps should follow logically and projections and scales should be chosen to suit the size and shape of the area to be mapped. Both scales and projections will inevitably be changed for specific purposes, but the maximum uniformity should be aimed at and, within a section of, for example, economic distribution maps, the scale should be uniform to enable comparability. Loose-leaf format facilitates revision.

Now that cartographers are so closely involved in the modern technical aspects of map plotting, the Hunter-Penrose-Littlejohn Limited *Graphic arts technicians handbook*, now in its fifth edition, will be found useful, also the *Jena review* and the *Hunting Group review*. Hunting Surveys Limited also issue regular 'Information sheets'. Some further useful references follow:

M Archambault *et al*: *Documents et méthode pour le commentaire de cartes (géographie et géologie)* (Paris, Masson, 1965-).

S Augustin and R Schoonman: 'Automation and design', *Penrose annual,* 1964.

E D Baldock: *Manual of map reproduction techniques* (Ottawa, Queen's Printer, 1964).

E D Baldock: 'The technical performance of map revision', *Kart nachrichten,* V, 4, 1963.

B Bannerjee and H R Betal: 'Perspective on the nature of map projection', *Geographical review of India,* XXVII, 2, 1965.

J Beaujeu-Garnier: *Practical work in geography* (New York, St Martin's Press; Arnold, 1963).

T Berlese: *Topografia generale* (Padua, Ceram, 1951).

M W Beresford: *History on the ground: six studies in maps and landscapes* (Lutterworth Press, 1957). Field studies in England.

S Bertram: 'The universal automatic map compilation equipment', *Photogrammetric engineer,* March 1965.

D P Bickmore and A R Boyle: 'An automatic system of cartography', *International yearbook of cartography,* V, 1965; Address to the ICA Technical Symposium, Edinburgh, 1964.

T W Birch: *Maps, topographical and statistical* (Oxford, Clarendon Press, second edition 1964).

R G Blakesley *et al*: 'The planning databank challenges the surveyor and mapmaker', *Surveying and mapping,* March 1967.

Roger Brunet: *Le croquis de géographie* (Paris, Société d'Edition d'Enseignment Supérieur, 1962). Sketchmap technique; block diagrams; statistical presentation.

William Bunge: *Theoretical geography* (Lund, C W K Gleerup, 1962). Lund Studies in Geography, Series C, General and Mathematical Geography, no 1).

John Bygott: *An introduction to mapwork and practical geography* (UTP, ninth revised edition 1964, revised by D C Money).

C F Capello and M L Chionetti: *Elementi di cartografia* (Torino, Giappichalli, three volumes 1958-60).

C C Carter: *Land-forms and life: short studies on topographical maps* (Christophers, 1959, repr 1961, revised by M O Walter).

W Chamberlin: *The round earth on flat paper* (Washington, DC, National Geographic Society, 1947).

Heather Child: *Decorative maps* (Studio Publications, 1956). Mainly modern maps, with some historical examples.

R J Chorley and P Haggett: 'Trend-surface mapping in geographical research', *Transactions,* Institute of British Geographers, XXXVII, 1965.

W G Clare: 'Map reproduction', *The cartographic journal,* December 1964.

M C Collins: 'The production of 1:50,000 contoured maps using high level photography and analytical triangulation', *Report of proceedings,* Conference of Commonwealth Survey Officers, 1963 (1964).

Co-operative Society for Geodesy and Cartography: *Review of geodetic and mapping possibilities* (1957).

Bruce Cornwall and A H Robinson: 'Possibilities for computer animated films in cartography', *The cartographic journal,* December 1966.

R E Cramer: *A workbook in essentials of cartography and mapping* (Dubuque, Iowa, Brown, 1963).

F M Crewdson: *Color in decoration and design* (Wilmette, Illinois, 1953).

W C Cude: 'Automation in mapping', *Surveying and mapping,* September 1962.

Alexander d'Agapeyeff and E C R Hadfield: *Maps* (OUP, second edition 1950). Map making, map reading and the history of maps.

C L Dake and J S Brown: *Interpretation of topographic and geologic maps, with special reference to determination of structure* (McGraw-Hill, 1925).

R O Davis: 'Problems of map maintenance' (ICA Technical Symposium, Edinburgh, 1964).

E R De Meter: 'Automatic contouring', *Nachr karten u vermessungswesen,* V, 9, 1964.

E R De Meter: 'Latest advances in automatic mapping', *Photogrammetric engineer,* November 1963.

G F Delaney: 'Problems in cartographic nomenclature', *Canadian cartography,* 1962.

O M Dixon: 'Sheet lines for topographic maps', *Surveying and mapping,* March 1966.

O M Dixon: 'Sheet designations for topographic maps', *Surveying and mapping,* December 1966.

C Duncan: 'Cartography and geographical studies', *Cartography,* 1963.

Bernard Dupuisson: *Procédés et méthodes des levés topographiques aux grand échelles* (Paris, Eyrolles, 1954).

G H Dury: *Map interpretation* (Pitman, third edition revised 1967). Includes chapter on cartographical appreciation by H C Brookfield and G H Dury.

Max Eckert: *Die kartenwissenschaft* . . . (Berlin, Walter de Gruyter, two volumes 1921 1925). A classic work and a major source book for later studies.

A Elwood: *Essentials of map-reading* (Harrap, fourth edition 1964).

G K Emminizer: 'The cartographic manuscript for the revision of maps', *Kart nachrichten,* V, 4, 1963.

W Filchner *et al: Route mapping and position locating in unexplored regions* (Basle, Birkhauser, 1957).

Nicola Franchi: *Elementi di cartografia* (Firenze, Istituto Geografico Militare, 1950).

J K Fraser: 'Canadian Permanent Committee on Geographical Names', *Geographical bulletin,* May 1964.

G J Friedman: 'How far automation?' *Surveying and mapping,* December 1964.

H Fullard: 'The problems of communication between editors and users of atlases' (ICA Technical Symposium, Edinburgh, 1964).

H Fullard: 'Revision of small-scale topographic and atlas maps', *Kart nachrichten*, V, 4, 1963.

R T Gajda: 'Automation in cartography', *The cartographer*, May 1965.

D W Gale: 'Register control in map reproduction', *The cartographic journal*, December 1965.

L S Garaevskia: *Kartografiya* (Moscow, Geodezizdat, 1955).

Alice Garnett: *The geographical interpretation of topographical maps* (Harrap, third edition revised, two volumes 1953). With twenty-five maps in folder.

A C Gerlach: 'Technical problems in atlas making' (ICA Technical Symposium, Edinburgh, 1964).

E Ghianda: *Topografia pratica* (Genova, Vitali and Ghianda, 1952).

C Godfrey: 'Cartography today', *Contemporary review*, July 1964.

J B Goodson and J A Morris: *Contour dictionary: a short textbook on contour reading with map exercises* (Harrap, third edition 1960).

Maitland Graves: *The art of color and design* (McGraw-Hill, 1951).

David Greenhood: *Mapping* (University of Chicago Press, 1964).

D L Griffith: 'Developments in cartographic techniques', *Report of proceedings*, Conference of Commonwealth Survey Officers, 1963 (1964).

S V Griffith: 'Contouring problems on general purpose maps', *Surveying and mapping*, 1952.

C B Hagen: 'Maps, copyright and fair use', *Bulletin*, Special Libraries Association, Geography and Map Division, December 1966.

V G W Harrison: 'Co-operative research in the graphic arts', *Penrose annual*, 1964.

D W Harvey: 'Geographical processes and the analysis of point patterns: testing models of diffusion by quadrant sampling', *Transactions*, Institute of British Geographers, 1966.

L I Haseman: 'Rapid military mapping', *Surveying and mapping*, June 1964.

A G Hodgkiss: 'Cartographic illustration: aims and principles', British Society of University Cartographers, 1966.

M Hotine: 'Rapid topographic surveys of new countries', *Surveying and mapping*. 1965.

Eduard Imhof: *Gelände und karte* (Zürich, Erlenbach-Rentsch, 1950 third edition and revised 1968). An outstanding work, finely illustrated.

Eduard Imhof: 'Isolinienkarten', *International yearbook of cartography*, 1961.

Eduard Imhof: *Kartographische Geländedarstellung* (Berlin, Walter de Gruyter, 1965). In German, but the illustrations would in any case be intelligible.

Eduard Imhof: 'Tasks and methods of theoretical cartography', *International yearbook of cartography*, 1963.

H W Kadan: *Kartographie: praktischer leitfaden für kartographen, kartolithographen und landkartenzeichner* (Leipzig, Fachbuchverlag, 1955).

R C Kao: 'The use of computers in the processing and analysis of geographical information', *The geographical review*, October 1963.

H A Karo: 'Modern developments in cartography and revision techniques', *Report of proceedings*, Conference of Commonwealth Officers, 1963 (1964).

H A Karo: 'Progress in international cartography', *Surveying and mapping*, Summer 1964.

Kartographische studien (Haack festschrift, 1957) *see* under 'Lautensach'.

J S Keates: 'Modern map design' *The scotsman*, 31 July 1964.

J S Keates: 'The perception of color in cartography' *Proceedings* of the Cartographic Symposium, Edinburgh, 1962.

M H Kilminster: 'The drawing and production of maps', *Cartography*, March 1962.

G Krauss: 'Difficulties of maintaining topographic maps and possibilities of overcoming them' (ICA Technical Symposium, Edinburgh, 1964).

Hermann Lautensach and Hans-Richard Fischer *editors: Kartographische: studien Haack festschrift* (Gotha, Geographisch-Kartographische, 1957). Twenty five papers on 'General geography' and 'Special and regional cartography', in a supplementary volume to *Petermanns geographische mitteilungen*, to commemorate the 85th birthday of Hermann Haack.

N P Lavrov: 'The perspective of the automation of the techniques of compilation and preparation to reproduction of geographic and topographic maps' (International Geographical Congress, London, 1964, Section IX, Cartography).

André Libault: *La cartographie* (Paris, PUF, 1963).

G B Littlepage: 'Revision of government charts and maps', *Kart nachrichten*, V, 4, 1963.

A K Lobeck: *Block diagrams and other graphic methods used in geology and geography* (Amherst, Mass, Emerson-Trussel Book Company, second edition, 1958).

B Lockey: *The interpretation of Ordnance Survey maps and geographical pictures* (Philip, seventh edition 1965).

G Lundquist: 'Production control' (ICA Technical Symposium, Edinburgh, 1964).

D Macdonald: 'The presentation of relief on maps', *The Australian surveyor*, March 1963.

R McLean: 'Must the artist disappear from mapmaking?', *Penrose annual*, 1964.

D H Maling: 'Recent trends in map use and presentation', *Penrose annual*, 1964.

T Margerison: 'Drawing for a computer', *New scientist*, 1 July, 1965.

A C Marles: 'Photomechanical processes on plastics', paper and discussion presented to the Fifth meeting of the Symposium of the British Cartographic Society, Swansea, 1965, on 'Current developments in cartographic practice and photomechanical techniques', reprinted in *The cartographic journal*, June 1966.

S F Marriott: *Map reading for the countrygoer* (Rambler's Association, revised edition 1964).

F J Monkhouse and H R Wilkinson: *Maps and diagrams: their compilation and construction* (Methuen, 1952; second edition, University Paperbacks, 1963, repr. 1964, 1966).

A H Munsell: *A colour notation* (Baltimore, Munsell Color Company, tenth edition 1946).

A M Parreira: *Manual de topografia* (Lisbon, Mocidade Portuguesa, 1954).

G Petrie: 'Numerically controlled methods of automatic plotting and draughting', *The cartographic journal*, December 1966.

Erwin Raisz: *Mapping the world* (New York, Abelard-Schuman, 1956).

Erwin Raisz: *Principles of cartography* (McGraw-Hill, 1962).

I D Reid and A F A Learmonth: 'Applications of statistical sampling to geographical studies, with special reference to cartographic representation of sampling error' (Canberra: Department of Geography, School of General Studies, Australian National University, *Occasional papers*, no 5, 1966).

F Reignier: *Les systèmes de projection et leurs application à la géographie, à la cartographie, à la navigation, à la topometrie*

(Paris, Secret d'Etat aux travaux publics, aux transports et au tourisme, two volumes 1957).

W W Ristow: *Three-dimensional maps: an annotated list of references relating to the construction and use of terrain models* (Washington, Library of Congress, 1964).

J A Roberts: 'The topographic map in a world of computers', *The professional geographer*, November 1962.

A H Robinson: *Elements of cartography* (Wiley, second edition 1963).

A H Robinson: *The look of maps: an examination of cartographic design* (University of Wisconsin Press, 1952).

R R Rollins: 'Printing management for maps and charts', *Surveying and mapping*, June 1964.

A Rosenfeld: 'Automatic imagery interpretation', *Photogrammetric engineer*, March 1965.

K A Salichtchkev: *Cartography* (Moscow, 'High school' Publishing House, 1966). In Russian.

Axel Schou: *The construction and drawing of block diagrams* (Nelson, 1962).

P C Sen Gupta: *How to map the earth true to size and shape* (Calcutta, Barman, 1954).

W A Seymour: 'Revision', *Report of proceedings*, Conference of Commonwealth Survey Officers, 1963 (1964).

J V Sharp, *et al*: 'Automatic map compilation', *Photogrammetric engineer*, March 1965.

R B Simpson: 'Radar geographic tool', *Annals* of the Association of American Geographers, 1966.

P Speak and A H C Carter: *Map reading and interpretation* (OUP, 1964).

A I Spiridonov: *Geomorphologische kartographie* (Berlin, Dt Verlag d Wissenschaften, 1956).

J A Steers: *Introduction to the study of map projections* (ULP, eleventh edition 1957).

A C Stephenson: 'Modern survey methods', *Geographical magazine*, April 1960.

L V Strees: 'The development of a precise automatic cartographic plotter', *Nachrichten aus dem karten und vermessungswesen*, 1963.

H Stump: 'Map revision', *Kart nachrichten*, 1963.

W R Tobler: 'Automation in the preparation of thematic maps', *The cartographic journal*, June 1965.

W R Tobler: 'The geographic ordering of information: new opportunities', *Professional geographer*, July 1964.

F Topfer and W Pillewizer: 'The principles of selection', *The cartographic journal*, June 1966.

C Travers: 'Cartographic maintenance requirements' (ICA Technical Symposium, Edinburgh, 1964).

J Tricart *et al*: *Initiation aux travaux pratiques de géographie: commentaires de cartes* (Paris, Société d'Edition d'Enseignement Supérieur, second edition 1960). Survey of the kinds of maps and mapping techniques.

United Nations: *Modern cartography: base maps for world needs* (New York, Department of Social Affairs, 1949). English and French editions.

United States Department of the Army: *Map reproduction* (Washington DC, 1956).

United States Department of the Army: *Topographic symbols* (Washington DC: US Printing Office, 1961).

R L Voison: 'Production control' (ICA Technical Symposium, Edinburgh, 1964).

R L Voison: 'The revision of non-Government maps', *Kart nachrichten*, 1963.

M Wood: 'Visual perception and map design', *The cartographic journal*, June 1968.

J W Wright: 'Three types of reconnaissance mapping', *The cartographic journal*, June 1965.

P Yoeli: 'Analytical hill shading and density', *Surveying and mapping*, June 1966.

P Yoeli: 'The mechanisation of analytical hill shading', *The cartographic journal*, December 1967.

A Zvonarev, *editor*: 'New problems and methods of cartography', Leningrad, *Nauka*, 1967, Academy of Sciences of the USSR Geographical Society. In Russian. Eleven contributions by specialists.

TRAINING IN CARTOGRAPHY

It was a saying of Erwin Raisz that ' a cartographer is a geographer who can draw a map besides '. In times past, map-making has been an art practised in the great cartographic houses of Blaeu, Mercator, Justus Perthes, Bartholomew and the rest, the techniques involved being passed on to younger members of the family and other apprentices. Until the first decade of this century, cartographers still learned by example and experience and, of course, to a certain extent, some of them still do. The mapping requirements of the first world war, however, and even more of the second, revealed the weaknesses of lack of co-ordination and of systematic training; courses in cartography at university level began to be advocated. The technical advances of the past few years have increased the pressure of demand for such courses.

Since the initiation of the Board of Ordnance, a civil establishment of surveyors and draughtsmen was maintained; cadets were engaged at an early age and gradually learned all the techniques of surveying and mapmaking. A two-year training scheme was in operation for youths recruited at about sixteen years of age and a second scheme gave a short course in drawing and surveying to young men recruited at eighteen or over. In 1946, the department, then a completely civilian surveying service, planned large-scale training schemes to create the necessary staff of some four thousand, to cope with the vast amount of work waiting to be done. Pre-entry qualification is now at least three GCE passes, preferably in English language, mathematics and geography. The current syllabus contains three parts: fifteen weeks basic drawing instruction, nine weeks advanced drawing instruction and two weeks elementary field survey instruction. On completion of the six months preliminary course, students are posted to the production section. Training continues for at least four years, including a written examination and frequent practical tests. Training then becomes more specialised and all work is strictly supervised to ensure the maintenance of high standards.

The Institut Géographique National maintains three grades of courses: course A is a two-year graduate course in theoretical and practical surveying and mapping, leading to the qualification of Geographical Engineer; course B is for post-high school graduates, who complete two years of theoretical and practical training to obtain the qualification of Technician in Geography; and course C, for selected high school pupils, is also a two-year course for the qualification Technician in Geography. The École National des Sciences Géographiques also has for many years given instruction in the fundamental features of geodesy, topography, photogrammetry and cartography. The National Cartographic Centre at Tehran initiated training courses in 1965, conforming with the A, B and C courses of the institute.

In June 1963, courses were begun at the Istituto Geografico Militare in Florence, for the training of cartographers and technicians; this was a combined effort by the Italian Cartographic Association, the Touring Club Italiano, the Institute, 'De Agostini' in Novara and the Ente Italiano Rilievi Aerofotogrammetriei. A journal is published every four months, the *Bolletino dell 'Associazione Italiana de Cartografia.*

The Aerial Photographic Unit of the University of Cambridge is a unique organisation, which, for more than twenty years, has worked to provide aerial photographs for teaching and research. The cartographic course set up in 1964 at the Oxford College of Technology has had the benefit of close co-operation with the Cartographic Department of the Clarendon Press. The one-year course was extended to two years in 1968 to allow for study in greater depth, including a six-week period of practical experience in a cartographic department. The course is recognised by the Department of Education and Science as ' advanced '. An 'A' level GCE in geography is required as a minimum pre-entry qualification, together with five 'o' level passes including english, mathematics and a science. The course is not intended to produce qualified cartographers and draughtsmen, but to give suitable and interested young people an intermediate stage of training. The geographical content of the course reaches first-year degree level; lithography and printing, map appreciation and design, and the interpretation of British and foreign maps are included. The final stages of the course cover all the processes of map reproduction and the diploma award is based on course work, a relevant thesis and on practical projects, such as the published atlas of Banbury. A foreign language study is also part of the course.

Cartographic training at the Survey Production Centre requires a pre-entry qualification of at least five 'o' level GCE passes, including geography, mathematics and english. This is a year's course, including all the basic features of map production on cartography, process photography, lithography, elementary photogrammetry. After the one year theoretical training, the students are employed for a second year on series of graded production tasks. The students were able to take advantage of the diploma course in cartography at Kingston College of Further Education.

Fairey Surveys and Hunting Surveys have given much time and thought to the training of cartographers capable of working in all parts of the world, especially in Asia and Africa; also in training local personnel in these countries. The urgency of the need for professional cartographers throughout the world has led steadily to the establishment of systematic academic training. The Cartographic Department of the Swiss Federal Institute of Technology in Zürich was one of the first to offer academic courses in map production. Similar courses were instituted in the United States and in the USSR. A chair of Cartography was established at the Budapest Eötvös Loránd University, with the view of maintaining an adequate supply of highly-trained cartographers to assist in the national mapping programme; many of the students hold also degrees in geography, geology, history or economics.

57

Discussion of the subject at international level came with the establishment of the Commission of the International Cartographic Association 'Training of Cartographers', 'to collect information on the different systems of training of cartographers now used in all member countries; to collate this information according to the various technical and professional levels and age groups of those under training; and to make this information available in as concise and convenient a form as possible.'

In Britain, in 1963, the working party set up by the Cartographic Sub-Committee of the British National Committee for Geography, first named 'Training of Cartographers' soon enlarged its brief and adopted the title 'Education in Cartography'. Its task has been to examine education in cartography in the United Kingdom in general and, in particular, to prepare for any consideration of the subject by the International Cartographic Association and to consider what range of studies is desirable in the education of cartographers. Defining the scope of the training, the working party agreed that it ranged from the study of the information collected by surveyors, of any kind, to the final reproduction of maps and charts by any means. The group estimated that between 3,500 to 4,000 cartographers were currently at work in Britain, with an annual intake of about 250, the majority being civil servants; most of them were draughtsmen, some eight hundred were reproduction staff and the rest professionals of various kinds. 'A Report by a Working Party of the Cartography Sub-Committee of the British National Committee for Geography' was produced in August 1965. Great variety of training was available to date; many university departments were already offering some kind of course in the subject, but the attention given to the technological aspect was minimal. The report suggested a basic list of subjects pertaining to cartography, which should figure in any training course from post GCE to post-graduate level, namely geography, measurement, design and reproduction. Pre-entry qualification for such courses should therefore include about five GCE subjects, including, preferably, geography at 'A' level and english language, mathematics, a science and a second language among the other subjects. A diploma in cartography was suggested, probably requiring a two-year full-time course; an MSc in cartography might follow first degrees in suitable subject disciplines. A Summer School in Practical Cartography was organised at the Department of Geography at Glasgow University, in 1964, with the co-operation of John Bartholomew and Sons Limited, Edinburgh. From this period of discussions emerged the Society of University Cartographers, with the aim ' to promote a high standard of cartographic illustration in universities ' and ' to evolve a training scheme for new entrants to drawing offices in Geography and allied departments '.

An Education Committee of the British Cartographic Society was set up in December 1966 to consider the scope of training courses at present in existence, to provide a forum for discussion of the aims of the various colleges and to consider further action. The Royal Geographical Society held a one-day forum on the training of cartographers in September 1967; it was convened by the Education Sub-Committee of the British Cartographic Society, under the chairmanship of J R Hollwey, and the members attending represented some of the chief organisations interested: Southampton College of Technology, Luton Technical College, Kingston College of Further Education, Oxford College of Technology, Redditch College of Further Education, Waltham Forest Technical College, the Ordnance Survey, the Survey Production Centre, the Military Survey, the Directorate of Overseas Surveys, Fairey Air Survey etc. Among the items discussed were the current pre-entry qualifications, the content of current and envisaged courses, the need for a national qualification and other similar points, and it was agreed to approach the Department of Education and Science to consider a national qualification in cartography, of agreed standard and syllabus. Following a resolution on 'National Diplomas in Cartography, Surveying and Photogrammetry', passed at the 1967 Commonwealth Survey Officers Conference, the British Cartographic Society is taking part in active discussions with interested organisations in Britain; while it is obvious that much remains to be done before a national qualification can be established, positive progress is at last being made.

Meanwhile, since the mid-1940s, the Department of Geography at the University of Glasgow had been developing its programme of teaching and research in the field of topographic science. At that time, it was usual for students in honours and general degrees in geography to undertake a fair amount of field traversing and plane-table surveying, the results of which were submitted as part of the examination. It soon became apparent, however, that with rapidly increasing technical developments and the demand for specialist cartographers in many fields, such superficial knowledge was not enough. Several geography departments began to offer cartography or mathematical geography as an optional third year subject— Edinburgh, Hull, Leeds, Leicester, London, Manchester among them —at Aberdeen, surveying and the history of cartography were offered. The Department of Geography in the University of Liverpool organised a short course in cartography in September 1965; part of the course was concerned with general topics, but emphasis

59

was on the work of a university cartographic department and the production of maps for books and journals. Glasgow, Cambridge, Newcastle, Bristol and Swansea offered surveying options.

The expansion of the Geography Department of the University College of Swansea was also the culmination of years of development. The new building, in 1956, included accommodation for a large general laboratory, a smaller honours laboratory, workshop and cartographic room, a map library and rolled map store, in addition to the more usual accommodation. An important step forward in British geography was this completion of appropriate accommodation and equipment for cartography and photogrammetry. A set of four photographic rooms houses a De Vere Horizontal Camera, a Leitz Reprovit 35mm copying unit, a Sirius dyeline copier and standard photographic equipment. Photogrammetric equipment included a Santoni Stereocartograph III, a Zeiss Stereotype, Zeiss and Bausch and Lomb Multiplex equipment, a Watts Radial Line plotter, and Zeiss Slotted Template equipment. One-year post-graduate diplomas in cartography were soon organised.

The University of Glasgow has made remarkable progress in courses for surveying and cartography since 1945. Second-year courses were soon offered to all honours students and two honours options in surveying and photogrammetry for BSC students having a background of science subjects. Until 1963, primary attention was given to the organisation of the first degree courses; provision of equipment was a costly business, especially for the photogrammetry course. In 1963, two post-graduate diploma courses began in cartography and photogrammetry, open also to graduates of other universities. Special equipment of great value has been installed for these courses, including survey instruments, plotting machines, photomechanical and photographic processing equipment. The department now has two Multiplex, a Wild A6 plotter, a Zeiss SEG IV rectifier and a new Wild B8 Ariograph. A Galileo-Santoni Stereosimplex IIC, a Kern PG-2L plotting machine and a Cambridge Stereocomparator were also obtained. Equipment in the Reprographics Laboratory includes a vacuum frame, whirler, diazo printer and photo-nymograph. Further details of the courses offered and of the equipment in the department are given in the article by G Petrie and J S Keates, of the University of Glasgow, 'Topographic science at the University of Glasgow', *The cartographic journal,* June 1968 and in the 'Courses in topographic science' brochure available from the university. Recently, some changes have been introduced in the Department of Cartography and Photogrammetry, in the light of experience; the introductory course common to both diplomas has been dropped and special short courses in cartography are given for photogrammetrists and in photogrammetry for cartographers; short courses in data processing and air photo-interpretation for both diplomas have been added and a short course in electronic measurement for photogrammetry students only. The diploma course in photogrammetry is recognised by the Natural Environment Research Council for the award of its Advanced Course Studentships. Practical work has been a feature of the courses throughout; specialised thematic maps have been made of various areas of Scotland and, in collaboration with glaciologists and geomorphologists, studies have been carried out on glaciers in Alaska and Iceland.

By this time, then, it has become axiomatic that cartographers need to be more widely educated and versatile; many of the students undertaking specialist cartographic courses already hold degrees in geography. Three kinds of experts are needed to prepare modern maps: cartographers and surveyors, draughtsmen and technicians. Much field work in different parts of the world is necessary before a cartographer can consider himself fully experienced. 'Revision' and discussion programmes will continue to be helpful. A one-week residential summer school in practical cartography for university cartographers, for example, was organised by the Department of Geography, Glasgow University, in September 1964; and the fourth meeting of the Symposium of the British Cartographic Society, held at the University College of Swansea, in September 1965, was devoted to the subject of 'The training and education of cartographers and cartographic draughtsmen', under the Chairmanship of D P Bickmore. Papers were given by M J Stansbury on 'The cartographic course at the Oxford College of Technology'; by K H Stephens on 'Cartographic training at Survey Production Centre'; by J Cridland on 'Basic training of Ordnance Survey draughtsmen'; these are reproduced, together with the ensuing discussions, in *The cartographic journal*, June 1966. Further useful references include:

P V Angus-Leppan: 'Cartography in the universities', *Cartography*, VI, 1, 1966.

R H Hardy: 'Current and future status of university surveying and mapping education in the United States of America', *Surveying and mapping*, June 1965.

R A Harris: 'Training of cartographic draughtsmen', *Canadian cartography*', 1964.

B Moriarty: 'Current status of cartographic education in American colleges and universities', *Professional geographer*, 1965.

P B Newlin: 'Proposed surveying and mapping curriculum at the University of Arizona', *Surveying and mapping*, June 1963.

A H Robinson: 'The need for state cartographers', *Surveying and mapping*, September 1963.

A H Robinson: 'The potential contribution of cartography in liberal education', *The cartographer*, May 1965.

2

INTERNATIONAL
MAPS AND ATLASES

' I've seen it on the map, but where is it really? ' NOEL
COWARD *Cavalcade*.

IT IS ONLY relatively recently that international agreement has been
reached on such matters as the prime meridian line, geographical
nomenclature, specifications for geodetic control, standard systems
of measurement and methods of presentation of topographic
features. International co-operation in geodetic surveying began
towards the end of the nineteenth century, gradually increasing in
scope, until in 1948 the Economic and Social Council of the
United Nations spent a large part of its Sixth Session in discussion
of these topics, acting on representations from such diverse countries as Brazil, Denmark, France and the United States, and
encouraged by such organisations as UNESCO, the Pan American
Institute of Geography and History, the International Council for
the Exploration of the Sea and the International Geographical
Union. Means of further encouraging national mapping programmes
were discussed with the aim of simultaneously developing closer
liaison with international programmes and the dissemination of
information on the new cartographic methods and techniques. A
follow-up meeting of experts was held in March 1949 to work out
more practical details.

Most of the topographical and cadastral mapping in the world is
now undertaken by national governments, though a great deal is
still done by non-governmental organisations. Governments also
publish at smaller scales maps based on large-scale original surveys,
either for general reference maps or for specific purposes. With the
development of closer communications between all the countries
of the world, however, it was inevitable that a map at a uniform
scale should become desirable, and such a project was first proposed
by Professor Albrecht Penck at the Fifth International Geographical
Congress at Berne in 1891. The scale suggested was 1 : 1M, at which
major natural and cultural features could be shown. The congress

63

approved the project in principle, but such a programme could be undertaken only with the backing of national mapping agencies; in 1909, the British government invited interested countries to a conference in London to discuss preliminary technical points. Leading members at this conference were Col S C N Grant, then director-general of the Ordnance Survey, Sir Charles Arden-Close, head of the Geographical Section, General Staff, J Scott Keltie, secretary of the Royal Geographical Society and Professor Penck himself. Meanwhile, discussions had taken place in 1895 and 1899 and final agreements on the technical aspects of the map were completed at the Paris Conference in 1913. It is interesting to read Professor Penck's own views, printed under the title ' Construction of a map of the world on a scale of 1 : 1M ' in *The geographical journal*, 1893.

It was agreed that the national mapping agencies should be responsible for the publication of relevant sheets and that the sheets should conform as nearly as possible with agreed specifications. A certain amount of mapping was completed during the thirty years or so following the 1913 agreements; Great Britain had undertaken 136 sheets of India and 132 sheets of Africa, while France and Germany had been mapping in eastern Asia. The Washington Congress in 1905 agreed that the United States should prepare a similar map of the American continent. Much disorganisation was caused, however, by the two world wars, and, at a meeting of the Seventh Assembly of the International Geographical Union held in Lisbon in 1949, a new Commission on the IMW was established with revised terms of reference. For the full recommendations and discussions, the *Report of the Commission on the International Map of the World 1 : 1,000,000* should be consulted (New York, American Geographical Society, 1952). Also in the *Report* are given the details of the existing state of publication and comments on specifications. A Central Bureau had been established at the Ordnance Survey offices in Southampton, to act as liaison between the governments concerned and to assist them in co-ordinating publication of the maps in the standard form; in 1949 the Commission examined what changes were desirable in the Bureau organisation. Because of the 1914 war, the first *Annual report* of the Central Bureau was not published until 1921, when information on progress in all countries had been reported and the resolutions of the 1909 conference, amended in Paris in 1913, were reproduced. Thereafter, *Reports* were issued annually until the 1938 *Report*, which included an account of the state of publication of the map to date, by Major Sanceau, then secretary of the Central Bureau. During the 1939-45 war, the offices of the Central Bureau at Southampton were destroyed, with the whole stock of IMW maps and the Central Bureau was transferred to the new Ordnance Survey offices at Chessington in Surrey. A short *Report* was issued covering the period 1939-48, asking for information from all countries on the progress of the Map. Forty one countries were in formal membership of the IMW project, but, among all the other difficulties encountered, not the least was the lack of maintenance of the subscriptions by some countries.

At the 1949 Lisbon Congress, then, a number of important topics were due for discussion: the organisation of the Central Bureau, any changes desirable in specifications and the most fundamental point of all, the function of the Map after fifty years of world changes. With regard to the Central Bureau, agreement was easily reached and it was also agreed that its functions should be widened to include the presentation of recommendations for revision of specifications regularly at international congresses on mapping and the organisation of special meetings to take any international action necessary at any time. The continued need for the IMW was agreed without hesitation, but it was emphasised that much work was required before the Map in fact met this need. For the first time, the transference of the Central Bureau to the United Nations was put forward as a serious proposition, in the light of the increased role of the United Nations in the co-ordination of the cartographic services. For example, seeing that the ICAO, FAO and UNESCO, besides other specialised organisations, published much of their information in cartographic form, the question was seriously debated whether any duplication of effort could be avoided.

The 1913 specifications had emphasised relief, hydrology, communications and population centres; these specifications had been published in 1914, as ' *Carte du monde au millionième* ' *Comptes rendus des Séances de la Deuxième Conférence Internationale.*

By 1949, it was estimated that the number of sheets required to cover the world would be 2,122. 405 sheets had been finished to date, of which only 232 had been compiled in accordance with the full specifications so far agreed; among these were the 107 sheets making up the *Map of Hispanic America,* compiled by the American Geographical Society. During the 1949 discussions, debate ranged between continuing the policy of decentralisation of the IMW to each nation, decentralising production to each nation, but maintaining a central cartographical establishment for co-ordination, and location of both organisation and production at a central cartographical establishment. (A map showing the status of publication in 1952 was included in the *Report on the United Nations Regional Conference for Asia and the Far East,* II, 1952.)

Another main line of discussion concerned the relationship of the IMW and the ICAO Aeronautical Chart. Professor Penck had originally thought of each sheet as 'a region in the frame of its natural surroundings', to be used for 'varied purposes of administration, navigation and commerce'. But he could not have envisaged such dramatic changes in 'natural' environments throughout the world, to such an extent that mapping at this scale might prove inadequate. The conclusion was that the two separate series were needed, because the IMW was a geographical map, incorporating as much detail as was possible on the small scale, whereas the Chart, designed with the specific needs of visual navigation in mind, carried the minimum of detail and that in a specialised form. It would be possible, however, to use the same sheet lines. Projections also could be standardised; at that time, the IMW was drawn on a modified polyconic, whereas the Chart used the Lambert conformal conic projection between the equator and 80 degrees latitude, and the polar stereographic projection beyond. In different countries, the need for both maps differed; most countries agreed that both series were necessary, but Canada, for example, asserted that there was no possibility of producing the IMW for the whole of its vast, sparsely populated areas. Most countries agreed that the role of the IMW as a base map was largely superseded, that the scale was too small to show really helpful detail, yet too large for effective generalisation, but that too much labour and cost would be involved in changing the layout of the map. Discussions continued, therefore, but certain advances were made. By September 1953, the transfer of the General Secretariat from the Ordnance Survey to the Cartographic Office of the United Nations was complete and the administration of the IMW was included in the vast aim of co-ordinating national cartographic services. *Annual reports* from 1954 have been published by the United Nations, Department of Economic and Social Affairs, New York, in English and French. Accounts are given of the progress of the administration and state of mapping, with index maps showing existing coverage by new or revised sheets. Bibliographies of selected official documents relating to the IMW appeared in the *First progress report* prepared by the Secretariat for *World cartography,* IV, 1954; and a first supplement to the bibliography was published with the *Report* for 1961.

At the IMW Conference in Bonn, 1962, new and more definitive specifications were agreed (United Nations: Technical Conference on the International Map of the World on the Millionth Scale, in two volumes: I, *Report and proceedings;* II, *Specifications of the IMW on the Millionth Scale,* UN., 1963, with illustrations and maps). The Lambert conical conformal projection was agreed as desirable for all new sheets outside the polar regions, with the existing modified polyconic projection left as optional, while the polar stereographic was to be used for polar regions. Agreement was reached also on sheet lines, to be basically four degrees north-south by six degrees east-west; also on major points of relief presentation and conventional signs—all set out in the *Report* above mentioned. The new resolutions were more flexible and therefore more useful. In effect, the specifications had a twofold aim: to provide, by means of a general purpose map, a document which would enable comprehensive study of the world to be made and which could therefore be of considerable assistance in planning for economic development; and to provide a comparable series on which thematic maps of the world could be based. The conference decided not to include vegetation cover on the sheets of the IMW, but to leave this as the subject of a special series. Members also agreed to continue any co-ordination possible between the IMW and the World Aeronautical Chart and to encourage the completion of the IMW sheets in the developing countries. Conventional signs were revised to bring them into line with the latest cartographic techniques; the number of contours and spot heights was increased and the one hundred metre contour was made compulsory.

A word should be said concerning the World Aeronautical Chart, which has been mentioned so far only in its relation to the IMW. The Chart, also at the scale of 1:1M, was devised by the Aeronautical Chart Service of the United States Air Force, and when the International Civil Aviation Organisation was set up, the Chart was put at its disposal. From the beginning, the present World Aeronautical Chart series developed and has proved of such practical use that many countries have given it priority over the IMW 1:1M. Features on the earth must be shown in such a way that they can be easily identified from the air. International boundaries are important, but not internal ones; the size and shape of settlements are the vital features and roads and railways are important as landmarks. Place names are of low priority, but additional information such as radar aids and technical air traffic control data are vital, also, of course, the position of airfields.

The primary use of the 1:1M map, as a kind of reconnaissance map, is not so urgent as it was sixty or seventy years ago. Aerial photography has made possible the rapid mapping of topographical features and large areas of the world are now covered at the scale 1:250,000. It should not be thought, however, that the IMW project has been a failure; achievements have been direct and indirect. The project gave considerable stimulus to mapping at the turn of the century, particularly in those countries hitherto unmapped at even small or medium scales. Secondly, of permanent importance have been the *Map of Hispanic America,* by the American Geographical Society, the *Map of Brazil* by the Club de Engenharia of Rio de Janeiro and other maps completed as a definite part of the series. The International map of the Roman Empire, *Tabula Imperii Romani,* 1 : 1M, was initiated by O G S Crawford at the International Geographical Congress of 1928, at which a Commission was set up to give the project further consideration. A congress was held for this purpose in London, at which it was decided that the preparation of the Map should be undertaken by the governments interested, that the Central Bureau of the IMW should co-ordinate the project, assisted by a Permanent Council of four members and that the IMW should be used as the basis of the Map. A progress report was issued by the Ordnance Survey in 1933; a further congress was held at the Royal Geographical Society in 1935 and the resolutions were published by the Ordnance Survey with the title ' International map of the Roman Empire ' (London Congress, 1936). Sheets were contributed by Italy, France, Egypt, Great Britain and Germany. Thirdly, the sheet lines have been used by many government mapping agencies as bases for other series. Fourthly, thematic mapping at 1 : 1M, for example, of geology, land use, vegetation and population, has enabled comparative study of these aspects increasingly throughout the world.

It must be remembered that the IMW was intended for indoor use, at leisure, for study or planning; also that the chief reason for uniformity of scale and specifications was to enable an immediate understanding of the size and character of another country. Two critical replies can be made to the latter assertion—that the IMW is no longer a sufficiently large-scale map on which to plot the complexities that exist today in parts of every country and also that, now that so much of the earth's surface has been covered by at least medium-sized topographical maps compiled by the national mapping agencies with the aid of aerial photography, the greater need at present is for base maps on which to compile the requisite thematic maps.

The production of the Land Map and Air Chart, 1 : 250,000 (Joint Operations Graphic), is an ambitious attempt to cover the world in a new series, designed to satisfy the requirements of ground and air forces in joint operations. Begun in 1966, the Map and Chart have the same topographic printing, but the ground version has heights and contours in metres, the air version in feet; the air version has also the full amount of detailed information needed in aviation. A paper describing the work at the Survey Production Centre and the technical details of production, presented to the

Symposium of the Cartographical Society, September 1967, was prepared by H A Bennett, L A Hogg and A A Hockyer; this paper, with figures and a printed specimen of the map, is included in *The cartographic journal* for December 1967.

The arguments indicated above and some others were taken up independently in the mid 1950s by the Department of Cartography, State Office of Geodesy and Cartography, Budapest, under the direction of Professor Sándor Radó and were subsequently discussed with the Geodetic and Cartographic Services of Bulgaria, Czechoslovakia, the German Democratic Republic, Rumania, Poland and the Soviet Union. After several years of preparation, a uniform world map at 1:2,500,000 began publication in 1964. This map is planned to cover the earth's surface in 244 sheets; it is considered to provide a valuable map in itself and to serve as a base map for applied maps of any kind. The same sheet and reference systems are used as for the IMW 1:1M series. Much thought has been given to the choice of projections; between 64 degrees south a conic projection is being used and the polar regions are being drawn on azimuthal projections. This is a general geographical map, dealing with physical and political aspects, using uniform conventional signs and a standardised colour scheme. The official languages of the map are Russian and English. Within the map, all geographical names are given in the official language of the territory concerned, in a Latin script or in an internationally accepted transliteration. It is a complex map, which nevertheless preserves clarity. Twelve colours are used, with land lettering in black and sea lettering in a bolder blue than that used for the water areas. Seven coloured layers, greens through yellow to browns, are separated by brown contour lines at 1-100, 100-200, 200-500, 500-1,000, 1,000-2,000, 2,000-3,000, 3,000-4,000 metres. Contours are inserted also at 300, 700, 1,500 and 2,500 metres. Sea depths are shown by five coloured layers and ten bathymetric lines between 100 and 7,000 metres, with important depths stated in exact figures. Settlements symbols distinguish six groups, between ten thousand and one million inhabitants; red hatching shows towns of more than three hundred thousand. Grey is used for railways, in two groups, and red for roads, in three groups. Canals are also shown. A practical approach is shown in the selection of data; for example, the distinction is made between perennial, seasonal and underground rivers. The first sheets to appear were those for London, Madrid and Rome, published by the East German Department of Geodesy and Cartography, with index sheets; they were drawn by Cartographia of Budapest in 1962. With them was issued a four-page explanatory leaflet, edited by the Hungarian State Office of Geodesy and Cartography, Budapest,

1964. Many more sheets have appeared since then and an interesting feature of the publishing programme is that groups of sheets cover large contiguous areas: three sheets of western Europe, for example, compiled by the Cartographic Service of the German Democratic Republic, ten sheets of southern and central North America, compiled by the Cartographic Service of Hungary, eight sheets of south and central Africa by the Cartographic Service of Poland. The South American sheets drew considerable attention on the occasion of the first Latin-American Regional Mexico Conference of the International Geographical Union, when not only the published sheets but the next sets of preliminary pilot prints were on exhibition. Further details are to be found in an article by Professor Sándor Radó, 'World map, scale of 1:2,500,000', in the *International yearbook of cartography*, VI, 1966, with an index map attached to the rear cover. Pergamon Press is the United Kingdom agent for these maps.

The United Kingdom has been responsible for the mapping of much colonial territory and has contributed in personnel and methodology to the mapping of many countries, particularly of India, Egypt and Siam. The importance of regular topographical survey and revision to a national economy was officially recognised in 1905, with the establishment of the Colonial Survey Committee, which concentrated mainly on tropical Africa. The Directorate of Colonial Survey, established in 1946, now known as the Directorate of Overseas (Geodetic and Topographic) Surveys, became in 1964 a part of the Ministry of Overseas Development, which in turn became the Department of Technical Co-operation; the headquarters of the Directorate is at Tolworth, Surrey. The Directorate's work was in this way acknowledged as part of Britain's programme of aid to developing countries; its resources are available under a technical assistance scheme to developing countries, both dependent and independent. The Directorate helps developing countries overseas, at their request, in basic survey and mapping and co-operates with local survey departments so that maps may be produced as rapidly and as economically as possible, frequently being able to cross national boundaries, if necessary, in order to plan geodetic survey and air photography on a wider basis. The maps themselves are compiled at Tolworth; there will be occasion to mention many of them in the course of chapter 3. The Directorate also acts as a centre for the collection and dissemination of technical information on surveying and mapping and it maintains comprehensive map and air photo libraries of overseas territories.

Maps are produced from air photographs, usually on the scale of 1:50,000; unproductive areas are mapped at 1:125,000 and the more populated areas at 1 : 25,000. Except in terrain where contours are considered essential, the first maps are one- or two-colour editions showing planimetric detail only, after which the second contoured editions are usually produced, adding detail provided by the local survey department. Many series have now been printed in five or six colours and specifications have been drawn up with individual countries for the maintenance of permanent national series. In many areas, thematic mapping also is well advanced; special investigations in these cases may be undertaken by the Directorate's Forestry and Land Use Section or by the Geological Survey Department. The *Annual reports* are central sources of information on the work of the Directorate. Each includes a summary of the year's work, analysed under headings such as air photography, field survey, computing, mapping, land resources investigation and training. The aim always is to establish a permanent framework of survey stations which can be maintained by local surveyors. A quarterly *Newsletter* is also issued and a *Catalogue of maps,* compiled in 1960, is supplemented by monthly lists of additions. For further information, F Dixey: *Colonial geological surveys: a review of progress during the ten-year period 1947-56* (HMSO, 1957) and K M Clayton, ed: *Guide to London excursions,* section 28 (20th International Geographical Congress, London, 1964) are recommended sources.

France, Portugal, Italy, the Netherlands and Germany have also been assiduous in mapping territories under their jurisdiction; many individual series have been initiated and, according to circumstances, maintained, as will be apparent in the relevant sections of chapter 3. The *Atlas des colonies françaises: protectorats et territoires sous mandat de la France* should be remembered in this context. Prepared under the direction of G Grandidier, it was published in Paris by the Société d'Editions Géographes, Maritimes et Coloniales, 1936; interleaved text accompanies the maps and there is a gazetteer. There had been a previous *Atlas colonial français,* by P Pollacchi (Paris, L'Illustration, 1929). The *Atlas de Portugal Ultramarino* was issued by the Ministério das Colonias, Lisbon, in 1948. Then there are the *Grosser Deutscher Kolonial Atlas,* compiled by P Springade and Max Moisel (Berlin, 1901-11), the *Deutscher Kolonial Atlas,* by the same compilers, published in 1936 by Fritz Lange, Berlin, and the *Wirtschafts Atlas der Deutschen Kolonien,* by M Eckert, published in Berlin in 1912. The *Atlante del colonie Italiane* by Mario Baralta and Luigi Visintin was published by the Istituto Geografico de Agostini in 1928 and the *Atlas van Nederland,* issued by Martinus Nijhoff of the Hague in 1938, in addition to the maps and atlases of individual territories to be mentioned in the next chapter.

Various other official bodies issue world series for special purposes. The War Office publishes a world map at 1 : 500,000; the Institut Géographique National also publishes a world map at this scale and produced a *Carte général du monde*, 1 : 1M, 1963-65. The chief world series of the Department of Geodesy and Cartography, Berlin, is at 1 : 2,500,000. Other government departments maintain 1 : 1M world maps, such as the Army Map Service; and some of the more lately developed national mapping agencies have begun map series on a world scale, such as that of the Survey of Kenya, Nairobi, at 1 : 250,000, 1960-.

The great geographical societies play an influential part in mapping all parts of the world. The American Geographical Society of New York maintains a series of maps of the land areas of the world at 1 : 5M. All sheets of the series covering Africa, Europe, Asia and Australia are constructed on a projection system devised by O M Miller at the request of the Army Map Service some years ago. The system consists essentially of three conformal oblated stereographic projections connected by ' fill-in ' projections in such a way that the latter are conformal at their edges and join the basic projections with no discontinuity of scale or angles. The society has produced numerous other maps of various parts of the world, many of them mentioned elsewhere in this text. The Cartographic Department of The Royal Geographical Society prepares maps of all kinds for the guidance of expedition parties or in demonstration of lectures, monographs or articles. The map supplements of the National Geographic Society have been notable for many years for their clarity and excellence of production.

Series of sheet maps, wall maps and charts are prepared and maintained by almost all the great commercial cartographical houses of the world. The selection mentioned here is determined by availability in Britain; they range from the scientifically accurate maps suitable for advanced study or planning to the increasing numbers of ' popular ' maps so useful to tourists, the best of which, while being presented in a more pictorial fashion, are still basically accurate. Indeed, many of them are actually based on the scientific surveys carried out by other surveying bodies. Fairey Surveys Limited is a fine example of the latter. From the early days of its foundation in India in 1923, Fairey Surveys and its associated companies—Fairey Air Surveys of Rhodesia Limited, Salisbury; the Air Survey Company of Pakistan Limited, Karachi; Fairey Air Surveys of Nigeria, Limited, Lagos; Fairey Surveys of Zambia Limited, Lusaka; Air Survey Company of India Limited, Dum Dum

—have worked in almost every part of the world, constantly adapting and improving methods of rapid and accurate reconnaissance surveys of all kinds, especially for town planning, irrigation and conservation projects and transport developments. The Hunting Organisation is a vast group of co-ordinating companies, of which the group most relevant to our work is the Hunting Surveys and Consultants Limited concerned with mapping, land survey and vertical photography; applied geology and geophysics, land-use and agricultural consultancy and the photographic services, each group consisting of two or more associated firms. The *Hunting Group review* includes information on all projects, in preparation or planned, with very useful photographs, diagrams and maps; and the *Information sheets* are devoted to detailed descriptions of the procedures involved in one assignment. The Cartographic Department of the Clarendon Press, which became a separate department in 1952, is mentioned here as a reminder only, for its pioneer work and individual maps and atlases are commented upon in other chapters of this text. The uniformly high standard of cartography issued by this department has resulted in a demand for its products throughout the world.

Of long-standing repute also is the Geographical Institute of Edinburgh, traditional headquarters of the firm of Bartholomew, founded in 1797. Six generations of cartographers, George John senior, who established the Geographical Institute, John, John George, noted for his development of the method of layer colouring, a third John, himself an eminent geographer, and, at the present time, John, Peter and Robert Bartholomew, have made many outstanding contributions to map-making; *The Times Atlas of the world,* other thematic atlases and British map series are mentioned elsewhere. In the educational field, the Geographical Institute has produced, for many parts of the world, numerous school and university atlases and map series, often in the vernacular. Bartholomew's world reference map series is in universal use; this is a series of detailed topographic maps, frequently revised, which present continents and regions of the world at larger scales than can frequently be offered in the average atlas. Each map is issued in a practical fold, with varnished covers. In the annual *Catalogue* of Bartholomew maps, an index diagram shows the world coverage by these maps. The whole continent of Africa is shown on the map at 1 : 10M, revised in 1967, with contour colouring, roads in red and railways in black. Boundaries show up well in purple. 'Africa, central and east', 1963, covering the Cameroun, Kenya and territories to the Congo is at 1 : 4M; 'Africa, central and southern' is a new map, 1968, at 1 : 5M, extending from the Congo to the Cape,

73

with an inset of Madagascar on the same scale: 'Africa, north-east', another new map at 1:5M, covers the territory from Tunisia to Somalia; 'Africa north-west', revised in 1966, also at 1:5M, extends eastwards to include Niger and Cameroun and southwards to the north of the River Congo; 'Africa, south and Madagascar', at 1:4M, revised in 1967, extends from Cape Town to the northern end of Lake Nyasa, overlapping 'Africa, central and east', and including Madagascar as an inset. 'North America', at 1:10M, revised in 1967, includes the Aleutian Islands, Greenland, the West Indies and Central America. The map of South America, at the time of writing, is withdrawn for revision. 'South-east Asia' at 1:5,800,000, was revised in 1966; it extends from Hong Kong to Java and from Burma to New Guinea. 'China', revised in 1967, is at 1:4,500,000; it includes Mongolia, Korea, Formosa, northern Laos and Vietnam. 'Japan', 1967, is at 1:2,500,000 and includes insets of Tokyo at 1:500,000, Okinawa at 1:1M, and Iwo Jima at 1:250,000. 'India, Pakistan and Ceylon' is available at 1:4M, revised in 1966, including Afghanistan, Nepal, Bhutan and the Andaman and Nicobar Islands. 'Israel' has just been revised, 1968, at 1:350,000, with inset maps of Negev and the Sinai Peninsula, and the 'Middle East' at 1:4M, 1967, covers the whole of the area bounded by Turkey, Iran and Arabia. Europe and Asia are available on one sheet, at 1:15M, revised in 1966. 'Europe, central', at 1:1,250,000, 1963, carries relief hill-shading and is intended mainly as a motoring map, indicating detailed road classifications and numbering; it extends from London to Venice and from Berlin to the Riviera. 'Europe, central and south-east', a new map produced in 1968, at 1:2,500,000, extends from the Gulf of Danzig to Crete and from Lausanne to Istanbul. 'Europe, western', at 1:3M, covers the area from Spain to Italy to southern Scandinavia and there is also a political map of Europe at 1:5M, 1967, on which international boundaries are clearly shown in different colours and railways and shipping routes are entered in black, but no roads. 'France and the Low Countries' are issued in two sections, north and south, at 1:1M, both revised in 1964, and there is a separate map for Scandinavia, at 1:2,500,00, 1965, which includes Denmark, Norway, Sweden and the Baltic. 'Australia', 1966, is at 1:5M, and 'New Zealand' at 1:2M, with inset plans of Auckland, Christchurch, Dunedin and Wellington at 1:200,000, also of some of the island groups. The Bartholomew maps of the British Isles are mentioned in chapter 3. There is also a 'World route chart' on the Mercator Projection, at 1:45M, 1967, showing railways, shipping and air routes against political colouring. Maps and atlases are also prepared on request for a large number of other firms and organisations both in Britain and in other countries.

Messrs George Philip and Son Limited began as a Liverpool business in 1834, transferring to London in 1856. The firm always had a strong interest in education and, with the passing of the 1870 Education Act, the publishing programme expanded to meet the requirements of elementary education. In 1902, when the London Geographical Institute came into being, a number of educational and reference atlases, wall maps, globes and textbooks had established themselves on the market. Sir John Scott Keltie and Sir Halford Mackinder edited textbooks for the firm. Through the years, several other firms have been absorbed into the business, notably Smith and Son's globe-making in 1916, Edward Stanford's map printing and publishing house in 1947 and Georama, illuminated globe and map firm, in 1948. Philip issues Stanford's general maps; these are political reference maps, usually having coats of arms or national scenes as a border, covering the world or individual continents, designed mainly for school, office or home use. As in the case of the Bartholomew maps, examination of the firm's maps produced during more than a century reveals the progress of geographical and cartographic thinking and technical developments.

In addition to atlases, to be mentioned later, Philip has specialised in wall maps. The large wall map of the world is drawn on Winkel's Projection, with the Atlantic in the centre; the smaller world map is on Gall's Projection, showing main shipping lanes. There is also an ' International world map ', with political colouring showing up vividly against a dark blue sea, and the principal towns marked. Physical maps of the continents or large regions are available at various scales, mounted on cloth, either with rollers or dissected to fold. A series of smaller wall maps of countries—Scotland, for example—is available separately; they show county or state divisions, communications and shipping routes and towns by symbols graded according to population. In the smaller series of physical maps, layer tinting is used without hill-shading and political boundaries are shown in red. A newer development is the series of ' Graphic Relief Wall Maps '; South America, for example, shows up very dramatically by this method, the curving edge of the Andes contrasting with the greens of the Amazon basin. The sets comprising the ' Comparative wall atlases ' are interesting; sets of five maps show land relief, political divisions and communications, four diagrammatic climate maps on one sheet, natural vegetation and density of population for each continent.

The first plastic relief and relief model maps to be produced in Britain were by the Clarendon Press, Oxford. Of the world series planned, the map of Great Britain and Northern Ireland, at a scale of one inch to sixteen miles, was the first to become available in 1964, in both flexible and rigid plastic. ' The Holy Land ' was also issued in 1964, at a scale of six miles to one inch; Biblical names are in different colours to indicate whether they occur in the Old, or the New Testament. Philip's ' Plastic relief model map of the world ' is very finely executed; measuring fifty-one by thirty-seven inches, it gives a particularly evocative representation of the great mountain chains of the world, the major river valleys and stretches of desert. Invaluable in both geography and history teaching, it cannot fail to assist in the understanding of movements of peoples, development of communications, etc. A ' Plastic relief model map of Europe ' is also available and a great variety of regional maps at various scales is issued, all set out in Philip's educational catalogue. Philip acts as agent for the Breasted-Huth-Harding historical maps and for the Denoyer Geppert world history wall maps, as well as issuing a splendid variety of special maps for the teaching of history, scripture and current affairs.

Geographia issues a series of wall maps, both special-purpose and general. Among them are three political maps; one on the Mercator Projection, mainly for commercial use, with a separate gazetteer-index; a map showing shipping routes very clearly, designed for industrial and administrative use; and the third, on the V der Grinten Projection, showing railways and shipping routes. Two predominantly economic maps are printed in French and German, one having statistical data in English on the reverse. The Cartographic Department of Nelson, from 1956, now removed from Edinburgh to London, and the later Cartographic Section of the Pergamon Press at Oxford, have so far become noted chiefly for atlas production.

Westermann maps are available in Britain through the Grant Educational Company Limited, Glasgow. From its foundation in 1838, the Georg Westermann Verlag, Braunschweig, has ranked with the great cartographical publishing houses of the world. More than a thousand skilled workers and specialists are now kept currently at work, using the latest reproduction and printing processes, mainly in the service of education at all levels. Among the most important parts of the publishing programme are the geographical and cartographical productions, sheet maps, atlases and teaching aids of all kinds. The house journal, *Westermanns monatshefte*, includes information of the firm's productions and of other research developments. Most of the Westermann maps are available in up to twelve languages; English, French, Dutch, Danish, Norwegian, Swedish, Finnish, Spanish, Italian, Arabic and Persian, in addition

76

to the original German. Wall maps of the world are drawn at the scales 1 : 15M and 1 : 18M; continent maps at 1 : 6M, with the exception of the map of Europe, which is at 1 : 3M. Regions of Asia are mainly at 1 : 4M, of America and Africa, at 1 : 3M and European countries at 1 : 900,000, 1 : 600,000 and 1 : 300,000, but new maps are constantly being added. All the wall maps are mounted on strong linen, with rollers and securing straps and strong hanging cord.

Special maps illustrating world living standards, economic alliances, world climates, vegetation belts, density of population, etc, use effective symbols, clear keys and explanatory legends. Relief featuring is bold, using layer colouring imaginatively to emphasise continuity of mountain chains, deep sea trenches and areas of lush vegetation. River systems are particularly well traced and, on the larger scale maps, hill-shading clearly indicates the relief detail. The size and, for the larger centres of population, the shape, is suggested by different symbols. A particularly useful series comprises detailed town plans, usually containing an inset of an especially important part; for example, New York, with an inset of Manhattan. Intended for teaching purposes—or for sheer fun—the ' Westermann pictorial wall maps ' were introduced a few years ago; a modern version of the ' Here be dragons ' idea, with the difference that these maps are based on accurate data. The pictorial map of Africa, for example, admirably suggests the vast drainage systems of Niger, Zambezi and Nile and vividly points the contrast between the arid wastes of the Sahara and the teeming life of tropical Africa. The Ahaggar highlands in the Sahara, frequently overlooked, are unmistakable and the great variety of flora and fauna indicated is astonishing. As the publisher notes, such a map forms an admirable transition to the two-dimensional map and is excellent for project work.

The Esselte Corporation, Stockholm, is a private mapping firm, which had great influence on Swedish map production and on international co-operation. For many years, the firm has maintained an active exchange of technical experience with the leading map printers in other Scandinavian countries and in Great Britain, America, Switzerland, Germany and France, and has itself taken the lead in glass engraving and the use of plastics in mapping techniques. The firm maintains most of the official maps, as well as producing educational atlases, thematic and tourist maps. Scribing techniques have been further improved by the design of new engraving tools, which are also exported to other countries. Esselte did much to encourage the formation of the International Cartographic Union and co-operates with other leading cartographic

77

publishers in international work, notably *The international year-book of cartography*.

The publishing houses of Artaria and Freytag-Berndt have been issuing educational wall maps since 1894, including general, physical and thematic maps. Methods were completely revised following the second world war and the latest techniques are now used. The maps are striking, using contour lines, distinctive hypsometric tints and spot heights for relief representation and clear drawing of river systems, roads and other main features. The scales adopted range from 1:100,000 to 1:300,000 for the home country, to 1:1,500,000 for north, central and western Europe, the whole of Europe at 1:3M and the other continents at 1:6M. A physical map of the world is available at 1:22M and a trade and communications map of the world at 1:25M. Recent maps show particularly fine hill-shading, which is done on Astralon or with an Aerograph air-brush. Details of the publication programme and some plates of examples from the map sheets are given in the article by Fritz Aurada: 'Entwicklung und Methodik der Freytag-Berndt Schulwandkarten', in the *International yearbook of cartography*, VI, 1966.

Two great Italian cartographic houses have made incalculable contributions to world mapping, both general and thematic. The Touring Club Italiano, founded in Milan in 1894, has since worked unceasingly to produce scholarly atlases, maps of Italy, Europe and other countries, a great number of travellers' guides, monographs, yearbooks and periodicals. The world atlases are mentioned later in this section, while the national and regional maps are referred to in their relevant places. Of great interest is the article by Manlio Castiglioni, 'Les récentes publications cartographiques du touring club Italiano', in the *International yearbook of cartography*, I, 1961.

The Istituto Geografico de Agostini of Novara was founded in Rome in 1901. The years have shown the development of cartographic techniques and the improvement of instruments; staff are trained at the institute in all the stages of map production, from first planning to printing and all the processes involved in revision. The institute was one of the pioneers in the photographic setting of names, particularly effective in revision and in translation into other languages. Plastic relief maps are now a speciality, for which separate catalogues are available. Wall maps for schools are produced in effective new relief styles. For further details of the work of the institute, an authoritative source is the article by Umberto Bonapace: 'La production cartographique de l'Institute Géographique de Agostini: buts et problèmes actuels', in the *International yearbook of cartography*, III, 1963.

The physical wall maps produced by the Hermann Haack Verlag,

Gotha, have become noted for the boldness and clarity of their design. The firm will be mentioned several times in later pages. Passing reference should also be made here of the cartographic work of Mairs Geographischen Verlag, though the majority of their work concerns Europe and, particularly, Germany. The Hallwag political maps have become familiar during recent years; they are fully coloured to show *de facto* boundaries. Relief is shown by hill-shading, with spot heights where desirable. Communications, including shipping lanes, pipelines and such special features as ruins or oases, are clearly depicted and town stamps are graded to give an indication of population. A world map at 1 : 33M was issued in 1964 on an equal area projection. The map is striking in appearance, mainly political in purpose, but showing relief in light grey shading. International boundaries, roads, railways, shipping routes and power lines are shown, also, in some areas, particularly interesting antiquities. Three insets are included, one on each of the polar regions and the third a map of the world on the Mercator Projection, showing time zones. For reference regarding current world affairs, it is a most effective map. The firm founded by André Michelin has specialised in guide books, including maps, and road maps, with tourists' interests chiefly in mind. A series of maps of various parts of Africa have recently been introduced.

Two examples of recent innovations must suffice to indicate the ingenuity with which cartographic organisations have attempted to meet contemporary requirements. To provide a regular topical map service on an international scale, the Hungarian State Institute for Geodesy and Cartography in Budapest began production in 1965 of a map service called *Cartactual,* with Professor Dr Sándor Radó as editor-in-chief. Quarterly looseleaf issues containing twelve map plates, with groups of thematic maps, sketch maps, etc on various topics, have been published since 1965, tracing changes in geographical information throughout the world, such as modifications of administrative boundaries, expansion or change in transport systems, changes in the numbers of inhabitants of settlements, foundation of new settlements, new geographical names and other such actual data required to keep maps up to date. The sets of maps are single sided, which facilitates quick reference and practical handling. The first issues were in monochrome, each map being carefully drawn with clarity and simplicity, with adequate documentation. With volume II, 1966, red colour was introduced to show changes, black remaining for existing detail. A particularly valuable feature is included, one which is fortunately increasing in map publishing, namely, the inclusion, not only of the source of data depicted, but also of the date. The second example demonstrates another pressing current issue, the potential relation between land and air mapping. In 1966, the Joint Operations Graphic Land Map and Air Chart, 1:250,000, was designed for use by ground and air forces, the same topographic detail appearing on both map and chart, with suitable variations. Production, by the Survey Production Centre, is planned for about six sheets a month. Involved in the production are the Cartographic Group, the Geodetic Office, the Map Research and Library Group and the Photographic and Printing Sections. Full details regarding specifications and a specimen section of the printed map are given in an article by H A Bennett, L A Hogg and A A H Lockyer, 'The land map and air chart at 1:250,000', in *The cartographic journal*, December 1967.

References to the IMW are to be found in all issues of *World cartography*. Particularly important is 'The International Map of the World on the Millionth Scale and the international co-operation in the field of cartography', in the issue for 1955, pp 1-12. Frequent references are also made in the reports of the United Nations Regional Conferences for Asia and the Far East. New issues are reported in *The cartographic journal*. Other particularly useful references to world mapping include:

G R Crone: 'The future of the International Million Map of the World', *The geographical journal*, March 1962.

R A Gardiner: 'A re-appraisal of the International Map of the World (IMW) on the Millionth Scale', *Empire survey review*, XVI, no. 120; reproduced also in the *International yearbook of cartography*, 1961.

A R Hinks: 'The International One-in-a-Million Map of the World', *Royal engineers' journal*, XVII, 1913, 77-86.

United Nations: *Modern cartography: base maps for world needs*. A summary of world mapping and details of mapping procedures.

WORLD ATLASES

By the beginning of the nineteenth century, the atlas form as we know it was in existence, gradually improving both in accuracy and in cartographic presentation. The classic concept of an atlas was essentially that of a collection of maps brought together for a purpose, with sufficient accompanying legend to make the maps intelligible and sufficient conformity of format to give an impression of unity to the production. The maps should be arranged in logical sequence, just as in the case of high quality books, and should include detailed tables of contents and an adequate index. Great attention should be given to the style of binding, to the paper and the type faces to be used, also to the style of spelling of place names. In recent years, the tendency to include explanatory text has increased, sometimes amounting to a considerable proportion of the whole work and occasionally bound separately in booklet or monograph form. Graphs, diagrams, statistics, photographs and bibliographical references are also common features of contemporary atlases; looseleaf form is frequently adopted, to facilitate frequent revision of those plates most needing updating. The increasing skill in cartographic techniques outlined earlier has enabled projections and scales to be used more effectively, symbols to be more evocative of information and it is accepted that the fullest documentation should be included, marking the authority of the work and enabling the atlas form to be in the fullest sense an invaluable reference tool.

The large general atlases are intended mainly for location and reference use, but, during the past decade particularly, increased interest in road travel, new approaches to regional planning, etc have introduced new features into the pages of general atlases. Noticeable also is the higher standard of educational atlases, reflecting the more advanced and practical approach to teaching in schools. It is perhaps in these atlases, in which the pressure is most felt to include as much detail as possible on relatively small scales, that improvements in techniques have been obvious; for example, the insertion of names to required size and style, by means of the photonimograph or other precise method, which enables more names to be included without blurring the topographic detail and also enables them to take a more active part in presenting, for example, comparative sizes of settlements.

Discussions of the International Geographical Union Special Commissions and of the International Cartographical Association have encouraged agreement on standards between countries; legend matter is increasingly presented in more than one language and the co-operation between leading map publishers in different countries has been one of the most interesting features of publishing trends in recent years. As with all kinds of published materials, the purpose of the atlas determines decisions in all these features; those intended for wider popular use are sometimes virtually graphic encyclopedias of geographical knowledge. Three-dimensional relief representation and the lavish use of illustrations make such works of the greatest interest to non-experts and to young students. It is evident, therefore, that two trends have been operating simultaneously in recent years: the atlases intended for scholarly use have become even more authoritative and expert, while the growing interest of laymen the world over in world affairs, and the increase in communications and travel have encouraged the production of atlases designed to become generally familiar. Indeed, the term 'atlas' seems to have become of such universal appeal that there have been a number of recent instances of works published as 'atlases' or 'atlas-histories', in which the atlas element is subsidiary to the text and illustrations, and this is surely to be deplored, especially when taken to an extreme.

Since 1817, Stieler's *Atlas of modern geography,* produced in Gotha by Justus Perthes, has been noted for its scholarship, balance and technical excellence, and the successive editions have effectively demonstrated the development of improved techniques. The new International edition, by Hermann Haack and others, was a thorough revision of the Centenary edition, completed in 1925. The fifty-seven instalments, each containing two accurate, up-to-date and detailed plates, included twenty-four completely new sheets. Relief is shown by brown hachuring and names are printed in the language of the country. Its chief use is locational; the excellent index contains about three hundred thousand entries.

In the category of world atlases, three productions are by general consent pre-eminent today, *The Times atlas of the world,* the *Atlante internationale della Consociazione Turistica* and the *Atlas mira.*

The first of the great world atlases in the modern tradition was *The Times atlas of the world,* published in sheets in 1920 and 1922, under the direction of Dr John George Bartholomew. The Mid-century edition of *The Times atlas of the world,* compiled by Bartholomew for The Times Publishing Company in five volumes, was completed in 1960: I *World, Australia, East Asia,* 1958; II *Southwest Asia and Russia,* 1960; III *Northern Europe,* 1955; IV *Southern Europe and Africa,* 1956; V *The Americas,* 1957. The aim of the atlas is clearly stated: 'to serve as an international atlas for general, official and library use'. The complete atlas consists of topographical and political maps, with historical and reference data. Geographical authorities throughout the world contributed to the accuracy of the maps, many of which are double-page spreads. The general style is typical of the House of Bartholomew; paper of fine texture, spacious layout, layer colouring for the relief and a judicious range of tints, green, buff, brown and grey, with white for permanent snow, blue for glaciers and black for lava sheets. The treatment of names is international in approach and the lettering is legible, especially the small names in Times Roman. The final maps were printed by deep etch photo-offset, by which sharp outlines and excellent registration have been achieved. The most significant areas are on the scale of $1:1\text{M}$. Each map area is portrayed on the projection and by the cartographic techniques most suited to it; and a unique feature at this date was the nine-plate coverage of the USSR on the scale of 1 : 5M, based on the *Atlas mira.* The five indexes total some 195,000 names, a most useful gazetteer index; in 1966 was published a new edition of *The Times index-gazetteer of the world,* which, though more extensive than the combined indexes of the Mid-century edition, indicates references to the atlas where applicable. The atlas has been the standard reference atlas of the English-speaking people, but the rapid changes in all parts of the world during the past decade have rendered a new edition inevitable. A one-volume atlas at a lower price was published by The Times and John Bartholomew in 1967, named the Comprehensive edition; it was based on the Mid-century edition, with the aim of bringing within the reach of a wide public an up-to-date atlas capable of aiding an understanding of world and even ' extra-terrestrial ' events. Space-saving has been achieved by some rearrangement of individual maps, but mainly by the use of thinner paper, printing on both sides of the plate and putting the indexes on the endpapers. Almost all the Mid-century edition plates have been retained, updated, a few new plates have been added and a number of plates, especially from volume I, replaced. A twenty two-page section includes introductory articles covering world minerals, energy, food and climate and extra-territorial subjects, such as artificial satellites. A useful nine-page glossary of geographical terms has been added immediately before the index. A revised edition, especially for the home market, was issued in November 1968. In an enthusiastic review of this atlas, it is compared favourably with ' certain highly touted recent atlases (which) have been bulked out with colour photography and other unnecessary data better fitted to a picture magazine than to an atlas '. To the discerning reader, it must be apparent in various places in this text that the present author completely concurs with this sentiment, however objective a view is attempted. A revised edition is due in 1969.

The latest available geophysical information gained from the researches during the International Geophysical Year has been incorporated and preliminary sections of the atlas are devoted to the solar system, the galaxies and the moon. The preliminary sections occupy forty-eight pages, followed by 121 double-page plates and 272 pages of index and other information. Scales range from 1 : 500,000 to 1 : 60M or 1 : 65M for world air routes, world climatology and world oceanography. World physiography is shown on 1 : 58M, with insets for structure and seismology on 1 : 110M. Regional maps of Britain are at 1 : 850,000, with a map of London and environs at 1 : 100,000. Spelling of names follows the guide lines of the Permanent Committee of Geographical Names and the US Board of Geographical Names. An interesting experiment is the presentation of the names of countries in capitals with red in-filling and, in spite of the vast amount of detail shown on the maps, the lettering does not usually obscure it. The Bartholomew clear colours are used; relief is shown by contours and layer colouring and the shapes of the main towns and cities are effectively indicated. The star charts are particularly clear and legible.

A second great world atlas is the *Atlante internazionale della Consociazione Turistica,* a comprehensive reference atlas first published in 1929 and substantially revised in 1938. The ninth edition, 1955-56, reissued in looseleaf form for frequent revision, was particularly important and marked the sixtieth anniversary of the Club's foundation. More than 170 plates, with large-scale insets, were each drawn and engraved by hand and printed by the most modern lithographic techniques. Parts of Italy are on scales as large as 1:250,000; most of Europe and the United States are on scales of 1:3M, other areas on smaller scales. Relief is by light buff hachures and hill-shading, with frequent spot heights. A brief summary of the source material is shown on the back of each plate and useful notes, glossaries and an index of some two hundred thousand entries form a second volume. Place names are given in the official language of each country. The atlas has been used as the base for other world atlases, notably for the two-volume *Gyldendals Verdens-atlas,* Copenhagen, 1951.

The publication of the Soviet world atlas, *Bolshoy Sovetskiy atlas mira,* 1937-40, planned in four volumes, was an outstanding event in the development of Soviet cartography, on which a specially organised research committee had worked. The two volumes completed by 1939 were withdrawn immediately on the outbreak of war. They covered the Soviet Union in great detail and a third volume was intended to cover the rest of the world; volume IV, *Climatology,* seems to have been published, but it is certainly difficult to come by a copy in Britain. The first volume consisted of eighty three pages of physical, economic and political maps of the world and eighty five pages of maps of the Soviet Union. The maps of the USSR showed comprehensively the achievements of the Soviet economy, advances in the study of natural environment and the development of science and culture. About half the map plates were devoted to economic topics, reflecting the importance attached to such maps by Lenin. A map of the financial relationships between the countries of the capitalist world was included, also maps showing the sources of raw materials and markets and a world map of transportation, indicating the financial ownership of railways and shipping lines. Already, in this first major atlas, the pre-eminence of Soviet cartography is revealed, as well as the depth of preparatory scholarship that has become the hallmark of Russian advanced studies. A translation of the titles and legends of the first volume was issued by A Perejda and V Washburne, in 1940 (Ann Arbor: Edwards). One of the notable features of the Atlas was the uniformity of scales and projections and general documentation.

A new atlas, edited by A N Baranov and others, was issued in 1954, by the Chief Directorate of Geodesy and Cartography, Moscow. The USSR is covered comprehensively at scales ranging from 1:1·5M to 1:5M and other countries of the world are well represented. The emphasis is on relief maps, plus general political and communications maps. Economic maps were not included, as these could be found in the *Geografischeskii atlas*, published the same year. Maps of areas outside Russia are arranged by continents and each section includes general maps of politico-administrative divisions, communications and physical features, followed by detailed regional maps on scales similar to those of the Russian maps. Relief is shown by coloured hill-shading in perspective; a few of the administrative maps have the hill-shading without the layer tints. The cartography is admirable, showing to especial advantage in the portrayal of complex river systems and deltas, or in areas of high and tumbled relief. Nearly all the major towns and areas of particular interest have been given inset maps. Certain maps are particularly fine, such as those of Iceland, Spitsbergen and Ethiopia. Detail on the maps of Britain is sometimes outdated, especially in respect of railways. Foreign place names have been transliterated into Russian. The index comprises more than two hundred thousand entries. In 1962, the State Publishing House for Geodesy and Cartography, in connection with the Ministry for Geology and Natural Resources of the USSR, brought out a new edition of the great atlas, with additional economic maps, those showing natural resources and power being especially valuable. The section covering the USSR itself forms, in effect, a national atlas.

A medium-sized *Atlas mira*, edited by S I Shurova and others, published in Moscow in 1959, consists of political maps at scales of about 1:3M for the European USSR, larger scales for a few special areas and elsewhere at smaller scales. As in the great atlas, it is well documented and has an index of some sixty-five thousand entries. A small *Atlas mira*, in a third edition, 1958, is a handy atlas of political maps at smaller scales and including less detail, edited by I M Itenberg and others. A pocket edition of the great atlas is available, published in 1955. Even this little atlas contains 165 pages of maps and 136 pages of detailed statistics and gazetteer. Still another *Atlas mira* is a military atlas, a *World complex atlas for the officer*, published in Moscow in 1958; composed of 214 pages, it is intended as an aid in the studies of political and military geography, including descriptive geographical maps of the Soviet Union and other countries and a series of military historical maps.

A new *Soviet world atlas,* 1967, was produced under the auspices of the Chief Administration of Geodesy and Cartography of the USSR. 250 full colour maps of the physical and administrative geography of the world include a number of new maps emphasising political divisions.

The Instituto Geografico de Agostini in Novara issued in 1959 a fifth revised edition of the *Grande atlante geografico,* edited by L Visintin; a beautiful atlas this, comprising 232 plates on astronomy, geology, physical, economic and political geography, climate and population at scales varying from 1 : 50M to 1 : 1,500,000. The numerous city plans are most useful. Special features include fine distribution diagrams and block diagrams and the clear, imaginative symbols used throughout. Brief text notes accompany the maps and there is an index of about one hundred thousand entries. A great variety of atlases is produced by this institute, including several others under the direction of L Visintin : the *Atlante mondiale,* the *Nuovo atlante geografico moderno* and many regional atlases.

The *Czechoslovak world atlas,* published in 1965 by the State Publishing House, Prague, is another fine production, including a historical section. Block diagrams are a feature and some folded plates give additional width west to east, as for the Mediterranean and North Africa map extending from Spain to Syria.

In the autumn of 1961 a world atlas was published simultaneously in Denmark, Finland, Germany, the Netherlands and Sweden, while French and Spanish editions were in preparation. This international venture was sponsored by the Cartographical Institute, Bertelsmann Verlag, in Gütersloh, which established a special department for the purpose, which has remained in existence in readiness for further work. *Der grosse Bertelsmann weltatlas,* under the general editorship of Dr Bormann, is an excellently produced atlas, based on several years of work on maps, documents and technical publications from all parts of the world, in conjunction with international geographic and scientific institutions and including the most recent results of geographical and scientific research work, such as that of the International Geophysical Year. The precepts of Alexander von Humboldt have been followed in that geographical surface forms and relationships are given particular prominence, but political, industrial and cultural data are also presented in detail. The maps, in eight colours, were produced by a special method which offered improved possibilities for colour gradation. The Atlas is not overburdened with text, but a valuable preface in German,

French and English discusses in general ' The atlas—its development and possibilities ', followed by a detailed introduction to the plan of this atlas, the scales and projections employed, contents and principles of nomenclature adopted and a key to languages and names. A fully comprehensive index gives place names in the language of the country; the official geographical terminology for rivers, mountains and other geographical forms is also given in the relevant language. Further editions were issued in 1963 and 1964 and the *McGraw-Hill international atlas,* 1964 (dated 1965), was based on this original, also under the direction of Dr W Bormann.

Three other important atlases have been produced by the Bertelsmann Cartographic Institute: the *Bertelsmann Hausatlas,* 1960, 1962, the *Kleiner Bertelsmann Weltatlas,* 1964, and the *Bertelsmann Atlas international,* 1963. Details of the work of the institute and descriptions of the four atlases, including a specimen map plate of each, is given by Dr Werner Bormann, in his article 'Aus der Arbeit des Kartographischen Institutes Bertelsmann ', in the *International yearbook of cartography,* IV, 1964.

The Oxford general atlases, prepared by the Cartographical Department of the Clarendon Press, should be mentioned together, as they have been all part of a well-considered plan. The maps are compiled with the co-operation of geographers, geologists, economists, historians and statisticians in Oxford and elsewhere, and the cartography is of the highest quality, in the van of new techniques. At the end of the second world war, the need for a new atlas became pressing; *The Oxford atlas,* edited by Brig Sir Clinton Lewis and Col J D Campbell, with the assistance of D P Bickmore and K F Cook, in 1951, which went into a fifth reprint with revisions in 1963, was virtually a new atlas. A fresh approach and new layout were necessary to reflect the altered emphasis in the strategic, political and economic spheres. In planning the layout, uniformity of scale sequence was aimed at, to enable comparison between the countries of each continent, and much thought was given to projections. A valuable feature of each map is a footnote giving full particulars of the projection used and indicating any incidence of scale errors and the corrections to be applied to them. The general section of the atlas contains eighty eight pages of plates in colour, followed by a section of twenty four colour plates dealing with important natural and human distributions. In the revised reprint of 1952, edited by Sir Clinton Lewis and J D Campbell, a section of distribution maps was included, edited by Professor Linton of Sheffield University; in this section, the more detailed maps are confined to areas where the topics chosen were known with sufficient accuracy to warrant representation at large scales. Most of the maps are in six colours; the range of layer tints, printed by half-tone, were limited to those which could be produced by four colours, light blue, yellow, brown and red. Contour lines were omitted. One of the biggest problems in atlas design is the selection of material for inclusion on road classification; this atlas attempts to distinguish between all-weather and fair-weather roads. English forms of place names have been given when these are in common use and geographical terms have been translated, except when the local form is the more familiar. More than fifty thousand entries are included in the gazetteer.

Derived from *The Oxford atlas* are *The concise Oxford atlas, The Oxford school atlas, The Oxford home atlas, The shorter Oxford school atlas* and *The little Oxford atlas. The concise Oxford atlas* is in its second edition, with a revised reprint in 1961; 120 pages of six-colour maps include some of special interest, such as those of Roman Britain and of the visible hemisphere of the moon, also an important section of special topic maps of Britain. *The Oxford school atlas,* in a third, revised and enlarged edition, 1960, reprinted in 1963, contains 112 pages of six-colour maps and a thirty two-page gazetteer. In this new edition, the 1:1M maps of Britain show roads as well as railways; new maps include a new map of the Mediterranean and southern Europe, the Middle East, and China, and the section of world maps has been extended to twenty pages, incorporating a series of economic maps based on the second edition of *The Oxford economic atlas of the world (qv). The Oxford home atlas of the world,* of which a third revised edition was published in 1963, is the general edition of *The Oxford school atlas,* giving the essential geographical and topographical information. *The shorter Oxford school atlas* gives the broad essentials of world geography; the third edition was revised in 1960, reprinted in 1963, when it contained sixty four pages of six-colour maps, with a short gazetteer-index. *The little Oxford atlas,* the general edition of *The shorter Oxford school atlas,* in a second edition with revision in 1963, is an excellent production, international in scope, but giving good space to Britain.

All these atlases include topographical maps showing striking photo-relief effects; the hill-shading technique developed by the Cartographic Department of the Clarendon Press produces an evocative three dimensional impression. General atlases published by Oxford University Press branches abroad are also available in this country. These include *The Canadian Oxford atlas of the world, The Canadian desk atlas, The Canadian Oxford school atlas; The Oxford Australian atlas,* in library and school editions, and *The shorter Oxford Australian atlas; The Oxford atlas for New*

Zealand, in library and students' editions; *The Oxford school atlas for Pakistan,* in library and school editions; *The Oxford college atlas for Ceylon,* in Sinhalese. Some of the atlas maps have been enlarged to wall maps.

The *Atlas général Vidal-Lablache,* first produced in 1894, with further issues in 1909, 1918, 1922, 1933, 1938 and 1951, is a fine example of the influence of the French school of geography, led by Paul Vidal de la Blache and Emmanuel de Martonne. The 1951 edition (Paris: Armand Colin) carries an increased number of maps in two parts, ' cartes historiques' and ' cartes géographiques', with an index-gazetteer of thirty one pages. The integration of geographical and historical studies has been one of the chief features of French geographical thought; in this atlas, for successive periods, the distribution of states, towns, commerce and population is presented, with brief but effective commentary, in such a way as to stimulate the consideration of these factors in interaction. The same publishers issued an *Atlas historique et géographique Vidal-Lablache,* of which a new edition appeared in 1960.

The *Atlas classique,* by Pierre Gourou (Hachette, 1966), is an interesting atlas of ninety six pages of maps and diagrams, in two parts, the first dealing with France and French territories, the second with the rest of the world. The *Atlas général Larousse,* published in Paris in 1959 (distributed in Britain by Harrap), contains seventy two full or double plate maps in six colours, a further number in two colours and 183 city plans, together with statistical tables, commentaries and a fifty five thousand name gazetteer. Historical maps comprise fifty five pages and thirty articles by eminent historians are a useful feature. The *Atlas internationale Larousse* (second edition, 1957), with text in French, English and Spanish, contains nearly seventy folding maps.

The *National geographic atlas of the world* originated from January 1958, when a new series of map supplements was introduced with the *National geographic magazine,* uniform in size and designed for assembly in a looseleaf binder. The society published the atlas in 1963, with Melville Bell Grosvenor as editor-in-chief, and a second, enlarged edition appeared in 1966. Emphasis is on the United States of America, but reasonably detailed and balanced coverage is given to the rest of the world. Arrangement is by geographical area, each section being preceded by a brief description of the country. The aim was to interest laymen as well as scholars, students and businessmen. Scales range from 1 : 30,560 to 1 : 11,404,800 and most of the maps are double spreads. Projections vary according to the purpose and extent of individual maps and cartography is of a uniformly high order and generally pleasing aesthetic appearance. Native spelling is used for those countries having a Roman alphabet; the transliteration of names in other countries is based on the systems prepared by the United States Board on Geographic Names and the British Permanent Committee on Geographic Names. A useful list of addresses of information centres is included, also a comprehensive index of some 127,000 entries. A detailed source of information is the article by Athos D Grazzini: ' Problèmes que présente la préparation de *l'Atlas mondial de la National Geographic Society* ', *International yearbook of cartography*, 1965.

The *PWN atlas of the world* has been issued by the Polish Scientific Publishers since 1963 with the relevant parts of the *PWN universal geography*. It is in four main parts, comprising eight fascicles in a plastic portfolio. Five hundred original climatic, demographic and economic maps are at scales ranging from 1 : 1,250,000 to 1 : 10M. The index contains some 150,000 geographical names.

New editions of medium-sized library or reference atlases have appeared in quick succession since the last war, many of them attempting with frequent, even annual, editions, to keep pace with rapid economic developments and political changes. Some of the best of these in Britain are published by the House of Bartholomew and George Philip, both publishers planning the range to suit different needs. The largest of the Bartholomew world atlases is *The citizen's atlas of the world,* of which a tenth edition appeared in 1962. The majority of the maps are printed in band colouring for clarity and are designed primarily to show settlements and boundaries, both internal and international. There are thirteen pages of physical maps covering the world by regions and eighteen pages of special subject maps. Every second double page is presented as a single spread of a large unbroken map. The introductory section gives such useful information as dates in the history of exploration, world areas, population, heights and a glossary of geographical terms. The index, of some 100,000 entries, refers to country or state and the direct position on the specific map. The maps are mounted on hinged guards, allowing the double pages to lie flat, and the fact that each volume has been hand-bound and designed for hard wear makes the atlas especially suitable for library and educational use.

The fifth edition of Bartholomew's *Advanced atlas of modern geography* (Edinburgh, Oliver and Boyd, 1960) was a complete revision, followed by a sixth edition in 1967. The plates were printed by a special colour process, making for exceptional clarity, precision and accuracy, and particular attention was given to physiographic features. The *Edinburgh world atlas of modern geography,* in a fifth edition, 1963, aims ' to give a realistic picture of the world and its problems as they appear today'. The volume contains a hundred fully coloured pages of maps, besides the introductory ones in black and blue. Population density is shown in relation to vegetation, climate and physical features. Two interesting early maps of the world are reproduced in the edition: Ruysch's map of 1508 and Wright's map of 1599. The House of Bartholomew has done much successful experimentation in the use of new and developed projections to suit special purposes, such as the Atlantis projection to show world airways. Also published for them by Oliver and Boyd was *The comparative atlas of physical and political geography,* in its forty eighth edition in 1964, an atlas including eighty pages of maps and a thirty seven-page index. In *The new comparative atlas* (Edinburgh, Oliver and Boyd, 1964), the maps have been entirely redrawn and have a brighter, clearer appearance. Layer colouring is now used, without contours, representing 'a just compromise between an overall impression and a highly accurate representation of relief'. Lettering is bolder and the index is improved. There are five good economic maps of the British Isles. Bartholomew's *The regional atlas of the world,* 1948, appeared in a revised new form as *The Columbus atlas: or regional atlas of the world* in 1954. 160 plates give a reasonable world coverage, with particularly full treatment for the United States. Plans are included for the principal towns and, in the fifty thousand name index, population figures are added. The House of Bartholomew also produces maps for a great many other publishers.

The library atlas, by Harold Fullard and H C Darby (Philip, 1938; ninth edition 1967), is in two parts, general and economic. The general section, at scales ranging from $1:2\frac{1}{2}$M to $1:8$M, was published previously as *The university atlas;* a new colour scheme has been used, with finely drawn hill-shading. Thirty two pages of updated economic maps and diagrams include distribution maps of major products throughout the world; diagrams and maps of world occupations, transport, imports and exports; economic development and production, land use and agriculture, minerals and industry for Great Britain and for other selected areas. Climatic graphs are included for more than two hundred representative stations throughout the world. The index comprises over fifty thousand entries, conforming in general with the rules of the Permanent Committee on Geographical Names and the United States Board on Geographic Names. The Atlas is handled in the United States by the Denoyer-Geppert Company of Chicago.

The twelfth edition of *The university atlas* (Philip, 1967), edited by G H Fullard and H C Darby, has incorporated modern developments in mapping techniques most successfully; relief is shown by lighter shades of layer colouring, with hill-shading. The general plan and format of the atlas have remained unchanged since the eighth edition of 1958 was redesigned to include post-war survey information, but improvements have been made in the contents of many of the maps and this atlas has been widely acclaimed as one of the best for educational purposes. New Zealand, for example, receives adequate treatment and the section on the new African states is particularly useful. Additional detail has been included, in line with current geographical analyses; for example, the vegetation map of North America indicates the various types of trees, as well as indicating the northern limit of the different species. A complete contrast is revealed in the equivalent map of Australia, which now includes the boundaries of artesian basins. The limits of maximum glaciation are shown on the geology maps of the British Isles and Central Europe. Ocean currents have been added to the world temperature maps, and, where relevant, seasonal streams and lakes are distinguished from permanent ones. Fourteen projections have been used. Hammer's Equal Area Projection for the world political maps has been used instead of the Mercator used previously. Four maps remain on the Mercator Projection, showing its value for areas in low latitudes and for navigation purposes; others are on a conical projection with two standard parallels if necessary. Physical features are depicted by contours, layer colouring, hill-shading and spot heights, according to the terrain. A plate such as no 102, India, Pakistan, Ceylon and Burma, shows dramatically the whole range of relief features; the southern cliffs of the Himalaya are vividly drawn, with the heights coloured blue, mauve and white, and the long streamers of high country at each end of the chain are well brought out, enclosing on the west the Khash and Sandy deserts and, on the east, the valley of the Irrawaddy. Lettering has been skilfully placed, so as not to be lost in the hill-shading. In addition to a map of the solid geology of the whole world, there are more detailed maps of the British Isles, France and central Europe, a structure map of the world and maps showing volcanoes, centres of volcanic activity and earthquake zones. The climate maps are supported by a series of climate graphs at consistent scales, and explanatory notes. Distribution of population is shown on an equal area map of the world, supplemented by larger scale maps of individual regions. Evidence of much recent research is apparent on the maps of the polar regions and of the oceans. Due importance is given to these vast areas, which are so influential to life throughout the world, but which, mainly from lack of knowledge, have not been adequately emphasised in many previous atlases. Significant of the growing appreciation of international uniformity, height and depth keys now show metres as well as feet; and square kilometres and centimetres have been added to keys to population and rainfall maps. The index also has been revised and now includes some fifty thousand entries.

Cassell's *New atlas of the world,* 1961, edited by Harold Fullard, containing 130 maps by George Philip, is a completely new work in the long line of authoritative atlases published by the House of Cassell since the first *Universal atlas* of 1891. It is especially useful for commercial interests and has an index-gazetteer of some seventy thousand entries. *Stanford's Whitehall atlas,* 1962, and the *Caxton world atlas,* 1960, edited by W Gordon East, also use Philip maps. The latter atlas was designed to be used in conjunction with a descriptive text, county by county, and it contains many illustrations and sketch maps.

Of Bartholomew ' handy' atlases, the *Graphic atlas of the world* should be mentioned. This general purpose reference atlas, which appeared in an eleventh revised edition in 1960, comprises ninety six pages of coloured maps, six pages of factual world geography and a gazetteer of over thirty thousand entries. The *Compact atlas of the world,* in a fourth edition, 1957, includes 128 pages of maps, mainly in political colouring, in addition to a series showing the climates of the continents; races, population, world economy and vegetation. Sixteen pages of introduction and a four thousand name index complete a useful general reference atlas. The *Handy reference atlas of the world,* in a sixteenth edition, 1954, includes physical maps for each continent; the *General atlas: physical and political geography,* third edition (Edinburgh, Oliver and Boyd, 1965), has eighty coloured map plates; and *The pocket world atlas,* 1966 edition, consists of political maps, with additional maps covering relief, climate, vegetation, ocean currents, time zones and communications, with an index of about four thousand entries.

Small general reference atlases by Philip each have their particular features. The *Modern world atlas,* 1960, is a handy illustrated atlas for general use, containing twenty four pages of index. The *Contemporary world atlas,* edited by Harold Fullard in 1960, is a comprehensive physical and political atlas, with a photographic section illustrating racial types, human activities and food production. The *Record atlas,* edited by George Goodall (twenty sixth edition, 1965), is useful for following world affairs, shipping and air routes: *The standard reference atlas,* edited by George Goodall and Harold Fullard, 1956, presents general geography and statistical information. The *Falcon world atlas,* 1961 and the *Practical atlas,* both edited by Harold Fullard, contain mainly political maps for general reference; the latter has a fuller index. The *Standard world atlas* (Philip and Rand McNally) gives a greater proportion of maps to the United States and Canada.

The Nelson *Concise world atlas,* edited by J Wreford Watson, 1961, is a small, well-planned atlas. Six introductory plates are given to astronomical geography and the structure of the earth and, among the forty eight pages of maps, the British Isles have a full treatment, which, with the photographs, diagrams and accompanying text, comprise a concise geography of Britain. The maps are of outstanding clarity, using light layer colouring, with hill-shading for relief representation, and clear typography. Special editions for Canada have been issued since 1958. A *Junior atlas* was published in 1962 and *Nelson's school atlas* has the same content as the world atlas, but is bound less durably.

The Hammond series of world atlases is designed mainly for use in American homes, offices, libraries and schools. All contain one or more sections in common. Each includes a set of forty eight coloured plates entitled ' Hammond's world atlas and gazetteer '. The *New international world atlas* (Hammond, 1958) celebrated the sixtieth anniversary of the firm; its scope includes ' The modern, medieval and ancient world : a comprehensive collection of modern post-war maps, factual data and a complete collection of historical maps covering two thousand years, with informative text '. This is a set of forty-six coloured plates called ' Hammond's historical atlas ', spanning the period from 3,000 BC to 100 BC, ' The cradles of civilization ' to ' The world of the United Nations and the cold war 1945-60 '. Social and economic statistical tables, a glossary of geographical terms, plans, an illustrated world gazetteer and an index of more than a hundred thousand entries all make this an informative reference atlas for British as well as American libraries. The *Globemaster world atlas* is of less value in Britain, having very little additional material other than the basic sections. *Hammond's advanced reference atlas* comprises four sections, each with its own title page : world geography, including a set of coloured photographs selected by Erwin Raisz to illustrate the major land forms of the world; ' Hammond's historical atlas '; ' Hammond's American atlas ', containing a wealth of material on American geography and history; and the common set of political maps. *Hammond's ambassador world atlas* was published in 1954, re-vised in 1956, and *Hammond's library world atlas* is another collection of sections also published separately: an extended version of Hammond's world atlas and gazetteer, containing additional maps of the countries of South America and the United States; an index of the United States, giving 1960 population figures; an illustrated gazetteer of the United States and possessions; an illustrated geography and gazetteer of the world, in colour and black and white; the races of mankind; the set of world distribution maps; an index of the world; finally, an illustrated gazetteer of Canada, including coloured maps of each province. *Hammond's diplomat world atlas,* 1961, is popular in presentation, intended for the non-specialist, but reliable, having taken its sources from many official mapping agencies, statistical offices and geographical societies. World maps are included, separate maps of the the polar regions and the continents, with particular emphasis on the United States and Canada. The atlas is well documented, having a separate index of the cities and towns of the United States and of Canada, tables of social and economic data for all countries of the world, a glossary of geographical terms, the races of mankind, notes on map projections and world statistics. A section ' Illustrated geography and gazetteer of the world ' is intended as providing background knowledge of world events.

Also from America come *Collier's world atlas and gazetteer,* 1953, including maps of individual states and countries, accompanied by text, especially useful for economic conditions; and the Rand McNally series, of which the most generally useful is probably the *Collegiate world atlas* 1961. 178 pages of coloured political maps and thirty three pages of historical maps are accompanied by a 120-page reference guide and an index of more than twenty five thousand entries. World population statistics are included, with 1960 census figures for United States cities. The *International world atlas* is a smaller atlas on similar lines; a section on world discovery and exploration is included, also an historical gazetteer, with an index of some eighty thousand entries. The *Life pictorial atlas of the world,* 1961, is also published by Rand McNally; the work includes six hundred pages of coloured maps and illustrations, thematic maps, charts and statistics. A fully coloured wall map of the world is issued in conjunction with this atlas. Two more popular Rand McNally atlases are *The readers' world atlas* and the *Worldmaster atlas.* The Rand McNally *Cosmopolitan world atlas* was revised in 1964. Philip produced a new census edition of this atlas in 1953 for sale in the United Kingdom and Eire; it is useful as a reference work especially for America and it includes a glossary of map terminology.

The Prentice-Hall world atlas, edited by Joseph Williams (Stanford University, second edition, 1963), was printed by the Geographisches Institut und Verlag in Vienna. It presents a systematic geography of the world, world economic maps and physical maps of the continents, with separate maps of their geographical regions. Most of the projections used are of the equal area interrupted

Eckert sinusoidal type and the relief representation is multicolour shaded, so that a three-dimensional effect is achieved. The economic maps are so orientated that the Americas appear in the centre, clearly showing the relationship of the Americans to the rest of the world with regard to commodities and trade. The latest available population figures are given for major political divisions; the index has been revised, but is still not adequate.

Most countries of the world have become expert in the production of handy world atlases, mainly intended for use in general libraries, schools or in the home. Most of them naturally have a bias towards the representation of the home country; some stress the mapping of economic and commercial factors for business planning purposes. For example, with special emphasis on Spain and Spanish interests, the *Atlas universal Aguilar* (Madrid, Aguilar, 1954) is particularly useful for areas of interest to the Spanish world. Spain is shown at a scale of 1 : 1M. In addition to maps, there is text on general and regional geography and photogravure illustrations. The layer-coloured physical maps are clear, but some of the general maps on smaller scales present rather a crowded appearance. Another world atlas with a comparable aim is the *Historisches-Geographisches Kartenwerk,* published in Leipzig by the Verlag Enzyklopädie. The intention of the atlas is to show the economic and cultural development of the earth's chief regions by surveying contemporary facts of land use and industry, mines and other industries, against the historical background and factors of educational, social and religious problems. The first folio to be completed was that sub-titled *Indien Entwicklung seiner Wirtschaft und Kultur,* 1958, containing sixteen plates on which are arranged some eighty six maps and an eighteen page explanatory text. This is an interesting venture, but some of the maps are rather over-crowded. Atlases with such aims as these in mind are especially suitable for looseleaf production, either produced all together, with subsequent revision as most needed, or in separate fascicules as they are completed.

The *Haack grosser Weltatlas,* under the direction of Rudolf Habel (Gotha, VEB Hermann Haack, Geographische-Kartographische Anstalt, 1967-) should comprise about 118 pages when completed. This atlas is intended as a reference work for location rather than as an analytical source. The majority of sheets are topographical, with some thematic maps included, ranging in scale from 1 : 300,000 to 1 : 20M; the regions of Germany are at 1 : 300,000 or 1 : 750,000. Layer colouring, with light hill-shading, is used to indicate relief features, urban areas are picked out in red on the more general maps and in an orange tint on the larger scale sheets and transport lines

96

are strongly marked. The sheets carry a grid system, plus latitude and longitude co-ordinates. Each issue is composed of four pages, each with title page in five languages, and an index map showing relationship with the whole sequence. The centre pages contain the map and the fourth page carries the text. Full documentation is included regarding sources and authorities consulted, which, in itself, is useful; when completed, this work will be very valuable, especially for East Germany, on which up-to-date information is not so easily come by.

Also in looseleaf format is the *Grosse Elsevier atlas,* in two folders with explanatory text, maps and diagrams, 1950-; another is the *Nouvel atlas mondial géographique et économique de tous les pays,* edited by Eugène Th Rimli, 1956-, published by Editions Rencontre Lausanne; Zurich, Stauffacher-Publ. By the same publishers is the *Neuer Welt-atlas,* 1958-, which also includes numerous illustrative plates interspersed with the maps.

For more popular use, the *Encyclopaedia Britannica atlas* appeared in a revised edition in 1954, under the direction of W Yust, with G D Hudson as geographical editor, and later reprints, with minor corrections, in 1959 and 1962. For reference, it includes physical and political maps, geographical comparisons, a glossary of geographical terms, geographical summaries and a gazetteer-index. In the illustrated encyclopedia section, tables for trade, population and resources are given for all countries. The maps were based on Rand McNally's *Cosmopolitan world atlas,* second edition, 1951. In 1965, the atlas was redesigned as the *Encyclopaedia Britannica international atlas,* adopting the maps produced by the Istituto Geografico de Agostini, used also in such atlases as the *Wilhelm Goldmanns grosser Weltatlas,* 1963. The editors were Ruth Martin, J V Dodge and M B Mitchell. In five sections, the first comprises ten colour relief maps, with added hachures. Fold-out maps are used for certain areas; emphasis is on America and coverage for Great Britain, for example, is quite unbalanced. The second section includes thematic maps of the world, with explanatory text. 215 countries are included in the ' Geographical summaries ' section, which quotes statistics by region òr by political units, on such topics as population, land resources, livestock, etc. The ' Geographical comparisons ' section attempts to cite examples of the ' highest, deepest, longest ' characteristics. The final section comprises a glossary of foreign names and the gazetteer.

The Reader's Digest great world atlas, published by The Reader's Digest Association Limited, under the direction of Professor Frank Debenham, 1962, and 1965, follows an unconventional and imaginative plan; in four main parts, ' The face of the world ', ' The countries of the world ', ' The world as we know it ' and an index in two sections, ' The British Isles index ' and ' World index '. The aim of the editors has been to stimulate interest in the universe ' from the centre of the Earth to the outermost limits of space ' and to demonstrate man's place in it. Particularly interesting plates show each of the polar regions and another set of three plates depicts the Indian, Atlantic and Pacific Oceans, with their surrounding land masses, thus evoking a more exact appreciation of the relationships of countries than is often given by the traditional atlas plates. The first section amply demonstrates the application of relief model techniques in map-making, emphasised by the bold use of colour. The second section contains conventional maps of individual countries and regions by Bartholomew. The third section employs diagrams, diagrammatic maps and figures to present a multitude of interesting topics, many inevitably simplified. Particularly successful are the double-page spreads of the world's chief minerals, of major explorations and the plate entitled 'Around the world ', showing six views of different parts of the earth as if seen from twenty-five thousand miles out in space.

The Pergamon general world atlas (Oxford, 1966), edited by S Knight, comprises ninety one pages of maps printed in six colours. Some interesting new features have been introduced into an atlas of this kind. Heights are given in metres and names are cited in vernacular form throughout. Physical and human geography are shown on different maps, and inset maps emphasise some interesting topics, such as the St Lawrence Seaway. The British Isles maps comprise physical, political and thematic maps, based on the Ordnance Survey ' Ten-mile map '; land use is included. The rest of the world is represented by physical and human maps, with short descriptions, especially in the case of the thematic maps. The gazetteer contains some twenty thousand entries, with a separate section devoted to Great Britain. Names are given in English as well as in the vernacular form and northings and eastings are stated. The larger Pergamon world atlas was completed in 1968 by the Polish Scientific Publishers, Warsaw, and the Polish Army Topographical Service, in collaboration with the Cartographic Department of the Pergamon Press; printing was also done in Poland. Experts in many fields have contributed to the two hundred pages of topographic folding maps and more than four hundred thematic maps and charts, all printed in at least ten colours. Production is in looseleaf form, issued in a binder designed to contain some 520 pages. It has been possible to include more complete detail on East European and Asian countries than hitherto. The major part of the world is covered by mapping shown at a scale of 1:2,500,00, with specially significant areas

98

shown at 1:250,000. Plans of world cities form a useful feature, and names and terms are used in the vernacular, with English equivalents for the most important of them. The latest available statistical data and census information have been incorporated. The maps of Great Britain have a land use background and other areas are shown in two styles of mapping, showing physical and human geography. The gazetteer includes over 150,000 entries, with cross references to important features and changes of place names which have occurred through the years. The chronological span of the atlas ranges from Babylonian times to the latest available lunar maps based on space rocket photography. The publishers stress that the requirements of businesmen have been much in their minds in the planning of this atlas. From the British angle, a number of critical comments could be made; some of these have been quite neatly expressed in the article ' Up the Poles ', in *The Sunday Times*, August 18 1968, 13.

The Van Nostrand *Atlas of the world* (Princeton, NJ, 1962) is a small work in paper covers. Short articles of the encyclopedia type on the countries of the world and tables of geographical facts and statistics are accompanied by a section on map projections and sixty pages of maps on various small scales. The index contains over thirteen thousand entries. The publishers seem to be proud of the fact that they are creating ' a new development ', changing atlases ' from mere collections of maps into universal combinations of world atlas, illustrated introduction to general geography, and comprehensive source of geographical and statistical information . . .'. Another encyclopedic reference work is Van Nostrand's *Atlas of western civilization*, by Frederic Van der Meer and T A Birrell, in a second edition, 1960, illustrated by about a thousand photo-gravures and fifty three annotated maps. Each map is designed to present a particular aspect or epoch of European cultural history, with emphasis on centres, points of contact, trends, etc.

Hamlyn's new relief world atlas (Paul Hamlyn, 1967) claims to be the first atlas to use the technique of shadow relief almost through-out the 106 plates; the relief model technique extends to the ocean bed as well as the land areas. Following a conventional section on map projections and one on space exploration, including one of the Mariner IV shots of Mars, are relief and vegetation maps for all the continents and smaller maps depicting climate, historical features, soil, land use, cultural aspects, etc. The more currently significant regions of the world are drawn on larger scales in ochre and blue only. This is an interesting and individual atlas; the policy followed regarding place names is conventional, adopting the agreed forms suggested by the Permanent Committee on Geographical

Names, and the sans serif typeface makes for clarity. *Newnes' pictorial knowledge atlas,* edited by Peter Finch, 1958, features ' view-through ' maps. Ninety pages of world maps are accompanied by a 228-page illustrated reference guide and index of some twenty seven thousand entries, also an illustrated section of five hundred explanations of geographical terms. *Goldmann's grosser Weltatlas* (Munich, 1955 and 1963) is in effect an illustrated encyclopedia on astronomy, geology, geography and climate, so much does it contain of explanatory text and diagrammatic illustration. Scales vary; central and eastern Europe are at 1 : 1M, the remainder of Europe at 1 : 2·5M, the Americas mainly at 1 : 5M and one map of Australia is at 1 : 10M. A feature is the small maps, diagrams and tables printed in the margins or on the reverse of the map plates.

The *Penguin atlas,* edited by J P Keates, is an English version of a Swedish production, printed in Sweden (Harmondsworth, Penguin Books, 1956) and is a handy pocket atlas comprising eighty pages of clearly printed coloured maps, mostly locational and administrative. The maps are inevitably on very small scales. There is a fifteen thousand name index. There is also the pocket edition of the *Atlas swiata,* 1958, adapted from the *PWN atlas of the world,* for the English market, by the Cartographic Department of the Pergamon Press, 1967-68. It contains much geographic information, tabulated under useful headings. Statistics, especially economic and population, are listed under separate countries. The maps are chiefly political; there is an index-gazetteer and a bibliography. *The world geographic atlas : a composite of man's environment,* privately printed for the Container Corporation of America in 1953, was designed and edited by Herbert Bayer. Explanatory text is illustrated by photographs, maps and diagrams. The *Winkler Prins atlas* is another of the encyclopedic type, published in a third edition by Elsevier of Amsterdam in 1957, with text, in Dutch, on morphology, climate, flora, fauna, population and descriptions of individual countries, with innumerable illustrations, maps, diagrams, an index and a bibliography. The *Ginn world atlas* was produced by Richard E Harrison and the staff of Ginn in 1963. The Atlas comprises sixty four pages of general physical and political maps and with it is published a forty seven page ' Workbook ', printed in two colours only, in which the surface features are very well represented.

Only the most important of atlases designed for school use are mentioned here. For further examples, reference should be made to the author's *Reference material for young people* (Bingley, 1967), of which a second edition is now in preparation. During recent years, constant improvements have been seen in the content, design and reproduction of educational atlases, encouraged by several sources, not least by the meetings of the Council of Europe committees discussing the standards and the format of geographical textbooks and maps. The increased emphasis placed on the use of maps and atlases at all levels of education has created a demand for a greater variety, as well as for higher standards. In many countries there exists at least one long-established firm which has specialised in educational atlases and which can therefore contrive to produce and revise them at reasonable prices. Since 1957, the Internationales Schulbuchinstitut, Braunschweig, has encouraged improved standards in the atlases used in European schools, with particular emphasis on the co-ordination of contents and on the definition and uniformity of terms and symbols. The arrangement of educational atlases follows usually one of two alternatives; either the first section of the atlas deals with the homeland, gradually leading the pupils on to further knowledge of other countries, or general and thematic maps of the world begin the sequence, followed by maps of continents and regions. Full use is now being made of the three-dimensional relief representation method, which is more evocative of ideas in the young student mind than contour lines. Clarity must be a main objective and every aspect of production must attempt to stimulate interest and imagination.

The Oxford junior atlas was an entirely new atlas when it was published in 1964 (1965), containing forty eight pages of seven colour maps and a seven-page gazetteer. It was designed for use in primary schools, so that simplicity and clarity were the aims throughout. Each map presents simplified information against a background of geographic landscape—relief, altitude and surface cover. Most of the maps are double-page spreads, without insets, as these were considered confusing for young children. Maps of continental areas are all at the scale of eight hundred miles to the inch, and a series of seven regional maps covers the world at twice this scale. A series of eight maps of the British Isles at 150 miles to one inch portrays such topics as weather conditions and population distribution; there are also six regional maps at sixteen miles to the inch. Three maps of Bible lands are included. Careful instructions are given in using the atlas; place names have been reduced to a minimum and the index locates places by verbal description. On the maps, names stand out boldly, but children should be warned that in some cases names have been cut in two by the edge of the page. The imaginative relief representation by layer colouring and hill-shading makes this an evocative atlas for use by children, and political boundaries show up well in black.

George Philip and Son specialise in educational materials. *Philip's modern school atlas* and the smaller *Philip's new school atlas* are both in their sixty fourth editions, edited by Harold Fullard. The latest editions have improved layout and new colouring, which makes a more immediate impact; hill-shading has been introduced into the maps of the continents. In the former atlas, more roads, in red, have been inserted, especially on the larger scale maps. In the latter, some completely new double-page spreads have in mind new examination requirements; a new map of world geology is included, also new regional maps of the Low Countries, the Rhine Valley and Switzerland, and Central European Russia. Both have adequate indexes. Local editions are prepared for Bristol, Cumberland, Durham, Hampshire, Hereford, Lancashire, Northampton, Nottingham and Nottinghamshire, Southampton, the East Riding and North Riding of Yorkshire, Wales, Edinburgh and south-east Scotland, and for Greater London. Philip's *Secondary school atlas* is also constantly revised; it contains more than ninety physical and political maps, maps and graphs of temperature and rainfall and maps of natural vegetation. Details are included for the principal countries and cities of the world.

Four atlases, all excellently produced by Verlag Ed Hölzel in Vienna are the *Österreichischer Mittelschulatlas* (Kozenn-Atlas), Lautensach's *Atlas zur Erdkunde,* the *Nouvel atlas général* (Bordas) and the Faber *School atlas.* The first of these would seem to have been the original; in the introduction to the seventy fifth edition, its origin is traced back to a Slovak geographer, Blasius Kozenn, who began this successful line of atlases in 1863. The Faber *School atlas* is an interesting production, designed specially for teaching by E Hölzel and edited for the English-speaking market by D J Sinclair, in its latest edition, 1966. The 154 maps give most areas reasonable coverage for the size of the atlas, with emphasis on the British Isles and Europe. The fourth edition includes eight pages of new maps showing world structure, vegetation, climate, population and economic and political status, with a gazetteer of some twenty thousand entries. Key areas such as the Indus Valley and Central Japan are particularly well executed, in the style of the parent atlas. The *Nouvel atlas général—la France et le monde,* by Pierre Serryn and others (Bordas, 1962), is a useful general purpose atlas in clear colours, comprising physical and thematic maps, the economic maps being especially useful; for example, the economic map of China.

The leading American school atlas is *Goode's world atlas,* in its eleventh edition, completely revised by E B Espenshade, jr (Rand McNally, twelfth edition 1964), previously named *Goode's school atlas.* It includes a pronouncing index of more than thirty thousand names. General physical and political maps are followed by other thematic maps of the world and major regions. Vegetation, land use and economic maps are generally on a continental basis. The maps are clear, especially in the expressive use of symbols; towns, for example, are shown by graduated circles and squares to indicate population. There is an extensive index.

Gyldendal of Copenhagen first introduced the interesting idea of producing a graded series of atlases for systematic study; *Atlas 1, Atlas 2* and *Atlas 3* were designed for progressive use in Danish schools. *Atlas 3,* edited by W F Hellner, was published in 1964 and is a particularly fine work. Three types of maps make up the atlas, physical, thematic and a series of town plans. Relief is shown by layer colouring in bright shades, with hachuring and hill-shading in some areas. Scales vary. There is a natural emphasis on Denmark and Scandinavia generally, but coverage of the rest of the world seems rather unbalanced; admittedly, it is difficult to present Africa adequately in two pages! The thematic maps and, particularly, the town plans, are the most valuable sections. First, Copenhagen, Stockholm, Oslo, Helsinki, Reykjavik, Aarhus, Odense and Aalborg at 1:200,000; then selected cities such as New York, London, Berlin, Paris and Moscow at 1:500,000. The lettering, which has been done by hand, is remarkably legible. A similar family of atlases is being produced in Britain jointly by Collins and Longmans. The contents of these new atlases have been based on the requirements of practising teachers; the publishers have aimed at producing a graded series from which teachers can select the material most appropriate for their pupils, not only according to age, but considering interest and ability. *Atlas 2,* the first to be completed and now in its second edition, 1967, is for the upper primary or junior secondary school stage and is intended to be ' a multipurpose atlas combining simplicity with comprehensiveness and an attractive appearance '. The fifty four pages of maps are so arranged that study begins with the homeland, then progresses to western Europe, the other continents and the world as a whole. *Atlas 3, Atlas 4* and *Atlas advanced* are mainly for CSE, fifth form and ' O ' level, and for sixth form, college and first year university respectively. The last two were published in 1968. The advanced atlas attempts to combine international scholarship with the latest cartographic techniques and the result is a reference work of great clarity covering the development and resources of the world. Large-scale general maps combine hill-shading with layer colours and contours; settlements were selected according to local importance and insets of urban areas, at 1:1M, are frequently included. Seventy two pages of thematic maps include a forty-page section on the world, country by country, presenting graphs of the world's produce. There is a comprehensive index of about thirty thousand names. Collins-Longmans also publish the

Study atlas, in its sixteenth edition, 1967, thoroughly revised to incorporate recent political and economic developments. The tenth edition was completely reconstructed and the general format has since been retained, with updating. The maps of the economy of the British Isles have been redrawn on the basis of the 1951 Industrial Census Report. Exercises are included, which are also revised with successive editions. The *Visible regions atlas* (sixteenth edition 1967) includes all the contents of the *Study atlas* except the exercises. The *Progressive atlas,* a new atlas designed to introduce young children to maps and atlases and the *Clarion atlas,* now in its eighth edition, 1967, should also be considered for school libraries. The latter, comprising forty eight pages of seven-colour maps, with sixteen pages of text, is specially bound in reinforced manilla. Special editions are issued for Wales, Malawi, Malta, South Africa in English and Afrikaans, and Rhodesia.

The policy, just mentioned, of producing atlases in several editions for use in different countries has been an interesting recent trend. Collins-Longmans publish also the *Pathfinder atlas* specially for West Africa, the *Ghana atlas,* the *Junior atlas* and the *New secondary atlas* for Malaysia and the *Primary atlas* for Zambia. Nelson has published the *Kenya junior atlas,* edited by P J H Clarke, 1964. George Philip publishes special editions for Malaysia, Australia and various parts of Africa, while editions of various physical and political maps are available in Arabic, Afrikaans and Rumi.

Several of the atlases already mentioned, notably the latest editions from Oxford, Bartholomew and Philip, include physical maps reproduced by the new developments in relief representation. For educational purposes and for popular use, these methods are particularly attractive and evocative. Professor Frank Debenham, an enthusiastic advocate of the new methods, demonstrated their potentiality in his *The world is round* (Macdonald in association with Rathbone, 1958) and in Harrap's *3-D junior atlas,* which he edited in 1955 (third edition 1963). The revised edition contains thirty pages of maps and a brief index; world vegetation and communications are treated first, followed by maps of the British Isles, then, progressively by maps of the countries and regions of Europe, Asia, Africa, the Americas, Australia, New Zealand and the world oceans. The *Relief form atlas,* by A Cholley, is an English edition of an atlas published by the Institut Géographique Nationale, based on air photographs taken in surveys of France and of French territories overseas. The photographs are presented with contour maps of the same areas and, in some cases, stereopairs of air photographs have been superimposed, one in red, the other in blue, so as to produce a three-dimensional effect when viewed through the spectacles provided. For some of the contour maps, contoured models have been photographed from two positions, so that they can be viewed in the same way. As an aid in developing the technique of reading from contour lines and in the study of land forms, such an atlas has a special value. Yet another interesting idea is effective in *The global atlas*: *a new view of the world from space*, published by Simon and Schuster of New York in 1958. Series of graphic relief maps were produced from sculptured models by Geographical Projects Limited, London, and a new method of colouring was designed to portray relief and vegetation or land use as they might appear if viewed from a point above the earth. This method is particularly effective in some areas and probably, especially to the layman, is more evocative of actual conditions than are the traditional methods.

In addition to the *Nouvel atlas général* mentioned above, Harrap is agent in Britain for *Meyers Duden-Weltatlas,* a comprehensive reference atlas comprising 174 pages of full-colour maps, 140 pages of tables and statistics and an index of some thirty four thousand names; and for *Meyers neuer handatlas,* a smaller atlas of 108 pages, of which thirty six pages are plates of full-colour maps and an index of about twenty thousand names, the rest being sketch-maps, tables and explanatory material.

The *Schweizerischer Mittelschulatlas* (Zurich, 1898-), though not readily available in Britain, is so important and has been so influential that it must be given a brief account; the atlas provides an excellent example of the development of Swiss cartographic method. The main source for the foreign topographical maps until 1910 was the Stieler atlas. For the revisions, 1927-32, directed by Professor Imhof, the principal source was the *Grande atlante del Touring Club Italiano* and since then, the 1954 *Atlas mira, The Times mid-century edition* and the national mapping services have been used as basic sources. The atlas was redesigned in 1955 and again, in a thirteenth edition, for the Konferenz der Kantonalen Erziehungs-direktoren, 1962, one outstanding feature being a new method of relief representation evolved by Professor Imhof. A high proportion of the maps has been revised and redrawn, incorporating new data and using new reproduction methods. The general effect of the topographic maps is as expressive as a relief model; contours are in brown, with lighter tints for the higher land and blue-grey shadow on the south-east slopes. Names in bold black stand out without obscuring the detail. The overall plan of the atlas has been reconsidered in the latest editions. Each area is now shown on a single map, combining the natural and cultural features. Switzerland is drawn on the scale of 1:1M; every important region is shown at least on scales of 1:15M, some on larger scales, such as Scandinavia at 1:6M. Political divisions and other thematic maps follow on smaller maps. The plan is one designed particularly for educational use, beginning with the home country and working outwards to the rest of the world, but the whole work is at a more mature standard than is the case with most school atlases. German, Italian and French editions are available. A detailed examination of the atlas is made by Professor Eduard Imhof: 'The Swiss Mittelschulatlas in new form', in the *International yearbook of cartography*, IV, 1964, which includes specimen map plates and a bibliography.

Another beautifully conceived atlas designed mainly for geography students and teachers is the *Geography atlas for secondary schools* (Moscow, Glavnoe Upravlenie Geodezii i Kartografii, 1954), edited by Y V Fillipov, which was revised in a second edition, 1959. It is in Russian, devoting about half the contents to the Soviet Union, with a balanced treatment of the rest of the world, introduced by a brief text. Political, geological, physical, climatic and economic maps of the world, continents and individual countries are included; more detail on all aspects of the geography of the Soviet Union is given, with particular attention to soil conditions and individual agricultural and industrial topics. In the new edition, two new maps were added: 'The Soviet Union—regions of the biggest construction sites of the Septennium 1956-65' and 'The economic-administrative regions of the Soviet Union'. The Directorate also issues many useful medium-sized world atlases, designed mainly for school use. Full details of these may be obtained through Collet's Russian Bookshop, London. The most important of these is probably the *Geographical atlas for teachers in secondary schools*, edited by A I Semenov, 1955, which comprises 192 pages of maps, including economic maps of the USSR.

The *Geographical atlas of the world for schools*, published by the State Publishing House for Geography, Warsaw, 1964, is also a fine production, of some 120 pages of maps and text, with an adequate index. The *Bos-Niermeyer schoolatlas der Gehele Aarde*, edited by Dr F J Ormeling, went into a forty third edition, published by J B Wolters of Groningen in 1964. In a large format, compared with others for school use, it contains eight pages of preliminary matter and 172 pages of maps, plus twenty eight pages of index separately pocketed inside the back cover. Of the 254 maps, the majority are thematic maps in two colours, blue and black, which effectively feature geology, climate, mineral resources, population, administrative areas, etc. These follow, in sequence, full-colour general maps of the same areas. The section devoted to the Netherlands is naturally particularly well detailed, showing balance and an imaginative use of colour. The rest of the world is treated less fully, but the more general maps give a clear impression of major relief features, in conventional style, with hill-shading added, and skilful selection of both material and place names has resulted in an evocative atlas for study purposes. A New school atlas by Esselte of Stockholm, 1967, was designed to meet new requirements in geography syllabuses. Emphasis is given to the study of environment in geography teaching; new types of maps have therefore been designed, using colour to indicate the principal characteristics of the environment, including relief and vegetation. Against this background, symbols for industries and other economic resources are introduced. The general physical maps in the atlas use a near-conventional range of hypsometric tints, on which is superimposed a series of vignetted layer colours to indicate detail.

The great general encyclopedias frequently include a major atlas section, as well as maps in the text. Particularly valuable is the atlas volume in the new edition of *Der Grosse Brockhaus,* 1960, which is arranged in five sections. Numerical and statistical information on the world and mankind is presented by a variety of graphical methods; then 280 plates in colour form the atlas section proper, including many double-page spreads covering topography, geology, vegetation, climate, anthropology and economic factors for the major regions of the world; the third section is a striking collection of plates illustrating typical land forms, settlement patterns and human life in various parts of the world; a small group of historical maps deals with some particularly important themes, such as the spread of western knowledge of the world; finally, there is an index of some fifty five thousand entries, including cross references from the current Russian and Polish place names in areas administered by the USSR and Poland to the German names. The chief maps are layer coloured, but most of the regional maps show relief very effectively by hill-shading in a brownish-grey tint. Scales are small; Germany and much of Central Europe at 1 : 1·25M, with the regional maps at 1 : 550,000 and parts of the United States at 1 : 4M.

As mentioned above, only a relatively few of the vast numbers of modern atlases have been mentioned; it may therefore be salutary to recall the words of one of the greatest living authorities on cartography, Dr Eduard Imhof—' It is only once in every two or three decades that a basically good and new world atlas is put out by some big cartographic publisher.'

REFERENCES

W G V Balchin: 'Atlases today', *Geographical Magazine,* April 1960, illus, diags.

C B Colby: *Mapping the world: a global project of the Corps of Engineers, US Army* (New York, Coward-McCann, 1959).

G R Crone: 'Atlases and maps', *British book news,* July 1964.

G R Crone: ' The future of the International Map of the World', *The geographical journal,* March 1962.

G R Crone: *Maps and their makers: an introduction to the history of cartography* (later relevant chapters) (Hutchinson University Library, 1953, 1962, 1966).

C B Fawcett: 'A new net for a world map', *The geographical journal,* September 1949, figs.

Irving Fisher and O M Miller: *World maps and globes* (New York, Essential Books, 1944). An excellent small book, to its date, having emphasis on map projections, with useful illustrations.

H A Karo: ' Progress in international cartography', *Surveying and mapping,* September 1964.

J J Klawe: ' The limits of editorial freedom in school atlases ' (ICA Technical Symposium, Edinburgh, 1964).

H E Mindack: ' Compilation aspects of a new world map', *The cartographer,* March 1965.

A H Robinson: ' The future of the International Map ' (International Geographical Congress, 1964, Section IX, Cartography).

W H Van Atta: ' From cross staff to satellite: surveying the world ', *Surveying and mapping,* December 1964.

3

NATIONAL
AND REGIONAL
MAPS AND ATLASES

MOST COUNTRIES NOW have an established national cartographical agency, but some areas, for one reason or another, still remain unmapped at medium or large scales. There are, however, few parts of the world that are without photographic coverage and maps based on air photographs. The speed of development and change in many areas has resulted in the need for efficient revision programmes and this operation also is greatly helped by aerial photography. In T W Birch: *Maps, topographical and statistical,* pp 10 and 11, is a useful table showing the percentage coverage by the official map series of several countries at 1:253,440 and larger scales. Styles of map design vary considerably from one country to another. Sometimes it is the character of the terrain itself which requires differences in the use of colour or specific symbols. Certainly, improvements in equipment and new technical developments are revealed in the changes introduced into successive editions of long-established series just as clearly as in illustration reproduction methods of typographical design. At the same time, beyond these obvious reasons for differences in map design, in indefinable ways the 'ethos' of a nation reveals itself in the prevailing forms of map design, as in the nation's literature, music or art.

The United Nations Commission on National Atlases was founded at Rio de Janeiro in 1956 to help and encourage the creation of national and regional atlases in individual countries. Important aspects of the commission's work have been the increase in unification and standardisation of the contents and legends of the principal maps, the compilation of a bibliography of national and regional atlases and of relevant literature, the extension of international contacts and co-operation in studying the problems of complex atlases. Meetings of the commission have been held in Moscow,

1958; Stockholm, 1960; Budapest, 1962; London, 1964; Paris, 1966, etc. The commission has a working party on industrial symbols.

A national atlas presents a synthesis of knowledge on the geographical elements that characterise the country; frequently, historical maps are also included to suggest reasons for the individual development and explain existing characteristics. In the purist sense, a ' national ' atlas must be sponsored and prepared by government agencies, inevitably with the assistance of any academic or private cartographical organisations that may exist in the country. The definition may be made clear by contrasting the ' national ' atlas sheets of Britain prepared by the Ordnance Survey at 1:625,000 and published by the Ministry of Housing and Local Government with *The Atlas of Britain and Northern Ireland,* which has a similar aim, was compiled with the help of many of the same experts, using the same data, but was published by a private organisation, the Cartographic Department of the Clarendon Press, Oxford.

The national atlas is also, one may say, a ' prestige ' publication, displaying, some to a greater degree than others, a national pride in beauties of scenery or industrial achievements. The base maps prepared for such an atlas have also been invaluable in the economic or regional planning of a country and the volume itself inevitably stimulates interest in both laymen and experts, providing a basis for further cartographical and other work. A later idea was the integrated planning of a series of atlases, either on a regional or a systematic basis. The International Geographical Union Commission on National Atlases has extended its interest to regional atlases and has given valuable assistance to organisations engaged in the preparation of integrated atlases. All national and regional atlases need the most careful planning, to be of maximum usefulness, particularly in the choice of projections and scales, to allow of adequate comparisons, cartographically and bibliographically; they give a fine opportunity for the display of craftsmanship, scientific accuracy and imagination. In many cases, the paper has been specially made for the purpose and close attention given to style of relief representation, the use of colour and symbols and the inclusion of necessary documentation, frequently including explanatory text and bibliographies. The atlas must suggest unity, yet each map sheet or group of sheets must have distinct individuality. A few general references may usefully be cited here, before treating of each country separately:

G R Crone: *Maps and their makers: an introduction to the history of cartography* (Hutchinson University Library, third edition 1966, chapters XI and XII).

N L Nicholson: 'Survey of single-country atlases', *Geographical Bulletin*, 1952, no 2.

K A Salichtchev, editor: *Atlas nationaux: histoire, analyse, voies de perfectionnement et d'unification* (Moscow/Leningrad, Edition de l'Académie des Sciences de l'URSS, 1960-64).

K A Salichtchev, editor: *Regional atlases: tendencies of development, subject matters of the maps on natural conditions and resources* (Moscow/Leningrad, Publishing House 'Science', 1964).

K A Salichtchev and J G Saouchkine: *Les cartes sociales et économiques aux atlas complexes* (Editions de l'Université de Moscow, 1968). In Russian.

Griffith Taylor, editor: *Geography in the twentieth century: a study of growth, fields, techniques, aims and trends* (Methuen, third edition 1957, chapter XXVI).

E L Yonge: 'National atlases: a summary', *Geographical review*, 1957, no 4.

ATLANTIC OCEAN, NORTH SEA

During recent years, much research has been undertaken on both the seabed and the waters of the Atlantic and North seas. Some of the results have been published in atlas form, such as the *Biographical atlas of the North Atlantic*, published by the American Geographical Society, 1959-60. The *Atlantic Ocean atlas of temperature and salinity profiles and data from the International Geophysical Year of 1957-58*, by F C Fuglister, was issued as the first of a series by the Woods Hole Oceanographic Institution, 1960. The *Atlas zur Schichtung und Zirkulation des Atlantischen Ozeans*, compiled by G Wüst and A Defant, embodied the results of the first Deutsche Expedition auf dem Forschungs-und Vermessungsschiff *Meteor*, which will run to many volumes of text, charts and figures. From August 1965, the research ship *Meteor* has been working on the Central Atlantic ridge. The research was directed by Professor Karl Brocks of Hamburg University and Dr Otto Meyer of the German Hydrographic Institute, leading a team of research scientists from twelve other institutions, mainly geophysicists and geologists. Special echo-sounders, nuclear magnetographs and gravimeters were used to investigate the structure of this submarine chain. This kind of research is typical of the projects being currently carried out in the world's oceans. One aspect being kept in mind was new evidence concerning continental drift. Also, the

111

behaviour of electricity in the atmosphere and its relationship with climate was studied by means of radio probes sent as high as the stratosphere and tracked by radar equipment on board the *Meteor*. Other special studies were undertaken at the point of intersection of the geographical and magnetic equators.

A second Atlantic ridge expedition was begun in 1967 in the area west of Gibraltar between the Azores and the Canary Islands, gathering information on marine geography, geology and biology. On this occasion ninety two research scientists and technicians from the Federal Republic, Spain, Great Britain, Portugal, Norway and France took part. The principal work was the measurement of the Great Meteor Bank discovered in 1938 by an earlier ship *Meteor*. For purposes of comparison and contrast, other marine ridges in the Atlantic and the continental shelves of Portugal and North Africa were also examined in detail. Thousands of samples were taken, which are now being carefully analysed at the Institute of Oceanography at Kiel and thirteen other German institutes for oceanographic research. In due time, it is hoped that a second atlas will appear incorporating the results of all this analysis.

Bermuda and Ascension Island are mapped by the Directorate of Overseas (Geodetic and Topographical) Survey, the former at 1:2,500 and 1:10,560 and the latter at 1:25,000 and 1:50,000, usually showing relief by contours at 100 feet intervals and spot heights.

The new island of Surtsey was quickly mapped at 1:10,000 by the Landmaelingar Islands, published in 1964 with a short descriptive history; it would be profitably studied together with *Surtsey, the new island in the North Atlantic,* by S Thorarinsson (Iceland, 1964) in Icelandic and English. Fine photographs and maps are included.

The North Sea, by L E J Brouwer, and *The law of the continental shelf,* by the Right Hon Lord Shawcross, were published as part V of the World Land Use Survey *Occasional Papers* (Bude, Geographical Publications, 1964).

A very useful fishing atlas of the North Sea—*Atlas Rybacki Morski Instytut Rybacki Gydnia*—compiled by A Klimaj St Rutkowicz (Warsaw, Państwowe Przedsiebiorstwo Wydawnictw Kartograficznych, 1957), maps the herring and mackerel catches in the North Sea and adjacent waters, including maps of specific fishing grounds, currents, temperature, salinity, etc, making altogether thirty six maps, with text in English, Polish and Russian.

North Sea, a BP map showing North Sea oil and gas licence areas, at approximately 1:1,500,000 was issued in 1966.

EURASIA

EUROPE : When treating of ' Europe ', it is always necessary to define the area under consideration. A vast literature has grown up on the evolution and interrelationships of the Eurasian and Mediterranean lands and, in the case of cartographic works, the area included varies according to the date of the map and the purpose for which it was compiled. In *The geographical journal,* September 1960, is a particularly succinct article by W H Parker, ' Europe: how far?', which provides an excellent starting point for anyone interested in the map of Europe; the point in question is, to use the author's own words, the ' controversy and confusion ' through the years concerning the geographical, cultural and political line of division between Europe and Asia, illustrated by well-conceived sketch-maps.

Bartholomew publishes a map, ' Eurasia ', at 1 : 15M, revised in 1966, which gives a clear picture of the whole complex area, showing contour colouring, roads in red, railways in black and boundaries in purple. Bartholomew's ' Political map of Europe ', at 1 : 5M, 1967, includes the Mediterranean. Railways and shipping routes are indicated, but no roads. ' Western Europe', at 1 : 3M, 1966, extends from Spain to Italy and to southern Scandinavia. ' Central Europe ', at 1 : 1,250,000, 1963, covers the area from London to Venice and from Berlin to the Riviera; relief is shown by hill-shading; and ' Europe, central and south-east ', at 1 : 2,500,000, 1968, extends from the Gulf of Danzig to Crete and from Lausanne to Istanbul. Bartholomew have also published ' Europa ', 1 : 4M, 1966, for BP, which includes an inset of Turkey at 1 : 4M, and on the reverse is ' Central Europe ' at 1 : 2,500,000.

Stanford's ' General map of Europe ' includes also the Mediterranean and the Near East to the head of the Persian Gulf, at 1 : 6M, with a pictorial border showing coats of arms. The Hallwag ' Political map of Europe ', 1 : 5M (Philip) extends from Scandinavia to the Mediterranean, including the Urals, the Caspian Sea and part of Arabia. Motorways are particularly well distinguished. A map of Europe at fifty miles to one inch is also available and one of central Europe at sixteen miles to one inch; individual countries are mapped mainly at sixteen or eighteen miles to one inch, and individual cities of Europe usually at five inches to one mile. Philip issues what is probably the largest reference sheet map of Europe for publication,

' Commercial map of Europe ', in two sheets at fifty six miles to one inch.

The Institut Géographique National, Paris, published the ' Carte de l'Europe ' in 1963, at 1 : 1M, using the Lambert Orthomorphic Projection; relief is shown by contours and hill-shading, with woodland areas distinguished and roads clearly marked. The *Grosse Shell—atlas of Germany and Europe,* 1960, aims at the presentation of maps for any given requirement, at the most convenient scales. Maps at 1 : 4,500,000, 1 : 1,500,000, 1 : 500,000 and 1 : 700,000 are arranged in systematic order, with a number of city plans at larger scales; a revised issue is published annually by Mair Geographical Publishers. Brief comments are printed below the map face and concise sources of information are quoted. ' L'Europe moins la France ', directed by A Frayasse, consists of thirty seven maps (Paris, Colin, 1957, Cahiers de cartographie, échelles diversés). With the increased popularity of the touring holiday, publications such as the *Hallwag Europa touring* have proliferated. This is a particularly useful guide, including ninety two pages of fully coloured maps and eighty six town plans, but there are many others which might serve individual purposes as well. Of unique interest on this aspect of Europe is Wigand Ritter: *Fremdenverkehr in Europa* (Leiden, Sigthoff, 1966), which includes text maps and maps in a separate pocket.

One of the first attempts to portray the geomorphological variety throughout Europe was the *Physiographic diagram of Europe,* by A K Lobeck (Wisconsin Geographical Press, 1923), an ingenious diagrammatic map showing the principal physiographic features of the continent, with a descriptive text. It was published in large-scale and small-scale editions. 'A tectonic map of Eurasia ' at 1 : 5M was issued by the Geological Institute of the Academy of Sciences of the USSR, Moscow, 1966; in twelve sheets, the map is drawn on the Bonne Projection and has legend matter in Russian and English.

At the International Geological Congress in Copenhagen, 1960, and at the Russian Trade Exhibition in 1961, one of the outstanding scientific exhibits was a hand-coloured tectonic map of Europe compiled by Russian draughtsmen from copy submitted by the various national geological surveys. The map, at 1 : 2,500,000, was sponsored by the Sub-commission on the Tectonic Map of the World, of the International Geological Congress. It covers Europe, the adjoining parts of the USSR, North Africa and Asia Minor, and was edited under the direction of the late academician, N S Shatsky. The map was printed for the International Geological Congress in the autumn of 1962, in French and Russian editions, by the USSR Academy of Sciences, but publication awaited the completion of the explanatory text, in 1964, edited by A A Bogdanoff, M V Mouratov and N S Schatsky, entitled *Tectonics of Europe; explanatory note to the International Tectonic Map of Europe,* comprising chapters, in English or French, contributed by national representatives on the sub-commission. The edition was limited to 1,500 copies. Sixteen map sheets are enclosed flat in an elephant-size portfolio. Each sheet bears its own legend. The map was of the greatest importance, not only for the cooperation afforded by so many geologists from different countries and the new knowledge gained of Europe and of the methodology of tectonics, but, even more important, it was, in effect, a pilot project for a more ambitious, international work to come.

An 'International geological map of Europe', sponsored by UNESCO, comprising forty nine sheets at 1 : 1,500,000, is to be published within the next few years, in accordance with plans prepared by the Commission for the Geological Map of Europe. At the time of writing, the following maps are already available and a brief delineation of their coverage will serve to indicate the magnitude of the task and the detailed work involved.

A3 Atlantic Ocean west of Scotland with Porcupine Bank and Rockall Bank (Isle of Rockall, UK).

C1 Arctic Ocean north-west of Norway, including Greenland basin and Norwegian basin, with between them the eastern spur of the Jan Mayer ridge.

C2 Caledonian orogenic belt of middle Norway; also covers Swedish territory.

D1 Basement and Caledonian orogenic belt of northern Scandinavia in the border region of Norway, Sweden, Finland and the USSR; southern Arctic Ocean including Spitzbergen Bank and Bear Island (Norway).

B3 Basement and Caledonian orogenic belt of Scotland and North Ireland.

C3 Basement and Caledonian orogenic belt of southern Norway and south-west Sweden; cretaceous and tertiary of Denmark.

D2 Mainly basement of north Sweden and north Finland with parts of Norway and USSR.

D3 Basement of south-west Sweden and southern Finland; paleozoic and mesozoic shelf-sediments of Estonian and Livonian SSR.

These eight maps were available in 1967; the following six became available in 1968, a combined edition of UNESCO and the Bundesanstalt für Bodenforschung, Hanover.

B4 Caledonian and variscian orogenic belt of Ireland and England; basement of north-west France; mesozoic and kenozoic sediments of south-east England, northern France and Belgium.

C4 Basement of southern Sweden and the island of Bornholm and the Bohemian Massiv; variscian orogenic belt of Belgium, north-eastern France and Germany; mesozoic and kenozoic sediments of the epivariscian platform area of France, Belgium, the Netherlands, Denmark, Germany and Poland.

D4 Basement of the Bohemian Massiv, variscian orogenic belt of Poland and Czechoslovakia; mesozoic and kenozoic sediments on the platforms of Poland and the USSR; Alpine orogenic belt of the northern Carpathians.

B5 Basement of north-west and central France; variscian orogenic belt of northern Spain and central France; mesozoic and kenozoic sediments of the Paris, Ebro, Aquitanian basins.

C5 Basement of the Bohemian Massiv; variscian and Alpine orogenic belt of south-eastern France, Switzerland, Austria, Italy and Yugoslavia; mesozoic sediments on the platform of north-eastern France and southern Germany.

D5 South-eastern part of the basement of the Bohemian Massiv; variscian and Alpine orogenic belt of the Carpathians in Austria, Czechoslovakia, Poland, USSR, Romania and of Bulgaria and Yugoslavia; kenozoic sediments of the great Hungarian basin.

The maps are available separately or as sets.

An 'International Quaternary map of Europe' was begun under the direction of Dr H-O Grahle in 1964; this is a very complex map, on which geological and morphological features are combined, at the scale of 1:2,500,000, sponsored by UNESCO and the Bundesanstalt für Bodenforschung, Hanover. It is expected to be complete in sixteen sheets. This map, the outcome of many years of discussion and successive plans, was finally launched at the First Commission for the International Quaternary Map of Europe, held in Leningrad in 1932. Two sheets were published, but further work was interrupted by the second world war. The map is one of several earth science maps of Europe, using the same topographic base and sheet lines; thus it is easy to extend the map to cover large areas of the sub-Arctic, western Siberia, north-western Soviet central Asia, North Africa and part of the Middle East. Four sheets are published to date: 1 Iceland, with the coast of Greenland; 2 the central part of Norway from the Trondheim fjord to Tromsö, Sweden north of 64 degrees north and west of 20 degrees east and the Arctic area south of about 20 degrees north; 5 Ireland, the peninsula of Cornwall and Wales, the Isle of Man, the Hebrides, and parts of the western coastal areas of Scotland; 6 parts of Norway and Sweden, part of the Baltic, the Jutland peninsula and parts of Great Britain. An index map of the completed map showing available sheets, sheets in preprint and those in preparation is included in *Nature and resources,* September 1968. Also at 1:2,500,000 a ' Metallogenetic map of Europe ', in nine sheets, was undertaken by UNESCO and the Bureau de Recherches Géologiques et Minières, Paris, 1968-.

Cartographia of Budapest completed a ' Hydrographic map of Europe's surface waters ', 1:10M, in 1965. *Europe in maps: topographical map studies of western Europe,* by R Knowles and P Stowe (Longmans, 1968), is a series of topographical studies based on extracts from large-scale maps of European countries together with matching aerial photographs and explanatory text, enabling the reader to obtain a full understanding of the area. Natural features, agricultural regions, areas of old and new industry and urban and rural settlements emphasise the variety of European landscapes.

A ' Soil map of Europe 1:2,500,000 ' was published by FAO in 1966 in six sheets, with an explanatory brochure, the result of intensive work by the working party on soil classification and survey of the European Commission on Agriculture, under the chief direction of R Dudal, R Tavernier and D Osmond. Following a meeting of European soil scientists held in Ghent, 1952, to discuss the unification of the different systems of soil classification and nomenclature, the working party was created in affiliation with the Sub-commission on Land and Water Use of the European Commission on Agriculture, one of its tasks being to prepare a revised soil map of Europe. The present map is based on the compilation of material available in the different countries or which was prepared for this purpose by some sixty contributors from twenty three countries. The first chapter in the booklet describes briefly the soil groups which have been distinguished in Europe and how these soil units have been composed into soil associations. Environmental conditions are then discussed—physiography, climate, vegetation, etc. Considerable space is given to explanations of the soil associations of mountains, subdued mountains, high plateaux, low plateaux, hills and plains; and the third chapter describes the soil associations. The final chapter gives a picture of the soil survey activities at present carried out in those countries which are members of the European Commission on Agriculture.

Particular attention has been given in recent years to the social and economic integration of Europe and the vital issues involved have found expression in map production. For example, the College of Europe at Bruges issued an economic map of Europe, ' Orbis terrarum Europae ' in 1955, compiled by I B F Kormuss. The work comprises sixteen maps at scales ranging from 1 : 3,600,000 to 1 : 45M, which have brought together factual information on political and economic institutions, crop limits, population distribution, fuel and power factors, steel industries and transportation. A fine cultural map by P Dumont and J S Baltus was included, also thirty four statistical tables. The maps, with a descriptive brochure, were issued in a jacket, or each map was available separately, in English and four other languages. Similar ideas were inherent in the preparation of the *Atlas of social and economic regions of Europe,* published from 1963 by the Soziographisches Institut an der Johann Wolfgang Goethe Universität Frankfurt-am-Main (Verlag Lutzeyer, Baden-Baden, 1963-), under the direction of Professor Dr Ludwig Neundör-fer. This is an atlas concerning the non-Communist nations of Europe, which is being issued serially under the patronage of the Council of Europe and with the assistance of the Federal Republic of Germany and the German Research Society (Deutsche For-schungsgemeinschaft). Separate folders are planned to accommodate 100 maps, with screw fastening; texts and keys are in German, English and French. Two detailed gazetteers list all the administrative sub-divisions in Europe, with a general map of the regions covered by the atlas and a list of statistical sources used. It should be remembered in using such an ambitious work as this that difficulties are still encountered in the differences met with in the availability and comparability of the statistics involved. The various sheets of the atlas present a compilation of sources designed to enable the user to find answers to his own questions; more widely, it represents a basis for the social and economic planning of Europe in the second half of the twentieth century. The work is arranged in three main series : the first series deals with the elementary facts of the socio-economic structure, the bases of occupations, agriculture and industries, and the movement of population; the second considers all manner of factors relating to internal structure—school, church and health services, analyses of levels of education, religious denominations, age structures and causes of deaths, etc; thirdly, changes and trends in all sectors included in the first and second series are traced, particularly in the agricultural structures, decentralisation of industrial locations, new concentrations and changes in the population structure. In effect, the first two series present the basic facts which will govern planning and decisions in the second half of the century and the third series will keep the atlas constantly up to date in the interpretation and interrelation of the facts.

Maps frequently accompany the numerous reports and other publications issued by the United Nations, the OECD, the Council of

Europe, the Statistical Office of the European Communities, etc and in the mass of literature which has developed from all the deliberations on the Common Market. Sometimes, they are the result of individual research, as in the case of the 'Distribution of population in Europe' map, by Alfred Söderlund, 1957, and the 'Economic map of Europe' by W William-Olsson, in two sheets, 1953.

Important to English-speaking readers is the *Atlas of central Europe,* taken from the latest reprint of the *Bertelsmann world atlas* and made available in Britain by Murray, 1963. At the standard scale 1:1M, basic topographic data are presented in greater detail than in most atlases; new trends in geographical teaching and thinking have been kept in mind and some map sheets use a 'sampling' method in depth, rather than attempting an overall more superficial coverage. There is a comprehensive index of more than thirty seven thousand entries.

The *Atlas of western Europe,* prepared by the eminent geographer and cartographer, Jean Dollfus, is issued in English, French, German, Italian and Dutch (Paris, Société Européenne d'Etudes et l'Information; Rand McNally, 1963). A committee of six members advised on the atlas, one for each of the countries of the EEC. Twenty four coloured and three monochrome maps cover geology, physical features, political and administrative divisions, major mineral deposits, demographic developments, present-day populations and their occupations, energy, communications and agricultural and manufactural land use. Illustrations and brief commentaries by Bernard Basdeloup accompany the maps. There is an index of place names and their equivalents and it is claimed that the geographical names have been 'restored to their original forms'.

One of the earliest atlases to attempt to portray the tangled history of Europe was *An atlas of European history from the second to the twentieth century,* by J F Horrabin (Gollancz, 1935). The text really predominates in this work, beginning with a note on the geographical factors, followed by sections devoted to successive historical periods, each accompanied by a black and white sketch-map. In the *Historisch-geographisches Kartenwerk,* by E Lahmann (Leipzig, Wirtschaftshistorische Entwicklung, Verlag Enzyklopädie, 1960), a vast amount of research from the Bronze Age to present times is contained in fifty three pages of text and 191 maps. The British Isles, France and the Benelux countries are covered and the whole work is excellently documented with illustrations, diagrams, bibliographies and indexes. The USSR dominated states of eastern Europe are covered in the *Atlas Östliches Mitteleuropa* (Bielefeld, Velhagen und Klasing, 1959). Fine maps are devoted to history, geology, physical features and natural resources. The legends are in a separate volume and captions are given in German, English and French.

An atlas of European affairs (Methuen, 1964), with text by Norman T G Pounds and maps by Robert C Kingsbury, is built round the theme of the continuing importance of Europe in world affairs. Monochrome sketch-maps and facing text deal with many aspects of European politics and economics, population, language and religious groups, individual countries and the great co-ordinating bodies such as the European Community, OECD and NATO, the Coal and Steel Community and the European Economic Community. It is a useful reference work, but hardly an atlas. Many other small publications have attempted, with varying success, to portray European history and political developments; the *Oxford historical atlas of modern Europe,* for example, by R L Poole, 1902, and the more recent *Atlas of European history,* in which sixty four pages of six colour maps show the physical setting of the major periods in the growth of European civilisation; other examples include F A Freeman's *Atlas to the historical geography of Europe* and the *European history atlas* by J H Breasted, published by Denoyer-Geppert of Chicago in 1937.

A specialised atlas, the *Atlas of European birds,* compiled by K H Voous (Nelson, 1960), comprises a series of distribution maps showing the breeding range of 419 species of birds breeding in Europe west of the Urals. Zoogeographical regions and climatic zones are shown and excellent plates illustrate comparative studies of individual species. It is hoped that more of this kind of work will be undertaken.

At a recent international conference held in Bonn, forty scholars from twenty five countries discussed the preparation and publication of an ' Ethnological atlas of Europe '. This will be the first of its kind and publication is planned for 1969; it will contain such interesting aspects as changes in the design of agricultural implements, the materials used for everyday things, such as homes, customs, ritual festivals, etc. A *Volkskundeatlas* is also being planned as a work of historical ethnology, depicting conditions as they were at the turn of the century; in progress also is a *Historical atlas of town plans for western Europe,* of which the British contribution is being directed by Mrs M D Lobel.

The increase in road travel and holidays on the Continent has given rise to numerous road maps and maps designed for tourists. Maps accompany all the regular guides, with those of Michelin and Mair being of particularly good standard. *The resorts of western Europe* (Bude, Geographical Publications, 1962) was an interesting

120

experiment. The title seems misleading, as Great Britain is not included, although Ireland is, also Iceland, North Africa, the Balkans and Asia Minor. The compilers were G W S Robinson and M G Webb, and the twenty two maps, in black and blue, were drawn by A C Clark. More than 1,200 resorts are named and several thousand marked, being graded by symbols according to calculations of numbers of beds available in hotels and boarding-houses. Not only the size of the resorts, but popularity, quality, types of visitors and length of season are also indicated, so that close scrutiny is needed for interpretation of the full data included on the maps. The reverse of the sheets carries notes and a list of the major resorts, grouped by countries and linked by numbers with the maps. Particularly since about 1960, the number of road maps and atlases has increased and the (usually) annual revisions incorporate revision and refinements in information and techniques. Much ingenuity in folding maps has become evident, for the convenience of the car traveller. The Kümmerly and Frey maps are particularly clear and striking. In 1961, the firm issued a *Road and travel atlas of Europe* (available in Britain through Benn), containing ninety six plates of maps and fifty six pages of commentary and index, in German, English, French and Italian. The series ' European country and regional road maps ' is by now well known in this country, distinguished by its clarity, up to date information and fine use of shading in mountainous districts. A *Road map of Europe,* 1 : 500,000, is edited under the auspices of the Alliance International de Tourisme; the latest editions have a neutral base of grey relief, on which different classes of roads are clearly shown.

Collins road atlas of Europe began in 1965, containing 232 plates of six colour maps, produced and printed by the Esselte Map Corporation, Stockholm. The basic scale is 1 : 1M, with eastern Europe shown at 1 : 2M and large cities and particularly notable tourist districts at 1 : 500,000. The ' Carta stradale d'Europa ', 1 : 500,000, comprises a series of uniform sheets issued by the Touring Club Italiano mainly for the high-speed motorist passing through central Europe; the work forms part of a larger project in progress, the 'Alliance internationale de tourisme' series. The background, in grey relief, provides a neutral base against which the different classifications of roads are shown in colours. Legends are given in eight languages. The *Europa touring* series is well known, produced by Hallwag and distributed in Britain by George Philip and Son. Twenty five countries are mapped, illustrated and described; eighty six town plans were included in the last edition and the whole work is well indexed. The *Europa Shell atlas of Europe* contains 198 pages of maps at twenty four miles to the inch, including a great deal of detail. The Hallwag (Philip) *Autropa atlas,* at fifty miles to the inch, is also an excellent road atlas.

The collection of essays edited by R Brugmans and others: *Sciences humaines et intégration Européenne* (Leiden, Sythoff, 1962, Cahiers de Bruges, nouvelle série, 1) includes a fine folded map in a pocket. The maps in Leon Dominian: *The frontiers of language and nationality in Europe* (New York, Holt, for the American Geographical Society; Constable, 1917) are still evocative. Useful maps are also in L M Alexander: *Offshore geography of northwestern Europe: the political and economic problems of delimitation and control* (Rand McNally, for the Association of American Geographers; Murray, 1966), which includes an account of the 1964 European Fisheries Convention and refers to the drilling for oil in the North Sea.

British Isles: The official mapping agencies in the United Kingdom are the Ordnance Survey of Great Britain, the Ordnance Survey of Ireland, and the Directorate of Overseas Surveys. The National Committee for Cartography, making liaison with the International Cartographic Association, is the Cartography Sub-committee of the British National Committee for Geography, under the auspices of the Royal Society, appointed in 1961. This sub-committee represents all kinds of British organisations and individuals interested in cartography, geodesy, geography, geology, photogrammetry and printing technology; it aims to provide a forum for the exchange of information and discussion on all aspects of cartography, thus stimulating development. Working parties have been set up to consider the training of cartographers, the standardisation of cartographic terms and automation in cartography; reports from the working parties were included in *The cartographic journal,* December 1965. The sub-committee also has it in mind to act as a co-ordinating body, when acceptable, between the different British cartographic interests. Delegations have been sent to attend the First General Assembly of the International Cartographic Association, 1961, presenting the first national report on cartographic activity in Great Britain. A British delegation went also to the United Nations Technical Conference of the IMW 1:1M at Bonn in the same year and to other relevant meetings since then. The sub-committee was responsible for the arrangements for the International Cartographic Association meetings in London and Edinburgh in 1964 and for much of the preparatory work. On this occasion, a second national report was presented. Symposia on current cartographic problems are held from time to time, usually at one of the universities. The first symposium, held in November 1961, was notable as the first occasion when representatives of government and commercial mapping organisations met for discussion with academic cartographers.

Christopher Saxton, William Camden, John Norden, John Speed, John Ogilby and the Cary brothers all had the right ideas about regional mapping and foreshadowed the day when the whole of Britain should be adequately mapped for the benefit of all its citizens, but it was not until after the Battle of Culloden in 1746 had revealed the inadequacy of existing maps of the Highlands that a survey at a scale of 1:36,000 was authorised. The Ordnance Survey, then, in common with similar mapping agencies in other countries, had its origin in a military organisation. For many years, the military requirements dominated map production; but it was fortunate that there was available a young engineer, William Roy, of outstanding talents and vision, who was appointed in 1765 Surveyor-General of Coasts and Engineer ' for making and directing military surveys in Great Britain '. From this time on, Roy continued to advocate the beginning of work for a complete national survey, but his recommendations were not actually implemented until the ending of the American war in 1783. In 1784, the first base was measured at Hounslow Heath under the direction of General Roy. In November 1967, a plaque, erected by the British Airports Authority, was unveiled, commemorating the start of the first scientific triangulation of Great Britain.

Through 1787-88, a triangulation, under the general aegis of the Royal Society, was made to determine the difference between the meridians at the Greenwich and Paris observatories, and in 1791, the major triangulation of Great Britain was begun. William Roy died in 1790, but, again fortunately, the third Duke of Richmond, then Master-General of Ordnance, had become firmly convinced of the soundness of Roy's ideas and it was he who in fact established the Survey on an official basis under the Board of Ordnance, having its headquarters in the Map Office of the Tower of London until 1841, when the buildings were destroyed by fire and the department moved to Southampton. During this initial period, much progress had been made on the establishment of control points by triangulation and mapping England and Wales at the one inch scale. In 1824, work had been interrupted by the urgency of mapping Ireland for valuation purposes and the greater part of the resources of the new department was engaged on this survey at six inches to the mile for the next twenty years. Experience gained in the Irish survey decided the completion of the survey of the six northern counties of England and of Scotland at six inches to the mile instead of one inch to the mile. Work on the principal triangulation of Great Britain was resumed in 1838 and completed in 1852, an achievement of remarkable accuracy, which formed the basis of mapping in Great Britain for the next hundred years. The triangulation rested on two base lines, one on the shores of Lough Foyle, the other on Salisbury Plain, measured in 1827 and 1849 respectively. As the triangles were carried across country, the one inch map sheets appeared. The first four sheets, issued in 1801, covered Kent and part of Essex and London.

The next stage involved much controversy, but in 1858 agreement was reached that, with the exception of mountainous and moorland areas, the whole country should be mapped at the scale of twenty five inches to the mile. A pilot scheme had already been conducted in the Durham area. Scotland and the six northern counties were first covered at this scale and by 1895 surveys of the southern counties were completed also, thus establishing a control system and large-scale maps covering the whole country which surpassed any other national mapping at that time. In 1870, control of the department passed from the War Office to the Ministry of Works and, in due time, to the Board of Agriculture, now the Ministry of Agriculture, Fisheries and Food.

The South African war and the first world war, followed by financial difficulties, severely hampered the maintenance of accurate maps, so much so that a parliamentary committee, under the chairmanship of Viscount Davidson, was set up in 1935 to decide on the measures to be adopted and it is the *Davidson Report*, published in 1938, that shaped the work of the Ordnance Survey as we know it. A first recommendation was that a National Grid, calibrated to the international metre, should be used as a standard reference system covering the whole country; large-scale plans of urban areas were to be compiled and an adequate system of continuous revision implemented; the 1:2,500 series was to be completely revised on a single national projection, thus superseding the 'County' series drawn on the Cassini Projection. These and other decisions involved a new triangulation for the whole country, which, due to the second world war, was not completed until 1952. Secondary and tertiary plotting was completed during the next few years. The second geodetic levelling was completed between 1912 and 1921 and the third between 1951 and 1959. On Ordnance Survey maps, all heights are expressed in feet above mean sea level, the datum point being at Newlyn in Cornwall. Between 1915 and 1921, the Tidal Observatory at Newlyn recorded the sea level at hourly intervals, from which a 'mean' was calculated. There are three orders of Ordnance Survey levelling, differing only in degree of accuracy: geodetic, secondary and tertiary. Numerous 'bench marks' have been established throughout the country, marking precisely determined heights. A levelling checking programme was initiated in 1956 to make any necessary corrections in subsidence areas every five years, in normal areas every twenty years and in mountainous regions every forty years. Levelling computations are processed on punched cards, which form the content of *Bench mark lists*, which are available for sale.

Counterbalancing all the difficulties, aerial surveying techniques had been advancing in efficiency; indeed, the department itself has constantly improved the techniques involved, enabling it to cope with the rapid changes and developments immediately following the second world war. Map printing had been the subject of research for many years, notably in photolithography and photozincography. The long period of development of the Ordnance Survey reveals the problems inherent in the production of a national series of topographical maps for the use of a changing and complex society.

During the second world war, the headquarters of the Ordnance Survey moved to Chessington in Surrey. The Publication Division remained there and has been responsible for the storage and distribution of the maps; the production units returned to Southampton, where vast numbers of staff have been trained and employed on all stages of production. New premises in Southampton to house the whole organisation were completed in June 1968 and the transfer to Southampton of the map stocks, involving the removal of several hundred tons of maps, began. The new address is Romsey Road, Maybus, Southampton. Current output is estimated at some fifty million sheets a year and over eleven million maps are held in stock. All field survey operations are under the supervision of the Director of Field Survey. The Geodetic Control Division is responsible for horizontal and vertical control and for air surveys. New instruments and equipment have revolutionised revision and the making of new maps. Stations of primary triangulation are, in general, about thirty miles apart. Helicopters can now be used for transport in difficult country and tellurometers, electronic instruments used in pairs to measure the transit time of radio waves along a line, are used for remeasurement of base lines.

The present state of publication may be summarised as follows. The large-scale plans are on two scales, 1:1,250, approximately fifty inches to the mile, and 1:2,500, approximately twenty five inches to the mile, both carrying the National Grid. Plans at both scales depict in detail the area represented and, in addition, show all administrative and parliamentary boundaries, spot heights along the principal roads, the position and height above mean sea level of Ordnance Survey bench marks and the position of Ordnance Survey co-ordinated points. The 1:1,250 scale plans will eventually cover all towns with a population above about twenty thousand and a number of other urban areas, each plan representing an area of five hundred square metres, carrying the National Grid at one hundred metre intervals. Photomosaics were temporarily issued for selected towns pending publication of the fifty inch plans.

The 1:2,500 scale ' County ' series plans resulted from the original large-scale work of the Ordnance Survey, carried out in the second half of the nineteenth century. They have all been revised at least once since their first publication. The plans are numbered on a county basis. They are being replaced now by National Grid plans, which will eventually cover the whole of Great Britain except the mountain and moorland areas. Each plan represents an area one kilometre square, but, as a rule, two plans are paired into a sheet representing an area of two square kilometres. In those areas where 1:1,250 plans have been prepared, the 1:25,000 plans are produced by direct reproduction from the larger scale. For the purpose of area measurement, the plan is divided into parcels, each measured individually and given a reference number, which is printed on the sheet, together with the area in acres. The linking together of more than one feature into one parcel is termed ' bracing ' and the symbol ʃ is placed across the common division. There are other symbols, used to denote, for example, ' measurement pecks ', which indicate the boundary of a parcel when no suitable ground feature exists. The symbols and conventions used are described in detail and illustrated in a leaflet published by the Ordnance Survey, *Parcel numbers and areas on 1 : 2,500 scale plans,* together with a history of the recording of parcels from 1850. When Ordnance Survey plans based on the National Grid were introduced, a new system of numbering parcels was adopted and a further change was made at the beginning of 1959. Towns, villages and built-up areas, where it would be impossible to show all the parcels individually, are banded together and treated as a single parcel. The perimeters of such parcels as these have been shown at different times by colour bands, stipple, hatching or a τ symbol. The present method uses the ? sign at regular intervals along the perimeter.

The medium-scale maps include two series; the six inches to the mile, or approximately 1:10,560 and the map at 2½ inches to the mile. The original maps of the ' Six inch ' series were issued on full-size sheets, each covering an area of six miles by four miles. Full-size sheets are still available for London, the Highland districts of Scotland, most of the islands and various other areas of Great Britain. The new type map, compiled from the new twenty five inch plans, were issued on quarter-size sheets, each covering an area of five kilometres square. Pending completion of the latter a provisional edition of the 'Six-inch' series was issued on the basis of the latest revisions of the old style map, in monochrome, with contour lines in red. Photomosaics were also issued temporarily for areas not covered by either the new or the old type maps. Now, a post-second world war series at this scale covers England and Wales and much of Scotland on National Grid sheet lines. The remainder of Scotland is covered by pre-war county sheets, which are being steadily replaced by National Grid sheets. Most National Grid sheets are five kilometres square, but some coastal sheets are a little larger. All features on the six inch maps are to scale, except in built-up areas, where the width of streets may be distorted to make room for the insertion of street names. Most of the 1 : 1,250 survey and all of the six inch resurvey are carried out by stereo-plotting instruments, using ground survey for those points only that cannot be completed by air survey. Six inch town plans are square, five kilometres a side, and are gridded at intervals of one kilometre. Brown contour lines indicate height in feet above sea level. Public buildings are shown in solid black; other buildings are cross-hatched.

One of the Davidson committee's recommendations was for a new map at a scale of 1 : 25,000 or roughly $2\frac{1}{2}$ inches to the mile, to be derived from the basic twenty five inch survey. This series proved very popular, the scale being midway between the 'Six inch' and the 'One inch' series and the smallest scale at which it is possible to show roads and similar features without exaggeration and to include most minor topographical features. Between 1939 and 1945 a War Office series of maps at this scale covered the whole country, being derived directly by photographic reduction from the six inch quarter-sheets. The first post-war edition has been completely redrawn and named the Provisional edition, comprising 2,027 sheets covering the whole of Britain except the Highlands and Islands of Scotland. No further major revision is planned, but important changes, including the completion of motorways, by-pass roads and other large features are added whenever a sheet is reprinted. The detail shown makes the series valuable for professional and educational purposes; all sheets are available in coloured or outline editions. The 1 : 25,000 'Second' series will gradually replace the 'Provisional' series. Individual maps will be derived from post-war 1 : 1,250, 1 : 2,500 or six inch maps, whichever provides the latest survey at the largest scale. Each sheet represents an area of twenty kilometres east to west by ten kilometres north to south. Specifications have been changed to make the map face more attractive. Otherwise, the same detail is shown as in the 'Provisional' series, with the addition of public rights of way, when these are known. All sheets are available in coloured and outline editions. Sheet numbers are based on the National Grid and can be obtained by reference to the one inch ' Seventh' series maps, on which the bold ten-kilometres grid lines represent the area of a ' Provisional' series sheet. The exact number of a 1:25,000 map of a particular area may be determined by adding to the prefix letters given in ' The incidence of grid letters' box shown at the foot of the one inch sheet, the grid reference, to two figures only, of the south-west corner of the ten-kilometre grid square. The numbers of the 'Second' series sheets may be found in the same way, except that, as each sheet covers twice the area of a ' Provisional' series sheet, the numbers of both ten-metre squares covering the map must be quoted. A special map at 1:25,000, in twenty five sheets, published in 1967, covers the whole of the Greater London Council area and shows the Greater London, London borough and ward boundaries overprinted in red on a grey outline base map of the 1:25,000 ' Provisional' series.

Of the small-scale maps, the current edition of the one inch, 1:63,360, maps is the seventh, begun in 1952. The first edition of the one inch map was printed from engraved copper plates, with relief shown by hachures, similar in style to the Cassini map of France. The early hachuring was rather heavy, tending to obscure other detail, but the hachures were later drawn more finely and with great skill. Spot heights were indicated. Contours were first adopted about 1830 as a result of experience on the Irish six inch survey; they were introduced on the six inch and one inch maps of northern England, which completed the first edition. The contours were surveyed instrumentally at fifty and 100 feet, then at intervals of 100 feet to 1,000 feet. Above the thousand-foot line, the interval was 250 feet. While all the detail on the engraved sheets was in black, the contour lines were not always clear, but later sheets had the contours printed in brown from a second copper plate. The use of colour in general followed the introduction of lithographic printing and photozincography. In the third edition of the one inch map, completed in 1912 and known as the Fully coloured edition, relief was shown by hachures in brown and contours in red. In all, there were six printings, brown and red for the relief, blue for water areas, green for woodlands, burnt sienna for roads and black for names and all other detail. Experiments in methods of relief representation and style of lettering continued. Through the years, the sheet lines have also undergone considerable changes, related to the projection and central meridian employed.

The current edition of the one inch map, based on the National Grid, covers Great Britain and the Isle of Man in 189 sheets and provides the country's most authentic, accurate and up to date small-scale map. It shows all the physical features and other information possible at this scale and relief by contours at a vertical interval of fifty feet. The sheets are revised at the rate of about twelve a year, the revision cycle for each sheet depending on the speed of change in the area concerned. Completed motorways and other such important changes are specially surveyed and added whenever a sheet is reprinted. All sheets are available in coloured and outline editions and an 'advance information service' is maintained, so that users who need them may obtain photomaps of corrected plates pending the issue of a new edition. A particularly important new sheet on this scale covers the whole of Greater London, in a coloured edition only.

In collaboration with the Ordnance Survey, David and Charles of Newton Abbot are preparing to republish from 1969 in exact facsimile the later printings of the first edition of the one inch map. The original plates were revised from time to time, the most important change being the addition of the railway network. The reprint, therefore, shows the original edition with its cumulative revision. The reprint is edited by Dr J B Harley of the Department of Geography, University of Liverpool, who will provide introductory notes for each sheet. The maps, which will be available either flat or folded within covers, will be invaluable to scholars and specialists in many fields. The maps covering Scotland will be republished in the same way, when the England and Wales sheets have been completed.

The one inch tourist maps cover some of the most popular tourist areas. They are based on the 'Seventh' series, but they also include additional information of interest to tourists, and relief (with the exception of the Cambridge area, which has contours only) is shown by a combination of contours, layer colouring and hill-shading, to give an easily understood impression of the topography. Sheets so far available are for Ben Nevis and Glen Coe, Cairngorms, Cambridge, Dartmoor, Exmoor, Lake District, Loch Lomond and the Trossachs, New Forest, North York Moors and the Peak District. The tourist map of the Cairngorms, 1964, is of particular interest, dealing as it does entirely with an area of high relief. Layer tints are combined with hill-shading, with cliff symbols indicated, and experiments have obviously been made with colour separation, resulting in the merging of adjacent tints. Roads and footpaths are shown and eighteen symbols indicate objects of interest to tourists or such information as the position of Mountain Rescue Posts. Nature Reserves and Forest Park boundaries are also shown. These maps are excellent examples of the individuality of areas of different 'landscape', even within a uniform series. The tourist map of the New Forest, 1966, is another most interesting sheet, extending from Salisbury to the Isle of Wight and from Blandford Forum to Botley, based on the one inch 'Seventh' series sheets fully revised in 1965. It differs from all others in the series; like them, it includes a great deal of information; even the key to the map is divided by ten main headings, Roads and parks, Railways, Symbols, Water features, Heights, Abbreviations, Boundaries, Predominant vegetation, Tourist information and Selected places of interest. The special symbols relate to Ancient monuments, various sports and viewpoints. Relief is indicated by hill-shading only. 'Dartmoor' and 'Exmoor' were added in 1967 to the 'Tourist' map series. On these, relief is emphasised by layer tints; National Park boundaries are shown, but in a yellow shade, which in some places may be confused with the yellow in-fill of roads.

The 'County administrative areas map of England and Wales', based on the one inch 'Seventh' series, began in 1965. The scale is 1 : 100,000, about midway between the one inch and the half-inch to the mile, which gives clarity of detail, while still enabling most counties to be shown on a single sheet. These maps are replacing the half-inch county diagrams.

Pressure on resources, especially the demands of planning, has forced the abandonment of the 'half-inch' series; another factor in this decision has been the existence of the Bartholomew series on this scale, which is completely revised every two or three years. A small number of special district maps of popular areas continue, such as the sheet covering the Snowdonia National Park and the 'Norwich' sheet, which covers much of Norfolk and Suffolk and the whole of the Broads. Relief is shown by contours and layer colouring and the wealth of road information and cultural data make them extremely popular.

Regarded as the standard motoring map series of Great Britain is the 'Quarter-inch' series, which owes its origin to the War Office and the Geological Survey. Work on it began in 1859, but was discontinued in 1872, when every engraver was needed for the one inch map; and was completed by 1884 at the request and at the expense of the Geological Survey. The map became available to the public in 1887. After the first edition, the series has been derived from the one inch map sheets and revised with them. The original series comprised twenty five sheets for England and Wales and seventeen for Scotland; the number was reduced in following editions to eighteen in the fourth edition and seventeen in the current fifth edition. The sheets are being reprinted at intervals of one to three years according to a programme related to the extent of changes taking place in the areas covered. The reprinted sheets are either fully revised from the one inch revision or revised for communications only. As this manuscript goes to press, the latest index map has appeared, on the back of the *Publication report* for August 1968.

Although still known by the name ' Quarter-inch ', the fifth series, begun in 1957, was drawn at 1:250,000 rather that at the true quarter-inch scale, 1 : 253,440. Clarity has been the chief objective of this map series, which is designed for many kinds of user. Contours, with layer colouring and hill-shading, show relief and all kinds of road information are included. Coloured and outline editions are available for this series also. An additional sheet at this scale covers Wales and the border counties of England, similar in style to the other sheets at 1 : 250,000, but in addition a green tint is used for land up to two hundred feet above sea level and hill-shading for land above that line. Contours are indicated at two hundred feet intervals. *The Ordnance Survey Quarter-inch atlas of Great Britain* is a new looseleaf atlas including the seventeen sheets of the ' Quarter-inch ' fifth series, trimmed and folded into pull-out style, with an illustrated introduction and a gazetteer index. The Ordnance Survey is maintaining an index of purchasers, who will be advised annually of new editions of the component sheets, which will arrive punched ready for substitution in the atlas.

Other important maps produced by the Ordnance Survey include the 1 : 625,000 series, which is mentioned below as the national atlas of Britain, and the two Great Britain sheets of the International Map of the World 1 : 1M. In style, these sheets conform to the Bonn specifications of the United Nations, so that, in appearance, the map is quite different from the usual Ordnance Survey productions and is distinguished still further by a light grey border. According to the agreed sheet lines of the 1MW, Britain appears on parts of seven other sheets. Symbols and differing type styles indicate centres of population and the administrative importance of the major urban areas. Road, rail and air communications are indicated, also national and county boundaries. Relief is shown by a combination of layer tinting and hill-shading. Other Ordnance Survey maps on the 1 : 1M scale include a physical map, an international local aeronautical map and the historical map series below mentioned.

Ordnance Survey maps have from the outset included a careful treatment of antiquities, especially after the six inch survey of Ireland, in which antiquities required special consideration. The Irish scholar John O'Donovan assisted greatly in the outstanding success of this part of the work. Similar principles were applied to the large-scale survey of Great Britain from 1853, but treatment was uneven until in 1920 O G S Crawford was appointed Archaeo-logy Officer. Air photography made the identification of sites easier. The practice of publishing special maps of historical importance began with the facsimile of the ' Gough map ' in 1870; in 1914, Symonson's ' 1596 map of Kent ' was printed in facsimile. The historical map series proper, however, began with the first edition of the map ' Roman Britain ' in 1924, which had great success in this country and much influence abroad. The base used was the 1 : 1M scale physical map of southern Britain extending to Berwick-upon-Tweed. Relief was already shown by five layer tints at 200, 400, 800, 1,200 and 2,000 foot intervals and a black overprint set out the established and suspected Roman roads, towns, forts and other evidence of mineral or industrial activity. Towns were given their Roman and modern names when possible; tribal names were indicated in their operative areas. Hadrian's Wall was shown and there was an inset of the Antonine Wall in Scotland at the same scale. The two further editions, in 1928 and 1956, reflect the increase in our knowledge and the area covered by the map was extended, first north to Aberdeen and, in 1956, to the whole of Great Britain.

Four little-known historical maps were also compiled before the second world war : ' Neolithic Wessex ', ' The Trent basin ', ' South Wales ', based on the Quarter inch and ' England and Wales in the seventeenth century ', at 1 : 1M.

' Britain in the Dark Ages ' was completed in two sheets, 1 : 1M in 1935 and 1939, comparable with the second edition of the map ' Roman Britain ', with the historical details overprinted on a physical base. The map was widely acclaimed and a second edition was soon envisaged; this was recently finished, the chief difference being in the introduction of a second colour. All information relating to the Anglo-Saxons is in black and red, symbolising Pagan and Christian, and the Celtic features are in blue. Accompanying the new edition is a completely revised introduction, bibliography and index.

Between 1939 and 1951 the historical mapping effort of the Ordnance Survey was devoted to the preparation of a map in two sheets, 'Ancient Britain ', prepared for the occasion of the Festival of Britain and designed to show visitors the principal extant archaeo-logical remains. It was compiled on the 1 : 625,000 base map, showing modern topography; seven major periods were distin-guished between paleolithic and the Dark Ages, besides indications of museums, earthworks, Roman roads and canals and sites whose age has not been identified with certainty. A second, revised edition was published in 1965. The next map in the historical series was ' Monastic Britain ' in two sheets at 1 : 625,000 in a first edition,

1950 and a second improved edition, 1955. This is a most complex map demanding a great number of symbols to indicate a vast body of information. In 1962 'Southern Britain in the Iron Age' was issued, using the southern sheet of the physical map of Great Britain at 1 : 625,000. On this base were imposed twenty two symbols, using four colours, red, purple, brown and black, to show four cultural periods of Iron Age life. Modern names have been indicated when possible and a list of sites is given in the introductory text.

A map of Hadrian's Wall followed in 1964, compiled by C W Phillips, on the scale of two inches to the mile. Hill-shading has been added to the contours showing ground relief, as physical conditions were important in siting the sections of the wall. Those parts of the wall still extant are shown in black, the lost parts in red. Each of these maps deserves full description, if space allowed. It is hoped that more such maps will be prepared, but it must be remembered that these form, as it were, only a spare-time occupation for the Ordnance Survey, whose vast publishing commitments have been outlined above. Each sheet is accompanied by a booklet, which includes an introductory survey, bibliography and index.

Revision is a problem to be faced in all map-making. The Davidson committee recommended that Ordnance Survey plans should be kept up to date by a system of continuous revision. A detailed talk on this subject was given to the Royal Geographical Society by W A Symons and B St G Irwin on March 2 1964; this talk was reprinted in *The geographical journal* for March 1965, together with the discussion. The method of continuous revision was introduced in 1945. All changes are surveyed as soon as possible, plotted on a ferro-prussiate blueprint of the plan on plastic, and the new information is incorporated in the next printing or the next edition of the map; the Advance Revision Information Service allows the new material to be available from the field sheets to anyone requiring it.

The Cassini Projection was first used for England and several central meridians were adopted before the final choice was made. Scotland and Ireland were first mapped on the Bonne Projection and after 1918 were redrawn on the Cassini Projection for the Popular edition. A form of Transverse Mercator Projection is now used for all parts of Britain. The point of origin of this projection is 49 degrees north, two degrees west, but the unavoidable error away from the meridian is redistributed by making the scale true at about 180 kilometres east and west of the central meridian, thus making the actual error unnoticeable, even on the large-scale plans. This uniformity of projection, plus the development of the National Grid and reference system, has resulted in a truly integrated national map service. An informative pamphlet on this subject published by HMSO is *The projection for Ordnance Survey maps and plans and the national reference system.*

A system of letters and footnotes enables identification of the editions of the several maps. The term ' series ' has replaced the term ' edition ' for maps or plans of a common design or specification. ' Edition ' is now used to signify a version of an existing map or plan for which the whole or a substantial part has been revised. The edition of a map or plan is shown by a letter printed in the left-hand bottom margin and, in the case of large-scale plans, also in the top right-hand margin. The year of publication to which the letter applies is shown in a footnote in the bottom margin. ' Reprint ' denotes reprinting of an existing map or plan on which the changes are limited to the correction of previous errors and also, on occasion, the inclusion of changes to topographic or other detail not resulting from systematic revision. An additional footnote states the changes that have been incorporated in a reprint.

To trace all the changes in design, colouring and symbols on Ordnance Survey maps and plans would be a major task in itself. Summarising some of these points, it must be remembered that the early series were printed from engraved copper plates, which had an inevitable influence on the style. Relief was first shown by hachures, rather in the style of the Cassini map of France. Contours were introduced in 1830, first on the Irish six inch and one inch maps of northern England, completing the first series. All the detail on the engraved sheets was in black, so that the contour lines did not always show up sufficiently clearly; hachures in brown were later printed on the one inch map, from a second copper plate, but the use of colour in general followed the introduction of lithography, then of zincography. The ' Fully coloured ' third series of the One inch, completed in 1912, shows relief by hachures in brown and contours in red; blue was used for water features, green for woods and burnt sienna for roads, with names and other detail in black. Following Bartholomew's successful production of their quarter-inch map of Britain, in which relief was shown by layer colouring, the Ordnance Survey used this method on the half-inch, then on the other maps. Several experiments were made in relief representation; when the fifth (Relief) ' One inch ' series was issued, relief was shown by contours in brown, hachures in orange and hill-shading in grey, with layers in buff tints. However, this style was considered too elaborate and, in the sixth series (the ' New Popular '), contours only were used, showing up well in brown. The seventh series of the one inch map was first printed in ten colours and, in a few areas and in the Tourist series, hill-shading was used effectively.

There were two styles: in the 'North York moors' style, a single-colour shadow tone in blue was printed in half-tone over the layer colours and the layer tints themselves were carefully chosen so as to reduce the 'step' effect of the layers; in the 'Lake District' style, the layer system was not used, but, instead, two printings in purple-grey tones on the shadow side of the hills and one printing in yellow on the illuminated side were added to the standard base map. The number of colours has since been reduced from ten to six by combining some of the printing plates.

The Provisional sheets of the map at 1 : 25,000 were published in two styles. The original style uses blue for drainage features, brown for road in-fills and relief, and black for built-up areas, principal boundaries, road outlines and the various vegetation symbols. Buildings may be either solid or hatched. On the later sheets, black was replaced by grey for all symbols except boundaries, road outlines and railways. The contour interval for both styles is twenty five feet. The Provisional Outline edition without relief was published in grey.

Judged by clarity of detail, use of colour and lettering, and general design, the Ordnance Survey maps set a very high standard and the craftsmanship of the cartographical drawing staff ranks with the finest in the world. All entrants to the Ordnance Survey drawing office begin with an intensive period of training and are instructed in the methods used to prepare the fair drawings needed for ultimate printing. The work may be on paper, coated metal plates, glass, film or plastic. A short period is spent on the survey side to provide a background understanding of the work before joining a field group for practical work.

In 1967, the Ordnance Survey announced plans to make some changes as a beginning for adoption of the metric system. The grid used on all maps is a metric grid already, the intervals varying from 100 metres to ten kilometres according to the scale. Heights on new and revised plans at 1 : 1,250 and 1 : 2,500 scales were the first to be changed from feet to metres; areas on new and revised plans at 1 : 2,500 are to be shown in hectares instead of acres; and the contours on new six inch maps and on the 1 : 25,000 maps directly derived from them are to be drawn at a metric contour interval. Further changes will be made very gradually, as resources permit, and conversion scales will be shown in margins.

Transparent copies of Ordnance Survey maps and plans are supplied to order by the department, usually only to customers who are licensed by the department, acting on behalf of the HMSO, to reproduce Ordnance Survey maps and plans. Transparencies of 1 : 10,560 and smaller scale maps are available on tracing linen,

135

Astrafoil or Permatrace, with the impression either on the dull or glossy side and with either a forward or a reverse printing. Transparencies of any 1:1,250 or 1:2,500 National Grid sheets may be had on Permatrace or on Kodagraph, both of which are plastics with a matt surface on each side, providing an excellent drawing surface. These impressions are available in reverse printing only. It is, of course, an infringement of copyright to use the transparencies to make unaltered reproductions of Ordnance Survey sheets, without permission from the Ordnance Survey.

Emphasis must be given to the part played by the Ordnance Survey in the British mapping achievement as a whole, notably in the 1:625,000 sheets described below as forming the basis of a national atlas of Britain.

The Ordnance Survey allows discount on map publications, including Geological Survey maps, for educational use, on the understanding that these maps will not be either sold or given away. The maps are available from appointed retail agents in the main cities throughout the country.

Three pamphlets issued by the Directorate-General of the Ordnance Survey provide a guide to understanding the map series mainly prior to the second world war; *A description of Ordnance Survey small-scale maps*, 1947, reprinted with an addendum in 1951; *A description of Ordnance Survey medium-scale plans*, 1949, reprinted with corrections 1951; and *A description of Ordnance Survey large-scale plans*, 1947, all including numerous extracts from map and plan sheets. From 1885 on, details of the progress of the Ordnance Survey can be obtained from the *Annual report* and much historical and technical information appears in the *Reports of the Departmental Committee on the Ordnance Survey*, 1935-38. Early map publications are listed in the *Catalogue of maps*, 1904 and 1920. A *General information and price list* is frequently revised and advance notices of individual new maps or changes of policy are circulated as necessary. Each month, a *Publication report* is distributed to subscribers, giving exact information on new publications and revisions at all scales, any miscellaneous information, changes in regional organisation, agencies, etc. Levelling information can be supplied for most parts of Great Britain in the form of *Bench mark lists*, with indications of the latest available values. The same values may appear on the current map or plan, though in many areas the *Bench mark lists* may show more recent levelling. A full description of the six inch and twenty five inch maps is given by H St J L Winterbotham in 'The National Plans', in the *Ordnance Survey professional papers*, new series, 16, 1934. The latest edition of the *Ordnance Survey map catalogue*, 1968, covering medium and small-scale maps, is a clearly printed booklet in two colours, with a loose index at 1:1,250,00 showing the sheet lines of the 1:25,000 Provisional and second editions. Index maps in the text show the sheet lines of the standard ' One inch ' series, one inch Tourist maps, half-inch District maps, standard 1:250,000 series, the Wales and the Marches sheet, the IMW 1:1M, the 1:625,000 series and the administrative areas maps of Scotland and London. There are some notes also on Geological Survey, Soil Survey and Land Use Survey maps, maps of Ireland, of the Channel Islands and the Isle of Man; most of the entries are annotated and there is also a great deal of general information on lesser-known aspects of the Survey's work. A few other particularly useful references are:

T M Baker: ' The revision of the one inch to the mile maps (1:63,360) of Great Britain ', *Kartographische Nachrichten,* V 4, 1963.

Major-General Geoffrey Cheetham: ' New medium and small-scale maps of the Ordnance Survey ', *The geographical journal,* May-June (November) 1946).

K. M. Clayton, editor: *Guide to London excursions* (20th International Congress, London, 1964, section 27). Part of the first edition one inch map of the Keswick area is reproduced, showing admirably the effectiveness of the fine contour drawing; and, for comparison, on the opposite page, the same area in the 1957 edition, which uses contours and hill-shading.

Sir Charles Close: *The map of England or About England with an Ordnance Map* (Peter Davies, 1932). ill, maps. Two chapters are included on the history and work of the Ordnance Survey.

G R Crone et al: ' Landmarks in British cartography ', *The geographical journal,* December 1962.

Major-General L F de Vic Carey: ' The Ordnance Survey ', *Geographical magazine,* April 1960. ill.

J R B Dennett, L H E Hobbs and B F White: ' Cartographic production control ', *Cartographic journal,* December 1967.

R A Gardiner: ' How Britain's national maps are produced ', *Penrose annual,* 1964.

D L Griffith and J J Kelly: ' The quarter-inch to one mile map of Great Britain ', ICA Technical Symposium, Edinburgh, 1964; also in the *International yearbook of cartography,* V 1965, together with the discussion.

J B Harley and C W Phillips: *The historian's guide to Ordnance Survey maps,* published for The Standing Conference for Local History by The National Council of Social Service, 1964, based on four articles from *The amateur historian,* 1962 and 1963.

B Lockey: *Interpretation of Ordnance Survey maps and geographical pictures* (Philip, sixth edition, 1961). ill, maps.

' Making maps for Britain ', *The litho-printer,* October 1963.

R F Peel : ' Geomorphological fieldwork with the aid of Ordnance Survey maps ', *The geographical journal,* September 1949. References.

C W Phillips: ' The Ordnance Survey and archaeology 1791-1960 ', *The geographical journal,* March 1961.

C W Phillips: ' The special archaeological and historical maps published by the Ordnance Survey ', *The cartographic journal,* June 1965.

F M Sexton : ' The adoption of the metric system in the Ordnance Survey', *The geographical journal,* September 1968. A section of an Ordnance Survey map 1 : 25,000 is included, showing comparison of contours at twenty five feet with those at five metres.

W A Seymour and B St G Irwin : ' Continuous revision of Ordnance Survey plans ', *The geographical journal,* March 1965. Report of a paper, with discussion.

Dorothy Sylvester: *Map and landscape* (Philip, 1952). ill, maps, figs. Notes on the interpretation of Ordnance Survey maps and the application of these methods to regional survey.

C J Sweeney and J A Simson: ' The Ordnance Survey and land registration ', *The geographical journal,* March 1967.

D E O Thackwell: ' The continuous revision of the 1 : 1,250 and 1 : 2,500 plans of Great Britain ', *Kartographische Nachrichten,* V 4 1963.

The history of the retriangulation of Great Britain 1935-1962, written and compiled by officers of the Ordnance Survey under the authority of the Director-General, was published in 1968. This work gives a complete record of the achievements of the Ordnance Survey in this project, with detailed explanations of the methods used and the results obtained. It is expensive, but it is a definitive work, including twenty four half-tone plates, a bibliography and twenty diagrams in a companion case. An official history of the Ordnance Survey, to be published by HMSO, is in preparation by R A Skelton.

The Ordnance Survey is also responsible for producing the basic information on areas, for official purposes. This includes, for example, the provision and maintenance of the total areas of administrative counties and civil parishes, total areas of inland water and of foreshore areas. In mountain and moorland regions, for which 1 : 2,500 plans are not produced, measurements are made on the six-inch maps.

The map products of the Ordnance Survey have been a primary source of information for commercial and other map publishers, subject to normal copyright conditions, and it is also fortunate that there are in Britain a number of other experienced and long-established cartographical houses who have in many ways responded to the requirements of education and commerce. Chief among these are the Cartographic Department of the Clarendon Press, Oxford, John Bartholomew and Son, George Philip and Son, Geographia and the Cartographic Departments of Collins, Longmans, Nelson and the Pergamon Press.

The Clarendon Press became a separate department in 1952, though naturally maintaining the closest liaison with the Oxford University Press and, as has been indicated in chapter 1, has become noted throughout the world for consistent excellence in quality of cartographical work and for experiment and invention. Atlases and general maps produced by the Press are mentioned elsewhere, as is the series of ' Oxford plastic relief maps ' of which here the series 3, 4, 5 and 6 are particularly relevant. The maps have been published in association with Barclays Bank and the Educational Supply Association, Harlow, Essex, is the agent for them. The maps are available in two forms, an outline edition and a fully coloured edition. Series 3, 1964-, comprises seven maps of regional areas of Great Britain and Northern Ireland at the scale six miles to the inch, covering areas such as ' Wales and the Midlands ', ' South-east England ', ' South-west England ', etc. They are available framed or unframed, printed in six colours on rigid plastic. Topographical information is comprehensive, towns and villages are indicated, also roads, railways, with traffic density classified, and canal systems. Simplified land use is shown by distinguishing colours for woodland, farmland, rough pasture and built-up areas. County, rural and urban boundaries are shown, also National Parks. Average annual rainfall is indicated by small symbols. The National Grid has been included at ten-kilometre squares.

Series 4 includes ten maps of selected areas of Great Britain, also at six miles to the inch, printed in six colours—' Birmingham and the Potteries ', ' Cardiff and south Wales ' ' Liverpool—Leeds ', etc, completed in 1965. Basic colouring indicates land use; built-up areas are shown by pink hatching; road data are based on 1954 figures and rail data on figures for 1958-59; airports are also shown and rainfall figures are inserted. Series 5, ' Great Britain and Northern Ireland ', comprises the seven sections of series 3 mounted together in one piece; it is available framed or unframed. Series 6 began from 1964 a new series at $2\frac{1}{2}$ inches to the mile, based on the Ordnance Survey 1 : 25,000 map, beginning with ' The Oxford area '. The vertical scale is six times the horizontal scale. Choice of colours differs from the conventional shades. Blue, for example, in varying depths, denotes land use, purple is used for wooded areas, green for rivers and lakes and yellow for roads, stronger in shade for the roads carrying the greatest volume of traffic. The symbols for railways also indicate traffic use. There is a reference key to many of the principal buildings in Oxford.

Probably the most used of the printed maps issued by John Bartholomew and Son is the half-inch map of Great Britain, which dates from 1875 and has an interesting history. Specimen sheets of an experimental map showing relief by layer or contour colouring were compiled for display at the Paris exhibition of 1878; comment was favourable and, after further experiment with the maps for *Baddeley's Lake District guide,* 1880, the system was adopted for this series, being therefore the first topographical series in any country with this method of relief representation. Fifteen different colour sheets are used for the layers. Originally based on the Ordnance Survey map sheets, it has for some time now been continuously revised by the firm's own information service and is generally accepted as one of the most up to date maps, covering Great Britain in sixty two sheets. County boundaries are shown; roads are classified, showing also Ministry of Transport road numbers, and railways, canals, overhead power lines and a number of social features are included. Each edition of a sheet is planned to last for only three years, frequently a shorter period in development areas. In looseleaf book form, the ' Half-inch map ' series can be used for office or library reference. All maps are cloth backed, mounted on guards and bound together by four locking screws. Volumes are supplied with the latest editions of each half-inch map and are available as the ' Great Britain ' looseleaf, ' England and Wales ' looseleaf and ' Scotland ' looseleaf. Each volume contains an index map and county coloured maps of the area covered. The Bartholomew *Gazetteer of the British Isles* (ninth edition, reprinted 1966) should be used with these maps.

The Bartholomew sixth-inch map is designed for the motorist travelling long distances but who, at the same time, requires a considerable amount of detail. The whole of Great Britain is covered in eight sheets of uniform size: ' Home counties ', ' West country ', ' Midlands and East Anglia ', ' Wales ', ' Lancashire and Yorkshire ', ' South Scotland ', ' Central Scotland ' and ' North Scotland '. The scale is large enough to show all the essential road information. The Royal Automobile Association assisted in the compilation and design of the original map and their information service helps to keep the sheets up to date. Roads are classified and indicated by six different symbols for motorways, selected RAC routes, A roads, B roads, unclassified roads, tracks and indifferent roads. Short

140

distance mileages are brought out in purple and the road numbers are shown in red. Railways, stations, county boundaries and a number of social features are marked and relief indicated by means of background contour colouring from green to grey.

A national atlas of Britain had been under discussion for many years when, in 1935, a committee appointed by the Council of the Town Planning Institute recommended a commission to direct a national survey of the natural and economic resources of the country; the National Atlas Committee of the British Association also submitted proposals in 1939. In 1941, an advisory committee was established under the chairmanship of the Director-General of the Ordnance Survey, as a result of which two map offices came into being, in London and Edinburgh, under the aegis of the Ministry of Local Government and Planning, with the Department of Health for Scotland, to examine the mass of official data to be represented on successive sheets of such a national survey. In *Government information and the research worker*, edited by Ronald Staveley and Mary Piggott (Library Association, second edition 1965), the section 'Towards a national atlas' by S W E Vince, revised by W A Payne (p 174), tells the full story of the 1 : 625,000 'Planning maps of Britain'. (*Note* that the prices quoted are no longer accurate.) The maps cover geological and physical structure, land use, mining, industry, administrative areas, population, communications, public utility undertakings and many other features of the national life and economy, much of the material not previously mapped before.

The majority of the maps have been prepared by the Ministry of Housing and Local Government, with other departments and research organisations and published by the Ordnance Survey at the scale of 1 : 625,000, which is approximately ten miles to the inch. The original manuscript maps, maintained in continuous revision, are available for official use. The current list of published sheets and prices may be seen in the *Catalogue* of the Ordnance Survey. Each map is in two sheets, Scotland and England north of Kendal and Northallerton, and the rest of England and Wales. Explanatory texts are available for the following sheets: land classification, average annual rainfall, population, limestone, vegetation, the grasslands of England and Wales, local accessibility, vegetation reconnaissance of the Survey of Scotland. The first map in the series is the topographical base map, which serves as underprint to all the other maps except the map showing physical features. This latter map presents relief by clear, limpid layer tints at 200, 400, 600, 800, 1,000, 1,400, 2,000 and 3,000 feet, with spot heights. The inland hydrographical features show clearly in blue and it is probably the finest outline of the British river systems yet printed. The continental shelf is also indicated in great detail. The narrow gap between Anglesey and the mainland, for example, shows with exquisite clarity. The map of solid geology has been prepared and drawn by the Geological Survey on a larger scale than has previously been published covering the whole of Great Britain. The 'Land utilisation map' was prepared by the Land Utilisation Survey of Britain and the climatological maps have been supervised by the Meteorological Survey. An interesting map is that of 'local accessibility' which shows the hinterlands of the great cities from the point of view of their transport facilities and public utilities.

The 1:625,000 'Route planning map' has become particularly well known and it is understood that it will be revised annually. It is in two sheets, designed not for use in a car, but for previous study in home or office. The 1966 edition incorporated all major road changes in 1965. Motorways were shown in blue, A roads in red and B roads uncoloured. Dual carriage-ways are clearly distinguished and Ministry of Transport numbers indicated. Other features include sea and air ferry terminal points, giving the frequency and lengths of the journeys from those points; rail car ferry terminal points; a table of distances between major towns; National Parks and 'areas of outstanding natural beauty'; larger scale inset maps of a number of towns. The 1967 edition introduced further innovations, the most important being the portrayal of the Ministry of Transport primary route system or 'green routes'. Apart from the motorways, these roads are intended to be the main traffic routes in both urban and rural areas, connecting primary towns, eventually forming a network of through routes. There is the further distinction of 'super primary' towns. Primary routes are coloured green, the names of primary towns are boxed in green and the 'super primary' towns have a light green in-fill to the box. The blue formerly used for motorways is a lighter shade and the red in-fill has been limited to class I roads not classed as primary routes. The green tints previously used, for National Parks and 'areas of outstanding natural beauty' have been omitted and the buff tints used for hill areas have been changed to yellow. Inset maps of selected towns are still added and roundabouts are now shown on them. For the first time, legends are given in French and German as well as English.

On the map 'Administrative areas', north and south, boundaries have been revised to April 1966, at the time of writing. A series of population maps, all in north and south sheets, have been compiled, using census and other data. 'Population: changes by migration, 1921-31' and similar maps for the periods 1931-38/39 and for

1938/39-47 show by colour tints administrative areas and the net changes resulting from migration; symbols show the number of people by which the population of each administrative area changed during the period covered by the map. A second group of maps, 'Population: total changes, 1921-31', with similar maps for 1931-38/39 and for 1938/39-47 show the net changes resulting from births, deaths and migration. Symbols of proportionate size show the number of persons by which the population of each administrative area changed during the period. The map 'Population change 1951-61', published in two sheets in 1966, using the figures of the 1961 Census, employs symbols only. The base map is printed in grey and proportional circles indicate changes in population. Percentage decrease in population is shown by two shades of green; percentage increase by three shades of red. In addition, two sizes of dots show numbers below the limit of the proportional scale. The contrasting colours emphasise very clearly the trend of decrease in the urban centres, with an increase at the edges. 'Population density, 1951' shows in full colour seven types of population density ranging from 'dense urban' to 'virtually uninhabited'; 'Population of urban areas, 1951' represents, by solid red circles of proportional size, the population of all urban areas from the vast London conurbation to Scottish small burghs.

It is impossible as yet to evaluate the 1:625,000 map sheets as an atlas, but comparison between individual sheets is inevitable with those of *The atlas of Great Britain and Northern Ireland*, published by the Clarendon Press in 1963. This atlas is not a 'national' atlas in the purist sense of the term; it bears no official seal and received no financial assistance from the British government. However, many government departments, as well as numerous other specialist and research organisations contributed to individual parts, and it is a matter of the highest commendation that, after all the years of discussion, a work of such magnitude and excellence should, after all, have been produced by private enterprise. The standard scale adopted was 1:2M, which allows the whole of the United Kingdom to be shown conveniently on one page of reasonable atlas size. This is also the scale of the 'Transparent reference overlay'. Enlargements of scale have been made to 1:1M and 1:500,000 for regions and reductions to 1:8M for some distribution maps. All maps are on the same projection as the Ordnance Survey maps. The larger scale maps and the overlay are marked with the National Grid, which is used also for reference in the gazetteer.

In the words of the publishers, the atlas is 'a statement, on a general basis, of modern Britain's resources, physical, economic and industrial—a complete and ordered portrait of this country from the rocks beneath to the industry above'. Historical geography is not included. Much of the material was derived from the 1951 Census and from other fact finding surveys and some of the material used was hitherto unpublished. 1955 agricultural and fisheries figures were used, with 1948-57 averages, and 1954 figures for the majority of industrial sources, but the latest information possible was incorporated where this was particularly vital, as in the cases of atomic energy and air traffic. A fold-out section at the back of the atlas gives details of authorities and sources. While preserving the traditional form, an original approach is revealed, both in general design and in the treatment and colouring of individual maps, in the series of structural systems, for example, in the drift map, the map of coastal relief and the agricultural maps. Most of the base maps show relief in pale grey colouring, while the relief on the regional maps is shown by hill-shading, actual height of layer colouring and land use by superimposed colouring or dots. Two lists, of maps and of topics, form a subject index to the atlas and a twenty four page gazetteer follows the maps.

The atlas is in four main sections: physical features; agricultural and industrial resources; and demographic social and communications aspects, each section being followed by regional summary maps for the section; the last section is a series of general regional maps at 1:500,000. Great skill and imagination are evident throughout the atlas, in the placing of individual maps, sometimes 'pairing' especially comparable maps on facing pages, and in the insertion, where relevant, of inset maps, diagrams and tables of statistics. The atlas begins with a reference map showing the distribution of farmland, woodland, rough pasture and built-up areas. Then a series of eight maps each depicts one geological system where it outcrops on the surface—a most evocative series, enhanced by carefully chosen block diagrams and bore-hole diagrams. Submarine and coastal relief, from the western edge of the continental shelf to east of Denmark and south of Ushant, are shown on a double-page spread, based on contours and soundings on Admiralty charts and from observations assembled by the National Institute of Oceanography. Submarine deposits on the continental shelf are given and a series of 1:8M maps shows sea temperatures, salinity and tides. The map depicting the varied characteristics of Britain's coastal geomorphology is particularly valuable.

The section dealing with climate is interesting, not only for the wealth of information included, but for the representational methods chosen. A series of small-scale maps is accompanied by dispersion diagrams for temperature, rainfall and sunshine over a thirty year period—a method admirably suited to the presentation of a climate

144

such as that of Britain. An enterprising section is that covering water and soil, including river flow, summer water balance, water supplies and water authorities. The map showing vegetation on non-agricultural land presents information accurately reduced from one inch Ordnance Survey maps and a series of detailed maps shows an analysis of woodlands according to type of tree, using one hundred acre dots. Very important sections are those comprising the maps on agriculture, fisheries and industries. Almost every map calls for detailed analysis and, for students studying either any aspect of British geography or national and regional atlas design, it is suggested that a really close study of this atlas would provide a most fruitful starting point. The documentation of the atlas is so sensible and the cartographic presentation so admirable (with only the most minor reservations) that information and ideas are positively stimulated in the user's mind.

Density of agricultural labour and sizes of agricultural holdings are shown and vivid dot maps analyse individual crops and livestock. Import and export tables frequently accompany the maps. Similarly, sea fisheries are analysed by type of fish and the concentration of the industry is shown. For other industries, up to eighty percent of the total of British industry, each map shows distribution in terms of employment, based on information from the relevant authority. Some receive particularly close analysis; the coal industry, for instance, for which the maps are accompanied by diagrammatic statistical maps based on data received from the National Coal Board. The demographic maps show population changes; skilled and professional occupations are given separately, besides age groups, retail trade, department stores, administrative and parliamentary boundaries. Of special interest are the maps dealing with internal and external communications. A coastal shipping map was compiled from such sources as Lloyds List and the records of the Post Office ship-shore service. Overseas shipping and cargoes are effectively portrayed. Railway, road and air freight and passenger traffic data are based on daily returns and an interesting map shows telephone trunk traffic.

The *Concise Oxford atlas of the British Isles,* edited by D P Bickmore, is made up of the relevant sections and gazetteer in the *Concise Oxford atlas.* Six-colour topographical maps are followed by a series of maps giving historical data and general information on climate, scenery, population, communications and recreations. The thirty four page gazetteer includes a brief account of historical sites. In 1966, Collins-Longmans published *The Reader's Digest complete atlas of the British Isles,* including England, Wales and Scotland, Northern Ireland, the Channel Islands, Jersey, Guernsey

145

and associated islands, the Isle of Man and the Republic of Ireland. The atlas is divided into four sections: 'The nature of the land', consisting of relief maps as they would be seen from an aircraft; 'The country we live in', topographical maps at 1:380,160 and 1:443,520 based on Ordnance Survey maps; 'The fabric of a nation', a section of eighty seven pages of thematic maps; and 'Places in the British Isles', which is a thirty two entry gazetteer, followed by statistical 'Facts about the British Isles'. The atlas is well-produced and a vast amount of information is packed into its pages. It is more a graphic encyclopedia than a conventional atlas, produced for the family rather than for the scholar or specialist. Folded plates are used on occasion to give a wide spread, relief form maps are included in the first section, together with pictorial block diagrams. The three-dimensional panoramic maps give a vivid idea of the diversity of land forms in Britain. The approach to climate is also striking. Most of the thematic maps are on small scales and many of them are too generalised to be of value, but they are evocative of ideas and a mass of historical, governmental, religious, cultural and economic data is included. Interesting topics such as bird migration are presented, also sections devoted to insects and wild flowering plants. The map of birth-places of famous people may be rather ineffective, but the diagrams depicting the machinery of government are useful. The variety of cartographic forms certainly reflects the current period of experimentation shown, for example, in the map 'People on the map', which is an original diagrammatic map compiled by Dr T H Hollingsworth showing the voting population divided into existing parliamentary constituencies. The area of each constituency on the map is determined by the number of voters it includes and the shapes have been arranged in such a way as to retain their location, their boundaries with neighbouring constituencies and the rough shape of the British Isles. The result is thought provoking indeed. Numerous colour photographs are interspersed throughout the text and a feature is the information portrayed on topics not usually found in such an atlas, such as language and dialects, folklore, customs and buildings typical of individual areas. A time chart records notable events from pre-Norman Conquest to 1964.

Bartholomew's Survey atlas of England and Wales is probably the best library reference atlas for this area. Topographical maps at the scale of half-inch to the mile show relief by layer colouring and a section of maps covers geology, climate, population and land utilisation. A ' Short general index to towns and villages ' compresses about ten thousand entries and the prefatory matter includes a bibliography, ' The cartography of England and Wales ', arranged chronologically. *The desk planning atlas of England and Wales* consists of a series of some sixty sheets at various scales, in process of publication by the Ministry of Housing and Local Government since 1955.

Stanford's series of general maps includes a full-colour map of Great Britain at 1 : 633,600, with an inset map of northern Scotland, 1 : 750,000. Detail includes county boundaries, railways, canals, abbeys, battle sites, etc. Geographia issue a reference map of Great Britain on a scale of nineteen miles to the inch, showing counties in colour, railways and other communications, including canals, and a number of social features. Many other individual maps are issued— 'British Isles, main roads', nineteen miles to the inch; 'Great Britain: motorways map', at sixteen miles to the inch; 'England and Wales, main roads', at $12\frac{1}{2}$ miles to the inch; 'England and Wales, with counties in colour', $12\frac{1}{2}$ miles to the inch; 'England and Wales: commercial and political', at nine miles to the inch; also many county and district maps.

'British landscapes through maps' is a growing series sponsored by the Geographical Association and edited by Professor K C Edwards; each presents a detailed study, illustrated by maps and photographs, of a representative area of Britain within the compass of an Ordnance Survey map.

Previously, many individual efforts had gradually developed the art of geological mapping in this country, but William Smith is generally acknowledged to be the 'father' of modern geological mapping, with his large map of England and Wales, 1815, London to Snowdon section in 1817, county maps of Berkshire and Oxfordshire in 1819 and 1920 respectively and smaller maps of England and Wales in 1920, among others, all published by John Cary, making use of the Ordnance Survey maps when possible. The Geological Survey of Great Britain, instituted as the Geological Ordnance Survey in 1935, was foreshadowed in 1832, when De la Beche, Secretary of the Geological Society, was authorised to add geological colouring to the one inch Ordnance Survey map of Devon, with adjoining Somerset, Dorset and Cornwall. The present Geological map of Britain comprises five main series: the Twenty five miles to one inch, 1 : 1,584,000, in a fourth edition 1957, coloured or uncoloured; the Ten miles to one inch; 1 : 625,000 in two sections, Scotland and England north of grid line 500 kilometres north, and the rest of England and Wales, both 1948, with a second edition in 1957; the Quarter-inch to one mile, 1 : 253,440, maps of England and Wales in various editions, and similar maps for Scotland and Northern Ireland; the One inch to one mile maps, 1 : 63,360, of England and Wales in a series of coloured reprints with the National

Grid added, also of Scotland and of Ireland; and maps on a six inch scale only for the coal-fields and other areas of particular economic importance. The rest of the series is preserved in the Geological Museum and/or the regional centres as hand-coloured or monochrome manuscripts. Special colour-printed maps have been compiled for the London district and, in addition to the published maps, manuscript maps are available for inspection at the offices of the Geological Survey. Accompanying the map sheets are *Memoirs,* illustrated with sections, text maps and other diagrams, each providing a most valuable source of information for the region covered. *Annual reports* of the department have been published since 1896. *Sectional list* no 45 (HMSO) gives a complete annotated list of Geological Survey publications for sale, including the explanatory *Memoirs* to the one inch sheets. The Geological Survey and Museum issues frequently revised *Lists of Geological Survey maps,* also a *Quarterly publications report* and an annual *Summary of progress.* For further information, reference should be made to Sir Edward Bailey: *Geological Survey of Great Britain* (Allen and Unwin, 1952) and Sir J S Flett: *The first hundred years of the Geological Survey of Great Britain* (HMSO, 1937), which includes a bibliography. The Geological Survey and Museum now comes within the Institute of Geological Sciences. A 'Tectonic map of Great Britain and Northern Ireland' at 1:584,000 was published by the Ordnance Survey for the institute in 1966. Detailed special maps are also undertaken; for example the 'Hydrogeological map of north and east Lincolnshire', at 1:126,720, published by the institute in 1967, and the 'Map of the drift geology of Great Britain and Northern Ireland' was the subject of an article by K M Clayton in *The geographical journal,* March 1963, including maps and references.

The first contract given to a British university for the production of one inch geological maps has gone to the Department of Geology of Exeter University. The contract was made with the Natural Environment Research Council in 1966 and the department is working on a five year plan for a geological survey of south Devon, including part of Dartmoor. The area stretches from Haytor south to Ashburton, east to the north of Torquay and thence to Budleigh Salterton, originally surveyed at the end of the last century. The British Geomorphological Research Group, established by the Department of Geography, University College of Swansea in 1961, is currently working on the 'Geomorphological map at 1:625,000' under the direction of Professor D L Linton; it shows the surface forms of Great Britain as they are generally classified by British geomorphologists. Five regional editors have supervised the work on Scotland, Wales, northern, south-western and south-eastern England. The group also compiles *Current research in geomorphology* (Philip, 1964-). The publications of the Geologists' Association are most important, particularly the *Proceedings,* issued in four parts each year, embodying the research of members, and the *Geologists' Association guides,* of which more than thirty have been published concerning individual regions. *Monthly circulars* contain announcements of meetings and papers to be read, with full particulars and lists of maps, books and other source material suggested for consultation.

Local geological and topographical societies throughout the country have made incalculable contributions to the progress of British geomorphology; almost every county and many towns have formed a geological or similar society, chiefly engaged on local work. The files of transactions, proceedings and monographs are invaluable sources of information and maps; the *Transactions* of the North Staffordshire Field Club, for example, established in 1865, which includes a special geography section, are a particularly apt example: the Norfolk and Norwich Naturalists' Society specialises in valuable coastal studies, printed in their *Transactions;* the Hertfordshire Natural History Society and Field Club, since 1875, has been noted for its studies of the topography, geology, climate, fauna and flora of Hertfordshire, also published in the *Transactions,* with other occasional publications; the Devonshire Association is another well-established county society, 1862-, which has a separate geography section, specialising in the study and mapping of roads, harbours, forestry, geology and agriculture of the county. Many of these societies are, of course, also interested in archaeology and the early development of the area, with which we are not so concerned here. Local surveys are numerous; for example, the *Survey of Leeds,* undertaken by the Thoresby Society, 1961. Academic geology departments have also contributed invaluable work in all kinds of field studies; one of the most recent is the centre based on the University of Keele, which publishes *The North Staffordshire journal of field studies.*

The British Exploration Club has a particular interest in cave surveys; many other cave and pot-holing societies have been formed within the past thirty years or so, including the Wessex Cave Club, the Bradford Pothole Club, the Northern Pennine Club, the Hereford Caving Club and many others, whose original maps and plans made in the course of their explorations are most valuable. The Croydon Natural History and Scientific Society has compiled an *Atlas of Croydon and district,* which is periodically revised; a feature of the Darlington and Teesdale Naturalists' Field Club is its collection of information and maps concerning local topography, especially the field paths; the Ussher Society specialises in the geology and geomorphology of south-west England and the surrounding marine areas. Mention must be made also of the work of the Manchester Geological Association. A number of other organisations, for example, the Field Studies Council, have in recent years produced original maps of various aspects of the country, in the course of their work or in specific projects sponsored by government departments. An example of the latter is the Survey of Grasslands of England and Wales, conducted in 1940 for the Ministry of Agriculture, by Sir R George Stapledon and Dr William Davies. Twenty three types of grassland are distinguished, ranging from first-class rye-grass pastures to the heather and cotton grass of mountain moorland.

The War Office has its own Survey Service, forming part of the Corps of Royal Engineers, which is responsible for all maps needed in any part of the world by the Army and Air Force. Civilian map constructors and cartographic draughtsmen are employed for the work, or Army recruits sometimes join up solely for the survey work. Government hydrological surveys are of the greatest importance, especially having regard to the pressing need for the conservation of water and also to the increasing incidence of flooding noticeable in many parts of the country. The 'Kent rivers hydrological survey', carried out by the Ministry of Housing and Local Government, resulted in the Kent rivers area being one of the best managed and best gauged catchments in the country; the survey followed the usual pattern: a general account of the area, dealing with such topics as population, rainfall, run-off and groundwater conditions, the major catchments in greater detail, an analysis of existing and potential supplies and demands. 'The North Lancashire rivers hydrological survey' is another particularly advanced work; the maps are usually accompanied by illustrations and diagrams, with a bibliography.

Stanford's geological atlas of Great Britain and Ireland, first published in 1904, was issued in a second edition in 1907, adding maps and descriptions of Ireland. The third edition, by H B Woodward, 1914, included also the geology of the Channel Islands. T Eastwood revised the work for publication in 1964; in this edition, Ireland and the Channel Islands are again omitted. The atlas amounts to a handbook of geological information rather than a conventional atlas. Here the ten mile sheets are printed in black and white at twelve miles to the inch. Each section of the map is accompanied by some pages of text summarising the stratigraphy, with comments on economic geology and structure. Introductory chapters cover the

150

stratigraphical succession of the country as a whole, again with constant reference to economic products. The main geological maps of the counties are excellent and description is given of the scenery county by county, also of the scenery on all the major railway routes. There are eighteen pages of drawings of fossils and a separate chapter on economic products illustrated by maps, which seem not to match the standard of the rest of the work. Bibliographical references are restricted and need to be supplemented by more recent work.

The complete list of publications of the Meteorological Office is given in the HMSO *Sectional list* no 37. The *Daily weather report* was first produced by lithography in July 1868; weather maps were first included in March 1872. The *Report* now includes a weather map for the northern hemisphere north of latitude thirty degrees to forty degrees from the midday of the previous day to the day of issue, at a scale of 1 : 30M. Weather maps for the British Isles and adjacent parts of the continent for 1800 hours GMT on the previous day and 0000 hours GMT on the day of issue at a scale of 1 : 20M; weather maps for western Europe and the eastern Atlantic for 0600 hours GMT on the day of issue, at 1 : 20M. The *Weather map,* first published in 1916 to explain how weather maps were prepared and used, now in its fourth edition, includes reproductions of charts, as do numerous other publications of the Meteorological Office, the London Royal Meteorological Society, the Radcliffe Observatory, Oxford and the Liverpool Observatory. J A Taylor and R A Yates: *British weather in maps,* now in a second edition, provides a valuable analysis.

The Meteorological Office first planned a *Climatological atlas of the British Isles* before the second world war. On the recommendations of the National Agricultural Advisory Council in 1945, however, publication was held up to include additional information and maps with the interests of agriculture in mind, and the atlas was at last published in 1952. 220 maps were arranged in ten sections, dealing with all aspects of the climate and weather of the British Isles. Each section includes an introduction and bibliography, with tables and diagrams to supplement the maps. Most of the maps show average conditions for 1901-30, selected as being a standard period for climatological averages for all the meteorological services of the world. In some sections, data have been incorporated for later periods, covering as many years as possible. Particularly useful are the snowfall maps, maps showing thunder frequency, monthly average means of vapour pressure, relative humidity and saturation deficit.

Among other local meteorological societies, which have contributed to climatological knowledge throughout the country, may be mentioned the Chester Meteorological Society. It should be remembered that the majority of the organisations whose work has been mentioned above usually keep at their headquarters, or in their library if they have one, the original drawings of surveys, whether the material is published or not; these are in most cases available for inspection by accredited workers.

A coloured map of the 'Principal systems of farming in Great Britain', compiled for the Association of Young Farmers' Clubs by Sir L Dudley Stamp and Keith Buchanan, was the first general survey of current agricultural systems in this country. Eight colours on the map distinguish urban areas, hill sheep farming, stock rearing, dairying, cropping with livestock, intensive farming and horticulture. A twelve page illustrated booklet accompanied the map. Shortly after, in 1930, Sir Dudley Stamp initiated the Land Utilisation Survey of Great Britain. Field work was carried out mainly between 1931 and 1933, by teams of voluntary workers, many of them senior school children, and every acre of land in England, Wales, Scotland, the Isle of Man and the Channel Islands was recorded on about twenty thousand six inch Ordnance Survey maps. The results were reduced to the scale of one inch to the mile and published in 150 sheets for England and Wales and the more populous parts of Scotland. The work was organised on a county basis and the findings were eventually co-ordinated and published in a series of *Reports,* one for each county, under the title of *The land of Britain.* The whole operation was summarised by Sir Dudley Stamp in *The land of Britain: its use and misuse* (Longmans, 1948), now in a third edition 1962, which gives a summary of the work of the Land Utilisation Survey and includes chapters describing at length the far-reaching changes in land use mapping since 1948. A shorter summary, *The land of Britain and how it is used,* had been issued for the British Council in the previous year. This monumental work, which represented Britain's land use at the time of the second world war, has been incorporated into the national atlas series at 1:625,000.

A Second Land Use Survey of Britain, based on the Ordnance Survey 1:25,000 sheets, was begun in 1960 by the Isle of Thanet Geographical Association under the direction of Alice Coleman of King's College, London. Full details are to be found in the *Land use survey handbook,* by Alice Coleman and K R A Maggs, also in ' The first twenty-four published land use maps ', a special issue of *Panorama* (Isle of Thanet Geographical Association, 1964). A monograph, *Land use maps in eleven colours on the scale of 1:25,000* is obtainable from Stanford. The larger scale base maps were essential for this second survey, not only because of the more complex contemporary conditions, but because data in greater detail are called for now in geographical studies and in planning. At this scale, at least sixty four categories of information are represented clearly. The first-order categories can be interpreted at a glance, while the second order need closer attention in relation to the key. The whole of England and Wales will be covered by 843 maps. Six inch maps are still used in the field. The Ordnance Survey provides the film bases of the colours blue, orange, grey and black, on which the further land use colours can be imposed. The main groups of colour are grey for settlement, red for industry, orange for transport, black stipple for derelict land, lime green for open spaces, light green for grassland, light brown for arable land, purple for market gardening, purple stripes for orchards, dark green for woodland, yellow for heathland, moorland and all rough land, blue for water features and marshy country and white for naturally bare land such as rocky cliffs, sea beaches, roads and pavements. Each of these groups is sub-divided, and, in addition, symbols are used for further definition. Voluntary surveyors have completed the majority of the surveying, but special vegetation surveyors undertook the work in heath and moorland areas, with the aid of Nature Conservancy grants. The surveying work in Scotland was begun in 1964; again, the moorland areas are being left until funds are available for professional vegetation surveys. In co-operation with the Ordnance Survey, both topographic and land use maps on the scale of 1 : 25,000, are being published on the same system of sheet lines. Most of the original maps are available for consultation in the Department of Geography, King's College, the rest being in the possession of the county organisers. A series of *County memoirs* including interpretative commentary are to be issued as completed; the published maps, as completed, will be available from Edward Stanford, London. For further explanation, reference should be made to Alice Coleman: 'The second land use survey in progress and prospect', *The geographical journal,* June 1961. Miss Coleman also made a contribution to the discussion at the symposium on experimental cartography held at Oxford in October, 1963; her address is printed in section 6, 'The mapping of vegetation, flora and fauna', in *Experimental cartography: report on the Oxford Symposium, October 1963* (OUP, 1964). In 'Cartographical progress' in *The geographical journal* for March 1964, Miss Coleman contributed notes on 'Some cartographic aspects of the second series land use maps'.

Individual scholars or scientists have been interested in the classification and mapping of soil types in Britain since the seventeenth century, but it was not until 1919 that a Soil Survey Sub-committee of the Development Commission was appointed to discuss the need for a soil survey of Britain. Methods already in use in the USA and in Russia were studied, much interest was aroused and a Soil Survey Conference was held in 1926 at the Harper Adams Agricultural College. Further meetings established a standard system of soil description, outlined in the *Soil Survey handbook,* compiled by G R Clarke in 1940. Meanwhile, soil mapping had been in progress in Wales, south-east England, Oxfordshire and in the Lothians and Aberdeenshire. A Soils Correlation Committee was created in 1930, as there was still no central organisation to co-ordinate the work. The committee toured the areas being mapped and helped greatly to improve and standardise techniques. It was replaced by the Soil Survey Executive Committee in 1936 and in 1939 the Soil Survey of England and Wales was at last established, with Professor G W Robinson as Director. The surveyors, numbering six at that time, continued to work from two Welsh centres, and centres at Harper Adams, Reading, Long Ashton and Wye, under the supervision of Professor Robinson. In Scotland, the Macaulay Institute for Soil Research was founded in 1930 and work was also done in the West of Scotland College area.

In the reorganisation of 1946, the headquarters of the Soil Survey of England and Wales was transferred to Rothamsted and centres were established at the regional centres of the National Agricultural Advisory Service at Newcastle, Leeds, Ormskirk, Derby, Wolverhampton, Cambridge, Bristol, Reading, Wye, Starcross, Aberystwyth and Cardiff. Soil maps have since been steadily compiled in many parts of Britain at 1:25,000, published at the scale one mile to the inch; they show soil series or other suitable units, which are arranged according to major soil groups and sometimes also according to their parent materials. The drainage status of the soils is nearly always indicated. *Memoirs* have been published, describing the environment and soils, with comments on agriculture, forestry and other forms of land use. In many cases, folded copies of the maps are included in the *Memoirs.* The chief use of soil maps and other data is still in agricultural advisory work; detailed soil maps have, for example, been made of all the Experimental Husbandry Farms as well as of the farms attached to research centres and agricultural institutes. The value of soil surveys is particularly seen in areas given over to specialised products such as fruit and individual vegetable crops and, increasingly, the relevance of knowledge about surface and sub-surface soil is being realised. Full details of the year's work on a country or regional basis are given in the *Annual reports* of the Soil Survey of Great Britain, published by the Agricultural Research Council, London; the *Report* for 1960, no 13, is especially useful, as it contains an article, ' Soil survey in Britain ', by Alex Muir, which outlines the development of soil survey and mapping. In addition, a considerable literature has grown up on this subject, published largely in Ministry of Agriculture *Bulletins* and *Research monographs* and in the transactions and papers of local natural history and other learned societies. G R Clarke: *The study of the soil in the field* (OUP, 1957) forms an excellent introduction. The maps in the monumental *The agrarian history of England and Wales,* edited by H P R Finberg, in several volumes in progress by the CUP, will be most valuable.

A notable atlas was published in 1962; the *Atlas of British flora,* edited by F H Perring and S M Walters for the Botanical Society of the British Isles. Interesting for the subject-matter, the distribution of each of the three thousand vascular plants of Britain, it was also one of the first atlases to be compiled with the help of a data processing machine. The data were recorded on punched cards, which, when played through the machine, entered black dots in the appropriate places in the grid squares of the map. Transparent overlays in a pocket show factors of climate, topography and geology. A *Supplement,* edited by F H Perring, added a further seventy five maps (Nelson, 1966), including maps of microspecies, sub-species and varieties of hybrids. This supplement completed the ' Distribution maps scheme ' of the society before its collection of data transferred to the new Nature Conservancy Experimental Station at Monks Wood, Huntingdonshire; but the mapping schemes are in hand by the British Lichen Society, the British Bryological Society and the British Conchological Society.

A palaeographical atlas of the British Isles and adjacent parts of Europe, edited by L J Wills (Blackie, 1951), was a courageous attempt to interpret the known facts about the geography of past times. The accompanying text, illustrated with maps and figures, is in itself a valuable contribution.

The mapping of individual diseases in this country dates from the early nineteenth century—Dr John Snow's classic mapping of cholera deaths in the Golden Square area of London, for example, in 1848, and the investigations sponsored by the British Medical Association in the 1880s into the incidence of several major diseases in Britain. Some reports were published, but a set of maps compiled by Isambard Owen were not published. Between 1919 and 1923, Percy Stocks, medical statistician to the General Register Office, studied the evidence available in the hope of finding any common factors in the deaths from cancer, and several maps for parts of England and Wales were compiled during the course of these studies and in the years following. However, *The national atlas of disease mortality in the United Kingdom,* completed in 1962 by G Melvyn Howe on behalf of the Royal Geographical Society in association with the British Medical Association (Nelson, 1963), was the first major project of its kind in the United Kingdom, assisted financially by the Isaac Wolfson Foundation and the University College of Wales, Aberystwyth. The first volume covers the period 1954 to 1958. A second volume covers 1959 to 1963, using the 1961 Census material. Maps of population density and mortality from all causes are followed by fourteen maps of the chief causes of death, based on the International Statistical Classification of Diseases; the historical introduction is valuable and each map is accompanied by descriptive text. Full details of the background to and construction of the atlas were given in a talk to the Royal Geographical Society on October 28 1963, printed in *The geographical journal,* together with the ensuing discussion, in the issue for March, 1964; and in the June issue for that year appeared a review of the atlas by Dr Jacques M May, chairman of the National Geography Commission of the International Geographical Union, author of *Studies in medical geography* in three volumes, 1958-61, and chief editor of the *Atlas of diseases.* The atlas was an experimental work and an attempt to establish the value of such maps as research tools in these studies. The difficulties inherent in the recording of reliable statistics were highlighted during the compilation of the atlas; for example, the practice of recording deaths in hospital revealed no correlation between the disease and the factors in the life that may have caused it. This practice has since been modified to recording death in hospital only after a six month residence period, but even this record may be misleading and may seem to exaggerate the deaths from a particular disease within the radius of a large specialist hospital. Most of the work was done by Dr Howe, with the support of the Department of Geography and Anthropology, University College of Wales, Aberystwyth; no interpretation of the maps was attempted at the time.

Dr J T Coppock has become the British authority on the analysis and mapping of agricultural statistics. His article ' The cartographic representation of British agricultural statistics ', illustrated by sketch-maps and diagrams, in *Geography,* April 1965, is essential reading in this specialist field and his atlas, *An agricultural atlas of England and Wales* (Faber, 1964), is the present standard work. More than two hundred maps, with many pages of commentary, explanation and analysis, illustrate regional variations in owner-occupancy and tenancy; the extent to which part-time farming is carried on; geographical distribution of the larger holdings and the distribution of agricultural population. A valuable collection of statistics on the

156

distribution of crops and livestock is included. Much of the data was analysed by computer methods. An appendix by A Sentance of the Department of Numerical Automation at Birkbeck College, deals with computer programming for agricultural information.

A previous agricultural atlas of Wales, 1921, and of England and Wales, 1925, with J Pryse Howell as chief editor, was prepared by the Agricultural Economics Research Institute, Oxford, with the assistance of the Statistics Branch of the Ministry of Agriculture. Dot maps showed distribution of arable, grass, principal crops and principal classes of livestock. They were printed on thin paper, so that they could be superimposed upon other maps for correlation. A second edition, revised by M Messer, was printed and published by the Ordnance Survey in 1932, when all the maps were redrawn according to later statistics. There are also more specialised agricultural atlases, such as the *British atlas of potato varieties,* issued in 1965 by the Potato Marketing Board.

For many years there has been an increasing interest in Britain in the mapping of social data. The task of mapping is centred as much on making full use of existing statistics as on new surveys and the exploration of new sources of data. In addition to the published census figures, census data by Enumeration Districts are available from the Registrar-General. Many universities are in one way or another using these data. The Population Studies Group of the Institute of British Geographers was formed in January 1963 to consider the practicability of mapping census data for the whole of Britain and this group has been working on the preparation of a *Census atlas of England and Wales;* from 1966, maps at 1:2M began to appear, showing analyses of age groups, persons born outside the United Kingdom, numbers of persons per household, types of tenure, etc; population distribution, intercensal change 1951-61, population density and sex ratios were mapped at 1:500,000. Most of the maps are monochrome, with an occasional use of colour; they were published with the necessary key maps, maps of topography and administrative areas and a brief explanatory text. The special maps in the *Atlas of Britain and Northern Ireland* were based on similar data, as also are other atlases in preparation, such as the *Atlas of Northern England,* the *Atlas of London,* just published and the *Atlas of disease mortality,* already mentioned. ' Urban hinterlands—15 years on ', compiled by F W H Green, was published in one sheet covering England and Wales by the Royal Geographical Society, in 1966. In the same year, the Department of Geography, University of Nottingham, published an *Atlas of population change in the East Midland counties 1951-61,* consisting of twelve maps compiled by R H Osbourne.

Geographia has made a particularly notable contribution in the series of marketing maps entitled ' Sales manager's and population maps of Great Britain ', which show by symbols the business possibilities of different parts of the country, using census figures for 1951, amended where possible from the Registrar-General's 1956-57 estimates. They show the marketing hinterlands of the major centres, and accompanying each is a handbook which includes estimated populations and notes the main retail outlets for each centre. The ' Markets and media survey of Great Britain ' was an interesting pioneer survey, designed to facilitate the co-ordination and planning of sales and advertising. Sixty five marketing regions are distinguished, at a scale of ten miles to the inch, with statistical tables based on the 1961 Census. The ' Manufacturing and industry survey of Great Britain ' comprises twenty eight national maps, with graphs and regional maps, based on Board of Trade figures. Twenty eight categories of manufacturing industry are recognised in this survey and each of them is mapped on a separate fold-out sheet of Great Britain. On the reverse of each sheet, tables give the numbers and percentages of each category of industry within seventy one regions; the regions are then represented by tables and graphs, providing a breakdown of industry by numbers, also, in those areas in which more than thirty establishments are found in any one industrial group, a breakdown is added according to numbers of employees in the establishments. A graph of each region shows the relative size of each industrial group within the region. Marketing maps of the West Midlands, Liverpool and Merseyside and of Glasgow and Clydeside have been published separately, prepared with the co-operation of the Department of Geography, University of Birmingham. The maps are printed in black and red.

With the increased use of maps by the general public, an interesting new kind of atlas has come into being, carefully planned to present the maximum information for tourists and those following their hobbies, in handy form. An example is *The National Trust atlas* (London, 1964), in which fifty nine maps, based on the Bartholomew fifth inch to the mile maps, show places of historic, architectural or scenic interest. Double-page maps are followed each by its own gazetteer, so that the numbered sites can be quickly identified, in addition to a separate index. Individually marked are castles, cathedrals, houses and gardens open to the public, ancient monuments and sites, civic buildings, villages of interest, woods, commons, cliffs and beaches. With similar aims, the *Shell nature lovers' atlas of England, Scotland and Wales,* by James Fisher, was published by the Ebury Press and Michael Joseph for Shell-Mex and BP in 1966. Bartholomew's *British Isles pocket atlas,* in frequently revised editions, remains the basic all purpose pocket atlas.

There are numerous road maps of the British Isles. Bartholomew's *Road atlas of Great Britain,* specially designed for motorists and tourists at a basic scale of a fifth of an inch to the mile, is now revised every year. The Outer Hebrides and Orkneys and Shetlands are on a scale of one tenth of an inch to one mile. Relief is shown by contour colouring and spot heights. A double page is given to each area, so that a wide panorama can be studied at one time. A smaller Bartholomew atlas, the *Roadmaster motoring atlas of Great Britain,* in a third edition, is also designed for the needs of motorists. It contains fifty nine pages of maps at six miles to the inch, arranged in double pages. Contour colouring shows generalised physical features and six classes of roads are distinguished, indicating mileages, road numbers, ferries, etc. Names stand out clearly. An index map is placed inside the front cover and a mileage chart inside the back cover; there is also an index of four thousand names, selected with the business and touring motorist particularly in mind. A Bartholomew county map of England and Wales at ten miles to one inch, 1961, indicated road classifications; and a touring map of England and Wales, at twelve miles to one inch, 1962, included an inset map of long distance mileages. Bartholomew's ' British Isles motoring map 1 : 1м ' is a useful reference map, serving many purposes; contours are shown, also hill-shaded relief features, against which the network of communications and the siting of settlements shows up well. There is a clear inset map showing counties.

The new *Sunday Times Royal Automobile Club road atlas of Great Britain and Ireland* (Nelson, 1968) contains ninety three pages of full-colour road maps, at five miles to one inch, based on the Mobil sheet maps. The maps are large and clear, having the routes boldly marked; of several special features, a section entitled ' Beat the traffic ' comprises a series of maps showing alternative routes to the main, obvious ones, for cross-country journeys such as Birmingham-Llandudno, Manchester-Cardiff. The town plans are excellent, including a comprehensive guide to London. *The Reader's Digest AA book of the road,* published by the Reader's Digest Association in collaboration with the Automobile Association, 1966, was the first edition of a road atlas, based upon the Ordnance Survey quarter-inch maps, with the addition of information of interest to tourists. A section shows routes through ninety towns, also strip maps of key routes and various types of useful information are included, such as a section on first aid. Geographia are producing ' Super motorist maps ' at a scale of three miles to the inch, covering Great Britain in twenty seven sheets. Each map folds into a plastic waterproof cover, about $5\frac{1}{2}$ inches by nearly 9 inches. The key to the maps is on the back side of the cover, so that it cannot be consulted easily once the map is spread out. The maps are clearly printed in red, green, blue, yellow and black. Motorways existing and planned are distinguished in blue, as are mileages, national and county boundaries, and water features; the National Grid is added. Primary routes are shown in green, class 1 roads in red, class 2 roads in yellow, and others with a double black line; the black stipple used for continuously built-up areas is useful. Woodlands are shown by light green stipple and a vast amount of other information is indicated either by black or red symbols. Gradient signs and occasional spot heights are the only indications of altitude.

In a small country such as Great Britain, having many centres of population, the maps accompanying the place directories are particularly useful, also those issued with the innumerable guides compiled by local authorities. Special purpose maps covering individual areas are published by many departments and organisations. Some of the most interesting include the maps of inland waterways, such as Stanford's ' Inland cruising map of England ' and the special canoeing maps by the same firm, for England and Wales and for individual areas such as the Norfolk Broads and the Thames. The *Waterways atlas of the British Isles* was published by Cranfield and Bonfiel Books, 1966; maps at 1:506,880 show navigable canals and rivers, indicating derelict canals and including information on navigation. Useful maps are included in the ' Canals of the British Isles ' series, in progress by David and Charles of Newton Abbot.

Large-scale maps and charts are available for all the major British ports, the publications issued by the Port of London Authority being particularly comprehensive. Other maps and charts of ports and harbours are to be found in the *Digest of port statistics,* produced by the National Ports Council. ' The siting and development of British airports ', by K R Sealy, consists of ten sheets at various scales, some of which are included in the article of the same title, in *The geographical journal,* June 1967. Official air charts are naturally available for those who need them. Official rail and bus maps are constantly revised and readily available; and a ' British bus services map ' at approximately 1:20,000 was produced by Ian Allan of Shepperton in five sheets, 1965.

Some pioneering regional cartographic ventures include *The regional atlas of north-east England,* now being prepared at Newcastle University and the *Atlas of Banbury,* produced by students in training in the cartographic course at the Oxford College of Technology, in conjunction with the Department of Town Planning in the College. The students collected the data, processed it, statistically or otherwise, selected suitable data for inclusion in the atlas and completed the draft, and production of plates ready for printing by lithography. It is expected that such specific projects will increasingly be undertaken by students and such work may be compared with the *Lithofacies maps* . . . to be mentioned below. *East Anglian studies,* edited by L M Munby, are papers, surveys and maps prepared by tutors of the University of Cambridge Board of Extramural Studies, including studies of the river system of Norfolk, Colchester in the eighteenth century and the Suffolk landscape, many of them illustrated by line drawings (1968). The series *Yorkshire field studies* is published by Leeds Institute of Education, University of Leeds.

For individual areas, such sources as the ' West Midlands Green Belt map ', 1 : 63,360, published by the Midland New Towns Society, 1966, are invaluable, as are the maps in the Hydrological Survey and the urban planning studies of the Ministry of Housing and Local Government, and of the Water Resources Board, such as ' Water supplies in South-east England ' (HMSO for the Ministry of Land and Natural Resources, 1966), which includes fourteen folding maps and diagrams in a separate case. The maps in the publications prepared for the annual meetings of the British Association for the Advancement of Science are most informative—K C Edwards : *Nottinghamshire and its region,* 1966, etc—each presenting the city in relation to its hinterland. The Department of Geography, University of Newcastle-upon-Tyne, is publishing a series of *Planning reports,* such as, for example, *Northern region and nation: a short migration atlas, 1960-61,* by J W House and K G Willis, 1967; and in the monograph edited by S R Eyre and G R J Jones, *Geography and human ecology: methodology by example* (Arnold, 1966), maps, illustrations and diagrams demonstrate short studies of specific places in Great Britain and elsewhere, to exemplify regional method. From time to time, Bartholomew issue special maps, such as that of Yorkshire at the quarter-inch scale, 1962, showing detail of the Ridings. A one inch to the mile map of the Lake District was also issued in 1962, extending from Bassenthwaite Lake to Lake Windermere and from Hawes Water to Ennerdale Water; relief is shown by contour colouring and hill-shading, and roads are classified, including footpaths and tracks. A notable feature of the Bartholomew maps and plans is that each normally bears the date of issue or revision.

Innumerable maps of all kinds are available for the London area; the most important cartographical project on the area to date, however, is *The London atlas* (Pergamon Press, 1968-). This atlas is the third in the current trend of production of full-scale atlases of great cities, following those for Berlin and Paris, to be mentioned below. The idea originated with Professor Emrys Jones and the Department of Geography, London School of Economics, their aim being the presentation of conditions in the whole of the London region, not only for reference and study but as a basis for interpretation and policy. For full details of the production, reference should be made to Professor Emrys Jones' talk on 'The London atlas', presented to the Royal Geographical Society on February 22 1965 and reprinted in *The geographical journal,* September 1965, together with the discussion that followed the talk. A number of figures in the text give an idea of the detail in which this atlas has been planned. Compilation was facilitated by consultation with the County Planning Officers, the Ordnance Survey, The Ministry of Housing and Local Government and other such bodies as the Royal Geographical Society, the London Topographical Society, the London Climatic Society, the Council of Urban Studies and the Port of London Authority. Physiography, geology, resources, hydrology of the Thames Estuary, microclimatology and pollution are covered; a historical section traces the development of the settlement from the earliest times to 1963. The 1961 Census has been the source for social and economic data, plus information gained from the 1966 Sample Census; also the relevant departments in industry, transport, etc have provided primary material. The second part of the atlas is to become available in 1969.

To date, the best known general atlas of London has been the Bartholomew *Reference atlas of Greater London,* in many editions, of which the latest complete revision was the thirteenth in 1968. This edition contains, for the first time, town plans of the dormitory towns of Brentwood, Guildford, Harlow, Hatfield, Hemel Hempstead, Royal Tunbridge Wells, Sevenoaks, Welwyn Garden City and Woking. Taking in the whole of the Metropolitan Police area and outlying built-up areas, the atlas now covers more than seventeen hundred square miles from Windsor to Gravesend and from St Albans to Reigate. Postal districts, new London boroughs, county, local government and Metropolitan Police boundaries are marked. Three main scales have been used: two inches to the mile for the outer area, four inches to the mile for the Greater London area and ten inches to the mile for Central London from Westminster to the City. Seven general and administrative maps are included and the general reference value is increased by the introduction, which lists places of importance, entertainment and general interest, including more than a thousand addresses, with their telephone numbers, and by the index, which contains over sixty two thousand names with their postal districts and boroughs, adding many courts, streets and buildings too small to be shown on the map.

Many revised editions have been issued also of the Geographia *Atlas of London and suburbs* containing maps at three miles to the inch. The area covered is bounded by Barnet, Becontree, Croydon and Greenford. Streets are named and house or shop numbers are given at intervals along major thoroughfares. There is a comprehensive index. Geographia issue also a handy size *Central London atlas,* covering the City of London and the West End. Of the many special purpose maps of London, the Duplex map in full colour, 1963, is an interesting project; on one side a map shows motorists the best routes through London, based on information supplied by the AA, and on the other side is a quarter-inch contour-coloured motoring map of the Home Counties, reaching a radius of about fifty miles from London. A ' Central plan of London ' at $3\frac{3}{4}$ inches to the mile is a detailed street map extending from Hampstead to Denmark Hill and from Blackwall to Shepherd's Bush, first published in 1960. ' London, Westminster to the City ', at nine inches to the mile, first came out in 1963; it is an excellent reference map, as even very small alleys and courts can be located through the index on the back and there is a separate map of the London Underground. ' The London map directory: a street atlas of London and its surrounds ' has been issued by Geographia since 1922.

Maps of cities are increasingly being compiled as part of study projects, development plans, etc, some in loose sheets, others in some form of binding. Examples are *An atlas of Durham City,* edited by H Bowen-Jones at the Department of Geography, Durham College in the University of Durham, 1960, with an illustrated introduction. In *An atlas of Harrogate,* edited by J A Patmore, the twenty two maps and diagrams were drawn by A G Hodgkiss, P J Treasure and J Lynch, of the Liverpool University Department of Geography and the work was published by the Corporation of Harrogate in 1963. Photographs and reproductions of prints are included. The *Historical atlas of Cheshire,* a sixty four page production including thirty one maps, was edited by D Sylvester and G Nulty for the Cheshire Community Council, 1958; and *A Gloucestershire and Bristol atlas* was issued by the Bristol and Gloucestershire Archaeological Society in 1961.

The publications planned for the international geographical congresses and other major meetings are invariably sources of original maps; for example, the *Guide to London excursions,* edited by K M Clayton, with specially prepared maps of the drainage, geomorphology, climatology, settlements, population density and communications, mostly based on Ordnance Survey maps, and particularly interesting larger scale maps and sketch maps to illustrate specific aspects of the text; *The geography of Norden* is another fine example. The original maps drawn by such cartographic departments as that of the Royal Geographical Society are of great significance; also, the large-scale surveys carried out in connection with government enquiries, such as the ' South-east Survey '. Geographical theses normally include original regional or local maps; an example is the author's own doctorate thesis, ' The city-port of Plymouth : an essay in geographical interpretation ', in which there were included a great number of thematic maps based on current, often unpublished, statistics; in this case, the work assumed a particular interest, for it was completed just before the second world war and therefore became the only systematic account of the pre-war city.

Commercial cartographic firms are frequently called in to assist in surveying and mapping for new developments; such a service is Hunting Surveys and Consultants, with its many associated centres in England and Scotland. Recent aerial photographic surveys carried out in Britain have included the seven hundred square mile Kesteven division of Lincolnshire at 1:10,560, for general planning and revision of the $2\frac{1}{2}$ inch Ordnance Survey maps; and the maps of the Greater London Boroughs of Newham and Barking, at 1:2,600, for redevelopment purposes.

In addition to the examples cited above, every issue of the authoritative British geographical journals—*The geographical journal,* the *Transactions* of the Institute of British Geographers, *Geography,* the *Scottish geographical magazine,* etc—and many of the journals devoted to the earth sciences, to economic and sociological subjects, include maps and sketch-maps. *The cartographic journal* in each issue includes at least one article of interest on Britain and the map supplements are extremely fine. Many of the high level overseas journals frequently include maps embodying research on some aspect of Britain. The following are three of the many recent relevant works :

J T Coppock : ' The cartographic representation of British agricultural statistics ', *Geography,* April 1965.

J A Taylor, editor : ' Early crop production in the British Isles ', *Papers* presented at Symposium IX, March 1966 (Aberystwyth, Department of Geography, University College of Wales, 1966); ill, maps, plans, dia, bibliog.

J A Taylor and R A Yates : *British weather in maps* (Macmillan, second edition 1967); text maps, dia.

WALES : In addition to all the publications and map series mentioned above which include Wales, are the following special sources for the cartography of Wales. ' Wales and the Marches ' was a special sheet of the ' Quarter-inch ' Ordnance Survey map series, published in 1959, the first map to cover the whole of the Welsh region on one sheet at a scale of this order. It includes Liverpool and Bristol, the two English ports with which the economy of Wales is so closely linked, also Chester, Shrewsbury and Hereford. It is similar in style to the ' Quarter-inch ' fifth series, but in addition a green tint has been used for land up to two hundred feet above sea level and hill-shading in violet for land above that line, combined with contours at two hundred feet levels.

' Wales and the Midlands ', at 1 : 380,160, was the first of a new series of plastic relief maps prepared by the Cartographic Department of the Clarendon Press, Oxford, in association with Barclays Bank, planned to cover the British Isles. Rough pasture, farmland and woodland are distinguished. Roads are classified by width and by volume of daily traffic. An excellent introduction to the geography of Wales is *Wales in maps,* by Margaret Davies, published in a second edition by the University of Wales Press, Cardiff, 1958. The black and white maps, designed particularly for school use, are accompanied by a commentary which is itself illustrated by maps and diagrams, covering physical geography, history, agriculture, population and settlement, commerce and industry.

J I Jones was the compiler of two useful atlases, *A historical atlas of Wales* and *A geographical atlas of Wales,* both published by Hughes and Son of Wrexham in 1956. Another historical atlas, *An historical atlas of Wales from early to modern times,* compiled by William Rees, was first published by Faber in 1951; and in a second edition in 1959. About seventy monochrome plates are designed to illustrate in map form the continuous story of the country's development in relation to the physical background. Illustrations and diagrams are also used to good effect and full notes are appended to each map. The maps and plans in James A Taylor: *Wales: agricultural geography and land use* are most valuable. This work was a paper prepared for the International Geographical Union Inter-regional Conference at Mexico City in August 1966, on selected developmental problems, in rural Wales, printed in Spanish in volume 2 of the *Proceedings* and in an English version by the Department of Geography, University College of Wales, Aberystwyth, 1967.

In 1961, a vegetation survey of the Welsh uplands was begun, whereby some 2,700 square miles of moorland and mountain country were to be mapped at six inches to the mile. All species constituting a ten per cent coverage or more are recorded; all Forestry Commission plantations are included and a vegetation atlas will be published in due course. From such plotting of statistical analysis and comparison with topographical, climatic and soil maps, the factors controlling vegetation, such as slope, altitude and aspect, soil texture and drainage, geology, climate and exposure, can more easily be related. Soil maps of Wales are nearing completion; especially in north Wales, where the soil survey was begun by Professor G W Robinson; the map and memoirs are of high quality and detail. Current land use surveying in Wales is particularly interesting as an example of the value of such periodic surveys; comparison with the 1931 picture, for example, reveals very clearly the steady increase in the proportion of non-agricultural land. The mapping of derelict land in south, central and north Wales will correlate with these results. At the same time, afforestation is a dominant feature, showing more than seven per cent under the direction of the Forestry Commission compared with two per cent in 1931. Some land has gone to industry. Improved road facilities link both with industrial activity and with new forms of tourism and the greater attention to National Parks and nature reserves.

Maps, often original, are to be found in the journal of the University College of Swansea Geographical Society, *Swansea geographer,* in the volumes of the 'Land of Britain' series covering Wales and in the many excellent monographs on the geography of Wales, among others—E G Bowen: *Wales: a physical, historical and regional geography* (Methuen, sixth edition 1957); E H Brown: *The relief and drainage of Wales: a study in geomorphological development* (University of Wales Press, 1960); 'The physique of Wales', *The geographical journal,* June 1957, with illustrations, text maps, figures and references, and in the many published works, especially geological, of F J North. The study by Harold Carter, *The towns of Wales: a study in urban geography* (University of Wales Press, 1965) also includes interesting maps.

An historical atlas of Wales from early to modern times, compiled by W Rees (Faber, second edition 1959; reprinted 1967), comprises seventy plates. Maps and text have been revised since the first edition of 1951 and a map of early railways has been added. Interpretative text is linked to each atlas plate and special sections deal with 'Origins', 'The Dark Ages', 'The Anglo-Norman Conquest', 'From the Conquest to the Union' and 'Modern Wales'. Bartholomew has published an 'Historical map of Wales and Monmouth', compiled by L G Bullock. This is a brightly coloured pictorial map, illustrating the historical and cultural background of the principality, surrounded by the arms of fifty six of the counties, cities and towns. An interesting article in this context is by Emrys Jones and Ieuan L Griffiths, 'A linguistic map of Wales, 1961' in *The geographical journal,* June 1963, illustrated by a map and diagrams, and including references.

SCOTLAND: The first survey of Scotland was carried out after the 1745 Rebellion, which had shown up the lack of reliable maps of the country; under the direction of General Roy, the survey was completed in 1755, but was not published. General Roy himself stated that it was not to be considered as an accurate map, because of 'inferior instruments', but the experience presumably helped towards the efficiency of the first really systematic survey, begun in 1784, when General Roy was charged with the execution of the Ordnance Survey of Great Britain. Details concerning the early progress of the map of Scotland and local conditions may be found in the 'Report for the Select Committee on the Ordnance Survey (Scotland)', together with the *Proceedings* of the committee, Minutes of evidence, etc, printed by order of the House of Commons, July 1851 and reviewed in *The Edinburgh Review,* CXCIII VII 1852. The *Proceedings of Edinburgh Royal Society* for 1851 also carried reports of progress of the map. The first one inch map of Scotland was published in 1843 and the later programme of Ordnance Survey mapping followed the pattern for the country as a whole, except that the mountain and moorland areas were not included in the 'Six inch' scale series. These areas have since been covered by a new photogrammetric survey, for which two survey documents for each map sheet have been produced on air survey plotting machines, one for the detail and the other for the contours. The detail on these sheets has been scribed and the rock features drawn. The one inch maps of Scotland have at least an inch of overlap between two adjacent sheets. The contour interval is fifty feet. The colour scheme is the same as in those for England and Wales, except that the orange in-fill is extended to 'other' motor roads. The sheets of the mountainous and moorland areas do not need revision more frequently than every twenty five years, but in the lowland and east coast areas, revision is carried out as frequently as is required.

The Ordnance Survey 'Administrative areas' maps for Scotland, at a quarter-inch to the mile, are in nine sheets, 1950-66, showing the boundaries of parliamentary constituencies, counties, counties of cities, burghs, districts and civil parishes, overprinted in colour on a base map compiled from the 'Quarter-inch' fifth series. A special 'Administrative areas' map for Glasgow and district, at one inch to the mile, gives the same administrative information on the base of the 'One inch' seventh series. The Ordnance Survey map of the

Cairngorms covers an area of more than a thousand square miles, stretching from Grantown-on-Spey to the Forest of Atholl and Glen Cova. Layer tints and hill-shading give a striking three-dimensional impression of the mountains and valleys and a great deal of information of interest to tourists has been added. Another special tourist map was issued in 1967 for ' Loch Lomond and the Trossachs '.

The atlas of Scotland, published in 1895 by Bartholomew for the Royal Scottish Geographical Society, was among the first of modern national atlases. Bartholomew currently issues several maps of Scotland, most of them frequently revised. Among the most used is the ' County map' at the scale of ten miles to the inch; this is an administrative reference map, showing counties in background tints and classified roads, completely revised in 1966. The ' Scotland, touring map' is at the scale of twelve miles to the inch and was revised in 1967. Contour colouring is shown, roads are classified and numbered and county boundaries are shown in purple. This is a very clear map, folded so that it can easily be referred to in a car and it includes an inset map of long distance mileages. Several maps have been compiled with the tourist in mind. Among these, ' Map of Scotland of old ', compiled in 1961 by Sir Iain Moncreiffe and Don Pottinger, outlines the areas belonging to about 330 Highland clans and Lowland families; 174 clan crests are reproduced in colour, with the coats of arms of their chiefs and the map is available folded or unfolded in a carrying tube. Another ' Historical map of Scotland ' has been compiled and revised by L G Bullock. This, again, is a pictorial map, indicating the location of incidents and characters prominent in Scotland's history, the sites of battles, castles, etc. The coats of arms of 129 Scottish clans, cities and towns are printed all round the map. ' Scotland, south-east ', also published by Bartholomew, was compiled by Dr K A Steer in 1961 at a scale of quarter-inch to the mile. Contour colouring is shown and historic buildings open to the public are marked, also 185 of the principal antiquities prior to AD 1100. These sites are listed on the reverse of the map, with brief notes on each. The ' Pentland Hills map ', first issued in 1958, at the scale of 1½ inches to the mile, covers Edinburgh and the country south to Dunsyre and Eddleston. Contour colouring is given and tracks, field paths and other rights of way are clearly indicated. Bartholomew has published also a number of city plans, including the ' Edinburgh plan ' at four inches to the mile, with index and notes, the ' Glasgow plan ' at 2½ inches to the mile, with a large-scale inset of the centre of Glasgow and a street index on the reverse side, and the 'Aberdeen plan' at four inches to the mile, also carrying an index of streets on the reverse side. There is also the Bartholomew *Edinburgh atlas guide,* comprising thirty two pages of maps at 3½ inches to the mile, with index and notes; and the 'Edinburgh pocket plan', covering the area from Currie to Musselburgh at 3½ inches to the mile, including a selective index and a list of places of interest.

Geological reconnaissances were undertaken by John Macculloch as early as 1811 and by the 1830s he had completed a broad geological survey of Scotland, which was published, after his death, at the scale four miles to the inch. His more detailed survey work had appeared in the papers of the Geological Society and in his *Description of the Western Islands of Scotland*, 1819. Modern geological mapping is the responsibility of the Geological Survey of Scotland in Edinburgh. Good progress has been made with the production of soil maps, especially in the Lowlands, at one inch to the mile. In the Highland areas, there is less need for detailed soil mapping. The first agricultural surveys were made by the Board of Agriculture from 1794 onwards, maps of individual counties being under construction until 1813, by various cartographers working on different scales; for many counties, a soil map was prepared in conjunction with these surveys. *An agricultural atlas of Scotland*, compiled by H J Wood, was published by Gill in 1931. The dot method was used in the distribution maps; there are useful notes and an introduction treating of the agricultural regions of Scotland. Information for current maps is compiled by the Department of Agriculture and Fisheries for Scotland. Especially useful maps were included in the department's *Types of farming in Scotland* (HMSO, 1952). A 'Fishing map of Scotland' was compiled by Malcolm A Scott and published by Hardy Prothers (Alnwick) in 1962, at a scale of ten miles to the inch. A list of individual waters is included on the reverse, with notes.

A population map of Scotland, based on the 1961 Census, 'Scotland population distribution on the night of 23 April 1961', was compiled by a team of staff and research students in the Department of Geography, Glasgow University (Collins of Glasgow, 1965); it is most important for all planning as well as geographical interests. The manuscript sheets of the Ordnance Survey maps, 1:250,000 were used in compilation. (Refer: *The Scottish geographical magazine*, April 1966, p 53-54.) The Registrar-General for Scotland made available unpublished statistical material and the distribution of population has been shown in as much detail as possible at a scale of 1:500,000, using actual areas and numbers rather than proportional symbols. For rural population, units of twenty five persons have been used, shown by red dots of 0·5 millimetre diameter. For towns and villages of more than 200 population, solid red shows up the built-up area, and the number of persons is printed in hundreds. Burgh and county boundaries are shown in black, as is the 100 kilometre National Grid. Detail is shown against a grey relief background and the reverse side of the sheet carries an outline and index of administrative units. It is hoped that further special surveys of Scotland will follow. Two articles in *The cartographic journal,* June, 1967, are relevant in this context: ' Population enumeration on a grid square basis: The Census of Scotland, a test case ', by Jean Forbes and Isobel M L Robertson; and 'A cartographic analysis of the Glasgow 1961 Census ' by Mary Jelliman, both illustrated by maps and figures.

Scotland is particularly rich in the number and quality of estate maps which have survived. The Scottish Record Office contains a fine collection of large-scale manuscripts, engraved and lithographed plans, which have been valuable for studying the evolution of the Scottish rural landscape. A *Union catalogue of large-scale manuscript maps of Scotland* was begun under the auspices of a committee comprised of representatives from the Scottish universities, the Scottish Record Office, the Royal Scottish Geographical Society and the National Library of Scotland (HMSO, 1966-). The complete index is kept in the Map Room of the National Library; some three thousand main entries, arranged by county and parish, and more than three thousand cross references to surveyors and owners include items in public and private collections. In the case of the latter items, photographic reproductions are made if possible and are filed for reference. Entries consist of exact title, date of survey, the surveyor's name, scale, size and a short summary of contents.

In 1965 the Edinburgh Branch of the Geographical Association published an *Atlas of Edinburgh,* comprising thirty nine pages of maps, text and illustrations in a paper cover, with a spiral binding, the material having been provided by members of the branch, mainly by teachers of geography in Edinburgh schools. The work was organised and introduced by Professor J Wreford Watson, with a brief outline of ' The rise and growth of Edinburgh '. All aspects of physical features and climate, land use, social and statistical data are included; daily commuter traffic in and out of Edinburgh is shown. The maps are interspersed with photographs, diagrams, graphs and drawings, all annotated, and the skilful use of cartographic techniques makes this a useful teaching atlas and introduction to the geography of the area.

The Royal Geographical Society of Scotland has always been interested in cartographical matters and in November 1959, in discussion with representatives of the Geography Departments of

Edinburgh and Glasgow, it was decided to give a still greater proportion of space to the subject in the *Scottish geographical magazine*. A Cartographic Sub-committee, set up in 1960-61, created without delay a Cartographic Section to encourage interest in map-making among members and to provide a forum where professional academic and lay map-makers could exchange ideas and information. An exhibition, 'Mapping Britain', was organised by the society in co-operation with the Ordnance Survey, in May-June 1960, for which a descriptive catalogue was prepared. In addition, a group was established to consider the spelling of names on Scottish maps, especially Gaelic names, and another group has been discussing the qualities required in satisfactory cartographic lettering.

The largest Scottish universities show increasing recognition of the importance of cartography in relation to geography. The Universities of Glasgow and Edinburgh now have full-time lecturers in cartography and both have well-equipped photogrammetric laboratories. Post-graduate diploma courses in cartography and in photogrammetry were initiated at Glasgow in October 1962. Other original maps are produced from time to time in the publications of the Geological Societies of Edinburgh and Glasgow, the Royal Highland and Agricultural Society of Scotland, the Royal Scottish Forestry Society, the Geographical Societies of Edinburgh, Glasgow and Aberdeen and the Scottish Field Studies Association.

Important sources of economic maps are the volumes of the three statistical accounts of Scotland. *The third statistical account of Scotland,* in particular, compiled and edited by the four Scottish universities, dates from 1951. Each university has been responsible for the county volumes in its own region, the whole being published by Oliver and Boyd, under the aegis of the Scottish Council of Social Service. Geographia has published marketing maps of the conurbation of Glasgow and Clydeside, similar to the West Midlands pioneer map, prepared with the co-operation of the Department of Geography, Glasgow University.

The Automobile Association has issued a frequently revised road book of Scotland from 1938, including gazetteers, itineraries, maps and town plans, also Bartholomew's 'Road map of Scotland' at one inch to five miles. Chambers also publish a *Guide to Scotland,* which includes a number of coloured sectional maps. There are useful maps also in the *Scientific survey of south-eastern Scotland,* published in 1951 by the British Association for the Advancement of Science, the *West Highland survey: an essay in human ecology,* edited by F F Darling (OUP, 1955) and the Scottish volumes in the 'Land of Britain' series.

IRELAND

The British Ordnance Survey has no responsibility now for maps of Ireland. The chief survey officer, Ministry of Finance, Ordnance Survey in Belfast, has charge of maps of Northern Ireland and, for the Irish Republic, the assistant director of survey, Ordnance Survey in Dublin. The national committee is the Ordnance Survey in Dublin.

Topographic maps are in two categories, large- and small-scale. The twenty five inch plans, 1 : 2,500, cover Northern Ireland except for the mountainous and moorland areas. Large-scale maps have naturally been required, particularly in the Belfast area, and a special series at this scale has been prepared. These plans are to be used as a base for future revisions, both of the twenty five inch itself and the ' six inch ' scale series. In the latter series, 1 : 10,560, each sheet covers an area of about six by four miles, with relief shown by contours. Both series are in monochrome. The small-scale map series include a one inch to the mile series in eleven sheets, each covering an area of about thirty by twenty four miles. These sheets have been revised since 1951 and the revised edition is used as a base for the newly designed 'Half-inch' and 'Quarter-inch' series, 1 : 126,720 and 1 : 253,440 respectively. Contours are drawn at fifty and 100 feet, then at intervals of 100 feet to one thousand feet, thereafter intervals of 250 feet. A third series of the one inch began in 1961. A provisional edition of the new quarter-inch map is available, mounted and folded; emphasis is on relief and communications, names being limited to the more important settlements and topographical features. A particularly attractive map issued by the Dublin Office in 1964 was of Dublin at 1 : 25,000, set in its hinterland, with an inset of the City Centre; an index of streets and places of interest accompany the map. A map of Dublin airport at 1 : 20,000 was issued by the Survey in 1966.

A description of the Ordnance maps of Northern Ireland, prepared by the Northern Ireland Ordnance Survey, was published in 1952 by the Ministry of Finance, Belfast. For the Irish Republic, the *Catalogue of maps and other publications of the Ordnance Survey* was published in Dublin by the Stationery Office in 1949. Both publications have been kept up to date by supplements. A guide to the current output of the Dublin and Belfast Ordnance Surveys, together with other topographical maps, atlas maps, early maps of the country and of the counties, estate maps and town plans is *Ireland in maps: an introduction, with a catalogue of an exhibition mounted in the Library of Trinity College, Dublin 1961, by the Geographical Society of Ireland in conjunction with the Ordnance Survey of Ireland*, compiled by John Andrews (Dublin, 1961).

172

The Geological Survey, Dublin, previously a division of the Geological Survey of Great Britain, became independent in 1922. A geological and topographical map of Ireland at 1 : 1M was issued in 1928 and a series of geological sheets at 1 : 253,440 has since been prepared for the whole country. A map forms part of the *Handbook of the geology of Ireland* (Murby, 1924), by G A J Cole and T Hallissy. The National Soil Survey, a department of the Soils Division of the Agricultural Institute, Wexford, began a detailed survey of the whole country on a county basis in 1959, working at a scale of 1 : 10,560, with the published maps at 1 : 126,720. The method of colour printing is based on that used by the Netherlands Soil Survey Institute. More than forty colours have been necessary for each of those maps already completed. Accompanying *Memoirs* follow the accepted pattern; a brief introduction outlines climate, relief and geology, followed by a short discussion on soil development and the techniques of soil mapping and recording. The text is illustrated and maps are enclosed in a pocket. Two other series of soil maps are being published, showing soil suitability and soil drainage, also glacial drift maps as desirable. 'A glacial drift map of Co Limerick, 1 : 126,720 ', for example, was issued by the National Soil Survey, Dublin, in 1966. The article by J Lynch, 'Preparing and printing soil maps in Ireland', in *The cartographic journal,* June 1967, includes some informative details.

The basic field survey for the Land Utilisation Survey of Ireland, begun in 1936, was the six inch to the mile map, later reduced to the one inch scale and superimposed on an Ordnance Survey base. The first section to be published was the map and text of *The Land of Ulster: the report of the Land Utilisation Survey of Northern Ireland, the Belfast region.* by D A Hill (HMSO, 1948). A long-delayed report, *Land use in Northern Ireland: the general report of the Land Utilisation Survey of Northern Ireland,* edited by Leslie Symons (University of London P, 1963), provides a valuable summary, based on detailed field work carried out mainly by the Department of Geography in Belfast. In four sections, 'Environmental background', 'Land use', 'Structural problems in agriculture' and 'County analysis', it is fully documented, published with the maps and illustrated by photographs. Sir Dudley Stamp had compiled *An agricultural atlas of Ireland,* published by Gill, in 1931.

The maps accompanying other Irish official publications are of central importance. Some very interesting ones were compiled in the middle years of the nineteenth century, during the efforts to improve communications. A series begun in 1837 showed the factors affecting canal and railway use, with population statistics based on the 1831 Census. The 1841 Census (1843), a vital document in any study of the development of modern Ireland, included four maps showing the distribution of population, a separate map of Dublin and a number of graphs. A useful article dealing with the achievements of the following decade is ' Population changes in Ireland, 1951-61 ', by James H Johnson, in *The geographical journal,* June 1963, with maps and references.

There are also the maps accompanying the various planning reports : the *Location of industry in Northern Ireland,* 1944; *Planning proposals for the Belfast area,* 1945, with a second report in 1951; and the *Belfast regional survey and plan,* 1962 (1964), directed by Sir Robert Matthew. A map illustrates the published results of the survey *Economic development in Northern Ireland, including the report of the economic consultant, Professor Thomas Wilson* (Belfast, HMSO, 1965).

An interesting *Linguistic atlas and survey of Irish dialects* was published in 1958 by the Dublin Institute for Advanced Studies, containing three hundred plates, with an introduction. Original maps, illustrating other specific topics, are frequently to be found in *Irish geography,* 1947-, the annual publication of the Geographical Society of Ireland, founded in 1934 especially to promote geographical studies of Ireland; and in authoritative monographs such as T W Freeman : *Ireland: a general and regional geography* (Methuen, third edition 1965); R Common, editor : *Northern Ireland from the air: features of the landscape explained by a combination of geographical text, line drawings and air photographs* (Queen's University, Belfast, 1964, issued with a folding map of Northern Ireland, 1 : 500,000); the classic J K Charlesworth : *The geology of Ireland: an introduction* (Edinburgh, Oliver and Boyd, 1953); the British Association for the Advancement of Science : *Belfast in its regional setting: a scientific survey;* James Meenan and D A Webb, editors: *A view of Ireland: twelve essays on different aspects of Irish life and the Irish countryside* (Dublin, British Association for the Advancement of Science, 1957).

Bartholomew's ' Quarter-inch ' map series of Ireland covers the whole of the Republic of Ireland and Northern Ireland in five uniform sheets : Antrim-Donegal, Dublin-Roscommon, Wexford-Tipperary, Cork-Killarney and Galway-Mayo. Revised editions are prepared approximately every three years. Relief is shown by contour colouring, the counties are outlined in purple and as much detail as possible at the scale is included. Emphasis is given to the road system; they are classified and numbered, from motorways to quite minor roads, which may interest the tourist. Other maps from Bartholomew, compiled chiefly with tourists in mind, are the ' Touring map of Ireland ' and the ' Historical map of Ireland ', compiled by L G Bullock. On the former, contour colouring is shown, all roads are classified and numbered and an inset map shows the chief touring areas and mileages. The latter is a pictorial map illustrating sites of historical interest, the areas controlled by individual clans and the coats of arms of forty five towns. For reference purposes, there is also a county map at the scale ten miles to the inch. Among the most useful guides to Ireland is the *Shell guide,* first issued by the Ebury Press in 1962, compiled by Lord Killanin and M V Duignan, well illustrated with maps. The Automobile Association issues *The AA road book of Ireland: with gazetteer, itineraries, maps and town plans* and *The contour road book of Ireland: a series of elevation plans of the roads with measurements and descriptive letterpress,* compiled by H R G Inglis, which was revised and enlarged by R M G Inglis in a second edition 1962.

Entries for maps are included in the national bibliography, maintained on cards in the National Library in Dublin; and the Stationery Office in Dublin issues an annual list of government publications, a number of which include maps. *A bibliography of County Galway,* by Mary Kavanagh (Galway County Libraries, 1965), also includes maps.

ISLE OF MAN: The Isle of Man is covered by the standard Ordnance Survey maps at one inch to the mile and smaller scales. Other useful maps appear from time to time illustrating articles in the publications of the Isle of Man Natural History and Antiquarian Society and in the reports of the various government boards concerned with agriculture and fisheries, forestry, mines and lands, tourism, highways, harbours and airports. In the 'Land of Britain' series, the volume for the Isle of Man was contributed by N Pye and E Davies; informative maps are also included in J W Birch: *The Isle of Man: a study in economic geography* (CUP, for the University of Bristol, 1964).

CHANNEL ISLANDS: The Ordnance Survey publishes on its own behalf only the three inch map of Guernsey, the most detailed map of the island available, fully revised from aerial survey material in 1966. For all other maps and plans of the Channel Islands, the authorities in Guernsey, Jersey, Alderney and Sark are responsible. Bartholomew issue on one sheet, contour-coloured road maps of Jersey at about $1\frac{1}{2}$ inches to the mile, of Guernsey at about $1\frac{3}{4}$ inches to the mile, of Alderney at one inch to the mile; and, without contours, Sark, at approximately two inches to the mile. Herm and Jethou are mapped at one inch to the mile. G Dury contributed *Channel Islands* to the 'Land of Britain' series, in 1950, and H J Fleure wrote the Guernsey volume in the ' British landscape through maps ' series, 1961.

SCANDINAVIAN COUNTRIES : The definition of ' Scandinavia ' varies. *The geography of Norden* . . ., edited by Axel Sømme (Oslo, F W Cappelens, for the Norwegian National Committee of Geography, 1960; Heinemann, 1961; New York, Wiley, 1962) includes Denmark, Finland, Iceland, Norway and Sweden and this is the grouping used here. In this fine publication is included a section of high quality coloured general and thematic maps, in addition to maps and diagrams in the text. The work is an excellent example of co-operation in geographical studies, being the united effort of thirteen eminent geographers from the leading universities and colleges of the five countries, to mark the occasion of the Nineteenth International Geographical Congress, Stockholm, 1960.

The Bartholomew ' World ' series reference map of Scandinavia, 1 : 2,500,000, covers Denmark, Norway, Sweden and the Baltic; it was issued in 1957, latest revision 1965. Contour colouring is shown, boundaries are marked in purple, roads in red and railways in black. Stanford's ' General map of Scandinavia ', also at 1 : 2,500,000, includes insets of Svalbard, Jan Mayen Island and Bornholm Island. Available in a paper edition, on varnished cloth, with rollers, or folded on cloth, it is decorated with coats of arms in a pictorial border. Geographia issue several maps covering the Scandinavian countries, also a *Touring atlas* covering Sweden, Norway, Denmark, Finland and Iceland, showing roads, railways and canals very clearly; nine town plans are a useful feature. A map of Denmark and southern Scandinavia, at sixteen miles to the inch, also emphasises roads and railways.

' The northern countries ' is the title of a general map of Denmark, Iceland, Norway and Sweden, published in 1952 by Fritzes Hovbokhandel, for the foreign ministries of these countries. ' Northern Scandinavia ', a map published by Cappelens, covers the area from the Gulf of Bothnia to the North Cape, at 1 : 1M. It is a fully coloured map, using hill-shading to indicate relief and has printed, in English on the back, information likely to be of use to tourists. The Russians have carried out a great deal of surveying and scholarly work along the shores of the Baltic. One useful work has been the *Atlas of the ice of the Baltic Sea and adjoining regions,* which consists of sixty three pages of maps, diagrams and statistical tables, published in 1960 by Gidrometeorologicheskoye Izdatel'stvo, Leningrad.

Thematic maps of the countries have also been produced by a number of specialist agencies; for example, 'A map of the northern wood industries', at 1:1,500,000, was published in 1965 by AB Svensk Trävaru-Tidning. Information about them is to be found in such journals as *Scandinavica: an international journal of Scandinavian studies; Scandinavian agricultural research,* journal of the Scandinavian Society of Agriculturists; *Scandinavian forest literature,* compiled by the Scandinavian Forest Union's Secretariat, etc. Reference should also be made to such fundamental works as G H T Kimble and Dorothy Good, editors of *Geography of the Northlands* (American Geographical Society, special publication no 32; New York, Wiley, and Chapman and Hall, 1955) which is well illustrated by maps; W R Mead: *An economic geography of the Scandinavian states and Finland* (ULP, second edition 1964), covering Denmark, Norway, Sweden and Finland; and A C O'Dell: *The Scandinavian world* (Longmans, 1957).

Norway: The Geographical Survey of Norway, Norges Geografiske Oppmåling, at Oslo, consists of a Geodetic Division, a Topographic and Cartographic Division, a Map Reproduction and Printing Division and a Cadastral Division; this organisation is also the national committee for cartography, representing Norway at the International Cartographic Association. The chief map series comprises 1:25,000, 1:50,000, 1:100,000, 1:200,000, 1:250,000 and 1:400,000. The sheet lines of the basic series, the 1:50,000, which has covered the country on a revised survey carried out between 1955 and 1965, are based on international geographic co-ordinate lines. Two colours only are used, blue for drainage and black for all other features. Important cities and some other areas of particular significance have been mapped at 1:25,000 as well as at 1:50,000. The 1:100,000 series, 'Rektangelkart and Grateigskart', published between 1940 and 1955, covers most of Norway, but it has been superseded by the new 1:50,000 series. The sheet lines of the 'Rektangelkart' are based on a kilometric grid. Relief is shown by contour lines and/or hachures, the contour interval being thirty metres, and three colours have been used, blue for open water, an olive shade for relief and black for all other features, including relief detail and drainage. The sheet lines of the 'Grateigskart' are based on the Oslo Meridian and equatorial parallel lines. Relief is shown by contours only, at thirty metre intervals. In this series, four colours have been used, blue for drainage, brown for relief, red in-fill for major roads and black for cultural features and for symbols indicating types of vegetation.

The 1:200,000 is the 'Fylkeskart' series, covering mainly the central and southern regions of the country. Each fylke may be covered by from one to four sheets, depending on the size of the area concerned, and the sheets may be coloured or monochrome. On the monochrome sheets, relief is shown by hachuring, while on the multicoloured sheets, brown shading is used, with blue for open water, green for woodland and black for all other features, including drainage. The 1:250,000 series, the 'Landgeneralkart', is also based on the Oslo Meridian and the equatorial parallel lines. Relief is shown by contour lines in brown at fifty metre intervals and some sheets have also relief shading in brown; blue is used for drainage, red for roads and black for other features.

Eighteen sheets of the 1:400,000 series cover the country; this series was made from one of the early surveys, carried out between 1863 and 1910. 'Norges Kommunev', 1:1M was prepared in two sheets, north and south, by the Norges Geografiske Oppmåling, in 1965, being mainly an administrative map, and another 'Norge' 1:1M was published by J W Cappelen in 1966.

Norway in maps, by Tore Sund and Axel Sømme, provides a thorough introduction to Norwegian topographical maps (Bergen, A S John Greig's Boktrykkeri, 1947). Nineteen maps at 1:50,000 in a separate folder, also maps and sketch-maps in the text, cover all types of Norwegian landscape, industrial regions and special subjects, of which farming, fishing and forestry are naturally of first importance. Tables of conventional signs are included, and the English equivalents of the geographical terms used are valuable.

The publications of the Geological Survey, Oslo, are in a single, numbered series, of which the serial, *Norsk geologisk tidsskrift,* 1905-, is a central source of geological information for the country. 'Norges geologiske undersokelse', 1:250,000, was published in four sheets between 1949 and 1952; in addition, a geological map of Norway forms part of a geological map of Scandinavia, 'Geologisk översiktskarta över Norden', 1:1M, published in 1933 by the Generalstabens Litografiska Anstalt, Stockholm.

The superb atlas volume in *Norge,* published by J W Cappelen, Oslo, in 1963, in four volumes, was the result of close co-operation between the Norwegian Geographical Institute, the Army Map Service in Washington and the cartographic firm, Esselte, in Stockholm. The first volume of the work includes articles on climate, population and industry, etc together with a systematic description of Norway region by region. Volumes 2 and 3 form a topographically arranged encyclopedia of Norway, and the fourth volume 'Norway in maps', deals with early maps and the development of mapping, with a fifty seven page section covering contemporary Norway at a scale of 1:400,000. Excellent town plans are included, for Oslo,

Bergen, Trondheim, **Kristiansand** and Stavanger and environs, at 1 : 150,000. The whole work is beautifully produced, including some magnificent colour photographs, and is well documented, having an index-gazetteer of some sixty thousand entries, a list of surces and a glossary of Lapp terms used.

Another cartographic venture which has achieved international acclaim is the Cappelen ' Road and tourist map of Norway ', in five double sheets, at scales of 1 : 325,000 for south and central Norway on three sheets, and at 1 : 400,000 for the northern part on the other two sheets, as follows : southern Norway, to the Bergen railway line; central Norway to Alesdund-Røros; Møre and Trøndelag; Nordland, with Lofoten and Vesterisen; Troms and Finmark. Each sheet overlaps adjacent sheets and also parts of neighbouring countries are included to show the communications links with Sweden, Finland and the Soviet Union. The map has been based on the most recent data from the Norges Geografiske Oppmåling and from experts throughout the country. Six colours are used in printing. Road networks are classified and numbered in accordance with the new system brought into effect in June 1965. The key, printed on the reverse of the sheets, includes about sixty symbols indicating hotels, pensions, youth hostels, camping sites, notable scenic points, ferries, etc, in English, French and German. Steamer routes are shown on all except the most southern map sheet. Paper for the maps was specially manufactured to obtain a high folding quality; the sheets are available either flat or folded within a durable cover.

Cappelen also issue a map of southern Norway at 1 : 1M, on which relief is shown by hill-shading, with heights and depths given in metres. Much information is included of interest to tourists either on foot or in a car. Bartholomew is the British agent for these maps, also for a series of mountain maps of Norway, fully coloured, showing roads, ferries, foot-paths, mountain huts, hostels and camping sites; for the ' Greater Oslo plan ', showing central Oslo at 1 : 13,000, with the Oslo region on the reverse, at 1 : 30,000, with index and legend; for the *Road atlas of Norway,* which consists of twenty two pages of maps in six colours, at scales of 1 : 325,000, 1 : 1M and 1 : 4M and ten town plans at 1 : 35,000; and for the Cappelen 1 : 1M map of northern Scandinavia.

Geographia has issued a map of Norway in two sheets at 1 : 1M for the Royal Norwegian Auto-Club; the southern sheet includes an inset of Oslo fjord on a larger scale. *The Royal Norwegian Auto-Club atlas* comprises eighty pages of contoured maps. Separate maps are available for Oslo fjord at a scale of three miles to the inch, for eastern and western Norway, at eight miles to the inch, and for

179

southern Norway at four miles to the inch, all showing contours, roads, railways and other major features.

In 1950, Axel Sømme compiled an *Atlas of Norwegian agriculture;* and a settlement map in fifteen sheets, based on the 1950 Census, was published by the Bureau of Statistics. Original maps are to be found in the quarterly *Norsk geografisk tidsskrift,* which includes articles in English or with English summaries, and in such scholarly publications as *Norwegian cirque glaciers,* edited by W V Lewis (the Royal Geographical Society and John Murray, 1960 RGS research paper no 4) and V H Malmström: *Norden: crossroads of destiny and progress* (Van Nostrand, 1965). A guide-book to Norway, prepared for the International Geographical Congress in 1960, was issued as a special number of *Norsk geografisk tidsskrift,* 17 1959-60.

Maps are included in the national bibliography of Norway, *Norsk bokfortegnelse,* which has a long history, but since 1952 has been compiled by the Norwegian Department of the university library at Oslo, Norge Geografiske Oppmåling has issued annual reports on the work of the Survey since 1885; it carries English abstracts. A *Catalogue* of the official map series may be obtained from the Geodetic Institute, Oslo.

Norway is responsible for the mapping of Spitzbergen. The latest map has been completed in four sheets at 1 : 500,000, 1964-, by the Norsk Polarinstitutt, Oslo; contours are shown at 100 metre intervals, with indications of spot heights and glacial information. The Royal Geographical Society issued a ' Map of Spitzbergen ' in 1962.

Sweden: In Sweden there are three great cartographic bodies, the Swedish General Survey, Rikets allmänna Kartverk, in Stockholm, which is responsible for the national maps of Sweden and consists of an Economic Map Division, a Topographic Map Division, a Photogrammetric Division, a Geodetic Division and a General Division; the Esselte Map Service, Stockholm, directed by Dr Lundqvist; and at the same address, under the same direction, AB Kartografiska Institut and Generalstabens Litografiska Anstalt. The Swedish Cartographic Association, Kartografiska Sällskapet, Stockholm, forms the national committee representing Sweden at the International Cartographic Association.

The chief national map series are the Photo map of Sweden, 1 : 10,000, the Economic—or General land use—map of Sweden, at 1 : 10,000 or 1 : 20,000, the Military map of Sweden, at 1 : 50,000; the basic topographic scale is 1 : 50,000, and there are series also at 1 : 100,000, 1 : 200,000, 1 : 400,000, 1 : 800,000 and 1 : 1M. In addition, a new general map at 1 : 250,000 has been compiled by photographic methods. The Photo map was composed of mosaics prepared from rectified aerial photographs and has provided the base map for the Economic map and the Military map. The sheet lines of the Economic map, begun in 1937, follow the kilometric grid lines and the sheets use three colours, green for the picture of the land obtained by photogrammetry, with yellow substituted for surveyed arable or horticultural areas, and black for all other features. The only parts of the country not to be covered by this map are those of sparse population density and difficult natural conditions in the centre and north-west. The map is constructed on the Gauss Conformal Projection. In all, about twelve thousand sheets will be needed to complete the series, which, it is estimated, will be achieved by 1970. The chief purpose of the map is to provide accurate cartographic data for planning in agriculture and forestry. For reference, Sweden is divided into sections fifty kilometres in width, numbered from one to thirty two from south to north and lettered A to N from west to east. One grid square, therefore, represents one sheet of the 1:100,000 topographic map or four of the 1:50,000 map sheets; the latter are in turn distinguished by the geographical notations, NW, NE, SW and SE. Further sub-division is added as necessary and all sheets are headed by the local name. The final revision policy is still under discussion; some sheets, for example, those covering the Stockholm area, have already been revised.

The Military map shows relief by contours at five metre intervals.

Compilation of a new 'Topographic map' at 1:50,000 began in 1954, designed to replace the 1:50,000 draft map, which served to provide provisional information pending more accurate mapping at this scale, and the older topographic series at 1:100,000 and 1:200,000. On this map also contour intervals are at five metre intervals. Five colours are used, blue for water features, browns for relief, green for woodland, grey for built-up areas and primary roads and black for all cultural features. The topographic map at 1:100,000 covers the south of Sweden to latitude 61 degrees 30' north, having the sheet lines based on a kilometric grid system. North of this parallel, the topographic map at 1:200,000 is based on geographical co-ordinate lines. Both maps were at first printed in two colours, blue for water features and black for hachured relief and all other features. The later sheets have used four colours. Experiments have been made with relief shading; forests are shown by green stippling, sea and lake surfaces by blue stippling. Hydrographic details are included, so far as the scale allows, and greater detail is being added, in harbours, for example, and by showing sea lanes, lighthouses and other navigational aids. Contour lines are at fifteen metre intervals above the tree line. Many of the northern sheets show also arable and meadow land, the boundaries of villages, Crown lands and forests. In addition, a number of sheets at 1:100,000 have been compiled for the northern area, based on the sheet line system of the 1:200,000 series. The more recent sheets include five metre contours in brown, with every fifth contour indicated more boldly. Spot heights are also given. Types of buildings are distinguished and they are drawn to scale where possible.

Twenty five sheets comprise the 'General map of Sweden', 1:400,000. Relief above the tree line is shown by contours at fifty metre intervals and by shading below that line. Blue is used to show drainage, brown for the relief shading and for the in-fill of main roads, brown hatched in black for the built-up areas, green for woodland, grey for sandy areas, orange screen to show boundaries and black for all other features. Seventeen sheets make up the 1:500,000 'Hypsometric map' series. Contour lines are at thirty three, sixty six and 100 metres up to five hundred metres, thereafter at 100 metre intervals. Hypsometric tints are used on all the sheets, some including as many as twenty three different colour tones. The whole country is covered by three sheets each of the 1:800,000 and 1:1M general maps. Five colours are used on these, blue for water, brown for relief, green for woodland, orange screen to mark boundaries and black for all other features. The current intention is to issue general maps only at 1:250,000 and 1:1M. The map at 1:1M, first published in 1932, was redrawn in 1958; revision is planned at five year intervals.

A new series at 1:250,000 was begun in 1961 by the Generalstabens Litografiska Anstalt, planned to cover the country in forty eight sheets. These sheets also are plotted on the Gauss Conformal Projection. Relief is shown by hill-shading. A useful reference is 'The official maps of Sweden: the modern series' in *World cartography,* VII 1962. A system has been evolved in Sweden by which complete air survey is carried out within every seven years; this source of information is used by all professions, as well as by the mapping agencies.

Swedish cartographers have given much attention to thematic mapping. Professor T Hägerstrand at the University of Lund is actively engaged in the modern methods of geographic data information systems, particularly with regard to land use planning. Sten de Geer produced the first effective maps of population distribution in 1919, at the scale 1:500,000. Twelve maps, based on the 1917 Census, were bound into atlas form. For the first time the dot method was used, combined with a brown layer colouring, which brought out the relative intensity of settlement. One dot represents one hundred people, groups of dots represent communities of between two hundred and five thousand and a globe symbol denotes larger towns. In the case of small towns, the dots are arranged in ordered rows, to indicate the built-up areas, while the village communities are shown by irregularly placed dots. White patches clearly show uninhabited areas. Factors having an influence on the distribution of population are indicated, such as the upper limit of cultivation, hydrographic features, industries, transport, shipping routes and fishing grounds. Since de Geer's original mapping, four revised maps have been prepared, at scales ranging from 1 : 500,000 to 1 : 2M. Since 1954, G Lundqvist has used proportional squares instead of spheres for symbols of urban population. The co-ordinate system map of population, 1960, is linked to this series of population maps because it gives absolute numerical statements for unit areas of one square kilometre. In collecting the population data for each unit area, census registers, with marginal markings, are used, also the official co-ordinate system, or grid, devised by the Topographical Survey Office in 1938, and various maps of reference.

The Esselte Map Service is currently working on a new large-scale population map of Sweden, at 1 : 100,000, which will amount to about a hundred sheets. These maps will serve as a base for future maps and for use in planning. Meanwhile, in 1963, Esselte published a map, prepared by the Central Bureau of Statistics, at 1 : 1M, showing the changes in population between 1951 and 1960 within the communes and parishes and the extension of the localities in 1960. On the same scale, published in 1964, was a map showing the distribution of population in Sweden in 1960, using the symbols dots and squares; international and administrative boundaries are also shown. ' Some trends in Swedish map reproduction ', compiled by C M Mannerfelt and Gösta Lundqvist (Generalstabens Litografiska Anstalt Esselte AB, Stockholm, 1952), is a very useful booklet.

Sveriges Geologiska Undersökning, Stockholm, issues four categories of publications. Series ' Aa ' refers to the geological map sheets at 1 : 50,000, with the accompanying descriptive bulletins, which began in 1862; series ' Ba ' denotes miscellaneous maps on various scales, also with bulletins, from 1865; series ' c ' includes reports and special bulletins, dating from 1868; and from 1900, monographs have been published from time to time, which are known as series ' ca '. Geological maps are also prepared by other geological organisations, notably by the Geological Society of Stockholm and the Geological Institution of the University of Uppsala.

The *Atlas över Sverige* was first prepared in 1900. In 1953, a new edition began to appear in looseleaf form in a leather covered case, published by Svenska Sällskapet för Anthropologi och Geografi, Stockholm (Kartografiska Institutet; Generalstabens Litografiska Anstalt). This is a magnificent atlas, giving a clear picture of the country's natural and cultural geography, population, economy, trade and port traffic, also including some historical sheets, such as the sheet of ' Mediaeval Sweden ', with the index of medieval geographical names. The fact that all the work of editing, drawing and reproduction has been done within one organisation has enabled the planning and execution of the atlas to be particularly successful. The complete atlas will consist of about 150 pages of maps and about three hundred pages of text in Swedish, with English summaries. Map titles and legends are also given in Swedish and English. A number of sheets are revised each year. For the atlas, Magnus Lundqvist prepared the maps of the forests of Sweden, making use of the records of the National Forest Survey investigation, also of figures collected from individual farmers on the use of arable land. This data was transferred to punched cards and figures for each category of information required were calculated, using IBM machines. The figures were plotted on the manuscript maps and coloured by hand. The cartography of the whole atlas is of a very high order, particularly in the use of colour; in recent sections, the Esselte strip mask method has been used to produce colour tones.

Two special atlases of Sweden are the 'Agricultural atlas of Sweden ', first produced by O Jonasson and others in 1939, with a second edition in 1952, published by the Landsbrukssällskapets Tidskriftsahtieblog, Stockholm; and *Sverige Nu : Atlas över folk land och näring,* edited by Hans W Ahlmann and others (Stockholm, A V Carlsons Bokförlags, 1949). The latter covers the major aspects of the geography of Sweden, each map with explanatory text in Swedish only. It is intended as a study reference atlas and for the general public. There are three groups of maps : dealing first with natural features; district maps covering the whole country at a basic scale of 1 : 1M, with relief shown by spot heights; and a series of thematic maps dealing with population distribution, agriculture, fisheries, industries, trade and communications.

For tourists and visitors, the *Royal Auto-Club atlas of Sweden* was issued in 1961 by Geographia. The basic scale is eight miles to the inch, with larger scale maps of the Stockholm and Malmö regions. Special emphasis is on communications and the major features of interest. Geographia also issue a map of Sweden at 1 : 2M, and another in two sheets at the scale of sixteen miles to the inch, which includes insets of Stockholm and Malmö at nine miles to the inch. A new tourist map of Sweden in ten sheets, the *Svenska turistkarten,* at 1 : 300,000, was published in 1967 by Generalstabens Litografiska Anstalt, who also maintain a ' Volkswagen servicekarta över Sverige ' at 1 : 1,750,000. Guide-books to Sweden were published on the occasion of the International Geographical Congress at Stockholm, 1960. Original maps are to be found also in the files of scholarly geographical periodicals, the *Geografiska annaler,* the two series of *Lund studies in geography* and *Ymer,* all of which publish quite a proportion of articles and summaries in English.

Maps are included in the national current bibliography, *Svensk bokforteckning,* taken over by the Bibliographical Institute in 1953, previously issued in various forms and titles since 1861. Maps are also listed in the quarterly journal *Globen,* Stockholm. A useful catalogue of maps, with text in Swedish and English, was issued in 1951 by the Rikets Allmänne Kartverk, Stockholm, and a catalogue of official maps published in Sweden may be obtained from the Swedish Survey Office. The publications of scholarly organisations, both containing original maps and giving information on new surveys and research, include those of the Geographical Association of Stockholm, the South-Swedish Geographical Society, Lund, the Geographical Society of Lund, the Geography Departments of Stockholm, Lund and Uppsala and the Swedish Society of Anthropology and Geography, Stockholm.

Finland : Turning to Finland, the *Atlas of Finland,* published by the Geographical Society of Finland in 1899, warrants first mention. Not forgetting the Bartholomew *Atlas of Scotland* already mentioned, Finland was the first country to publish a national atlas in the contemporary sense; it was not only a fine atlas in itself, but it inspired a new trend in atlas production. Issued ' to assist the people of Finland to know themselves and their country ', the atlas throughout gives an indefinable impression of enthusiasm and encouraged the twofold development which rapidly followed : the publication of detailed maps, officially sponsored, to make known the geographical, economic and social aspects of a country and, secondly, to prepare large-scale maps which should also be of service in planning resources or developments of any kind. A second edition of the atlas appeared in 1911 and the third, 1925-28, was issued in two volumes, thirty eight plates of maps and a volume of text. The fourth edition, compiled by the Finnish Geographical Society and the Institute of Geography of the University of Helsinki, under the direction of Leo Aario, was completed in 1960. The cartographic work and reproduction of plates, carried out by the General Survey Office, is of a uniformly high standard. This edition contains 445 maps at scales ranging from 1:1M to 1:9M, many of them on double-page sheets. Comparison between different geographical factors is facilitated, as related maps are grouped together. All aspects of the nature and economy of the country are covered, and, from the time of the appearance of the first edition, the atlas has been of special interest as a survey of a northern country in such authoritative detail. The meteorological section, for example, includes frost data and the sections on forests and water power are of great practical value. Population density and analysis are particularly well treated, as are agriculture and industry. The second volume, the text, entitled *Explanatory notes,* was published in English in 1962; legends and marginal information on the map sheets are in Finnish, Swedish and English.

The national committee for cartography is the National Cartographical Society of Finland, Suomen Kartografinen Seura ry, Helsinki. Map-making in Finland has a long tradition, but a new period of development began with the foundation of the Geodetic Institute in 1918, on the initiative of the Geographical Society, as soon as the country achieved independence. Primary triangulation, on which all maps were based, was completed and joins up with the triangulation of Estonia across the Gulf of Finland and with that of Sweden over the Gulf of Bothnia. *World cartography,* VII 1962, should be consulted; the first-order triangulation net, 1959, is shown, together with other index maps of gravity measurements and levelling and of the coverage by various kinds of maps.

Cartographic coverage of the country is the responsibility of the National Board of Survey of Finland, Maanmittaushallitus, Helsinki, which includes three divisions, the Geodetic Bureau, with a Division of Aerial Photography attached, the Cartographical Bureau and the Topographic Bureau. The main series are at 1:20,000, 1:100,000, 1:200,000 and 1:400,000. Four 'old' series include three in Russian at 1:21,000, 1:42,000 and 1:84,000 and one in Finnish at 1:42,000. The sheet lines of these four series are based on the international geographical co-ordinate lines. The Russian series cover the southern areas of the country; sixty nine sheets at 1:21,000 primarily for the immediate surroundings of Helsinki, 155 sheets at 1:42,000 for most southern Finland and twenty seven sheets at 1:84,000 for the south-eastern part of the country. The Finnish series comprises 132 sheets covering southern Finland. There are also now two series of economic maps at 1:100,000, road maps at 1:400,000 and large-scale topographic coverage of the larger cities and towns.

The 1:20,000 map is the base map for the country. Each sheet usually covers an area of ten square kilometres and compilation is from controlled photomosaics at a scale of 1 : 10,000. The sheets are of two types : those issued before 1951 are monochromes, while those published since that date are printed in five colours, blue for open water, brown for the contour lines at five metre intervals, red for roads, orange for cultivated areas and black for all other features. By reduction from them, economic maps are prepared at 1 : 100,000 and a general map at 1 : 400,000, except in areas of sparse population density, in which aerial photography has been directly used. The contour interval on the 1 : 100,000 sheet is ten metres. Each sheet covers an area of about 1,200 square kilometres and five colours are used, blue for drainage, brown for relief, green for woodlands, red for roads and black for cultural features. Black hatching indicates built-up areas. The 1 : 200,000 sheets each cover about 4,800 square kilometres; about 100 cover the country. Twenty three sheets of the 1 : 400,000 cover the country, each sheet including about 19,200 square kilometres.

Especially in northern Finland, the land has risen, so that in 1935 the Geodetic Institute began the second levelling of Finland, which revised the 1892-1910 first levelling and also completed new levelling in the far north, not previously covered. The whole operation was completed by 1955. More than ninety per cent of the country has been aerially photographed, usually at a scale of 1 : 10,000, by the Division of Aerial Photography in conjunction with the Defence Forces Topographic Service. Besides the economic maps at 1 : 100,000, there are two series at 1 : 1M, which overlap in places. One series is on the Gauss-Krüger Projection, in fifty four sheets covering most of the central and eastern parts of Finland; the other is on the polyhedric projection, again comprising fifty four sheets, covering the southern and north-eastern areas.

The first road map to cover the country was issued in 1932 in twelve sheets at 1 : 400,000. Many sheets overlapped; contours were drawn at intervals of twenty metres and five colours were used, blue for drainage, brown for relief, green for woodlands, red for roads and black for all other features. Several revised editions have appeared since then. In addition, a general map of the main roads has been available in successive editions at 1 : 1,500,000, which includes maps of all the cities. A new series of road maps, ' Suomen Tiekarta ', at 1 : 200,000, began in 1965, published by the National Board of Survey. The whole country is covered in fourteen sheets, clearly printed, showing four classes of roads and motorways, with distances between major towns quoted. Inset on the map faces are town plans and information of interest to tourists is given in Finnish and English. A road map at 1 : 500,000, showing sixty two city and town plans on the reverse, was issued in 1966.

187

The Geological Research Institute publishes rock maps and soil maps at 1:100,000. The maps prepared by the Geological Survey are at 1:1M, and each is issued with an explanatory bulletin. The latest geological map of Finland forms part of the geological map of Scandinavia, 1:1M, by the Generalstabens Litografiska Anstalt, Stockholm. Geological information is to be found also in the publications of the Institute of Geology, University of Helsinki, and of the Finnish Geological Society. The Board of Forestry has compiled maps of the state owned forests, based on aerial photographs. A coastal chart series at 1:50,000 is issued by the Board of Navigation; special charts have been made along the main shipping lines and a map series of inland waterways is at 1:40,000.

Finnish cartography has, during recent years, become noted throughout the world. In addition to the official activities already mentioned, the mapping and surveying firm Kunnallisteniikka Oy has become known as FINNMAP. The main part of its work consists of aerial photography and mapping at large scales, 1:500, for example, for special purposes. The firm includes a traffic engineering department which carries out traffic investigations and surveys for traffic, highway and road planning. Projects have been carried out also in many parts of the world; base maps for planning purposes have been prepared in Iran and Nigeria, and, for geological work, in Spitzbergen. In most cases, aerial photography is used, with, if possible, ground control. Stereoplotting in preparation for engineering works is a speciality of the firm, demonstrated in England, Ghana, Iran, Malaya, Sweden, Germany and Denmark.

A fine cartographic achievement was the *Atlas of the archipelago of south-western Finland,* 1960, edited, with the accompanying text, by Sten Segerstråle, and prepared under the auspices of the Nordenskiöld Foundation. Printing was done by the Land Survey Office. In two parts, the first comprises a number of systematic studies, the second the presentation of twelve typical skerry districts, the whole amounting to twenty eight map plates and 179 pages of text, to which are added superb photographs, diagrams and a bibliography.

In the Bartholomew ' World reference map ' series, Finland is included on the sheet for Scandinavia. Geographia issues a physical-political map of Finland at a scale of twenty miles to the inch, showing roads and railways, also a road map at thirty two miles to the inch, stating distances and with inset maps of Helsinki. George Philip has issued a 1:2,500,000 map for BP since 1965. ' Finland 1:1M ' was issued by the Survey Office in 1960, in one sheet, which carries also a gazetteer of towns. Large-scale maps of Helsinki became available, for example, at 1:30,000 by the Real Estate

Office, Survey Department, in 1963, with a map of central Hensinki at 1 : 15,000 on the back, and an index of streets. A guide-book to Finland was issued for the International Geographical Congress at Stockholm, in 1960; and *Fennia*, the journal of the Finnish Geographical Society, includes original maps and sketch-maps, besides giving information on new cartographical projects. Most of the articles have English summaries.

Maps issued by the Central Board of Land Survey are noted in the *Maanmittaushallituksen karttaluettelo*. Reference should be made also to the ' Report on cartographic activities in Finland ', prepared by M Kajamaa for the International Cartographic Association Technical Symposium at Edinburgh in 1964. A *Catalogue* of official maps of Finland may be obtained from the Finnish Survey Office.

Iceland : Iceland is frequently included in polar studies, in general cartographic works on the Scandinavian area and, occasionally, together with the Faeroes. The Geodaetisk Institut at Copenhagen maintains a topographical map of Iceland at 1 : 500,000. The latest new map of the country was published in 1966 by the Tourist Association of Iceland, at 1 : 750,000. The Bartholomew ' world reference ' series includes a map of Iceland at 1 : 750,000, layer coloured in six tones of green and brown, with contours drawn at 100, 200, 500, 1,000 and 1,500 metres and spot heights at greater altitudes. Symbols marking glaciers, hot springs and freshwater lakes are effective. Official Icelandic spelling of placenames has been adopted. There is also a BP road map at 1 : 600,000.

The cartography of Iceland, by Halldor Hermannsson (Cornell UP, 1931) concerns mainly early maps, but good map coverage to its date was included in N E Nørland : *Islands kortlaegning* (Copenhagen, 1944). Research articles on Iceland, including original maps or sketch-maps, are frequently to be found in the Swedish geographical journal *Geografiche annaler;* volume 41 formed the ' Guidebook to Iceland ' prepared for the International Geographical Congress held in Stockholm, 1960. All publications issued in Iceland since 1944 have been noted in the *Årbok,* which also includes publications in foreign languages relating to Iceland.

Denmark : In Denmark, there are two official mapping agencies, the Royal Danish Institute of Geodesy, Kongelik Dansk Geodatisk Institut, which is the national committee of cartography, and the Directorate of Land Registry, Matrikeldirektoret, both in Copenhagen.

The national series of maps of Denmark are published by the

institute at the following basic scales: 1 : 10,000, 1 : 20,000, 1 : 40,000, 1 : 100,000, 1 : 150,000, 1 : 200,000, 1 : 300,000, 1 : 750,000 and 1 : 1M. The geographical co-ordinates indicated on these maps refer to the Copenhagen Meridian and international parallels; at the same time, the sheet lines follow generally a kilometric grid. In the three larger scale series, relief is shown by contour lines, while the other maps mark only spot heights. The series at 1 : 10,000 is the ' Kvartblade ', having contour intervals at two metres. Three colours are used, blue for drainage, brown for the contour lines and black for all other features. The series 1 : 20,000, ' Malebordsblade ', each sheet of which covers about seventy one square kilometres, follows the same specifications as the 1 : 10,000. The 1 : 40,000, 'Atlasblade', has contours marked at two metres and uses five colours, blue for drainage, brown for relief, solid brown for primary roads and broken brown for secondary roads, green for woodlands, grey hatching for built-up areas and black for other features. Each sheet covers about 284 square kilometres. The 1 : 100,000 is the ' General-stabkort '; each sheet covers about 1,350 square kilometres and four colours are used, blue for drainage, brown in-fill for primary roads, green for woodlands and black for other features, including indica-tions of vegetation. The 1 : 150,000, ' Nyt faerdselkort ', is primarily a road map. The whole series comprises ten sheets, each covering an area of between nine and eleven thousand square kilometres, with some overlapping. Six colours are used, the differences being in the distinction between primary and secondary roads, in red and brown respectively. An orange overprint shows administrative boundaries. The 1 : 200,000 series has the same colouring as the 1 : 150,000; twelve sheets, with overlapping of adjacent sheets, complete the series. Three sheets make up the series at 1 : 300,000, the 'Automobil-kort '; colouring follows a similar pattern, except that a red tint is used to indicate large city areas. The three small-scale series are all single sheet maps. The 1 : 500,000 is similar to the 1 : 300,000, with the addition of national route numbers. The 1 : 750,000 is also similar, but only the major cities and towns are shown. The 1 : 1M map is in the style of the International Map of the World 1 : 1M, except that the sheet lines have been changed to include the whole country.

In 1947, the institute began a series of 1 : 250,000 ' Höjdekort över Danmark ', in which relief is shown by layer colouring in green and brown, with contours at ten, forty, eighty, 120 and 150 metres and depths of coastal waters at six, ten, and thirty metres. Towns are printed in red, indicating their extent and shape, size of population being indicated by the size and style of lettering. Four types of smaller towns are distinguished. Symbols for deciduous and coniferous forests, heather moor, meadow and marsh are overprinted; other information includes railways and main roads, airports, social features and interesting historic monuments. An inset map of Copenhagen, 1 : 100,000, shows the growth of the city. Greenland and the Faeroes are included, the former at 1 : 4,500,000, showing spot heights in metres, and ice conditions, the latter at 1 : 250,000, with relief shown by layer colouring at contour intervals of two hundred metres and settlements differentiated by size. In addition, a new series at 1 : 25,000 has been published by the institute during the past few years.

The Danmarks Geologishe Undersøgelse, Copenhagen, has issued geological maps at 1 : 100,000, with descriptive texts, since 1893. Maps are also frequently included in published reports and papers. The latest general geological map is included on the 1 : 1M map of Scandinavia published by Generalstabens Litografiska Anstalt, Stockholm.

The excellent *Atlas of Denmark,* begun in 1949, is edited by Niels Nielsen and published in Copenhagen by Hagerup. It is an example of geographical co-operation at its best, between geographers associated with the University Geographical Laboratory, the Royal Danish Geographical Society, the Carlsberg Foundation, the Danish government and a number of specialists and firms. The major aspects of the physical geography of Denmark are covered, with emphasis on glacial land forms, and block diagrams are effectively used in conjunction with the maps. A real attempt has been made to show how the underlying rock formations govern the landscape. The country has been divided into ten regions. Another main section of the atlas is given to aspects of population. Aage Aagesen's *Population maps* on a basic scale of 1 : 200,000, form the second volume of the atlas, published in 1961. Population distribution is shown in detail, one dot representing twenty five inhabitants in rural areas and proportional circles for settlements of more than five hundred persons. A map of Copenhagen is included at 1 : 100,000, also maps of the Faeroes and Greenland on smaller scales. Text, in Danish or English, provides background information regarding historical evolution, occupational changes, etc, together with comments on the ten regions. Air photographs add to the beauty of this fine atlas.

An atlas for schools, *Danmark-atlas med Faerøerne of Grønland,* edited by Johannes Humlum and Knud Nygård, was issued in 1961 by Gyldendal of Copenhagen. Twenty five very fine map plates and diagrams are complemented by fourteen pages of text. A motoring map of Denmark, South Norway and South Sweden, at 1 : 1M is published by Hallwag of Berne, and BP issued a road map of Den-

191

mark at 1:520,000, in 1966, with street plans of major towns on the reverse. Many guide series include Denmark, and Geographia issue a road map of Denmark at six inches to the mile, with legend in English. The Geodaetisk Institut maintains ' Kobenhavn— turiskort' at 1:15,000, with an index of streets. A guide of Denmark was prepared for the International Geographical Congress of 1960, held at Stockholm, which was issued as volume 59 1960, of *Geografisk tidsskrift,* journal of the Royal Danish Geographical Society, in which most contributions are in English or have English summaries. There are usually English summaries also in the research series *Folia geographica Dannica.* Original maps are to be found in the pages of *Geografisk tidsskrift,* 1877-, in *Kulturgeografi,* 1948- and in *Danmark,* a remarkable historical gazetteer by J P Trap, 1858-62, in a fifth edition by Niels Nielsen and others, 1953-, in which maps are included for every county, parish and city.

There has been considerable bibliographical interest in Denmark in early cartography and local maps, but the current output is inadequately listed. A *List* of official maps is available from the Institute of Geodesy.

THE NETHERLANDS & BELGIUM
The Netherlands: having taken a leading part in map-making in the sixteenth and seventeenth centuries, has of late years emerged again as one of the great cartographic nations of the world. The national committee representing the Netherlands at the International Cartographic Association is the Kartografische Sectie van het Koninklijk Nederland Aardrijkskundig Genootschap.

The national maps, published by the Topografische Dienst (Topographic Service), Delft, are in four main series, two at 1:25,000 and the others at 1:50,000 and 1:200,000. The sheet lines of all the maps are based on a kilometric grid system and the Greenwich Meridian and the international latitudes are shown at the margins. The original editions of the ' Chromo-topografische kaart des rijks' at 1:50,000 were drawn on the Bonne Projection, but later revisions used the stereographic projection. The second map at 1:25,000, the ' Fotogrammetrische kaart des rijke ', has always used the stereographic projection. The colour scheme in all the series varies little, blue is used for water features, brown for contours, green for woodlands over black vegetation symbols, pale green for meadow land, yellow for sandy areas and administrative boundaries, a pale pink shade for non-cultivable land, red for primary and secondary roads and black for all other features. Red and black are used for building symbols on the topographic sheets at 1:25,000 and tinted red is introduced to show built-up areas on the 1:50,000 sheets.

705 sheets of the topographical map at 1 : 25,000 cover the country; some outline base maps are also made available for those who need them in planning projects. The contour interval is 2½ metres. The photogrammetric map at 1 : 25,000 runs to 382 sheets. The 1 : 50,000 maps are available in 'full' sheets or 'half' sheets, some being published in both forms. Full sheets cover forty kilometres east to west by twenty five kilometres north to south; half sheets are twenty kilometres east to west by twenty five kilometres north to south. The contour interval is five metres. The 1 : 200,000 'Chromo-topografisches kaart des rijks' covers the Netherlands and adjoining areas in neighbouring countries in twenty three sheets, each of which is normally eighty kilometres east to west by forty kilometres north to south. A new series, 'Overzichskaart van Nederland', 1 : 250,000, has recently been completed in six sheets by the Topo-grafische Dienst, compiled from the 1 : 100,000 series on the Trans-verse Mercator Projection; this is a very clear, superbly printed map.

A Geological Survey was established in the Netherlands in 1903, operating under a number of titles until 1936, when the official title was confirmed as De Geologische Stichting, Haarlem. A geo-logical map of the Netherlands has been compiled at 1 : 600,000; the accompanying text is available in English. Other publications of the Survey and of the Geological Societies of Leiden and The Hague contain valuable maps, frequently local. 'Autokart Nederland', 1 : 250,000, was issued by NV Cartografisch Instituut Bootsma at The Hague, with twenty nine small town plans on the reverse. Various large-scale plans of Amsterdam have been published by the Depart-ment of Public Works, and an 'Amsterdam town map' at 1 : 15,000 appeared in 1966, prepared by the NV Cartografisch Instituut Bootsma, with a street index on the reverse.

The *Atlas van Nederland* is one of the most recent of the national atlases and a most exciting cartographic publication—a fine example of the high standard of the work of the Topographic Service at Delft, which is responsible for its preparation. In looseleaf form, in a very sturdy binder, it has been published by the Government Printing and Publishing Office from 1963. As long ago as 1932, a commission was appointed for the preparation of a national atlas, but it was not until 1951 that the Ministry of Education, the Royal Netherlands Geographical Society and the Society for Economic and Social Geography worked out detailed plans for it. An efficient organisation now exists, consisting of a board of trustees, an edi-torial committee and seventeen sub-commissions, one for each main section of the atlas. The atlas is expected to be complete, in 109 sheets, in 1973. The sheets so far issued represent all sections of the atlas, beginning with a sheet demonstrating the history of

7

cartographic techniques in the Netherlands by sixteen excerpts from maps printed from the sixteenth century to the present, compiled by Dr C Koeman. The basic scale used is 1 : 600,000, which enables the Netherlands to fit conveniently on double-page spreads, each mounted and folded. Contemporary social and economic conditions are mapped in detail, with additional inset maps and diagrams as required. Statistics and concise explanatory text are also added to some of the map faces and/or on the reverse, the whole being most carefully planned to make the greatest visual impact. The treatment of soils, as one would expect, is particularly well done, and the detail in which all the coastal parts are portrayed is cartographically very fine. The clear colours used throughout are striking, especially in the thematic maps and in the relief map, ' The Netherlands and surrounding countries ', at 1 : 500,000, in which eight layer colours are used, with contouring, to show land relief, and five colours to show sea depths and the tidal flats. Although so much information is included on the maps, the effect is not of overcrowding and much thought has been given to the selection of typefaces and size of type to achieve such clarity.

Three special atlases should be mentioned. The *Atlas van de Nederlandse beschaving*, by J J M Timmers (Elsevier, 1957), contains forty nine maps demonstrating the cultural history of the Netherlands. The other two reflect the agricultural interests of the country : *Landbouwatlas van Nederland*, thirty four plates with explanatory text, prepared by the Ministry of Agriculture, Fisheries and Food Supply of the Netherlands (NV Uitgevers-Maatschappij, W E Tjeenk Willink, in two volumes 1959) and the ' Dutch potato atlas ', comprising forty six looseleaf plates, compiled with descriptions, by J A H Esch and others (Wageningen, Veenman, 1955).

Belgium: The two agencies in Belgium providing topographical information are the Institut Géographique Militaire, Brussels, founded in 1831 as the Dépôt de la Guerre et de Topographie, which publishes maps at 1 : 20,000 and smaller scales; and the Service de Topographie et de Photogrammétrie, Brussels, which prepares plans at 1 : 10,000 and larger scales. The national committee representing Belgium at the International Cartographic Union is the Société Belge de Cartographie, Belgische Vereniging voor Kartografie.

By 1882, a topographical map of the country had been completed at 1 : 20,000; a reduction of this map at 1 : 40,000 and three other derived maps at 1 : 100,000, 1 : 200,000 and 1 : 320,000 were also available. In the 1920s, a new series of the map at 1 : 20,000 was begun, based on a new triangulation and executed with much improved techniques. The series was not finished, however, because, with changing conditions, especially after the second world war, the series was unsuited to modern requirements. In 1945, the Institut Géographique Militaire, then the Institut Cartographique Militaire, planned a new mapping programme, which included the preparation of a basic topographical map by aerial photogrammetry at 1 : 25,000, the 'Carte de Belgique', and four standard maps at 1 : 50,000, 1 : 100,000, 1 : 200,000 or 1 : 250,000 and 1 : 400,000, to be made by successive generalisations of the basic maps at 1 : 50,000. In fact, the 1 : 25,000 series replaced the older 1 : 20,000 series. It is drawn on the Belgian Lambert Projection and each sheet covers an area sixteen kilometres by ten kilometres. Contour intervals are at $2\frac{1}{2}$ metres and spot heights are marked. Six colours are used, blue for water features, brown for relief, green for woodlands, red for highways and paved roads, a grey tint for built-up areas and black for all other features. Marginal notes are in French, Flemish and English. The map should be complete in 241 sheets; a revision programme began in 1958, whereby the sheets are revised every six years.

The 1 : 50,000, based on the standard 1 : 25,000, has replaced the old 1 : 40,000 series. This series has only recently been completed and, in the meantime, a 'rush' edition was issued under the series title '1 : 50,000 type R', which consisted of a revised edition of the 1 : 40,000 maps, incorporating such new data as was possible from aerial photographs and some field checking. In the new maps, the Bonne Projection has been retained. Seventy four sheets complete the series, each covering an area of thirty two kilometres by twenty kilometres. The contour intervals are five, ten or twenty metres, depending upon the nature of the terrain. The edition issued for civil use is printed in five colours, blue for water, brown for relief, green for woodland, red for roads and black for other planimetric features. Marginal inscriptions are also in the three languages.

The new standard 1 : 100,000 series is based on the new standard 1 : 50,000 series, and this map should be complete by 1969. At present a 'rush' edition is available in twenty four sheets. The modern large scale plans at 1 : 10,000 and 1 : 5,000, designed primarily for use in town planning, building and engineering, are prepared by photogrammetric methods. Larger scale plans are also being made for certain localities. All the plans are multicoloured. A 'Belgique carte administrative', 1 : 500,000, published by the institute in 1964, is the latest administrative map to show province, arrondissement and commune boundaries.

Plans for the *Atlas de Belgique*, the national atlas of Belgium, were adopted by the Comité National de Géographie as early as

1937, but the work was necessarily delayed. The first of the loose-leaf sheets, at a standard of 1:500,000, were issued in 1954. The atlas covers all aspects of the geography of Belgium, incorporating the latest available data from government and other authorities; sketch-maps and diagrams add to the information included and commentaries in French and Flemish accompany each sheet. The maps of the *Atlas du Survey National,* begun by the Administration de l'Urbanisme, Ministère des Travaux Publique in 1964, supplement those of the national atlas. Several hundred of these maps, covering a wide range of subjects, are available for reference in the offices of the administration in Brussels. The maps are excellently produced and brief texts provide explanatory comment for each.

Belgium is included in the Bartholomew reference map series on the sheet 'France and the Low Countries, north', at 1:1M, revised in 1964. Shell began issuing a 'Benelux road map', 1:410,000, in 1960, prepared by the Mairs Geographischer Verlag, Stuttgart. A 'Carte géologique de la Belgique', 1:160,000, has been in production since 1920, by the Institut Géographique Militaire, also one at 1:1M first issued in 1932. Topographical, geological and other thematic maps are to be found in the publications of such organisations as the Société Royale Belge de Géographie, Brussels, the Société Royale Belge de Géographie d'Anvers, the Institut de Géographie, Brussels, the Institut de Géographie de l'Université, Ghent, the Institut de Géologie et de Géographie Physique, Université de Liège, and in the *Bulletin* and *Mémoires* of the Société Belge de Géologie, de Paléontologie et d'Hydrologie, Brussels, the *Annales* and the *Mémoires* of the Société Géologique de Belgique, Liège and in the *Revue de géologie,* 1920-, published by the Université de Liège.

Volume II 1951 of the *Mémoires et documents,* published by the Centre de Documentation Cartographique et Géographique, is devoted to *Belgique et Pays Bas*: *documentation bibliographique,* including maps. There is also the excellent *Bibliographique géographique de la Belgique,* compiled by M-E Dumont and L De Smet, published by the Brussels Commission Belge de Bibliographie, in Belgian and French, beginning in 1954. Four fascicules appeared between 1954 and 1956 in the *Bibliographie Belgica;* supplements have been issued at intervals as individual fascicules.

FRANCE & LUXEMBOURG

France: The national committee for cartography is Le Comité français de Cartographie, which was founded in 1957 in consultation with the chief cartographic organisations official and commercial. There are many organisations in France responsible for the acquisition of cartographical information and mapping programmes of various kinds. These include the Institut Géographique National; the Division des Travaux Topographiques; the Service du Cadastre and the Service Hydrographique de la Marine; the Bureau de Recherches Géographiques et Minière; le Centre de Documentation et de Recherches Cartographiques et Géographiques; the Centre d'Etudes Phytosociologiques et Ecologiques; the Direction Générale des Eaux et Forêts; the Direction de la Météorologie Nationale; the Institut de la Carte Internationale de Tapis Végétal; the Office de la Recherche Scientifique et Technique Outre Mer; the Service de l'Information Aéronautique. The information collected and recorded by all these bodies is invaluable and the cartography of published maps is of the highest standard.

The first map covering the whole of France was that of Cassini de Thury; and the first step in the compilation of a systematic series of maps of France was taken in 1802, when a Commission was set up to discuss conventional signs, metric scales and other practical issues. From 1814 on, a government commission was appointed to consider the project of a new topographic map of France. The resulting map, the ' Carte de l'Etat-Major ', 1 : 80,000, based on the triangulation by the survey engineers and on Bonne's Projection, gradually replaced the Cassini series. This map is in monochrome and does not show contours, but is heavily hachured. About 1900, a new series on the scale of 1 : 50,000 was initiated, but progress was slow. The first sheets to appear covered the important areas in the eastern part of France; until the mid 1950s, indeed, the only medium-scale map covering the whole country was the photographic enlargement of the early 1 : 80,000 to 1 : 50,000. From 1940, however, with the reorganisation of the department and the use of aerial photography and stereoplotting machines, progress has been more rapid. The function of the new institute was extended to include the Service du Nivellement Général and the various overseas survey services. The institute, therefore, provided geodetic, topographic and other cartographic materials, not only for Metropolitan France, but for the various overseas territories which looked to France for technical aid. Such concentration of effort had many advantages; a Groupe des Escadrille Photographiques was organised as part of the institute and also attached to the institute was the Ecole Nationale des Sciences Géographiques, which covers all the techniques of geodesy, topography, photogrammetry and cartography. The Photogrammetric Department of the Institut Géographique National is now one of the largest in the world.

All the national map series are now published by the IGN, the national agency responsible for the topographical maps at the scale of 1:10,000 or smaller. To understand the complex development of the various map series, in many ways comparable with that of Great Britain, it must be recalled that in the early nineteenth century, the Dépôt des Cartes et Plans was merged with the Dépôt de Fortifications and subsequently with the Dépôt de la Guerre, to which the survey engineers were attached. In 1887, the Dépôt de la Guerre was replaced by the Service Géographique de l'Armée, which, in 1911, was granted some independence, ceasing to be part of the General Staff, and thence in 1940 became reconstructed as the civilian organisation, Institut Géographique National, under the Ministry of Public Works.

The main task of the IGN in Metropolitan France is the production and maintenance of the 'Carte de France', in the 1:20,000 base map series and the 1:50,000 series. Both series follow the sheet lines system based on geographical co-ordinates shown in decimal grades. The former series is published in one-eighth sheets for areas south of the 54 grade line and in one-quarter sheets north of this line, while the latter is in full sheets. The sheets of the 1:20,000 series provide the basic map for all current cartographic work and also the essential cartographic data for economic projects. Cartographic standard is extremely high. The normal contour interval is five metres in reasonably level country and ten metres in the mountain regions. Four colours are usual, blue for water, brown for relief, green for woodland and vegetation cover and black for planimetric features and other symbols. A neutral tint was introduced after the series began, to distinguish important rocky massifs and also a screened blue was used for open water areas. The varied relief features of the country are shown in great detail, also roads, railways and waterways, mine locations and all kinds of cultural information. The sheets comprising the 'Région Parisienne' section also show built-up areas in red.

The monochrome map produced by the Service Géographique de l'Armée, the Plans Directeurs de la Guerre 1914-18, have been revised to correspond with the standards and specifications of the 'Carte de France', 1:20,000. The 'Carte de France', 1:50,000, is issued in five separate editions: 'Type 1922', 'Type spécial 1922', 'Type 1900', 'Edition oro-hydrographique', and 'Edition au relief'. 'Type 1922', sometimes called the 'Edition normale', carries full contours at intervals of five, ten or twenty metres, according to the nature of the country. The five colours used are blue for water features, brown for relief, green for woodlands and vegetation, grey for the rocky peaks of the mountain massif and black for all other planimetric features. 'Type spécial 1922' follows the same specifications as the above, with the addition of built-up areas indicated in red. ' Type 1900 ' has been systematically replaced by the ' Type 1922 ' series; eight or ten colours were used and the contour interval was ten metres. The ' Edition oro-hydrographique ' represents geographical and geological features; the ' Edition en relief ' is a three-dimensional series made of light cardboard material. Two other series, at 1.200,000 and 1 : 1M, are also issued as relief models.

The ' Carte de France au 10,000 ' maps are usually derived from the surveys or from the actual maps of the 1 : 20,000 series. The majority of sheets have been issued in monochrome and they provide a useful supplement to the 1 : 20,000 series in areas of particular density or social features or especially complex relief.

Three editions have been issued of the ' Carte de France et des frontières au 200,000 ': ' Type 1880 '; ' Type 1912 ' and ' Type 1912-42 '. The series was designed to replace an earlier monochrome map of France and is based on the 1:80,000 series. Eighty one sheets cover the whole of France, most of them being in the ' Type 1912-42 ' edition. The contour interval is forty metres and six colours are used, blue for water features and screened blue for open water; brown for relief, green for woodlands; red for roads and black for all other cultural features.

Twenty two sheets make up the ' Carte de France et des pays limitrophes au 500,000 '. The sheet lines are derived from the IMW 1 : 1M. The colouring system is very similar to the major series: blue for water features and screened blue for open water; a different relief representation, being in grey-green with added hachures; green for woodlands, red for the main roads and black for all other cultural features.

The recently completed ' Nouvelle carte de France ' in 292 sheets at the scale of 1 : 100,000, also published by the Institut Géographique National, is intended as an intermediate map between the large-scale series and the numerous road maps now available. Eight colours are used to show road details and all features of general interest. Mechanographic punched card equipment for processing the geodetic information has been in operation for some time. The geodetic card catalogue is arranged by geodetic points, each point being represented by two punched cards containing all relevant information—name, type, co-ordinates, height, system of projection, date of calculation, 1 : 50,000 sheet number, maps of the point on the sheet and the code of the survey on which the point was determined. For each sheet in the 1 : 50,000 series, a card is also prepared, giving details of each geodetic survey which determined points shown on the sheet. An alphabetical dictionary of all geodetic points, giving full details, is in continuous preparation on punched cards, corrections being made as they occur. An excellent new series at 1:250,000 has recently been published, based on the 1:100,000 map and printed in eight colours. Relief is shown by contours and hill-shading and the road network is exceptionally clear. For further details on these services, reference should be made to Robert Perret: 'Essai d'une carte structurale de la France', in *Mémoires et documents,* III 1952, of the Centre de Documentation Cartographique et Géographique and to the article 'L'atelier mécanographique de l'Institute Géographique national' in the *Report of the Second United Nations Regional Cartographic Conference for Asia and the Far East.* The file of the annual publication *Exposé des travaux de l'IGN* gives a complete record of the work of the institute.

The Service de la Carte Géologique de la France has issued the 'Carte géologique de la France', 1:1M, in four sheets. Geological details have been superimposed on the black outlines of the early topographical sheets at 1:500,000 and on the 1:50,000 sheets of the 'Carte de France' since 1945. The maps are periodically revised. *Mémoires* have been prepared in connection with the geological maps since 1893 and an informative *Bulletin* since 1889-90. A number of geological societies in France issue publications in which original maps frequently appear. A Douai groundwater map at 1:50,000 has been compiled by the Bureau de Recherches Géologiques et Minières, derived from the Douai hydrogeological maps of 1963. It is in the nature of a pilot project, for similar maps will be prepared for other areas, when the test of time may have suggested improvements.

The Service de la Carte de la Végétation de la France has prepared a vegetation map of France, 1:200,000, directed by P Rey, published by the Centre National de la Recherche Scientifique. The maps are accompanied by *Memoirs* and a set of 1:1M maps by which other natural features closely associated with vegetation may be related. Professor Henri Gaussen has prepared some of the maps in the series, using the vegetation classification system devised by him, which he demonstrated also in the vegetation maps of the *Atlas de France* (see below). The 'Carte forestière de la France', which is being issued in sheets at 1:100,000 by the Service de L'Inventaire Forestier National, has adopted the same sheet lines as the 1:100,000 topographical map. The forestry information is interpreted from aerial photographs and the map is printed in seven colours. L'Office National Météorologique succeeded the Bureau Central Météorologique as the principal centre for research and documentation in meteorology. La Société Météorologique de France is also a most scholarly body and its journal, *La météorologie,* frequently contains original maps and charts. The *Annales de géographie* and the many journals published by the geographical societies throughout France also include invaluable maps or sketch-maps illustrative of their texts. The maps in the *Cahiers de l'Institut d'Aménagemant et d'Urbanisme de la région Parisienne* also warrant special mention. Sketch-maps are also a particularly valuable feature of French academic monographs.

A map at 1:500,000, 'Zones de peuplement: industriel ou urbain', was published by the Institut National de la Statistique et des Etudes Economiques in 1962. In nine sheets, this map shows the spread of urban and industrial population as revealed by the 1954 population figures. The base map is printed in grey, showing the boundaries of the communes, and different colours are super-imposed to show the various numerical groups. 'France adminis-trative et statistique' at 1:500,000, published in 1964 by Girard et Barrère, is a clear map in four sheets, showing towns classified by population into ten divisions, and the communications network of roads and railways. A revised edition of the 'Carte des chemins de fer français' at 1:800,000, was issued by the Institut Géographique National in 1965 in four sheets. The 'Michelin road map of France' began in 1922. In 1950, the whole map was completely redesigned and revised. The scale remains the same, 1:200,000, but a number of details—lettering, for example, and the width of roads—have been changed to achieve greater clarity and accuracy. Each sheet takes about eighteen months in preparation, for much research is done and each sheet is carefully documented. Other commercial map publishers include Blondel La Rougery, founded in Paris in 1904, who are the official publishers for the Air Ministry; Girard et Barrére, whose history as a map and globe-making house goes back to 1780; and Taride, also of Paris, who have been issuing tourist maps, guides, a great variety of popular maps and globes since 1852. The BP booklet of maps of the roads of France includes eight general sheets, one sheet covering the Paris region, one sheet on Corsica and town plans of Paris, Bordeaux, Lyon and Marseille (Paris, Société Française des Pétroles BP, 1966). Harrap issues a particularly useful coloured wall map of France, by A Rouby, 'La France physique, économique et touristique', 1967, at the scale of one inch to fifteen miles; coloured slides accompany the map.

The *Atlas de France* is one of the great national atlases, first published in looseleaf form between 1934 and 1945, with a second edition 1953-59, by the Comité National de Géographie, printed by the Service Géographique de l'Armée (Paris, Editions Géogra-phiques de France). Some plates in the second edition were reprinted from the first, but by more modern methods; new plates show types of agriculture, livestock distribution, electrical energy and density

201

of railway traffic. Emphasis is on economic topics. Field work was of a very high quality and the whole production demonstrates the best of French cartography. Interesting new techniques include the use of proportional spheres to denote urban centres of more than five thousand population; but it should be noted that population data are from the 1946 figures. Actual figures are printed by the names of the chief towns and cities. Interesting plans of medieval towns are included. Scales range from 1 : 1,250,000 to 1 : 8M.

The *Atlas de la France de l'Est* was prepared by l'Association pour *l'Atlas de la France de l'Est* and edited by La Librairie Istra, Strasbourg (Nancy, Les Editions Berger-Levrault, 1960-). Seventy looseleaf maps, accompanied by explanatory notes, in a binder, give a more detailed treatment of the region than was possible in the national atlas. The region covered approximates to Alsace-Lorraine, and Professor Henri Baulig, in the introduction, comments on the unity of the region from the population aspect. Its frontier situation and the independence of spirit of its people are stressed. Another useful regional atlas is the *Atlas du Nord de la France*, prepared by the Institut de Géographie de l'Université de Lille (Les Editions Berger-Levrault, Paris, 1961-). Seventy four plates, revised at intervals, with text, are issued in a screw binder. The *Atlas de Normandie*, published by l'Association pour *l'Atlas de Normandie* is in the first stages of publication. The complete atlas will cover administration, physical geography, agriculture and industry, population, communications and social geography.

A most interesting work, *Atlas aérien: France*, published by Gallimard, consists of five volumes, each containing up to 350 air photographs and pictorial maps : *1* Alps, Rhône, Provence, Corse, 1955; *2* Bretagne, Loire, Sologne et Berry, entre Loire et Gironde, 1956; *3* Pyrénées Languedoc, Aquitaine, Massif Central, 1958; *4* Paris et la vallée de la Seine, Ile-de-France, Beauce et Brie, Normandie, de la Picardie à la Flandre, 1962; *5* Alsace, Lorraine, Morvan et Bourgogne, Jura, 1964.

Several special subject atlases of France have been published, original in conception and design. The *Atlas industriel de la France*, prepared by F Anisensel and others under the direction of Robert Giry for La Documentation Française, was completed in 1959 (1960). 1954 Census data were used, divided into two main parts, consumption of energy and industries. Notes and statistics help to elucidate the maps. *Atlas de l'industrie textile* comprises twenty eight plates showing centres of production and processing in the wool, cotton, silk, jute and cordage, man-made fibres and other such industries, showing also sources of imports and exports. The accompanying text and statistics make this a most useful reference work. The *Atlas hydroélectrique de France* covers the economic geography of water power. A general map is followed by index maps of the five major areas of hydroelectric concentration, namely, East, Brittany, Central Massif, Pyrenees and Alps. Thirty three sectional maps are based mainly on the official 1:200,000 series or on the 1:400,000 map of Girard and Barrère. Sources, dams, conduits and plants are clearly shown and also included are illustrations and descriptions of power plants.

The unique *Atlas de France vinicole* was produced under the general sponsorship of several wine growers' committees and syndicates and was published under the patronage of regional syndicates representing the vineyards of Bordeaux, Bourgogne, Côtes du Rhône, Champagne and Coteaux de la Loire, in five volumes, with a sixth volume, ' Les eaux-de-vie de France—le cognac '. Photographs and reproductions of engravings are included and the descriptive text is given in French, German, English and Italian, also, where relevant, in Spanish and Portuguese.

The *Atlas historique et culturel de la France,* by Jacques Boussard, was published by Elsevier in 1957. Each section, ending with the technical era 1914-55, contains maps and photographs, reproductions in offset lithography, text and notes on the illustrations. A bibliography and index complete the reference value of this scholarly atlas. Another useful reference work for sociologists, economists, geographers and historians, the *Atlas historique de la France contemporaine 1800-1965,* was prepared by P A Bouju and others in 1966. 461 maps are divided into eight sections: administration, population, economics, politics, religion, education, cultural life and French abroad. Still in preparation is the series *Les atlas linguistiques de la France par régions,* under the direction of Albert Dauzat. In publication since 1958 by the Centre National de la Recherche Scientifique, this atlas continues the work of Gilléron and Edmont in the *Atlas linguistique de la France.*

Paris, the second of a new genre of regional atlases, has been prepared under the direction of Madame Beaujeu-Garnier, professor at the Sorbonne. The contents are to include a series of maps on the physical environment, about thirty six sheets on land use and transport, thirty six on economic activities, thirty depicting social data and about twelve showing the regional role of Paris. *A map book of France,* compiled by A J B Tussler and H H Allen (Macmillan, 1968), is an integrated unit composed of text, maps and diagrams on the physical, regional, economic and industrial geography of France, with the aim of encouraging independent work on the part of the student; it is in two parts, regional and systematic.

General topographie, by de Fontagnes (Paris, Colin, 1948), gives a full, illustrated account of the geodetic, topographic and cartographic operations used by the French for their maps and plans at that time. *Cahier de cartographie: France et France d'outremer,* by A Frayasse, comprises forty pages of concise text, published by Colin, 1957, and an article by P Flatrès deals with contemporary specific work: 'Cartographie de l'utilisation du sol dans le Nord: méthodes et résultats', *Hommes et terres du Nord* (Lille, 1966, no 2), including maps, with an English summary. *France et pays d'expression française,* by A Labaste, is available through Harrap, in a new edition 1965; besides splendid illustrations, there are 117 maps and diagrams in two colours. Numerous other relevant monographs could be cited, especially on local geography, most of them illustrated by original maps or sketch-maps.

The Comité National Français de Géographie began the *Bibliographie cartographique de la France* in 1936-37; maps are included in the *Bibliographie de la France,* also in the monthly trade list *Biblio.*

Luxembourg: The national committee for cartography of Luxembourg is the Administration du Cadastre et de la Topographie. Topographical maps of the Duchy have been prepared and published at 1:25,000 by the Institut Géographique National de France, with the collaboration of the Administration. This is a coloured series; there is also a monochrome series, 1:10,000. A 1:50,000 series in colour was begun in 1956 by the Institut Géographique National and a first revision of this series by photographic air surveys was completed by the IGN in 1964. A new map at 1:20,000, in colour, began to replace the 1:25,000 in 1964-65. The Service Géologique de Luxembourg was founded in 1937; the latest 'Carte géologique générale du Grand-Duché de Luxembourg' was published at 1:100,000 in 1966.

CENTRAL EUROPE

The central areas of Europe are naturally complicated to study from any aspect as a result of their complex history. Some very fine mapping has been carried out, both locally and nationally, at various times, and many maps have been compiled covering areas with varying boundaries. Two modern atlases are extremely useful. The *Atlas of Central Europe,* published by the Kartographisches Institut Bertelsmann in a second edition 1962 (Murray, 1963), had the aim of promoting the cause of European union. It covers Germany, the Low Countries, Switzerland and Austria, stressing the distribution of major features of production, industry, population and communications. Twenty pages in full colour, on a basic scale of 1:1M include some physical and other maps not found in any British or American publication. A good deal of modernisation has been applied, though the background colours are still perhaps rather heavy. Names throughout have not been translated and there is an index of more than thirty seven thousand entries. *The Atlas östliches Mitteleuropa,* a looseleaf publication, in two volumes, edited by Theodor Kraus, E Meynen and others, began in 1959 by Velhagen and Klasing of Bielefeld. The area covered stretches from the Baltic to the Moravian Gate and from the Elbe to the eastern boundary of Poland, cutting across the more usual regional boundaries. The boundaries of Germany are shown pre-second world war. Sixty eight map plates and four maps in a pocket are grouped under seven main divisions: natural conditions, historical and political factors, population and settlement, economy and communications. The second volume contains explanatory text in German, French and English. In the pocket are sheets of the pre-war 1:300,000 ' Übersichtskarte' covering the former German territories. Very fine documentation is given and the photographic plates are excellent.

The prepared sheets of the *Atlas des deutschen Lebensraumes in Mitteleuropa,* edited by Professor N Krebs, were not published after the war for obvious reasons, but in 1958 the section on morphology, in revised form, was published as *Landformen in mittderen Europa: morphographische karte mit reliefenergie,* by H Waldbaur (Leipzig, Wiss Veröff, Deutsch Institut Länderk). The map is at 1:2M, clearly showing the relation of geological data to the relief forms. The cartography and production are highly scholarly, eight graded colours showing the relief, with details emphasised in black, and having an explanatory text and bibliography. A great number of monographs and articles have been written on regions of central Europe, especially in French and German. Many of these contain informative, frequently original maps, such as, for example, Pierre George and Jean Tricart: *L'Europe centrale,* in two volumes (Paris, PUF, 1954) and, extending to include the Adriatic coast, Josef Wartha: *Die häfen der nördlichen Adria und ihre Beziehungen zur Österreichischen Aussenwirtschaft* (Universitäts verlag Wagner, 1959), an erudite contribution to the study of economic geography of central Europe and to port geography. For English-speaking readers, Alice Mutton's *Central Europe: a regional and human geography* (Longmans, 1961) is probably the best introduction. *Les républiques socialistes d'Europe centrale,* by A Blanc and others (Paris, PUF, 1967), covering Eastern Germany, Poland, Czechoslovakia, Hungary and Romania, is well illustrated with maps in the text and includes a bibliography.

Germany: The leading cartographical organisation in the Federal Republic of Germany is the German Society of Cartography, Deutsche Gesellschaft für Kartographie, Hamburg, which also forms the national committee to the International Cartographic Association. Germany has a long tradition of scholarly work in cartography; outstanding individual names readily come to mind, also the long-established Justus Perthes Geographische Anstalt of Gotha, until the division of Germany after the second world war, when it removed to Darmstadt. The Gotha Institute is now known as VEB Hermann Haack. The modern pattern of map publication has, however, become extremely complex. Even after the unification of Germany in 1871, the separate states continued to be responsible for their own mapping programmes, with little or no co-ordination. Basic differences existed between the mapping of north and south Germany and, with the added complications resulting from the second world war, some kind of co-ordination became essential. The Arbeitsgemeinschaft der Vermessungsverwaltungen der Länder was created to establish communication and a measure of co-ordination between the individual map services. The ten ' Länder ' are North Rhine-Westphalia, Lower Saxony, Schleswig-Holstein, Hamburg, Bavaria, Hesse, Bremen, Rhineland-Palatinate, Baden-Wurttemberg and Saar. The Institut für Erdmessung, founded in 1950, was later named the Institut für Angewandte Geodäsie and this is the central research body in geodesy, photogrammetry and cartography, which has prepared map series at 1 : 200,000, 1 : 500,000 and 1 : 1M. There is also the Bundesanstalt für Landeskunde und Raumforschung, which comprises the Institut für Landeskunde under the direction of Professor E Meynen, and the Institut für Raumforschung, which is concerned with regional planning. Individual ministries have their own cartographic departments, which issue specific publications and maps in hydrography, forestry, transport, etc.

Germany's tradition in private cartography has been long established. The most notable agency has been the Justus Perthes Geographische Anstalt of Gotha, whose influence extended far beyond Germany. With the division of Germany after the second world war, those cartographic agencies in what is now known as the German Democratic Republic either moved outside this area or became nationalised. Justus Perthes moved to Darmstadt and the Gotha Institute is now known as VEB Hermann Haack. East German mapping since the second world war has followed the Soviet pattern; some tourist and school maps are widely known, but the major part of the cartographic work is not generally available.

To the end of the second world war, the largest scale map to cover the greater part of Germany was the ' Topographische Karte ' at 1:25,000. For reconstruction work after the war, larger scale maps were needed and a topographical map series at 1:50,000 was undertaken and completed in 553 sheets, many already revised. In addition, a beginning was made on a series 1:5,000 for the areas most involved in economic and social planning. Changes and corrections are reported to the Land Survey Offices, where they are plotted on base maps to be incorporated on new revisions as they occur. Very full details are included in D S Rugg's excellent article 'Post-war progress in cartography in the Federal Republic of Germany', which includes also a bibliography, in *The cartographic journal,* December 1965. Other relevant references are G Krauss: 'Map series of the Federal Republic of Germany' and 'Critical remarks about the revision of the official topographic map series of the Federal Republic of Germany', both in *Kartographische Nachrichten,* V 4 1963. The established standard topographical series is now the 1:50,000 and the original 1:25,000 series has been revised. For a time there was a series at 1:100,000, but this has now been discontinued. As it is now virtually impossible to obtain the official current maps of the Democratic Republic, reference must frequently be made to the pre-war sheets of the 1:25,000 and the old 1:100,000 series for this area.

German cartographers have been foremost in the production of regional and thematic mapping. Before the first world war, geology, morphology and climatology had been mapped, followed by aspects of human geography. Between the wars, the main interest shifted to agricultural and industrial topics, notably the *Atlas des deutschen lebensraumes* begun in 1937, but never finished because of the 1939 war. The principal geological survey was at 1:25,000, but in the case of a few individual states, which have carried out their own surveys, the scale is 1:200,000. All the official geological maps of Germany are listed in the *Berichte zur deutschen landeskunde* (Bundesanstalt für Landeskunde und Raumforschung, Bad Godesberg, supplement no 4 1961). The body of the work is an index, with entry by topographical title of the sheets, of all the sheets published in the 1:25,000 series. Full bibliographical details are given, together with the names of map collections in which they are included. Similar lists are entered for smaller scale series and for geological maps published by academic or commercial organisations. The whole work amounts to a review of the history of German geological mapping, as well as a current check and location list.

In response to the increasing demand for information on the availability of ground water for population and industry, the Bundesanstalt für Landeskunde und Raumforschung prepared a 'Hydrogeological map of the Federal Republic of Germany' published between 1953 and 1960 in fourteen sheets at the scale of 1 : 500,000. Larger scale maps showing greater detail and explanatory text are part of the publication. The 'Topographische Übersichtskarte', 1 : 200,000, published by the Institut für Angewandte Geodäsie has been based on the previous 1 : 25,000 and the 1 : 50,000 series. The scale was chosen instead of the 1 : 250,000, so that extra detail could be shown. The map is printed in eight colours; relief is shown by contours and hill-shading and a road classification is indicated.

Many geography departments had also been preparing maps of specific areas or topics and all these projects were the forerunners of the great ten volume *Deutscher Planungsatlas* begun in the early 1950s, initiated by Professor Kurt Brüning and published by the Akademie für Raumforschung und Landesplanung at Hanover in collaboration with the respective state authorities. First published were the volumes for Schleswig-Holstein, Lower Saxony and Bremen, Hesse, Bavaria and Berlin. The *Atlas von Berlin,* 1962, volume IX of the series, was a work of major importance, not only as part of the atlas as a whole, but in the pattern it set for a new development in city atlas planning, followed by the atlases for Paris and London already mentioned. The *Atlas von Berlin* was originally intended to be a separate geographical atlas, but, for financial reasons, it was decided that it should take its place as one of the 'Planungsatlas' series. Because of the political situation, comparable statistical material was not always possible to obtain. In outline, the atlas includes topography and general situation, administration, landforms, population, settlements and housing, economy, transport and communications, public services, cultural amenities, health and the overall structure of the city. In such an atlas, the information included can be specific, concerned with individual industries, for example, and the 'flow lines' of traffic. A building developments plan was incorporated. Diagrams are included as necessary and an explanatory text accompanies the map sheets. A full analysis of the atlas is given by Georg Jensch: 'Ein themakartographischer Kommentar zum *Atlas von Berlin*', in the *International yearbook of cartography,* IV 1964; Werner Witt contributed two articles to the second and third volumes of the *Yearbook* respectively, 'Deutscher Planungsatlas' and 'Regionalatlanten in der Bundesrepublik Deutschland'.

The remaining volumes of the atlas are for the regions Rhineland-Palatinate, Baden-Wurttemberg, Saarland, North Rhine-Westphalia, Hamburg and a comprehensive volume on the German Federal Republic at 1 : 1,500,000. The volumes are not uniform in size, scale or in overall plan of contents. Differences in geographical setting, local factors and cartographical resources have made each an individual work, but a substantial part of each volume consists of a factual basic survey of available local resources for planning purposes.

Similar atlases have been published independently of the project. The *Atlas of Nordrhein-Westfalen* has been in publication since 1950 by the Ministerpräsidenten des Landes Nordrhein-Westfalen, Düsseldorf. On a basic scale of 1 : 500,000, the sheets show relief, climate, administrative divisions, aspects of population and economic development. A 'Ruhr coal-mining district atlas' has been prepared by the Association of Communities in the Ruhr coal district, which demonstrates specifically the factors of planning and development.

Die Bundesrepublik Deutschland in karten, a new official cartographical work on the Federal Republic, began publication in 1965, by the Statistisches Bundesamt, Institut für Landeskunde and the Institut für Raumforschung. The atlas is divided into five sections, administration, physical geography, population and social geography, economic geography and natural regions. Most of the maps are at 1 : 1M or 1 : 2M, titles are given in English and French as well as German. When completed, this will be the most comprehensive cartographical work on Germany. A very special and fascinating atlas is the *Atlas der deutschen volkskunde*, which has a complex publishing history. The Deutsche Forschungsgemeinschaft commissioned preparation of the work in 1930 and more than 100 sheets had been published when the second world war intervened. These sheets, mainly on the scale 1 : 2M, some on 1 : 4M, showed all manner of folk traditions throughout the country, mapped with great clarity and skill. There was to have been a commentary, but, so far as is known, this was not published. In 1954, the atlas material was considered again, under the direction of Professor Matthias Zender, and publication by N G Elwert Verlag of Marburg began in 1959. It was completed in 1966, comprising a set of forty eight maps, with 975 pages of commentary by Dr Günter Wiegelmann. Germany was again a pioneer in mapping such a subject; other countries, such as Finland, Austria, Sweden, Switzerland and Hungary, all proud of their folk-culture, have since undertaken comparable work.

Also in 1965 the *Atlas des deutschen agrarlandschaft* was completed, prepared jointly by the university departments of geography and the Deutsche Forschungsgemeinschaft and published by Franz Steiner Verlag of Weisbaden, 1962-65. The atlas is in five sections: the first examines specific topics for the whole extent of the Federal Republic; the second provides an analysis of the relationship between physical factors and agriculture; the third concerns agriculture round the great urban centres; the fourth shows social and economic influences on land use and, finally, a selection of the topics considered is mapped on larger scales. The cartography, carried out by the Institut für Landeskunde, is superb, and the text concise and clear—the whole work demonstrating the best of German scholarship and the erudite results that can be achieved by co-operation between academic geographers backed by official resources. The *Pfalzatlas* consists of individual sheets at various scales, 1965-67, by Willi Atter, published by Pfalzische of Speyer for the Gesellschaft zur Forderung der Wissenschaften. Two pamphlets of explanatory text have been prepared for use in conjunction with the atlas.

For East Germany, the *Klima atlas für das gebiet der Deutschen Demokratischen Republik,* published in 1953 by the Akademie Verlag in Berlin, provides a detailed analysis of the climatic factors. The main body of the atlas comprises forty eight maps at 1:1M, depicting all aspects of temperature, precipitation, etc. An introductory section contains maps of relief, soils and vegetation, and a grouping of physical and climatic regions. A set of five phenological maps and six examples of exceptional conditions form the third section of the work, together with seven pages of diagrams and a brief explanatory text. VEB Hermann Haack began publication in 1956 of a looseleaf *Agraratlas über das gebiet der Deutschen Demokratischen Republik,* entirely in German. A pocket inside the back cover of the binder contains reductions onto one sheet of each section of maps.

Commercial cartography in Germany is concerned chiefly with the production of educational maps, wall maps and atlases, or with road maps and tourist maps of all kinds. The Mair Geographischer Verlag specialises in road maps; the ' Deutsche Generalkarte ' at 1:200,000, published between 1952 and 1957, is a topographical map designed with the needs of the motorist in mind. Topographical content was based on the 1:25,000 map, revised in the field, especially those areas including main roads and built-up areas. The sheets are uniform in size and in cartographic style. Colouring is much softer than in the majority of previous German cartography and great clarity has been achieved, with relief shown by hill-shading. A ' Verkehrskarte Deutsche Demokratische Republik ', 1:500,000, was published in 1964. This is a coloured map, distinguishing five categories of roads—with distances quoted in kilometres—railways, canals and airports. A small inset map shows airways. More than ten million copies of road maps are now published annually in West Germany alone. The article by Volkmar Mair, ' Strassenkarten aus Mairs Geographischen Verlag ', in the *International year-*

book of cartography, 1963, gives full details. Town plans are also published by Mairs, by the Institut für Angewandte Geodäsie, by the Stadtvermessungs-und Katastenamt and also by local firms such as E v Wagner and J Mitterhuber. Lists of streets and public buildings are usually printed on the reverse. H König of Frankfurt specialises in street plans of the important cities, such as Frankfurt itself, Berlin, Bonn, Bremen, Cologne and Hamburg. The sixth edition of the official map, ' Eisenbahnen in der Bundesrepublik Deutschland ', 1 : 750,000, was published by Hauptverwaltung der Deutschen Bundesbahn in 1964. Seven categories of railways are shown, with stations, and relief is indicated by hill-shading. The Reise-und-Verkehrsverlag maps of Germany cover most of the country at various scales, the most significant areas at two miles to the inch. An interesting new study of German tramways, containing maps, is F Van der Gragt: *Europe's greatest tramway network: tramways in the Rhein-Ruhr area of Germany* (Leiden, Brill, 1968).

Bartholomew published a reference map of Germany, Austria and Switzerland in 1957, in two parts, north and south, at a scale of 1 : 1M. They are essentially road maps, showing through-roads and special motorways, distinguished from other main roads and German Alpine roads. Distances are given in kilometres. Contour colouring is shown and there are inset maps of Berlin and Vienna. The latest Bartholomew revision of Germany is included on the ' Central Europe ' map at 1 : 1,250,000. Relief is shown by hill-shading.

The Deutsche Gesellschaft für Kartographie, founded in Leipzig in 1837, includes nine special divisions, each of which deals with specific cartographic issues. The leading cartographical journal is the *Kartographische nachrichten.* The value of the *Forschungen zur deutschen landeskunde* (Remagen/Rhein, 1885-) can scarcely be overestimated; the splendid maps are by the Bundesanstalt für Landeskunde. Biennial publications of importance are the volumes prepared for the meetings of German geographers known as the ' Geograbentag '; the specially prepared maps and sketch-maps are of the greatest value.

The Bundesanstalt für Landeskunde is the centre for the documentation of German geography and cartography. In addition, a number of other bibliographical works make known research either completed or in progress, the *Geographisches Taschenbuch,* for example, the quarterly *Berichte zur deutschen landeskunde* and the series *Die deutschen landkreise: Handbuch für verwaltung, wirtschaft und kultur,* each volume of which concentrates on a particular area, including maps and bibliographies. New research

211

in Germany, including cartographical projects, is made more widely known by the monthly *Cultural news from Germany,* issued by Inter-Nationes, Bonn. The *RV katalog* and the *RV der kleine katalog: landkarten, reise-führer, globen, atlanten aus aller welt,* published by the Reise-und-Verkehrsverlag, Stuttgart, is probably the most complete trade catalogue of maps currently available. The main catalogue is a looseleaf publication, kept up to date by monthly supplements, in German, English and French. Emphasis is on German production, but the balance between countries is otherwise well kept. In addition, there are the catalogues of the other great commercial cartographic houses. *Westermanns geographische bibliographie,* for example, has been published in ten issues a year since 1955.

Austria: The central cartographic agency in Austria is the Bundesamt für Eich-und-Vermessungswesen (Landesaufnahme) in Vienna; and the national committee to the International Cartographic Association has been formed from the Austrian Geographical Society, Kartographische Kommission der Österreichischen Geographischen Gesellschaft.

The national topographical map series are in four main categories. Austria 1:25,000 and 1:50,000 and 'Übersichtsblatt von Mitteleuropa' 1:750,000. The sheet lines of these maps follow the Ferro Meridian and the international geographic parallel lines. The 1:25,000 series is published in sheets of two sizes. Contours are shown at twenty metre intervals and in some cases ten metre contours are also added. Five colours are used, blue for drainage, brown for relief, green for woodland; a blue-green shade shows the presence of glaciers and black is used for all other features. The sheets of the 1:50,000 series, each covering fifteen square feet, have been issued in two editions; a provisional edition covered the country pending the 'definitive' edition. On these sheets, all features, including relief shown by hachures, are in black, with the exception of woodlands, which are identified by green shading. Contours at twenty metre intervals are drawn on the definitive edition. Four colours are used, blue for drainage, brown for the contour lines and relief shading, green shading for the woodlands, as in the provisional sheets, and black for all other features. 1:200,000 and 1:750,000 series are also in black, except for the green woodland tint.

The 'Vorlaufige Ausgabe der Österreichischen Kart 1:200,000 mit Strassenaufdruck' was published in 1963, a new series on the same sheet lines as the Austrian Staff Map of Europe, drawn on a Gauss-Kruger Projection, with contours at 100 metre intervals

and hill-shading. Another new 'Karte der Republik Österreich' at
1:500,000 was published by the Bundesamt für Eich-und-Vermessungswesen (Landesaufnahme) in 1965. 'Österreich at 1:650,000'
is produced by Falk-Verlag in Hamburg and Freytag-Berndt und
Artaria, Vienna, began an 'Österreich Strassenkarte', 1:600,000,
in 1965. Five categories of roads are shown, with distances quoted
in metres. Relief is indicated by hill-shading and spot heights in
metres. Information of interest to tourists is the chief purpose of
the map, which has legends in four languages. A similar 'Turisten-
Wanderkarte Österreich', at 1:100,000, was first issued in 1963.
'A map of road conditions in Austria', at 1: 625,000, produced in
1962, is still useful, published by the Österreichischer Automobil-
motorrad und Touring Club, Vienna. A BP map of Austria showing
roads at 1:600,000 is produced for the firm by Freytag-Berndt und
Artaria; street plans for Vienna and seven other towns at 1:20,000
are on the reverse.

Many atlases have been produced in Austria, varying in purpose
and quality, from the *Physical-statistical atlas of Austria-Hungary*
by J Chavanne and others, 1877, to the group of regional atlases
begun after the second world war. The Kommission für Raumforschung der Österreichischen Akademie der Wissenschaften (Commission on Regional Research of the Austrian Academy of Science)
is the official agency responsible for the national atlas of Austria
and the printing and publication are in the hands of the Kartographische Anstalt Freytag-Berndt und Artaria, Vienna. Professor
Hans Bobek is the chief editor. The complete project is magnificent
in its vision and scholarship, designed to be a geography of Austria
in maps, logically arranged and carried out with the co-operation of
experts in every aspect. The work is divided into twelve sections,
each dealing with a special topic: the situation of Austria in central
Europe; physical environment; land use in the past and present;
settlements; population distribution; characteristics and trends;
economy; transportation; administration; culture; the complex of
social, economic and political factors and problems of regional
development control. A final section summarises the cartographic
problems involved in the preparation of the atlas and comments on
some of the maps. The scale of 1:1M was chosen as the standard;
1:500,000 is also used. Work on the atlas began in 1957 and publication began 1958-. Each volume is supplied with a transparent
overlay, showing administrative centres and boundaries and the
sheets are all excellently documented. In the first fascicule was
included a map of Vienna showing land use and styles of buildings,
at 1:50,000. Full details about the atlas, including specimen map
plates, are given by Hans Bobek in his article 'Der Atlas der

Republik Österreich ', in the *International yearbook of cartography,* IV 1964.

Several of the regional atlases provided both information and experience for the compilation of the national atlas. The *Atlas of Carinthia,* 1925, should be mentioned, also the *Burgenland atlas,* by Fritz Bodo and Hugo Hassinger, 1941, which particularly set the pattern and the ideals for those that followed. In the foreword to the latter atlas, Hugo Hassinger states that the aim of the atlas was to bring ' a scientific contribution to the knowledge of its south-eastern province, its people, their history and their relationship to the soil, their cultural and social reforms '.

The first of the post-war atlases to be completed was the *Salzburg atlas,* edited by Egon Lendl and others (Salzburg, Otto Müller Verlag, 1955). Scales vary from 1 : 50,000 to 1 : 3M and the cartography is very fine. 138 maps cover natural features, land utilisation, agriculture, industry, aspects of population, communications and cultural geography, both past and present. The text for use in conjunction with the maps, contributed by a number of experts, is itself illustrated by maps and diagrams. It amounts to a regional geography of the Salzburg region; the bibliographies are particularly useful. The *Atlas von Neiderösterreich* is in seven parts, published on behalf of the government of Upper Austria by the Upper Austrian Geographical Institute from 1958-. Statistics are based on the June 1951 Census. 140 plates cover the physical, political, historical and social geography of the region, at scales of 1 : 500,000 and 1 : 1M. The cartography is of very high quality and the thematic maps are particularly clear. A transparent overlay shows the administrative divisions included on each map. There are twelve sections, carefully planned: general map; soil and hydrography; climate, vegetation; fauna; administrative and ecclesiastical divisions; settlement and population; agriculture and forestry; industry and trade; traffic; nationalities and dialects; education, arts and science; Upper Austria on historical maps. A text of two hundred pages forms a regional geography in itself. An *Atlas der Steiermärk,* in German, 1965-, is being published by Steier-Märkischen Landesregierung, under the auspices of the Graz Akademische Druck und Verlagsanstalt, but a copy has not yet been examined. The *Österreichischer Mittelschulatlas,* in a ninety second edition 1966, should be remembered, for nearly fifty maps of Austria are included in it.

Original maps are frequently found in the academic works of German scholars or in theses. Ingo Kuhre's study *Der südöstliche Odenwald und das angrenzende bauland,* published by the Geographisches Institut der Universität, Heidelberg, in 1964, is a good example, or the *Festschrift* prepared by the Gesellschaft für Geographie und Geologie Bochum; the first publication of the Institute of Geography at the new Ruhr University—P Busch and others, editors: *Bochum und das mittlerer Ruhrgebeit* (Bochum, Deutschen Geographentag, 1965), with a number of maps in a separate pocket. Scholarly periodicals in the field, publishing original work, are numerous. Maps are included in the *Österreichische Bibliographie,* compiled by the National Library from 1946.

Switzerland: The Topographical Survey of Switzerland, Eidgenössische Landestopographie, Berne, was founded in 1838, but a special organisation, the Swiss cartographic association, Schweizerische Arbeitsgemeinschaft für Kartographie, forms the national committee for representation at meetings of the International Cartographic Association. Switzerland is very efficiently mapped and the standard and beauty of Swiss cartography is unsurpassed. The adaptation of mapping techniques and design to suit a terrain having a high proportion of mountain relief has given Swiss mapping a particularly individual quality.

Since 1935, the national maps have been co-ordinated in three basic series, at scales of 1:25,000, 1:50,000 and 1:100,000. There is also a series at 1:500,000. In the normal edition at 1:25,000, black contour lines are shown at ten metre intervals in the central areas and the Jura, at twenty metre intervals in the Alpine country. Hachuring, very skilfully executed, indicates the major rock faces and, at every 100 metres, the contour lines, photogrammetrically determined, pass through the rock hachures. Together with layer colouring, an accurate indication of height is achieved, also a striking picture of slope profiles. Yellow relief tones showing areas of sparse vegetation add to the dominant impression of the rocky heights. Eight colours are used in the printing. Six colours are used for the map at 1:50,000. The whole relief of the country is shown by twenty metre contours. On the more generalised map at 1:100,000, contours are at fifty metre intervals. These show rather faintly, but every two-hundred contour is drawn more boldly. Ten colours are used for this series, the additional shades being yellow and red to emphasise the major road network. In 1965, a new series of the national map at 1:100,000 began. All these maps reveal great care in the choice and placing of lettering. A wide variety of forms has been used through the years, improving with technical developments. Several models of photosetting machines are now used—Photosetter, Lumitype, Linoform, Monophoto, Hadego, etc. Glass engraving was introduced about 1953, with a resulting improvement in clarity. An excellent example of the artistry of Swiss mapping is the special map of the Aletschgletscher,

one of the most spectacular of the Alpine Glaciers, which formed
the Swiss contribution to the work of the International Geophysical
Year. It was made by the topographical survey in co-operation
with the Laboratory for Hydraulic Research, at 1 : 10,000, showing
contours at ten metre intervals, blue on the snowfields, otherwise
black or brown. The complex glacial geomorphology is exquisitely
drawn. An important article on Swiss cartographic methods is
by Paul Bühler: ' Schriftformen und Schrifterstellung . . .', in the
International yearbook of cartography, 1961. All the maps are
based on a national metric grid, on which are also based the sheet
lines of the *Atlas Siegfried,* also referred to as the *Atlas topographique de la Suisse,* which is composed of 462 sheets at 1 : 25,000
and 142 sheets at 1 : 50,000. In the publication, *Gelände und Karte*
(Zürich, Ergen Rentsch Verlag, 1950), Professor Dr Eduard Imhof,
one of the leading authorities on cartography today, shows the
representation of relief features on maps by reference to the system
and content of the national maps of Switzerland.

    Professor Imhof is also the chief editor of the Swiss national atlas,
Atlas der Schweiz, which began publication in 1965. Plans for such
an atlas, under discussion for many years, came to fruition in 1961.
The work is a collective enterprise and will be a monument to
Swiss research and cartography; the Union of Swiss Geographical
Societies, university geographers and many official bodies have
collaborated. The scientific stages of cartographic preparation have
been carried out at the Swiss Federal Institute of Cartography,
Zürich, the topographical survey of Switzerland is responsible for
reproduction and publication, the whole project being under the
auspices of the Swiss government. More than four hundred poly-
chrome maps are planned, amounting to about ninety large-format
double-page map plates, plus a section of regional maps. The plates
are of a size to take a map of the whole of Switzerland at the scale
of 1 : 500,000; the regional and thematic maps are on larger scales,
using the large-scale map of Switzerland recently completed. The
first sheet to be issued was a fine general map; contours are shown
in brown at two hundred metre intervals, light tints of green and
buff represent the lowlands, showing up magnificently the black
rock-work and the light blue of the snowfields and glaciers, all
with strong blue-grey shading. Railways in red with white in-fill
and settlements show very clearly. On the thematic maps so far
issued, a variety of carefully chosen symbols in colour are superimposed on a background of hill-shading in white and light grey.
The clarity of colour achieved on the map plates is largely the
result of a newly developed three colour process by which the
three standard colours are toned down in five progressive stages

216

by line screens only. All aspects of the geography of the country will eventually be covered and the atlas will demonstrate fully the contemporary idea of a national atlas, serving the requirements of education at all levels, and is the basis for regional and economic planning, in addition to its role in portraying the individual character of the country as accurately and completely as possible. Annotations are given in German, French and Italian; an additional English text is planned. Full details about the atlas are explained by Professor Imhof in his article 'Der *Atlas der Schweiz*', in the *International yearbook of cartography*, VI 1966, which includes also a bibliography.

Geological maps of Switzerland are produced by the Swiss Geological Commission in Basle; the Geotechnische Kommission der Schweizerischen Naturforschenden Gesellschaft, Zürich, has also published maps and explanatory booklets concerning the geomorphology and natural scenery of regions of Switzerland. The files of a number of scholarly journals include original maps on some aspect of the geomorphology of Switzerland—the *Beiträge zur geologie der Schweiz*, 1899-, for example, the *Eclogae Geologicae Helvetiae*, 1888- the *Geographica Helvetica* and *Basler beiträge zur geographie und ethnologie*. The complex structure of the Alpine regions in particular has attracted many geographers and geologists; three of many monographs including maps are J Früh: *Geographie der Schweiz* in three volumes plus an index volume (St Gallen, 1930-45); A Galon: *The Alps, Austria, Switzerland* (Polish Scientific Publishers, 1958); and *Schweizer heimatbücher*, which is the title of a series of monographs in progress on aspects of Switzerland. A 'Vegetationskarte der Schweiz', 1:200,000, is available, in a revised edition, 1963, by Bild-und Stadtplan, Zürich; and a new map is included in *Die industrie der Schweiz*, by Hans Boesch, 1954. There are many guides covering Switzerland, and the Swiss National Tourist Office also maintains a 'Schweiz offizielle strassenkarte' at approximately 1:526,300, including a table of distances, a good deal of tourist information and fourteen small town plans on the reverse. Kümmerly and Frey issue a general map of Switzerland at 1:520,000. An official map of Basle, 1:12,500, has been produced since 1961 by Wassermann of Basle, with a list of streets.

The classified current national bibliography, *Das Schweizer Buch*, 1954-, compiled by the national library, includes maps in series 'A', which is issued twice a month.

Czechoslovakia: The Czechs may be justifiably proud of their cartographic history throughout a variety of circumstances and political régimes. Frequent reorganisation or reappraisal has resulted from the complex factors in the country's history; after the first world war, for example, the Military Geographical Institute emerged, in 1919, and again after the second world war, two main periods of re-examination and progress can be distinguished, that of the 1930s and 1940s and, secondly, the completely new cartographical planning programmes that have been put into operation during the past decade or so. At present, the national committee representing Czechoslovakia is the Czechoslovak Scientific and Technical Society, Cartographic Committee, formed in 1965 on the recommendation of the conference of cartographers held in the previous year in Prague. Members are experts from the cartographic and topographic organisations, scientific institutions and universities; individual groups deal with 'Education and qualification of cartographers', 'Cartographic terminology' and 'Mechanisation and automation of cartographical production'. Conferences are held regularly; a particularly influential meeting concerning thematic cartography was held in 1966 in collaboration with the Slovak Academy of Sciences.

The crowning achievement of the first period was the first national *Atlas of Czechoslovakia,* published in 1935 by the National Atlas Committee of the Czech Academy of Sciences and Art, Prague. In this magnificently produced atlas, fifty five plates containing 442 maps covered the whole field of geography, demography, agriculture, industry, transport and communications, banking, trade, education and schools, edited by Jaroslav Pantoflíček and a team of experts. Many diagrams and graphs are included and annotations are printed in Czech and French. Scales are mainly 1:1M or 1:2M, with some maps at smaller scales. The execution of the main maps is impeccable and there are several dynamic maps on population changes and on aspects of industry and commerce.

In 1959, the Research Institute of Geodesy, Topography and Cartography of Czechoslovakia, in co-operation with the Pedagogical Research Institute, initiated a programme of enquiry to consider the co-ordination of cartographical publications. Knowing the longstanding interest of the Czech people in education and of their genius for providing imaginative incentives to learning, it is not surprising that the 1959 programme should have begun with an evaluation of the maps, atlases and globes used in primary and secondary schools, both those published within Czechoslovakia and foreign works. Co-operation was given by 181 schools, involving more than three hundred teachers and thirty thousand pupils, also parents and industrial organisations; a special study was also made of cartography at university level. The information compiled was processed at the Computing Centre of the Central Office for

218

Geodesy and Cartography and many of the findings were incorporated in new editions of the *Schweizerischer Mittelschulatlas* previously mentioned. The Office for Geodesy and Cartography published a sixty page volume of maps at 1 : 15,000, in 1963; many of these are town plans, with indexes of streets. A road map at 1 : 1M was issued in 1964.

The Military Geographical Institute has published national maps of Czechoslovakia at 1 : 25,000, 1 : 75,000, 1 : 200,000 and 1 : 750,000; the Land Survey Office also publishes general maps at scales of 1 : 100,000 and 1 : 1M and large-scale plans at 1 : 5,000 and 1 : 10,000 for the Planning Department. The Topographical map of Czechoslovakia at 1 : 25,000, covers the country in approximately six hundred sheets. The special map of Czechoslovakia at 1 : 75,000 is composed of 169 sheets of a map showing contours at twenty metre intervals and using four colours, blue for drainage, brown for the contours and relief shading, green for vegetation and black for all other planimetric features. An edition of this map prior to the second world war showed relief by hachures and distinguished only woodlands by a green tint, all other features being in black. A third edition of physical and administrative maps was begun in 1965. The General map of Czechoslovakia at 1 : 200,000 is in twenty nine sheets, each covering an area one degree square, and the Survey map of Czechoslovakia, 1 : 750,000, is composed of seven sheets.

The Geological Survey, Prague, has published geological maps in five series since 1921, but a new official state geological map in thirty three sheets, compiled from a new geological survey, was issued in 1964; each sheet comprises a map at 1 : 200,000, with legend, geological sections and stratigraphical diagrams, accompanied by a volume of explanatory text in Czech or Slovak. An 'Atlas of the regional geology of Czechoslovakia', consisting of seven maps at 1 : 1M, was published in 1966; the maps deal with general geology, tectonic, quaternary and residual deposits, mineral resources, metallogenetic, hydrogeological and aeromagnetic features. Also at 1 : 1M were the maps in a 'Climatic atlas of Czechoslovakia' in 1958, issued by the Hydrometeorological Institute of Czechoslovakia. This work comprises eighty nine beautifully produced maps, with text, of which there are résumées in Russian, English and French. A series of atlas maps, 'Economic atlas of Czechoslovakia', published between 1951 and 1954, attempted to show the economic characteristics of the country, with indications of the potential of natural resources, in relation to other factors. In 1965, a new series of maps of agriculture, industry, vegetation, climate, geology and population at 1 : 2,250,000 was begun by the Central Office for Geodesy and Cartography.

Publication of the 'Atlas of Czechoslovakian history' was delayed by the second world war. It was eventually published in 1965 by a commission of the Czechoslovakian Academy of Sciences, in collaboration with many scholars and historical organisations. The country's history from the earliest times is analysed in eleven parts; within each section, general maps are followed by maps linking social and economic developments with the physical background. Scales of the major maps range from 1 : 1M to 1 : 5M, while numerous sketch-maps, city plans and diagrams help to clarify a most complex collection of data. Place names are given in the accepted 1960 form and, in addition, a twenty thousand name register shows equation with any previous forms used.

Freytag-Berndt und Artaria, Vienna, published a 'Tschechoslowakei Strassenkarte', 1 : 600,000, in 1965. This is a very clear, well-printed map, showing five categories of roads, with distances quoted in kilometres, and road numbers; places of interest to tourists are indicated.

The outstanding current cartographic publication is, naturally, the new 'National atlas of Czechoslovakia', published by the Central Office for Geodesy and Cartography, beginning in 1966. When complete, this will undoubtedly be a most exciting work; sheets have so far appeared dealing with natural environment, population, industry, agriculture, transport and standard of living.

The following works are available for further details : J Kouba: *Contemporary cartographic works in the Czechoslovak Socialist Republic,* Central Office for Geodesy and Cartography, special publication July 1964; R Malivanék: *Mapping for economic and technical purposes in Czechoslovakia,* Central Office for Geodesy and Cartography, special publication July 1964; Marie Medková and Ondřej Rovbík: 'Atlas der tschechoslowakischen Geschichte ', *International yearbook of cartography,* VI 1966, which includes additional bibliographic references; K Pecka: *Organisation of work on the National Atlas of Czechoslovakia and The Atlas of Czechoslovak History,* Central Office for Geodesy and Cartography, special publication July 1964. Maps have been included in the Czechoslovak National Bibliography, *Bibliograficky katalog československé,* since 1958.

Poland: The recovery of independence in Poland in 1918 increased opportunities for the development of geographical studies, only to be hampered again in the course of the second world war. There are now two central cartographic bodies, the Glowny Urzad Geodezji Kartografii, Head Office of Geodesy and Cartography, established in Warsaw in 1945, and the Instytut Geodezji i Kartografii, which represents the country at the International Cartographic Association. The former is the authority responsible for co-ordinating and supervising all geodetic and cartographic activities, including the production of maps and atlases for educational use, as well as of all maps required in planning the economic development of the country. *Ad hoc* committees were created for special publications, such as the Polish National Atlas Committee, until 1958, when a Committee for General Cartography was established as a permanent consulting body of leading geographers and cartographers. The committee is concerned with all matters involving international co-operation in the field of cartography, with technical problems and with new projects and publications.

The standard topographical map series is at 1:500,000; maps of individual areas and towns are produced at various scales. The new beginning of planned economy in Poland co-ordinated by the Central Planning Office and the Head Office of Regional Planning was rationalised still further in 1949, when the state Commission of Economic Planning became the only body in control of economic planning, and regional planning was subordinate to them. Speedy mapping became of central importance; numerous studies of geographical environment and population were carried out, mainly by university departments, to meet the needs of regional planning. In 1953, the Geographical Institute of the Polish Academy of Sciences was created to initiate and co-ordinate research; there are three main divisions, physical geography, economic geography and cartography.

Plans for a 1:25,000 'Geomorphological map of Poland' had been outlined in 1946; from 1950 onwards, work on the project was undertaken by all the academic geography departments. Methods of geomorphological mapping were worked out and, as the work progressed, the geomorphological map itself became of use as a basis for economic purposes and in regional planning. The 'Land Utilisation Survey of Poland' 1:1M, modelled on the work achieved in Britain, began in 1946, at a scale of 1:300,000, based on the topographic maps at 1:100,000. Arable land, pasture land, forests, water areas and settlements are distinguished, on the principles of the Land Utilisation Survey of Britain. In 1955, the Geographical Institute of the Academy of Sciences began a more detailed land use map, based on accurate field work, at a scale of 1:25,000. In this survey, detailed property relationships, individual crop rotation cycles and further classifications of the permanent types of land use, such as forests and pastures, could be included. Distinctions were made between different types of settlement, transport and waste land. On the basis of this map more mature programmes of development will be planned. In 1952, work began on a detailed hydrographic map of Poland, carried out in co-operation with the Department of Physical Geography in the Geographical Institute of the Academy of Sciences. Similar research, involving constant charting and mapping, has been carried out on the regional and local climates of Poland, having regard especially to the effects on agriculture, industry and health resorts. Soil mapping has not been so systematically completed for the country as a whole. The ' Geoprojekt ' is a separate organisation, employing many geographers concentrating on the study of urban natural environments, in connection with town planning; much large-scale mapping has been involved. The Polish geographical ' genius ' has therefore shifted from purely physical studies to the detailed analyses of settlement patterns and the relationships between towns and their hinterlands, this area being termed ' powiat '. Physical studies continue, but always with an applied use in mind. For example, as the problem of water resources has become more urgent, greater refinements had to be made in the hydrographic studies; not only surface water, but underground water supply, water pollution and fluctuations in level, etc have been studied. Lakes in Poland are particularly important; geographers, in co-operation with hydrobiologists, backed by the resources of some fifty research institutions, have mapped and prepared a register of more than six hundred lakes, the work being published by the Geographical Institute of the Polish Academy of Sciences.

The 'Atlas of Poland ' has its own traditions, for one of the first ' national ' atlases was Professor Romer's ' Geographical statistical atlas of Poland ', 1916. Work on the current atlas began in 1948 and was published by the Institute of Geography of the Polish Academy of Sciences, Warsaw, in 1953. Documentation was undertaken within the Geographical Institute, including bibliographical work, discussion on Polish place names, geographical terminology and the critical evaluation of statistical and cartographical sources. Maps are at scales ranging from 1 : 2M to 1 : 5M or 1 : 6M. The *Atlas geograficzny Polski,* by Michal Janiszewski (Warsaw, 1957), includes detailed sketch-maps of special features.

Geographers in Britain know the bare facts about Polish geographical and cartographical achievement, such as the one aspect outlined above, but it is difficult to see examples of the works. Even this limited knowledge was impossible until the Institute of Geography of the Polish Academy of Sciences created a new journal, *Geographica Polonica* (Warsaw, Export-Import Enterprise Ruch, 1964-), in English, with the intention of making Polish geographical work more widely known. The journal has been issued annually at first, but with plans for making it semi-annual or even quarterly, and to extend coverage to include the other languages used at international geographical meetings. Translations or summaries of Polish works will be included, as well as articles written especially for the journal. Polish Scientific Publishers (PWN) are also including more English titles and annotations in their catalogues. These developments are all moving in the right direction, but more translations are needed of the scholarly geographical originals of journals and monographs produced by Polish geographers especially during the past ten to fifteen years—works such as Stefan Jarosz: 'The landscapes of Poland' (Warsaw, *Budownictwo i architektura,* 1956); J Kostrowicki: 'The geographical environment in Poland: the natural conditions for the development of the national economy' (PWN, Polish Scientific Publishers, second edition 1961); Stanislaw Lencewicz and Jerzy Kondracki: 'The physical geography of Poland' (Warsaw, Academy of Sciences, second edition 1959); A Wrobel: 'The Voyevodship of Warsaw: a study in regional economic structure' (PWN, Polish Scientific Publishers, 1960, with English summaries); all these works include maps. There are thirty maps in *All about Poland: an encyclopaedic guide* in process of publication at the time of writing; the work is in English, containing some five thousand entries contributed by outstanding scholars and specialists on all aspects of the life of Poland from the beginnings of the Polish state to the present day, with multicolour and black and white photographs and diagrams.

MEDITERRANEAN LANDS: Considerable research has been carried out during recent years on the Mediterranean Sea and the shores surrounding it. Mention must be made of the publications of the International Commission for the Scientific Exploration of the Mediterranean Sea, established in 1919, many of which include illustrative maps, frequently original: *Rapports et procès verbaux* and the *Bulletin de liaison des laboratoires* report on work in progress. FAO and UNESCO have also conducted research in the sea— for example, the report of the *FAO Mediterranean project,* including a map, published in 1959, with a second printing in 1961; and the *Bioclimatic map of the Mediterranean zone,* published jointly by FAO and UNESCO as part of the *Arid zone research programme,* 1963-1966, 1:5M, in two sheets, printed in full colour. Also part of the publication are four 1:10M maps of homologous regions in small format and explanatory notes. *Land use in semi-arid Mediterranean climates* was the subject of a UNESCO/International Geographical Union Symposium, held at Iraklion in September 1962, of which the published papers formed volume XXVI of the *Arid*

223

zone research programme, 1964, in English and French. Particularly useful are the papers on the classification and mapping of geomorphological and land use aspects of lands in Mediterranean-type semi-arid climates, with regional studies. Of central importance is *Mediterranea,* the trimestrial review published by the Centre International de Hautes Etudes Agronomiques Méditerranéennes, Paris, also the relevant publications of the Centre National de la Recherche Scientifique, such as *Océanographie géologique et géophysique de la Méditerranée occidentale: colloques nationaux . . .,* 1962. Parts of Mediterranean countries are mapped for road and tourist use by many cartographic agencies; Hallwag of Berne, for example, published a ' Motoring map of the eastern Mediterranean ' at 1:3M, in 1964. ' Vegetation sheets of the eastern and western Mediterranean ' at 1:5M are among the latest sheets published, in 1968. These examples are in addition to the regular hydrographic charts and maps maintained by the Hydrographic Department of the British Admiralty and of the Bureau Hydrographique International and of other similar bodies.

Spain: The cartographic body representing Spain internationally is the Seminario de Estudios Cartográficos Associacion Española para el Progreso de las Ciencias, Madrid. Three organisations are responsible for national map-making, the Instituto Geográfico y Catastral, founded in 1870, the Servicio Geográfico del Ejercito and the Servicio Cartográfico del Aire, of which the first issues topographical maps in three main series. The ' Mapa nacional de España ' a six colour map in more than a thousand sheets, at 1:50,000, began in 1875 and is the basic series, having contour lines shown at twenty metre intervals; the colours are blue for drainage, brown for relief, green for woodland vegetation, grey for cultivated areas, red for primary roads, black for other roads, railways and all other features. Modern techniques have gradually been introduced and scribing on glass for colour separation has been in use for several years. Lettering is made with a photo-type-setting machine and either photomechanical positioning is used, or sticking. Reference should be made to the paper by R N de la Cuevas, ' New techniques used in producing the national map of Spain, 1:50,000 ', given to the International Cartographic Association Technical Symposium at Edinburgh in 1964. There are two ' Conjuntosprovinciales ', at 1:200,000 and 1:400,000; the former comprises sixty seven sheets in three colours, while in the latter, one sheet is devoted to each province. A ' Mapa de España ' at 1:500,000 covers the country in nine sheets, and the 1:1M map is in two sheets.

A new series of maps intended to cover the whole country at

1:25,000, including the Balearic and Canary Islands, is in progress by the Servicio Geográfico del Ejercito. Most of these sheets are printed in from three to five colours, but some are monochrome. A 1:1M 'Mapa fisico' was issued in 1930 and the Instituto Geológico y Minero de España completed a new 'Mapa geológicó de España y Portugal' in 1959. The Instituto Nacional de Investigaciones Agronomicas published an interesting 1:1M 'Mapa vinicola nacional' in 1959. Two 1:1M population maps have been issued. One shows population density in 1950, based on the number of inhabitants in each municipality, and the distribution of population according to a modified form of the Sten de Geer method; the second map shows increase or decrease in the population of each municipality between 1900 and 1950. Similar maps based on the 1960 Census figures are being prepared for publication during 1969.

The Geological Survey, established in 1873, has prepared various geological maps from time to time, also explanatory maps in other publications, such as the *Memorias,* 1873- and the *Boletin,* 1874-. The Instituto Geográfico y Catastral has also prepared a 'Mapa geologico de España', 1:400,000, in two sheets, from 1957-. In 1928, *Explicacion del mapa geológico de España* was published; and the latest of the geological maps is the 'Mapa Geológica de España', 1:1M third edition in 1952, which includes also Portugal and Spanish Morocco. Another scholarly source of geological information and maps is the file of *Estudios Geologicos,* 1945-, the journal of the Instituto de Investigaciones Geologicas 'Lucas Mallada', Madrid. A fifth edition of the 'Mapa geológico de la Península Ibérica, Baleares, Canarias', 1:1M, was issued in two sheets in 1966; and a 'Mapa pluviométrico de España', amounting, with diagrams and text, to more than five hundred pages, was compiled by P M Quijano Gonzalez and published in Madrid in 1946. On the occasion of the Sixth World Forestry Congress in 1966, the Ministerio de Agricultura, Dirección General de Montes, Caza y Pesca Fluvial published a new 'Mapa forestal de España', 1:400,000. In Spain the Ministerio de Agricultura has carried out a comprehensive land use survey, based on aerial photographs at a scale of approximately 1:30,000, reduced to a published map at 1:200,000. The classification established by the Commission on World Land Use Survey has been followed with some minor changes.

In the Bartholomew 'World reference map' series, the map of Spain and Portugal is at the scale 1:1,250,000, in a revised edition, 1964. Relief is shown by hill-shading and roads show up well in red, railways in black. George Philip issue a road map of Spain and Portugal for BP, at 1:2,500,000, revised in 1966. Michelin publish maps of individual areas, usually cities and their environs, such as the sheet ' Zaragoza-Barcelona ' at 1 : 400,000, first issued in 1962. ' Madrid ' was produced in 1964 by the Department of Geodesy and Cartography, Berlin, in the world map series at 1 : 2,500,000 currently being produced by the eastern European countries. Relief is shown by contours and layer colouring, without hill-shading, using twelve colours. The appearance of the map is informative and pleasing, but the ' English ' used does not conform to current English-speaking usage.

The Departamento de Cartografia de Aguilar issued a *Nueve atlas de España* in 1961, which must remain a standard work of reference for Spanish geography. In 445 pages, the atlas comprises maps of the geology, physical features, climate, economy, population, communications, history and regional geography of Spain at scales ranging from 1 : 600,000 to 1 : 3,250,000. The physical maps are clear, relief being shown by hypsometrical tints from light green to brown. Provisional boundaries and roads are marked in red. A useful feature is the series of cartograms supporting the provincial maps, showing population density, agriculture and climate. Seventy three pages of geographical analysis are included, also illustrated with maps and diagrams, and there is an index of some thirty five thousand entries. The *Commercial atlas of Spain* was issued in 1963 by the Chamber of Commerce. This is a concise work of fifty seven maps, accompanied by twenty three pages of text and 182 pages of statistics. Fifty provincial and seven area maps illustrate the statistics. Modern cartographic techniques were used in portraying economic and commercial studies of all the municipalities in Spain, classified into four categories. Total statistics are included, also on the provincial maps are tables of the basic statistics of the province together with the corresponding percentage of the national figures. A great deal of basic data is included on many aspects of commerce and market research.

The first parts, about 100 plates, of the national atlas of Spain appeared in 1966, demonstrating what is obviously to be a work of outstanding scholarship and cartographic excellence, planned and published by the Instituto Geográfico y Catastral, 1965, in conformity with the recommendations of the International Geographical Union Commission on National Atlases. The basic scale is the national topographic scale 1 : 50,000; other regional topographic maps at 1 : 200,000 and chorographic maps at 1 : 500,000 covering the whole country are reductions of the 1 : 50,000 sheets. District maps and maps of Spanish overseas territories at various scales are included. Thematic map sheets so far seen include maps of geology, geophysics and geodesy, climate, water supply and use, so vitally important to Spain, transport and commercial centres. A separate volume contains the index and explanatory text.

Special emphasis is on Spain in the *Atlas geográfico mundial y especial España,* edited by S A Seix Barral, published in Barcelona, 1957. There is also the atlas containing twenty seven town plans, *Spanische Städte,* compiled by O Jürgens (Hamburg, 1926).

The issues of the *Boletin de la Real Sociedad Geográfia,* Madrid, are a valuable source of maps; also the maps in the four volumes of the *Spain and Portugal* volumes of the ' Geographical handbook ' series (Geographical Section, Naval Intelligence Division of the Admiralty, 1941-45), though dated, are still useful. Notable for their maps are J M Kleinpenning: *La región Pinariega: estudio geográfico del Noroeste de Soria y Sudeste de Burgos* (University of Utrecht, Geographical Institute, 1962) with maps in a separate pocket; J B Maurel: *Geografía urbana de Granada* (Zaragoza, Departamento de Geografía aplicada de Instituto Juan Sebastian Elcano, 1962); and Manuel de Terán, editor: *Geografía de España y Portugal,* in five volumes (Barcelona, Montaner y Simon, SA, 1952-55).

The General Directorate of Archives and Libraries of the Ministry of Education began publication in 1958 of a monthly *Boletín del depósito legal de obras impresas,* including maps, which superseded the *Boletin de la propriedad intelectual,* which had been issued since 1847.

Scores of atlas maps and wall maps show the whole Iberian Peninsula. In addition, there is the 1:1M 'Grandes Carreteras de España y Portugal', produced by Michelin 1963-; an inset shows the environs of Madrid at 1:400,000. Coloured wall maps, physical and economic, of Spain and Portugal, by L Solé Sabarís and P Defontaines, at a scale of approximately 1¼ inches to 100 kilometres, are available through Harrap.

Portugal: Portugal has had a proud history in map-making, and early in the modern period the Instituto Geografico e Cadastral was founded in Lisbon; this institute now forms the national committee at the International Cartographic Association. A Bureau of Geodesy, Bureau of Cartography and a Bureau of Photogrammetry are divisions of the institute. Within the past thirty to forty years, when increasing expansion and demand for accurate maps of many different kinds, both of Portugal and of Portugal's overseas territories, became urgent, six other mapping organisations have been established: the Mozambique Geographical Mission, 1932, and the Army Cartographical Service, 1932, the Timor Geographical Mission, 1937, the Angola Geographical Mission, 1941, the Overseas

227

Geography Centre in 1946 and the Overseas Astronomic and Gravimetric Mission in 1959.

The main series of topographic maps of Portugal, published by the Instituto Geográfico e Cadastral under the title ' Carta corográfico de Portugal ', are at four scales: 1:50,000, 1:100,000, 1:200,000 and 1:400,000. The sheet lines follow the geographical co-ordinate lines, using the Lisbon Meridian and the equatorial latitude. All the maps are printed in the following colours: blue for water features, brown for contour lines, green for vegetation, black for railways and for the detail of sandy and rocky coastlines. Cultural features are shown in black on the sheets of the 1:400,000 series, in red on the other three series. The 1:50,000 series, which began in 1900, covers the country in 175 sheets. The contour interval is twenty five metres and a monochrome edition of this series is also available for some areas. A series at 1:100,000 was compiled between 1856 and 1904 and a new series began in 1950, comprising fifty three sheets. Some copies of the old edition, covering the country in thirty seven sheets, are still available. Seventeen sheets make up the series at 1:200,000, having a contour interval of 100 metres; there are three sheets of the 1:400,000, with a contour interval of two hundred metres. The Geological Survey, Lisbon, was officially established between 1883 and 1937. The fourth edition of the ' Carta geológicó de Portugal ', 1:500,000, was completed in two sheets in 1952. There is also the Geological map at 1:1M. The file of the *Boletin des Sociedade de Portugal,* 1841-, is valuable in this connection, as well as all the incidental publications of the Geological Survey. A 1:600,000 ' Carta hipsometrica de Portugal ' was issued in 1955 and there is also the 1:1M ' Carta hipsometrica '. The Geological Survey issued in 1960 a new ' Carta mineira de Portugal ' at 1:500,000, in two sheets, with explanatory text, 1965. Many rainfall maps have been compiled, including the 1:1M map completed by A N Veiga Garcia in 1943. The Estacao Agronomica National is the organisation mainly concerned with the study of land use. A ' Carta ecologia . . .', 1:500,000, was published in 1952, together with a booklet analysing the various features. A 1:500,000 ' Carta distribuicao da populacao de Portugal ' compiled by O Ribeiro, using the figures of the 1940 Census, was published in 1951. Road maps have proliferated during recent years. The ' Carta itineraria de Portugal ', 1:250,000, completed between 1904 and 1924, has been revised on a regular programme since 1941. The BP booklet of the roads of Portugal, containing four sheets of maps at 1:600,000, was issued in 1966; inset are Lisbon and Porto and their environs.

Preparation of the *Atlas do Portugal* began in 1941, under the direction of Aristides de Amorin Ciráo. The second edition, compiled between 1958 and 1959 to commemorate the fifth centenary of the death of Prince Henry the Navigator, is a most comprehensive and carefully planned atlas and a fine piece of cartography (Coimbra, Instituto de Estudos Geográficos, Faculdade de Letras, 1960). About forty maps, at scales ranging from 1 : 1,500,000 to 1 : 5M, include geology, drainage, climate, agriculture, population, industries and trade, communications and dialects. There is also a section dealing with Portuguese overseas provinces. Each plate is accompanied by an explanatory text in Portuguese and English.

The files of such periodicals as the *Boletin da Sociedade de Geografia de Lisboa,* 1887-, the *Boletin do Centro de Estudos Geográficos,* Coimbra University, the publications of the Instituto de Climatologia e Hidrologia da Universidade do Porto and of the Observatório Central Meteorológico and the monthly *Bulletin de la Société Portugaise de Sciences Naturelles,* 1907- are invaluable for maps and figures of all kinds in illustration of research; similarly the work, especially in regional geography, of the great geographer Orlando Ribeiro and the monograph *Iberische Halbinsel* by Hermann Lautensach (Munich, Keyser, 1964), which has a separate map supplement.

The national bibliography of Portugal includes maps : the *Boletin de bibliografía Portuguesa,* compiled by the National Library since 1935, with a break between 1940 and 1945. A monthly classified bibliography, with the same title, began in 1955. Maps were also included in the bibliographic contributions on Portugal by Hermann Lautensach to the *Geographisches Jahrbuch,* which were collected and expanded for publication as *Bibliografia geografica de Portugal,* by Hermann Lautensach and Mariano Feio (Lisbon, Centro de Estudos Geográficos, Institute para a Alta Cultura, 1948).

Italy: Italy, with a superb tradition of pioneer mapping, is also one of the leading nations of the world in contemporary cartography. The Istituto Geografico Militare in Florence is the official agency for map publishing in Italy and, in 1963, under its auspices and with the co-operation of The Touring Club Italiano, the Istituto ' De Agostini ' in Novara and the Ente Italiano Rilievi Aerofotogrammetriei, the Associazione Italiana di Cartografia was set up, with headquarters at the institute, to represent Italy at international meetings of the International Cartographic Association. Courses have been organised for the professional and technical training of cartographers and technicians, and a journal, the *Bolletino dell' Associazione Italiana de Cartografía,* is published every four months. Meetings and symposia are held.

The modern mapping of Italy dates from the establishment of the institute in Florence in 1872; by the turn of the century, the whole of Italy had been surveyed and the *Carta topografica d'Italia* was being steadily produced at three basic scales, 1 : 25,000, 1 : 50,000 and 1 : 100,000. The sheet lines of the three series followed the geographical co-ordinate lines based on the Rome Meridian and the equatorial latitude. The majority of the 1 : 25,000 series were monochrome maps; a few sheets in colour used blue for water features and glaciated regions, brown for relief and black for all other features. The contour interval was twenty five metres. The same contour interval was shown on the 1 : 50,000 series, which was issued in two types, the difference being in the use of colour. Both used blue for water and glaciation, brown for relief, green for vegetation and black for other features, including hatching to indicate the rugged topography in the highest altitudes; but one type had, in addition, a grey hill-shading to emphasise orographic representation. In the first instance, the 1 : 50,000 series was prepared to cover all the regions of the country not yet mapped at 1 : 25,000, gradually extending its coverage to other areas. 277 sheets of the 1 : 100,000 were produced, covering the whole country; this series also was issued in the two types, with the same colour scheme, except that brown was used for the hatching that emphasised the outlines of the high country, together with hill-shading. In addition, work progressed on a ' Carta delle regioni ', 1 : 250,000 and a ' Carta corografica d'Italia e regioni adiacenti ', 1 : 500,000, again in two types, one using brown shading for relief, while the other showed relief by hypsometric tints; and a ' Carte d'Italia ', 1 : 1M, in three colours, blue for water, brown hachuring showing the relief and black all other features.

New aerial surveys have begun for a revision of the 1 : 250,000 map, which has become outdated, and these surveys are used also as a basis for the new 1 : 50,000 map begun in 1964 (1965-). This new edition of the 1 : 50,000 map will comprise 652 sheets, the sheet lines being keyed to those of the 1 : 250,000, which, in turn, is a multiple of the 1 : 1M sheet lines. There are still two editions, one using seven colours, including shading for relief, the other six colours without shading.

Italy being a large country to map, with long extension north to south, production of large-scale maps is a difficult operation; many communes and provinces have already carried out mapping of their own areas, particularly for planning projects of all kinds, at scales ranging from 1 : 2,000 to 1 : 10,000. The 1 : 10,000 has become the scale most frequently used by the non-governmental agencies, it being cheaper and quicker to produce and to maintain. The purposes of these maps are essentially technical or administrative, but it is realised that the surveys will prove useful as a basis for future work, so that more local surveys are planned, notably for Sardinia and Tuscany. The current publications programme of the Istituto Geografíco Militare includes also revised sheets for the topographical series 1 : 100,000, with province and township boundaries shown in violet; an archaeological map, 1 : 200,000, issued with a pamphlet for each sheet, 1949-; special maps for the Geological Service, the Forest Department, the Board of Agriculture, Board of Trade and the Air Force. A ' Carte d'Italia ' at 1 : 200,000 is complete in twenty four sheets for Italy and Sicily and a regular revision programme is in process. A land survey at 1 : 200,000 has recently been completed. The ' Carta aeronautica regionale d'Italia ', 1 : 500,000, was completed in eleven sheets in 1965. The ' Officio geologica cartogeologica d'Italia ' is available for most areas at 1 : 100,000 and a 1 : 1M map for the whole country is in two sheets. For a detailed analysis, especially of the topographic maps, reference should be made to the article by Professor Giovanni Mussio, ' Practical lessons of cartography in geography study ', in the *International yearbook of cartography,* II 1962; also authoritative is 'A brief summary of national activity in Italy during the past few years ', a paper delivered by C P de Divelec to the International Cartographic Association in Edinburgh, 1964.

A fine cartographical achievement is the ' Carta della utilizzacione del suolo d'Italia ' at 1 : 200,000 published in twenty six sheets by the Consiglio Nazionale della Richerche, Direzione Generale del Catasto and The Touring Club Italiano between 1956 and 1966. Sicily is included. However, this is rather a small scale at which to show detail, especially as Italian land plots tend to be small also. The map has been compiled on the topographical base of the ' Carta automobilistica d'Italia ', under the direction of Professor Carmelo Colamonico. Twenty one kinds of soil utilisation are distinguished. The National Institute of Economic Agriculture, Milan, published a volume of eleven maps, *Carta della utilizzacione del suolo d'Italia,* 1961. The Istituto Geologica published a ' Carta geologica d'Italia ' at 1 : 1M in 1931; a series at 1 : 100,000 is compiled, but not all are for sale. In 1961, a revised 1 : 1M ' Carta geologica d'Italia ' was issued by the Ministero dell'Industria e del Commercio.

In areas where special development projects are undertaken, co-operation is sought from many different organisations, sometimes including commercial surveying teams. One such area, for example, is Calabria, where during the past hundred years or so, agricultural conditions and therefore the standard of living have been deterior-

231

ating. In the early 1950s, the Italian government began planning a complete redevelopment of irrigation schemes, land improvement, river control works, afforestation, agrarian reform, construction of aqueducts, roads and railways and encouragement of the tourist industry and, in 1958, Hunting Technical Services of London were contracted to prepare a new geological map of the area at 1 : 25,000, with the associate company, Compagnia Aereo Richerche of Rome and the Servizio Geologico d'Italia; in the finished work, an overlay is provided, showing the position of likely landslides, unstable slopes and other phenomena of erosion. Another outstanding map, published by the Consiglio Nazionale delle Richerche, was the 'Carta della densita della populazione in Italia', 1:1,500,000, 1951.

The Touring Club Italiano publish the 'Carta automobilistica d'Italia', 1 : 200,000, begun in 1935 with thirty sheets and continuously revised; also the 'Carta generale d'Italia', 1 : 500,000, a four sheet map begun in 1959, showing relief by layer colouring and spot heights, communications and boundaries. Each map has its own index and the whole series is revised frequently. These two series supersede earlier series, the 'Carta d'Italia', 1 : 500,000, produced in 1937, in twelve sheets, printed in seven colours, and showing road classifications and distances, and the 'Carta d'Italia', 1 : 250,000, in sixty two sheets, also printed in seven colours, and giving road information, but with contours at fifty metre intervals. The club is constantly adding to the special purpose map sheets, for tourists, for skiing itineraries, etc, one of the best known being the 'Carta delle zone turistiche' issued for a few special areas such as the Naples district, the Riviera and some mountain areas. Well-illustrated monographs are also issued from time to time, all with the aim of making Italy better known both cartographically and culturally. Up to date maps of Florence are maintained, with special issues as required; for example, a map was immediately prepared showing the area affected by the floods of the winter 1966, overprinted on a 1 : 8,000 sheet, issued together with a transparency showing the strength and direction of the currents. The *Atlante fisico-economico l'Italia* is a finely produced atlas, by Giotto Dainelli, published by the Touring Club in 1940, consisting of 508 maps. The explanatory text, *Note illustrative,* appeared separately, with its own subject index.

'Italia 1 : 1,500,000', published by Litografia Artistica Cartografica, 1964, shows the existing road network and the vast programme of trunk roads under construction, and is therefore more up to date than the group of thirty five maps, each with its own gazetteer, of the 'Carta d'Italia speciale per automobilistici, ciclisti e turisti, 1 : 25,000', issued by the Istituto Italiano d'Arti Grafiche.

Italy and the Balkans are shown together on a reference map by Bartholomew, 1957- at 1:2M. Predominantly a road map, railways are also shown, and contour colouring is added.

Numerous scholarly periodicals emanate from the national and provincial geographical and other scientific societies and from the university departments, many of them including maps. The Società Geografica Italiana, founded in Rome in 1867, has contributed original work not only concerning Italy, but other parts of the world; probably the scientific work carried out in the Karakoram area has become the most widely known. Professor Roberto Almagià, one-time president of the Italian National Committee and holder of many other honours, made notable contributions to several branches of geography, particularly cartography. His works *L'Italia,* in two volumes, 1959, and *Il mondo attuale,* in two volumes, 1963, are among the best known of his monographs and, as co-editor of the *Rivista geografica Italiana,* he helped to make it one of the leading geographical journals in the world. *L'Italia nell'economia delle sue regioni,* by F Milone (Turin, 1955), gives an exhaustive treatment of Italian economic geography, containing a number of maps. The Associazione Italiana degli Insegmanti di Geografia, founded in Rome in 1954, under the direction of Professor Dr E Migliorini, has already made its influence felt in geographical teaching and research and is maintaining the high standard of map-making and the understanding of maps throughout Italy.

Geographical maps published in Italy are listed in a special supplement to the monthly issues of the national current bibliography, *Bibliografia nazionale Italiana,* taken over by the National Centre in 1958, but founded in 1886 as the *Bolletino delle pubblicazioni Italione,* compiled and published by the National Central Library in Florence. Maps are listed also in the *Bolletino* of the Istituto Geografico Militare and a *Catalogue* of the institute's own publications was issued in 1961.

The majority of the general maps of Italy include Sicily also. In addition, the maps in Ferdinando Milone: *Sicilia: la natura e l'uomo,* should be mentioned.

Malta: The Directorate of Overseas (Topographic and Geodetic) Survey has mapped Malta and Gozo at 1:2,500, showing contour colouring. Very useful maps were included in *Malta: an economic survey,* published by Barclay's Bank in 1963; and two studies, including maps, have been carried out by H Bowen-Jones for the Department of Geography, Durham College in the University of Durham—*Agriculture in Malta: a survey of land use,* 1955, and, with other authors, *Malta: background for development,* 1961.

233

Majorca: The best source for information about the mapping of the island group is J G C E Rey Pastor: *La cartografia Mallorquina* (Madrid, Medinaceli, 1960).

THE BALKANS: The Balkans is another 'area grouping', which changes in extent according to the context. In this section, the important map-making countries are Hungary, Rumania and, to a lesser degree, Yugoslavia.

Hungary: The Hungarian State Institute for Geodesy and Cartography, Budapest, was founded in 1954 to take responsibility for civil mapping, previously combined with military mapping; in 1967, the title was changed to National Office of Lands and Mapping. The Office co-operates with the Cartographical Sub-committee of the Hungarian Academy of Sciences, with the Hungarian Geodetic and Cartographic Society and with other institutions in the making of maps and atlases. The National Office of Lands and Mapping, Department of Cartography, forms the Hungarian national committee to the International Cartographic Association. The maintenance of high standards in all cartographic activities is ensured by the courses in civil geographical cartography offered by the Cartography Department of the Budapest Eötvös Loránd-University. In addition to the Diploma of Cartography, a number of staff engaged in high level cartography in Hungary also hold degrees in geography, geology, history, meteorology or other specialised sciences. The highly qualified staff explain the great developments in Hungarian cartographical work within the past decade. International map exhibitions have become traditional at Budapest, organised by the Hungarian State Office of Geodesy and Cartography; each year, a particular theme predominates, such as, in 1962, national atlases, road maps in 1963, tourist maps in 1964, wall maps in 1965 and school atlases and globes in 1966. Hungarian cartographers also maintain close contacts with the other leading cartographic nations of the world in co-operative projects; the new edition of the *Atlas international Larousse,* for example is being made in Hungary, also the cartographical work on the 'Map of China', with the characters of the new Chinese Latin alphabet, is being processed in Hungary for the Swedish Publishing House, Esselte. The world map at 1:2,500,000 had its origin in Hungarian initiative and individual sheets have been regularly released since 1965 for the cartographical institutions of socialist countries. The new serial publication, 'Cartactual', to be mentioned in more detail later, is published by the Hungarian State Institute for Geodesy and Cartography.

It is not easy, however, in this country to learn of the exact details of the national topographical map series, or to see copies of any Hungarian map publications, with the exception of the national atlas and the maps issued mainly for tourists. Hungarian publishers of guides and tourist maps have of late been increasingly eager to make their country better known to people overseas. Many interesting and striking productions have been exported by the Kultura Hungarian Trading Company for Books and Newspapers. The most useful of these include a road atlas of Hungary and a map at 1:520,000, both in four languages. Twenty three maps in the atlas are at the scale of 1:360,000, with seventy nine smaller maps of Hungarian towns. 'Hungary: a guide book', by I Boldizsár, consists of nearly four hundred pages, profusely illustrated by photographs and maps, available in French and German editions, as well as in Hungarian. Separate guide-books and a map have been issued for the city of Budapest. Another road map of Hungary at 1:12,500 was issued by Kartograficii of Vállalata, Budapest, in 1964, and another at 1:525,000 in 1965. Six categories of roads are shown, with distances indicated in kilometres and the legends in five languages. The same publisher issues town plans and, in 1963, brought out an administrative map of the country at 1:500,000. Freytag-Berndt und Artaria completed a road map of Hungary and Czechoslovakia for BP, at a scale of 1:1,500,000, in 1966.

A work of the greatest importance has been the 'National atlas of Hungary', recently completed by the State Office of Geodesy and Cartography, in co-operation with the State Planning Office, the Central Statistical Office, the various ministries and the Hungarian Academy of Sciences, the research institutions and many individual specialists. A 'Regional atlas of south-east Hungary' is in preparation, intended as the first of a series of regional atlases designed to assist in regional planning. In 1960, the 'Climatic atlas of Hungary' was prepared by J Kakas under the direction of Dr Frigyes Dési, director of the Meteorological Service. Seventy eight plates, comprising 130 maps, are in a looseleaf binder, with twenty pages of text in Hungarian and German, published by the Verlag der Ungarischen Akademie der Wissenschaften in Budapest.

The Hungarian Geological Survey, Budapest, was founded in 1872. A geological map of the whole country was issued in four sheets at 1:500,000 in 1914. The *Annals* of the Hungarian Geological Institute, 1872- and the *Annual reports*, since 1882, give full details of work undertaken, illustrated with maps and diagrams; there are also the *Geologica Hungarica*, geographical series, 1914-, and palaeontological series, 1928-, also *Acta Geologica*, begun by the Hungarian Academy of Sciences, Budapest, in 1952. 'Cartography' has been published bi-monthly since 1954 by the State Institute of Geodesy and Cartography. The publications of the university geographical departments and of the geographical societies provide sources of maps in illustration of research; the Hungarian Geographical Society, Budapest, has flourished since 1872, having special divisions for physical geography, economic geography, methodology of teaching, and cartography, with branches at Debrecen, Pécs and Szeged. The quarterly *Bulletin* of the society includes English abstracts of most articles. In addition, there are several outstanding monograph sources of original maps incorporating specialist material: the publication ' Studies in Hungarian geographical sciences ', for example, edited by Gyula Kiklós, and published by the Hungarian Academy of Sciences in 1960 for the International Geographical Congress at Stockholm in 1960. Four papers in English and three in French are well illustrated by photographs, maps and diagrams. The papers given at the two conferences held to discuss the influence of the Carpathians on weather conditions also include valuable maps; publication of the complete papers by the Hungarian Academy of Sciences, in 1953, included English summaries. The Academy also published the monumental work by Marton Pécsi and others, *Ten years of physico-geographical research in Hungary,* 1964, which is available in English.

Maps are included in the monthly classified current national bibliography, *Magyar nemzeti bibliográfia,* published since 1946 by the national library.

Rumania: The Directía Topografica Militara, previously the Institut Cartografic Militar, in Bucharest, the official organisation responsible for the mapping of Rumania, was established early this century. But, long before this, an administrative map, with statistical tables of Wallachia and Moldavia, at 1 : 420,000, 1835, provided a valuable document on this complex part of Europe. With the foundation of the Rumanian Society of Geography in 1875, geographical studies made steady advance. The society's *Bulletin* appeared in 1879. The Office of Geologists was also created, which became the Geological Institute in 1906. These and other research bodies carried out geographical studies, particularly in regional and physical geography, based on or illustrated by maps. A geological map at 1 : 1,500,000 was published in 1927. With the second world war, far-reaching political, social and economic changes took place and geographical research was directed to practical requirements to a greater extent than previously. The Geographical Research Institute was founded in 1944, reorganised in 1952 and again in 1958, when it became the Geology and Geography Institute of the Rumanian

Academy. The Geography Society, now the Rumanian Natural Sciences and Geography Society, with its forty five branches throughout the country, also carries out fundamental research and field work. To a greater degree than in other countries, geographers undertake the mapping and planning reports of towns and their environs, soil and agricultural surveys, etc, all co-ordinated by the Rumanian Academy. The development of socialist agriculture and the creation of new industrial centres have made essential all kinds of research on local climate, hydrology and ecology, utilisation of water resources, etc. A general topoclimatic map has been compiled as a basis for microclimatic studies and ' The economic map of the Rumanian People's Republic', at 1:400,000, has recently been produced by the Geography Department of the Institute of Economic Sciences, Bucharest. ' Population density of the Rumanian People's Republic', at 1:500,000, and ' Population distribution in the Rumanian People's Republic', at 1:1M have accompanied detailed study of urban development and the geography of settlement. The territorial structure of regional economies is naturally of basic importance to the state and a group of geographers and the Geology and Geography Institute of the Rumanian Academy has produced a map of land use in Rumania, at 1:500,000. A population map, in accordance with the specifications suggested by the Commission for World Population Map of the International Geographical Union, was based on the population census of 1956. The work, which is perhaps the most important single achievement of Rumanian geographers to date, was published in 1960—the *Monografia geografica a republicii populare Romîne,* in two volumes, including two atlas sections covering physical features, agriculture, industry and population (Bucharest, Editura Academiei Republicii Populare Romîne, 1960-61).

A new political and administrative map at 1:1,500,000 was issued by the Directía Topografica Militara, in 1963. In 1966, a new road map of Rumania and Bulgaria was published by the BP Touring service, at 1:1,700,000.

In 1965, *The Atlas geografic Republica Socialista România* was published by the Editura Didactică si Pedagogică, Bucharest; 109 maps, edited by Dr Victor Tufescu, present a great deal of information about modern Rumania. Geology, relief forms, hydrography, climate, agriculture, industry, mining, population, communications and international trade, are all covered, either on double-page spreads at 1:1,750,000 or single pages at 1:2,500,000, with a number of regional maps on larger scales. Diagrams and sketch maps are included as necessary, for example, to illustrate the air links between towns. Most of the country is covered also by small-scale regional maps at 1:4M. Air links within the country are shown on a sketch-map.

For a country with so complicated a history, the atlas, *La Roumanie: la terre roumaine à travers les âges* is particularly interesting. Compiled by Romulus Seisanu, it covers history and political geography, demography and economic aspects, with text in Rumanian, French and English. *Agriculture en Roumanie: album statistique* is another atlas illustrating agricultural statistics, published by the Ministère de l'Agriculture et des Domaines in 1929.

Iaşi University has carried out a programme of study to determine the best methods for drawing up maps necessary for the systematic planning of towns; several special maps have been devised at scales of 1:2,000 and 1:10,000, using a general geomorphological map as a base map.

Maps are included in the current national bibliography, *Buletinul bibliografic al cărtii,* from 1952; in 1956, the work was taken over by the newly formed Central State Library, since when monthly *Lists* have been issued, with annual cumulations from 1952.

Yugoslavia: Here is another country having not only a complex physiographical character, but also a dramatic political history. In Belgrade, there is the Savezna Geodetska Uprava, which publishes topographical maps, the latest of which is the map at 1:750,000, widely used for general purposes and in teaching. Other organisations also issue topographical and special maps of the country, chief of which is the firm Učila in Zagreb, which publishes a complete catalogue of its publications. A relief map of Yugoslavia at 1:50,000 is the most used series; regional maps have been compiled at scales ranging from 1:200,000 to 1:500,000. The Geo-Karta Institute in Belgrade is engaged upon extensive geodetic and cartographic work, also on maps and atlases for use in schools.

The Geographical Institute of the Serbian Academy of Science, founded in 1947, and the Geographical Institute of the Slovenian Academy of Science and Arts are the major centres of research, co-ordinated by the Yugoslav Academy of Science and Arts. Special organisations have been created for particular purposes, such as the Adriatic Institute, which is chiefly concerned with mapping and other research on aspects of the Yugoslav coastal region. Then there is the cartographic section of the State Publishing House of Slovenia, which maintains a useful map of Yugoslavia at 1:500,000 and a map of Slovenia at 1:300,000. Geological and geomorphological maps are to be found in the files of *Annales géologiques de la péninsule Balkanique,* 1889-, published by the Geological Institute of the University of Belgrade. *Geoloski vesnik* has been published by the Geological Survey since 1932 and *Memoirs* since 1933. A geological map was issued at 1 : 1M in 1931 and at 1 : 500,000 in 1953.

Učila published a tourist map of the Adriatic Coast in 1960 at approximately 1 : 390,000 and an *Auto atlas of Yugoslavia* in 1962, containing twenty seven pages of maps at 1 : 600,000, with twenty six town insets. A new road map of Yugoslavia at 1 : 800,000 was issued by Reise-und Verkehrsverlag of Stuttgart in 1965; five categories of roads are shown, with distances quoted in kilometres, and through routes of many towns are inset. Relief is shown by hill-shading and spot heights in metres and many tourist attractions are marked. A road map at 1 : 2M was issued in 1965 by Philip for BP. A map of the railway network was produced in 1962 at a scale of 1 : 800,000, by Yugoslavia Railways, Belgrade. Ljubljana, a main centre of attraction for tourists, was mapped in one sheet at 1 : 15,000 by Turisticao Drostvo in 1963.

The maps in the two volume Admiralty handbook, *Jugoslavia* (Naval Intelligence Division, geographical handbook series, 1944), though frequently dated, are still useful. Reference should also be made to H R Wilkinson : *Maps and politics: a review of the ethnographic cartography of Macedonia* (Liverpool UP, 1951), which includes sketch maps.

Maps are included in the *Bibliografija Jugoslavije: knjige, brošure: musikalije,* which includes publications in all the languages of Yugoslavia, 1950-. Other main sources of information are the publications of the Geographical Society of Belgrade, especially the *Journal,* 1912-, *Atlasi,* 1929-, *Karte,* 1931-; also the *Croatian geographical journal,* 1929- and the ' Geographical reporter ', 1925-. All these files have suffered breaks in issue, usually because of political difficulties.

In 1949, the various republican societies formed a Council for Collaboration and in 1952 a national geographical committee was set up to co-operate with the International Geographical Union.

Bulgaria: Little detail can be ascertained concerning the work of the Office of Geodesy and Cartography, Sofia. A geographical atlas for schools devoted twenty two plates to Bulgaria, and a road map of Bulgaria was issued at 1 : 800,000 in 1962. The State Publishing House for Cartography, Warsaw, issued a 1 : 1M sheet, layer coloured, of Bulgaria, in 1964 and, in 1965, a tourist map, 'Autokarte Bulgarien ', 1 : 800,000, compiled by the Kartproekt in Sofia, was published by Harptverwaltung für Fremdenverkehr beim Ministerrat. Other maps are to be found in the *Bulletin* of the Bulgarian Geographical Society, in the *Review* of the Bulgarian

Geological Society and in the publications of the Academy of Agri-cultural Sciences, founded in 1961.

The national committee representing Bulgaria at the International Cartographic Association is the Scientific and Technical Union of Bulgarian Geodesists, Sofia.

Maps are included in the classified national current bibliography, *Balgarski knigopis,* published irregularly since 1897 and annually since 1953 by the Bulgarian Bibliographical Institute.

Greece: There are three main cartographic agencies in Greece, the Topographic Service, the Greek Army Geographical Service and the Photographic Section of the Ministry of Public Works, all in Athens. The principal series of national maps of Greece published by the Army Geographical Service include two topographic series at 1:20,000 and 1:50,000; a 1:100,000 'Military' map series, a 1:200,000 'Strategic' map series and a 1:400,000 General map. The contour interval on the 1:20,000 map is ten metres and four colours are used in printing, blue for water, brown for relief, green for vegetation cover and black for all other features. Contours on the 1:50,000 series are at twenty metre intervals, except in areas where supplementary ten-metre contours are desirable. The colour scheme is the same as for the 1:20,000, plus red for roads. The sheets of both these series form units of the map at 1:100,000. Sheets of the 1:200,000 series cover one degree square, with con-tours at 100 metres, with hypsometric tints, and the 1:400,000 series sheets cover two degrees square with the contour interval at two hundred metres and hypsometric tints changing every five hundred metres.

One of the most useful of the sheets issued by the Topographic Service of the Ministry of Public Works is the '1:500,000 road map' series, which covers the country, with the exception of the Dodecanese Islands, in six sheets. Contours are shown at two hundred feet intervals; four colours are used in printing, blue for water, brown for relief, red for roads and black for other features. A second series is the 'Geographical' map series at 1:800,000. The sheet lines are based on the Athens Meridian and the international parallel lines; one Greenwich Meridian is usually given on the lower margin of each sheet. Two sheets cover the country, over-lapping in the region of the Gulf of Corinth. The colour scheme follows the usual pattern, with the addition of hypsometric tints.

The Hellenic Geographic Society, founded in 1919, has done much to encourage mapping in the country and acts as the national committee representing Greece at the International Cartographic Association. Other organisations issuing maps include the Institute for Geology and Subsurface Research, Athens, established in 1949, noted for its two publications, *The geology of Greece* and *The mineral wealth of Greece,* both published in 1951. A metallic-mineral map of Greece, 1:1M, was issued by the Institute in 1965.

'Ellas', 1:200,000, was published by the National Statistical Service of Greece between 1963 and 1965; this is a series of some fifty district maps which constitute an atlas of Greece, drawn on a conical projection, with the prime meridian at Athens. Contours are shown at two hundred metre intervals, with layer colouring and spot heights added. International and administrative boundaries are marked and vegetation, roads and railways are clearly delineated. Legends are in Greek, English and French. BP is issuing a 1:100,000 geological map of Greece at 1:850,000, with an inset of Athens and the peninsula to the south-east at 1:300,000.

Several atlases cover this historic region. Nelson's *Atlas of the classical world,* edited by A A M Van der Heyden and H H Scullard, 1959, is well known, not only for the maps but for the magnificent illustrations. A *Shorter atlas of the classical world,* compiled by the same editors, followed in 1962. This was a new work, not an abridgement of the former atlas. The term ' atlas ' in this case is really misleading, for the 225 pages of text plus notes and illustrations are accompanied by eight plates of maps, end-paper maps and a few sketch-maps in the text. One of the most important recent works on Greece is the *Economic and social atlas of Greece,* published in 1964 by the Centre of Economic Research, Social Sciences Centre, Athens. In Greek, English and French, 126 maps, compiled by Bernard Kayser and Kenneth Thompson in collaboration with Roger Vaternelle and Basil Coukis, are presented as ' a foundation upon which even more detailed investigations may be undertaken '. Scales are 1:2M or 1:4·2M. Texts and statistics support a great variety of maps; relief and rainfall maps, using layer colouring techniques, distribution maps, making use of shading and dot methods. All aspects of Greek geography are represented.

Apart from this atlas and the publications of the Hellenic Geographical Society, there has been surprisingly little geographical or cartographical work in recent years. Many of the maps in the Admiralty handbook *Dodecanese* (Naval Intelligence Division, geographical handbook series, second edition 1943) are still useful, also the maps in *Die Griechischen landschaften* in four volumes (Frankfurt, Klostermann, 1950-59). Some of the works in the ' Research monograph ' series of the Athens Centre of Planning and Economic Development include maps.

In May 1966, Hunting Geology and Geophysics were invited to make a photogeological study of the area east of Salonika and the

River Axios in north-eastern Greece, to locate any possible minerals; the survey involved a mountainous area of about eight thousand square miles. With the aid of aerial photography, two surveyors compiled the geological maps and prepared the report in eight months; these are being published by the Institute for Geology and Subsurface Research, Athens.

THE MIDDLE EAST: One of the chief difficulties in dealing with any kind of documents, historic or contemporary, concerning the whole complex area known as ' The Middle East' is that of defining the exact area under consideration and the fact that different names have been applied at different times to some of the same regions. An area of central importance and controversy throughout historical time, it is a ' bridge' or ' pivot' area, having links with three continents; documents and surveys continually need to include parts of it in discussions of Europe, Asia and north Africa and, while maps do this to a lesser degree, the connections should be borne in mind throughout.

The Bartholomew ' Reference map of the Middle East', 1:4M, 1967, covers Turkey, Iran and Arabia and the intervening territories. Contours are coloured; roads are shown in red, railways in black. Geographia issues a general reference map of the Middle East, Suez and the Mediterranean, showing railways and pipelines, with an inset of the Nile delta.

The *Historical atlas of the Muslim peoples,* by Roelof Roolvink (Amsterdam; Djambatan, 1957), is an essential reference work, emphasising the wide distribution of individual peoples; the maps are carefully designed and printed. The ' Ethnographic map of the Middle East', 1:5M, by the Academy of Sciences, Moscow, 1960, shows this factor more generally. Equally important for an understanding of the area is the *Atlas of the Arab world and the Middle East,* published by Djambatan-Macmillan in 1960; in New York, St Martin's Press made the work available. Within a small compass, the physical and cultural characteristics of the area are brought out, aided by photographs and explanatory text. The range covered is from Morocco to Iran, and end paper maps portray two vital elements of the area—' The spread of Islam' and ' The world of Islam in the Middle Ages', showing trade routes and main products. The treatment of individual areas varies according to the availability of factual information.

In the same year, the second in the Oxford series of regional atlases appeared, *The Oxford regional economic atlas of the Middle East and north Africa,* prepared by the Cartographic Department of the Clarendon Press and The Economist Intelligence Unit, assisted by geographers at the University of Oxford and elsewhere. Fifty four pages of maps include general reference maps of the more developed parts of the region, showing relief by layer colouring; these are followed by maps of the whole area at larger scales, dealing with physical geography, agriculture, minerals and oil, industries, transport and population. A feature is the specially compiled original maps of selected areas, including the vegetation map by Professor Gaussen of Toulouse, a new soil map by Dr A Muir of Rothamsted, a new ' Water balance map ' by Gordon Smith of the Oxford School of Geography, the map and statistics on the Suez Canal, the strip map of the Nile and of the Tigris-Euphrates, with hydrological notes; the relatively large-scale petroleum map of the Persian Gulf area; the ethnographic and population maps and the land use and irrigation maps, which are also of great value. Detailed supplementary notes, diagrams, a bibliography and gazetteer complete this well documented reference source.

The *Atlas of the Middle East,* published by the ' Yavneh ' Publishing House Limited, Tel Aviv in 1964, provides forty pages of maps, many of them thematic maps, with text in Hebrew only. Egypt, the Sudan, Israel, Syria, Iraq, Arabia, Persia, Turkey and Cyprus are included. *An atlas of Middle Eastern affairs,* by N J G Pounds and K C Kingsbury (Methuen, 1964, with a revised edition in University Paperback, 1966), presents a summary of geographical, political and historical information in a handy format, but it is scarcely an atlas. The maps, in black and white, are subsidiary to the text; they vary in quality and some of them are very small. The second edition is said to be revised, but statistics seem to be drawn from 1961 sources. The *Atlas of Islamic history,* compiled by H W Hazard (Princeton UP, 1951; revised edition 1954, in the Princeton ' oriental ' studies series), covers the Near and Middle East and assembles material vital to the understanding of the area which is otherwise scattered. The maps were compiled by H Lester Cooke and J McA Smiley, with explanatory text, gazetteer-index and statistics of population, conversion tables of dates and index of place names.

A vast periodical literature covers the area, in which maps are occasionally included. The United Nations *Economic development in the Middle East: review of the main economic developments in the Middle East relative to agriculture, industry, petroleum, foreign trade and balance of payments,* 1951- is published in English, French and Arabic. The *Middle East journal,* quarterly from the Middle East Institute, Washington, DC, and the new *Middle Eastern studies,* published by Cass, are all likely sources. The Europa directory, *The Middle East and North Africa,* 1948-, includes Egypt, the Sudan, Libya, Turkey, Syria, Israel, Jordan, the Yemen and Kuwait, Cyprus, Iran and Iraq, giving geographical, social and economic surveys of each country, illustrated by maps; a new edition is prepared every one or two years.

A standard work since its inception has been W M Ramsay: *The historical geography of Asia Minor* (Royal Geographical Society, supplementary paper 1890; reprint by photolitho by Adolf M Hakkert of Amsterdam 1962) is an excellent starting point for the understanding of this area. Other relevant monographs, containing maps, include:

Sir Reader Bullard, *editor*: *The Middle East: a political and economic survey* (Royal Institute of International Affairs, third edition 1958).

George Cressey: *Crossroads: land and life in south-west Asia* (Chicago, Lippincott, 1960).

W B Fisher. *The Middle East: a physical, social and regional geography* (Methuen, fifth edition, revised 1963).

Jean Gottmann, *editor*: *Etudes sur l'Etat d'Israel et le Moyen Orient* (Paris, Colin, 1959). A collection of articles which have appeared in various French periodicals during the previous twenty five years.

B A Keen: *The agricultural development of the Middle East: a report to the Director, Middle East Supply Centre, 1944-45.* (HMSO, 1946).

George Lenczowski: *The Middle East in world affairs* (Cornell UP, second edition 1956).

S H Longrigg: *Oil in the Middle East: its discovery and development* (New York, OUP second edition, 1961).

Royal Institute of International Affairs: *The Middle East: a political and economic survey* (The Institute, second edition 1954).

Doreen Warriner: *Land reform and development in the Middle East: a study of Egypt, Syria and Iraq* (OUP, for the Royal Institute of International Affairs, second edition 1962).

Turkey: The Turkish Geodetic Survey was established in Ankara in 1909. Topographic maps are maintained at various scales. A series at 1:800,000 covers Turkey in eight sheets, overlapping the adjoining areas of Belgrade, Bucharest, Salonika, Athens, Crete, Lebanon, Syria, Iraq, West Iran, the Caucasus and the southern Crimea. The main series of topographic maps is at 1:200,000, which is being revised, using the results of the photogrammetric survey being carried out for the 1:25,000 series. The latter series, which covers most of the country, is not available to the public.

The first scale at which a systematic geological series was issued by the State Department for Geology and Mining at Ankara was at 1 : 800,000; explanatory texts are available in French for the Istanbul and Izmir sheets This map is now replaced by a revised map at 1 : 500,000, published by the Institute of Mineral Research and Exploration, Ankara, 1962-64; the map is complete in twenty one sheets, each sheet being accompanied by explanatory text in Turkish and English. Other thematic maps are not generally available, but may be seen at the Ankara Institute. An annual *Progress bulletin* is issued by the Institute, also monographs and *Reports;* increasingly, those of the latter which are for publication are issued in English, French or German as well as Turkish. Istanbul University has prepared an economic map at 1 : 800,000; and forestry and climatic maps are issued by the relevant departments. A population map has been completed in accordance with the International Geographical Union Commission for a World Population Map, using the figures of the 1955 Census. A 1 : 1M map is available, and the 1 : 25,000 series has been initiated chiefly for recording information on population distribution. Most of the road maps published before 1958 or 1959 are out of date, as the country's communications have been changing so rapidly, but a road map at 1 : 1·625M is maintained by the Ankara Tourist Office. Hallwag of Berne has produced a road map of Turkey at 1 : 3M from 1962; and the BP Touring Office published a road map of Turkey at 1 : 1,850,000 in 1966.

The *National atlas of Turkey,* published by the Faculty of Letters, University of Istanbul in 1961, contains eighty seven maps and diagrams covering all aspects of the life of the country. This atlas is in Turkish and English, but the greater part of source material for Turkey is not available in other languages. The *Review* of the Geographical Institute of Istanbul University is an authoritative source for original and sketch-maps. Monographs with useful maps include the following :

Bernard Lewis : *The emergence of modern Turkey* (OUP, for the Royal Institute of International Affairs, 1961).

Reinhard Stewig : *Byzanz-konstantinople-Istanbul* (Kiel, Schrift Geogr Inst, 1964).

The economy of Turkey : an analysis and recommendation for a development program. Report of a mission sponsored by the International Bank in collaboration with the Government of Turkey (Baltimore, Johns Hopkins P, for the International Bank, 1951).

Iraq : The Iraq Survey Directorate is situated in Baghdad. The basic scale for topographical mapping is 1 : 50,000, aerial photography being used in revision. The latest drawing office techniques have been steadily installed through recent years. The Iraqi government is now devoting half of the oil revenues to the basic development of the country and the greatest concentration of surveying is in geology, soil and land use, with the rehabilitation of agriculture being the first objective. In this work, the United States Bureau of Reclamation and various United Kingdom organisations are continuing to assist.

Early in 1967, the government invited UNESCO to undertake an investigation into the geology of Iraq and to advise on training facilities in schools, universities, research and government geological services and the role of non-governmental institutions in geological research. Arnost Dudek, director of the Prague Geological Survey and Felix Ronner, associate professor at the Institute for Minerals and Applied Geology at Gratz led the investigations and prepared a *Report,* which is of fundamental interest to any student of Iraq. Outside the universities, the most important institutions engaged in geological research and mapping are the Department of Geology of the Ministry of Oil, the Groundwater Department of the Ministry of Municipalities and Works and the Department of Geology of the National Commission for Atomic Energy. A thorough review of existing geological information is being undertaken, but improvements need to be made in geological mapping and exploration. The various reports and maps are widely scattered among the different institutions and the most urgent action is required to centralise a documentation centre for geological information, reports, maps, etc.

Much of the great plain formerly irrigated by the Tigris and Euphrates now lies derelict, the soil ruined by salt. An ambitious dam construction programme is in operation to harness the water from the two rivers and make it available for irrigation. Air reconnaissance surveys are the first stage in this work, carried out under contract by a firm such as Hunting Technical Services, assisted by local surveyors and scientists in the field. Final compilation of soil and land classification information is made on the 1:50,000 base maps. The reports and maps of the Arid Zone Research Institute, University of Baghdad, founded in 1961, are invaluable, but not all publicly available; the institute is administered in six sections, soil and geology, climate and environment, use of water resources for agricultural progress, arid zone reclamation projects, agricultural engineering and arid land reform.

The *Atlas of Mesopotamia: a survey of the history and civilisation of Mesopotamia from the Stone Age to the fall of Babylon,* by M A Beek, translated by D R Welsh and edited by H H Rowley (Nelson, 1962), contains twenty two diagrammatic maps, photographs and drawings, with 147 pages of text and captions.

The publications of the Iraqi Geographical Society, Department of Geography, College of Arts, Baghdad, 1960-, are a useful source of sketch-maps, also the maps in the Admiralty handbook *Iraq and the Persian Gulf* (Naval Intelligence Division, 1944) are still useful, especially the folded map in the pocket. Other useful maps are to be found in:

Nurik Al-Barazi: *The geography of agriculture in irrigated areas of the Middle Euphrates Valley,* in two volumes, 1961-63.

W H Al-Khashab: *The water budget of the Tigris and Euphrates Basin* (University of Chicago, Department of Geography, Research paper no 54, 1958).

C J Edmonds: *Kurds, Turks and Arabs: politics, travel and research in north-eastern Iraq, 1919-1925* (OUP, 1957).

Wilfred Thesiger: *The marsh Arabs* (Longmans, 1964). Concerns the marshes at the junction of the Tigris and Euphrates.

The economic development of Iraq. Report of a mission organised by the International Bank at the request of the Government of Iraq (Baltimore, Johns Hopkins P for the International Bank, 1962).

There is no current bibliography of maps available.

Syria, Lebanon, Israel, Jordan: The Service Technique du Cadastre, Beyrouth, was founded in 1921, to produce surveys and maps for local use. A geological map of Syria, Lebanon and northern Palestine, 1 : 500,000, is maintained, issued originally in two sheets in 1942 by the Délégation Général au Levant France Combattante, Beyrouth, with explanatory text; and a 1 : 1M 'Carte géologique de la Syrie et du Liban', with text, was issued in a third edition in 1945. A *Bibliography of Levant geology, including Cyprus, Hatay, Israel, Jordania, Lebanon, Sinai and Syria,* compiled and arranged by M A Avnimelech (Israel Program for Scientific Translation, 1965) is a useful source. The *Atlas climatique du Liban* began publication in 1966 by the Ministère des Travaux Publiques et des Transports, Direction de l'Aviation Civile. Fifty maps, with commentary, comprised the first volume. The maps in the Admiralty handbook *Syria* (Naval Intelligence Division, 1943) are still useful, also the maps in the following:

C P Grant: *The Syrian desert: caravans, travel and exploration* (Black, 1937).

The economic development of Syria: report of a mission organised by the International Bank at the request of the Government of Syria (Baltimore, Johns Hopkins P; OUP, 1955).

R Mouterde and A Poidebard: *Le limes de Chalcis: organisation de la steppe en haute Syrée romaine* (Paris, Librairie Orientaliste Paul Geuthner, 1945); with plans and maps in a separate volume.

Jacques Weulersse: *Paysans de Syrie et du Proche-Orient* (Paris, Gallimard, 1946).

The Ministry of Education is compiling a national bibliography of maps.

In the area now Israel and Jordan, traditional and contemporary factors are almost inextricably mingled. The Survey of Palestine, founded in 1920, became the Survey of Israel in 1948, within the Ministry of Labour, Tel Aviv. The Survey now acts as the national committee making liaison with the International Cartographic Association. Topographic map coverage is at various scales, the standard maps series being at 1 : 20,000, composed of about 260 sheets, having contours inserted at (usually) ten metre intervals. The sheets covering those parts of the country which have been developing quickly have been under constant revision, using aerial survey and modern plotting techniques. These sheets have generally been used as base maps for maps at smaller scales. A later series at 1 : 10,000 was completed entirely by aerial photography and photogrammetry, and this, in turn, is used as base material. The contour interval is five metres. A special series of maps, 1 : 10,000, was necessary to cover the southern areas of the country, to assist its rapid development; these sheets show agricultural specifications and development schemes in addition to the basic topographical information. A 1 : 50,000 series covers the country in fifty six sheets; a series of 1 : 100,000 is in twenty six sheets and general maps at 1 : 25,000 and 1 : 250,000 have also been prepared. After having for some years performed all medium- and small-scale cartographic work by scribing on glass, the Survey has finally changed completely to scribing on plastic foils, mainly Astrascribe and Scribalon. A new publishing venture, *Cartographical papers,* began in 1965, in Hebrew with short English summaries.

A 'Geological map of Palestine', 1 : 250,000, with explanatory text, was produced by the Palestine Geological Survey, 1947; a revised map at the same scale, in two sheets, was published by the Survey of Israel in 1965, in Hebrew and English. Geological maps are now available at 1 : 50,000, 1 : 100,000 and a new series, 1968- at 1 : 25,000. The publications of the Geological Survey are naturally central documents, also the *Bulletin* of the Geological Society, while, for a review of the whole area, A E Day: *Geology of Lebanon and of Syria, Palestine and neighbouring countries,* is most useful (Beirut, American Press, 1930). A 'Soil map' at 1 : 500,000 was completed in 1966 by the Ministry of Agriculture, Soil Conservation Service, published in Hebrew and English; also a soil erosion map at the same scale and a soil conservation map at 1 : 20,000. A new edition of the 'Road map of Israel' at

1 : 200,000 was issued by Zvi Friedlander in 1965, based on material supplied by the Survey of Israel. Four categories of roads are distinguished, with distances quoted in kilometres, a table of distances and an index to places. The Ministry of Tourism produced a 'Touring map of Israel' at 1 : 750,000 in 1966; this sheet also includes on the reverse a table of distances, a town index and details of interest to tourists.

The *Atlas of Israel* (Jerusalem, Department of Surveys, Ministry of Labour and the Bialik Institute of the Jewish Agency, 1956-), edited by David Amiran, while in many ways modelled on other national atlases, reveals some particularly interesting features. The atlas has been published at the rate of two folders a year, each folder containing eight to twelve sheets; the maps are made available as they are completed, not necessarily in sequence. The small size of the country enables maps to be presented at larger scales than is usual; in fact, the format of the atlas was decided by the size of page which would accommodate four maps of Israel at 1 : 1M. Scales throughout the atlas range from 1 : 500,000 to 1 : 2M. The relative smallness of the area covered should have made the compilation of the atlas easier, but two factors must be borne in mind in studying it; first, the rapid development of some parts of the area has made revision of some sheets necessary before all were completed and, secondly, political changes and periods of instability have resulted in varying coverage and reliability of available data. The editors have included time sequence maps for several aspects of the geography of the country. Topographical variety throughout the country portrayed at large scales gave an opportunity for interesting experiments in cartographical representation. In 1965, the atlas was virtually complete, except for continuing revisions and new additional sheets. Fifteen sections, comprising altogether 102 maps, each deal with a particular subject. Text is in Hebrew, with tables of contents in English, and an English translation of the text is in preparation.

The twentieth anniversary Israel atlas, compiled by Zev Vilnay, 1968, contains twenty maps in full colour of both historical and contemporary Israel, with additional illustrations.

The number of recent planning proposals put forward for various parts of the country have made necessary much preliminary surveying and the preparation of large-scale maps. These are fine studies in themselves; for example, the 'Land use capabilities for irrigable land', at 1 : 500,000, prepared by the Soil Conservation Service, Ministry of Agriculture, in Hebrew and English, and the 'Land use capabilities for dry land farming', at the same scale, both published in 1966. The Israel 'Physical master plan' was published by

249

the Ministry of the Interior in 1964; this has been followed by a series of detailed surveys of individual areas, such as the ' Physical master plan of the Israel coastal strip ', by Elisha Efrat and Ehud Gabrieli, consisting of seventeen maps at 1 : 20,000 (Jerusalem, Ministry of the Interior, 1966). The narrow coastal plain has been developing fast, population continues to increase and the need is to preserve the coastal amenities, while at the same time making provision for well-balanced residential development. These maps, in full colour, distinguish eight types of urban land use and eleven other categories, from agricultural use to bathing beaches. Each map is accompanied by a short explanatory text in Hebrew and English. Later in the same year, the two compilers completed the ' Physical master plan of the northern Negev ', which comprised fourteen coloured maps at 1 : 4M, each with a short commentary in Hebrew and English, and a location map at 1 : 500,000. Agricultural land, a wide variety of minerals and the attraction of numerous historic sites makes this an area of rich potential, providing the lack of water supplies could be remedied. The ' Physical master plan of Jerusalem ', by Elisha Efrat, 1967, contained fourteen maps at 1 : 150,000, with short commentaries in Hebrew and English; this was intended as a basis for planning for the next fifty years, but the unsettled political condition of the area is bound to have repercussions on this aim. For further information, reference should be made to *The Israel physical master plan,* by Jacob Dash and Elisha Efrat (Jerusalem, Ministry of the Interior, 1964), which itself includes maps.

The Palestine Exploration Fund has, from 1865, carried out a vast amount of research in the area; published works include the *Survey of western Palestine,* the *Survey of eastern Palestine,* the ' Great map of western Palestine ' in twenty six sheets, ' Map of Palestine ' in twenty sheets, ' Map of western Palestine ', showing water basins in colour, photo-relief maps, a plan of Jerusalem and two sheets of sections, north-south and east-west. *Palestine exploration quarterly,* 1869- has been an important source of information. The *Bulletin* of the Israel Exploration Society, 1933- includes English summaries and the *Israel exploration journal,* quarterly, has been published in English and French since 1950-51.

The maps in the Admiralty handbook *Palestine and Trans-Jordan* (Naval Intelligence Division, 1943) are still useful, especially the separate folded map in the pocket. A map is included in the Barclays Bank DCO survey *Israel: an economic survey,* 1963; other relevant works including informative maps are:

J Ben-David: *Agricultural planning and village community in*

Israel: studies on the human implications of settlements in the arid zone (UNESCO, 1964).

G S Blake and M J Goldschmidt: *Geology and water resources of Palestine* (Jerusalem, Department of Land Settlement and Water Commissioner, 1947).

Efraim Orni and Elisha Efrat: *Geography of Israel* (Israel Program for Scientific Translations, 1964, Oldbourne Press; New York, Daniel Davey, 1964; from the Hebrew work published by Achiasaf Publishing House, 1963). The first comprehensive study in English, including numerous maps and a topographical map from the Survey of Israel, 1 : 500,000, inserted in the cover.

L E Taverner: *The revival of Israel* (Hodder and Stoughton, 1961).

A survey of Palestine, prepared in December 1945 - January 1946 from the information of the Anglo-American Committee of Inquiry, in three volumes, includes analysis of population, land tenure, detailed analysis of agriculture, production, climate and soil, animal industry, citrus production, irrigation and drainage, forestry and soil conservation, fisheries, trade and industry.

Maps are included in *Kirjath Sepher,* the bibliographical quarterly of the Jewish National and University Library; official publications are also recorded in the state archives and library's *List of government publications,* bi-monthly from 1953.

It will be convenient to mention here some of the cartographical works compiled on the ‘Holy Land’, for some of these include parts now in western Jordan. Incidentally, a most valuable sketch-map by C G Smith of Keble College, Oxford, ‘Israel's territorial gains as a result of the June war’, was published in ‘This changing world’, *Geography,* July 1968. The map shows also other boundaries, water supplies, oil fields and the locations of phosphates, potash works and manganese—a most useful reminder that it is through such prompt publications as this that geographers are enabled to keep abreast of their ever-changing subject matter.

The Oxford Bible atlas, edited by H G May in 1962, contains in handy form a wealth of information about the Holy Land and its neighbours from the sixteenth century BC to about AD 70. Twenty six maps in five colours, most of which are accompanied by a brief text, provide comprehensive surveys of their topics, such as the Exodus, Palestine after the Exile, the Cradle of Christianity and the Dead Sea Scrolls. A standard relief map of the Holy Land is frequently used as a base. Three maps are of particular interest to geographers: the natural regions of the country, vegetation in biblical times and the mean annual rainfall, while many of the maps are relevant to the historical geography of the eastern Mediterranean. The great ‘Atlas of the Bible’ by L H Grollenberg has been trans-

251

lated and adapted from the fourth edition by R Beaupère (Paris, Elsevier) and a shorter version, translated into English by Mary Hedlund, was published by Nelson in 1959. *The historical atlas of the Holy Land,* edited by E G Kraeling (New York, Rand McNally, 1959), was one of the many scholarly works inspired by new archeological discoveries. Indeed, the text begins with an account of the finding of the Dead Sea Scrolls. A concise outline of the historical geography of the area is attempted in forty pages of coloured maps, a number of monochrome maps, plans, charts, illustrations and text. The hill-shading in green, yellows and browns, especially on the 'Ancient Palestine' map, is unusual, but striking. Of the numerous other similar atlases, especially for children, the *Westminster historical atlas of the Bible,* by G E Wright and F V Filson, published by the Student Christian Movement Press, 1945 and the *Lands of the Bible: a golden atlas,* by S Terrier (Rathbone Books, 1957) warrant particular mention.

Jordan is now a sparsely populated and relatively unproductive region. A survey of land use was made in 1963, of which the results, compiled by S G Willimot and others, were published by the Ministry of Overseas Development under the title *Conservation survey of the southern Highlands of Jordan.* As part of a programme of evaluation of the water resources of the east side of the Jordan valley, initiated by the Central Water Authority of the Hashemite Kingdom of Jordan, with financial assistance from Great Britain through the Middle East Development Division and the Foreign Office, London, a geological survey of the area was carried out by Hunting Technical Services between 1962 and 1965. The object of the survey was to determine zones within the area where ground-water was likely to be present in sufficient quantity and at a depth which would permit economic exploitation of the water by pumping. Careful mapping of the rock formations was made, particular attention being paid to the water-bearing potential of the various formations. Contour maps were constructed on particular geological horizons to show the inclination, folding and displacement produced by faulting on the water-bearing strata underground. *The economic development of Jordan: report of a mission organised by the International Bank at the request of the Government of Jordan* (Baltimore, Johns Hopkins Press; OUP, 1957) contains useful maps; other studies of Jordan containing maps include:

K M Hacher: *Modern Amman: a social study* (Durham Colleges of the University of Durham, Department of Geography, 1960; research papers series, no 3).

G L Harris *et al: Jordan, its people, its society, its culture* (New Haven, HRAF Press, 1958; survey of world cultures series).

Guy Mountfort: *Potrtait of a desert* (Collins, 1965). An expedition to Jordan in 1963.

Raphael Patai: *The kingdom of Jordan* (Princeton UP, 1958).

Isaar Schattner: *The lower Jordan valley* (Jerusalem, Scripta Hierosolymitana, 1962), with maps in a pocket.

G A Smith: *The historical geography of the Holy Land* (Hodder and Stoughton (twenty fifth edition 1931). Includes Syria, Palestine, Jordan, Israel.

Cyprus: The mapping of the island of Cyprus is incomplete. One of the latest surveys was carried out mainly by aerial photography, for the ' Cyprus land use map ' at approximately 1 inch to 1 mile, under the direction of R R Rawson and K R Sealy in the Geographical Laboratory at the London School of Economics, with the aid of Dr D Christodoulou, a Cypriot who knows his own country and who prepared his PHD thesis on the evolution of land use in Cyprus. The published map (Geographical Publications, 1956) was at the scale of 1 : 253,440. An explanatory text was published in 1960, written by Dr Christodoulou.

During recent years, considerable progress has been made in raising the standard of living in Cyprus and much cartographic work has been directed to, for example, pasture research to improve agriculture, increasing the productivity of grazing lands and preparing an integrated programme of crop and animal husbandry. All uncultivated land required surveying, a task demanding more time than the local survey staff could give. Hunting Technical Services were called in and spent eighteen months mapping the land on which pasture seeding was possible and surveying the remaining land with a view to grazing or other utilisation. Air surveys at approximately 1 : 35,000 were used and, with the assistance of the local ground staff, a set of sixteen maps at 1 : 50,000 was produced, showing different kinds of pasture and crops. A reduction of the ' Pasture survey ' maps on one sheet was made at 1 : 250,000 and a report on the present state and future prospects of land use in Cyprus was published. Climatic conditions were also indicated on maps at 1 : 250,000, showing the ' climax vegetation zones '.

A geological map of Cyprus was completed by C V Bellamy in 1903. In *Cyprus, the survey of physical features, economy and industry*, revised by Barclays Bank DCO in 1965, there is an informative map; and there are maps in Sir Harry Luke's *Cyprus: a portrait and an appreciation* (Harrap, 1957).

A new series of topographic maps of Cyprus at 1 : 25,000 was begun by the Directorate of Overseas Surveys in 1960, totalling fifty nine sheets. The Department of Lands and Surveys published an administrative map, 1:253,440, in 1966, with an inset of Nicosia at 1:15,840. L M Bear, director of the Geological Survey Department, initiated a new 'Geological map of Cyprus' in 1963 at a scale of 1:250,000. Soil surveys are carried out by the Soil Survey Section, Ministry of Agriculture and Natural Resources, Nicosia; a 'Reconnaissance soil map', prepared by C G Soteriades and G C Grivas at 1:125,000, was drawn and printed by Fairey Air Surveys in 1961. The Department of Lands and Surveys has also issued a 'Forest map of Cyprus' at 1:253,440, 1964. The forest information was supplied by the Cyprus Department of Forests; state forests are distinguished on the map, with forest stations and locations of telephones.

Legal deposit of printed books, maps and other materials was made compulsory in 1949 and these items are entered in a register maintained by the administrative secretary, but no list is published, so far as is known.

Saudi Arabia: The Arabian Peninsula is still one of the least mapped parts of the world, in spite of considerable progress being made in the southern and eastern areas. The Directorate of Overseas Surveys map of the Protectorate of South Arabia is now in its fourth edition; at a scale of approximately 1:3M, the map shows administrative divisions and also the member states of the Federation of South Arabia. The Directorate has also recently produced a 'Photo-geological map of Western Aden Protectorate', approximately 1:250,000, in two sheets, 1967. Interest in mineral or oil resources has encouraged mapping of some areas. The United States Geological Survey has for some years maintained special map sheets at 1:500,000, and in 1958 issued a map of the Arabian Peninsula based on these sheets supplemented by aerial survey. In 1963, The United States Geological Survey published a new topographical map of the Arabian Peninsula at 1:2M, sponsored by the Ministry of Petroleum and Mineral Resources of the Kingdom of Saudi Arabia and the United States Department of State.

H von Wissmann's map of the Hadramaut, published in 1932, was the first reliable general map of south-western Arabia to supplement all the travellers' sketch-maps. A map of the north-western part was made by H St J Philby following his 1936 journey; this was published at 1:1M by the Royal Geographical Society. The society produced a map of South Arabia east of Aden in two sheets, 1958, based on the two above mentioned maps and incorporating information from various other route maps; the scale of this map is 1:500,000 and the sheets are printed in six colours. 'The

Arab world ' at 1 : 1,250,000 was issued in 1959 by the International Association of Friends of the Arab World, Lausanne.

The remote mountain and desert areas on the western side of Saudi Arabia, mined for silver and gold in the days of the Romans, are now being systematically surveyed. Geologists from the United States, France and Japan have mapped large areas and in 1935 large-scale aerial survey was begun, on a co-operative basis, by the French Geological and Mining Research Bureau in consultation with the Saudi Arabian Ministry of Petroleum and Mineral Resources; in addition, a survey consortium, consisting of Hunting Geology and Geophysics, the Arab Geophysical and Survey Company, Aero Service and the Lockwood Survey Corporation, worked on a series of geophysical maps which have formed the basis of subsequent detailed field surveys in suitable areas.

The *Arab atlas* was prepared and produced in Arabic by the Directorate of Military Survey, 1965, to give as thorough a presentation as possible of the physical, political and geological features of the United Arab Republic and the Arab countries; general physical, political and distribution maps of the world and of the continents are also included, with other general astronomical and physical data and a sixteen page index. Local pronunciation of the most important place names is a useful feature.

Valuable maps are contained in the *Oriental explorations and studies* of Alois Musil, published under the patronage of the Czech Academy of Sciences and Arts, 1926-28. Most of the published monograms by H St J B Philby on the area contain maps, for example *S'au'di Arabia* (Benn, 1955) and *The land of Midian* (Benn, 1957). The maps in the Admiralty handbook *Arabia* (Naval Intelligence Division, 1946) are still useful, especially the folded map in the pocket. *The economic development of Kuwait: report of missions organised by the International Bank at the request of the Government of Kuwait* (Baltimore, Johns Hopkins Press, for the International Bank, 1965) also contains useful maps; similarly, the following monographs:

Z R Beydoun: *The stratigraphy and structure of the Eastern Aden Protectorate* (HMSO, 1965); issued as a special supplement to *Overseas geology and mineral resources*.

Adolf Leidlmair: *Hadramaut: bevölkerung und wirtschaft im wandel der gegenwart* (Bonn, Ferd Dümmlers Verlag, 1961).

Jacqueline Pirenne: *À la découverte de l'Arabie: cinq siècles de science et d'aventure* (Paris, Amiot-Dumont, 1958).

S G Shiber: *The Kuwait urbanization* (Kuwait Government Printer, 1965).

Wilfred Thesiger: *Arabian sands* (Longmans, 1959).

K S Twitchell et al: *Saudi Arabia: with an account of the development of its natural resources* (Princeton UP, third edition 1958).

Iran: The documentation regarding Iran is still to a great extent looking towards the past, but current awareness is gradually increasing. The National Cartographic Centre, founded in Tehran in 1953, consists of a Photogrammetric Division, a Field Survey Division, a Cartographic Division, a Planning and Research Division and a Control Division; the centre acts as the national committee representing Iran at the International Cartographic Association. In the past, there has been some duplication of mapping owing to lack of co-ordination; topographic mapping has been the concern of the Ministry of Foreign Affairs, the Military Geographic Service and the Cadastral Office, the Ministry of Industry and Mining, the Department of Irrigation, the Ministry of Roads, the Anglo-Iranian Oil Company, various municipalities and the Seven Year Plan Agency. In 1953, the latter agency centralised the civilian cartographic activities and set up the National Cartographic Centre, which is responsible for the maintenance of the topographical maps of the country, the centralisation of cartographic activities and documentation and the provision of assistance to other government agencies.

First-order triangulation began in 1956; second-order and subsidiary triangulations are mainly local in character, since they relate to local projects. Modern methods of triangulation, plotting and reproduction have been introduced. Ozalid printing has been widely used, since the editions are not required to be large; greater experimentation has recently been introduced, new colours added to the maps, etc.

A political map, 1:2,500,000, published by the centre in 1963, is a clearly printed map showing administrative boundaries. A series of large-scale regional plans is in preparation, at 1:2,500. These maps are being published in groups, as they are completed, such as the map of Ahwaz in fourteen sheets, of Ciráz in twenty sheets; and Rasht in eight sheets. 'Iran, 1:3M' is a new series published by the Sahab Geographical and Drafting Institute, in sixteen sheets, in Persian and English, 1963-, beginning with mineral deposits, physiographic divisions, natural vegetation and river basins. Much work needs to be done on the geology of Iran. For the International Geological Congress in Mexico, 1956, BP published a portfolio of geological maps of the country at 1:1M and sections at 1:250,000, based on the work of the geological staff of the Anglo-Persian Oil Company 1909-33 and the Anglo-Iranian Oil Company 1933-51.

For the New Delhi Congress, 1964, the company, with the co-operation of the local survey staff, prepared a further set of sixteen more detailed maps of south-west Iran at 1 : 250,000. Each sheet is accompanied by sections and a few pages of explanatory notes. Limited numbers of these maps were made available through Edward Stanford. *Géologie du plateau Iranien,* by R Furon (Paris, Mémoires du Museum National d'Histoire Naturelle, 1941) includes useful basic maps; and in 1961 a soil map of Iran at 1 : 2,500,000 was issued by the Soil Department, Irrigation Bongah, Ministry of Agriculture, Tehran, produced with the aid of FAO.

The official guide to Iran contains maps and an Iran 'Highways map at 1 : 2,500,000' was issued by the Ministry of Roads in 1966; more informative road maps are beginning to appear from various publishers.

The publications of the Imperial Geographical Society, Tehran, contain original maps and sketch-maps, also *Iran: probleme einer-unterentwickelten Landes alter Kultur* (Frankfurt-am-Main; Deister-weg, 1962). Attention should be drawn to an excellent article by P H T Beckett and E D Gordon: 'Land use and settlement round Kerman in southern Iran', *The geographical journal,* December 1966, illustrated by a fine folding plate of maps and diagrams.

Interest in retrospective bibliography has been evident for some time in Iran and current documentation is making progress, but there seems as yet to be no complete current bibliography of map production.

Afghanistan: A new map of Afghanistan at a scale of approximately 1 : 2M was published in 1965 by the Afghan Tourist Association. Province boundaries are shown. Relief is represented by hill-shading and two classifications of roads are indicated. In the same year, FAO issued a 'Survey of land and water resources', consisting of thirty one maps and seven charts. Topographic and geological surveys are included, at various scales.

In 1964, a geological map of central Afghanistan, at 1 : 150,000, was published by the Institute of Geology, University of Milan, compiled from the results of Ardito Desio's expedition to the Hindu Kush in 1961; there is also available a 'Geological map of Afghanistan', 1 : 1M, compiled by the German Geological Mission in Afghanistan and published by the Geological Survey of the Federal Republic of Germany and the Afghanistan Geological and Mineral Survey. There are excellent maps in *La géographie de l'Afghanistan: étude d'un pays aride,* by Johannes Humlum and others (Copenhagen; Gyldendal, 1959, Scandinavian University Books), and in *Agricultural Afghanistan* by I N Vavilov and D D Bukinich (Pan-

257

9

Soviet Union of Applied Botany, 1929). The text is in Russian, with an English summary and English captions to the maps.

USSR: Since the end of the second world war and increasingly during the past decade, information concerning the Soviet people and their country has become more readily available in the west. In a very special sense does this apply to original mapping; much of the country's cartographical output is still in the ' secret ' documents class and the export of large-scale maps is not encouraged, but plenty of small-scale special maps are released to interest tourists.

The essential surveying and mapping needed for military purposes was carried out by the Corps of Military Engineers, but, following the 1917 revolution, when the entire national economy was reviewed, mapping of the whole vast area of the USSR was planned. The Central Geodetic Department was established in March 1919. Lenin himself drew up a programme for the production of topographic and special maps and regional atlases, planned with the co-operation of all relevant bodies, to serve the revival of the economy and the cultural morale of the people. In 1935, the Central Geodetic Department was reorganised as the Central Department of State Surveys and Cartography, to be followed three years later by a further reappraisal, when the department became recognised as the co-ordinating body for all geodetic, cartographic and aerial photographic activities, with the title Central Department of Geodesy and Cartography. By this time, the relevant research and academic institutions had prepared themselves to take part in the overall plan and, as technical developments have advanced, Soviet geographers and cartographers have not only taken advantage of them, but have in many ways improved upon them with characteristic energy and scholarship. With equally characteristic common sense, the training of the required technicians for this vast programme was faced from the beginning. The Moscow Ground Surveying Institute was developed and became the Moscow Institute of Geodetic, Aerial Photographic and Cartographic Engineers, and Chairs of cartography were inaugurated at the Universities of Moscow and Leningrad. Later, in 1928, the state Research Institute of Geodesy and Cartography became fully equipped to act as the headquarters of research in these fields, under the title Central Scientific Research Institute of Geodesy, Aerial Photography and Cartography. At the same time, the USSR has been taking an increasing part in international congresses and conferences and has formed a national cartographic committee to represent the Soviet Union at the International Cartographic Association.

The great task of establishing a geodetic network was completed in four stages. First, the astronomic geodetic network was fixed, then second- and third-order triangulations established a control for topographic mapping at 1:25,000 and 1:10,000. Finally, the control network for topographic mapping at 1:5,000, 1:2,000 and larger scales was completed. Systematic surveying began about 1924, fairly slowly at first, but speeding up with the improvement in techniques and precision instruments. Russian technicians themselves evolved improved levelling instruments. Aerial photography and photogrammetry has been used as much as possible; the Institute of Aerial Photography of the Central Geodetic Department of the People's Commissariat of Heavy Industry was formed under the Council of Labour and Defence, in 1929. More recently, the application of electronic computers has enabled more rapid processing of data, especially important in the case of special topic maps.

The first topographic map series to be finished was that at 1:100,000, for which more than twenty six thousand sheets were needed to cover the whole country. This series is the base for the series at 1:200,000, showing relief, hydrography, settlements and roads, and for another series at 1:300,000. A series at 1:500,000 is used mainly as an air chart. The topographic systems are, in general, all in line with the sheet reference systems of the International Map of the World 1:1M. The first edition of the 1:1M map for the Soviet Union was completed between 1941 and 1945 in 182 sheets; they were in two series, one showing relief represented by hypsometric tints, the other by contour lines, with forestry shown by green symbols, the standard green colour for low relief being replaced by grey. *The manual for the compilation and preparation for the publication of the state map of the Soviet Union on the millionth scale* was published in 1940. A second edition of the map was completed in 1951, in three series: one series showing relief by hypsometric tints, the second using contours and hill-shading, and the third an outline series printed in pale shades for use as a base for other special maps. The ' State geologic map of the Soviet Union ', the ' State soil map of the Soviet Union ' and the ' State geobotanic map of the Soviet Union ' have all been subsequently completed using this base. Twelve of these comparable maps are planned on other aspects of geomorphology, soils phytogeography, climatology and mineral resources. Full details of the series, accompanied by an index map, are included in *World cartography*, VII 1962.

In 1962, the Central Office published a good map at 1:10M, showing nationalities of the USSR and the linguistic groups, based on the 1959 Census. An administrative map, at 1:15M, including the whole country, was introduced by the Central Office in 1963, and a set for the regions was begun, each at a suitable scale. In 1963 also the Central Office began to issue a series of historical maps, beginning with ' Russia in the thirteenth century ' in two sheets at 1 : 3,500,000.

An unsurpassed analysis, to its date, of Soviet geography was contained in the first two volumes of the great Soviet Atlas of the world, mentioned in Chapter 2. The co-ordination of research work in the USSR enabled all the relevant scientific organisations to contribute their special parts, and the aim of the atlas throughout was to emphasise the interrelationship of relief, soils, vegetation, etc. Systematically, since the second world war, outstanding cartographic productions have appeared—the *Marine atlas,* the *Atlas of physical geography* and the *Atlas of the Antarctic,* also all described elsewhere, and a series of regional atlases, all carefully planned to be of the most efficient use in development programmes and in education. The *Atlas of Moscow Oblast,* 1933, and the *Atlas of Leningrad Oblast and the Karelian ASSR,* 1934, had already pointed the way. Both of these atlases presented a comprehensive picture of the region and they are still interesting for the high quality and originality of the cartographic techniques employed. The first of the new series to be completed was the 'Atlas of Belorussian SSR ', edited by S N Malinin and others (MinskMoscow, Akademiya Nauk BSSR and Glavoye Upravleniye Geodezii i Kartografii SSR, 1958). 140 pages of maps presented a balanced picture of the economy at scales between 1 : 1M and 1 : 4M. A series of maps summarises the factors of agriculture and industry, using the dot method for distributions and cartograms. Cultural and historical maps show important stages of the country's development. A conference on the co-ordination of these atlases was held in Moscow in 1961 and in that year the 'Atlas of Armenia ' was finished; this atlas was even more comprehensive and detailed and the cartography is quite beautiful. The Armenian Geographical Society led the team in its preparation. The *Atlas of the Irkutsk and Kiev regions* followed in 1962, also that for the Ukraine and Moldavia. The *Azerbaijan atlas* was published in 1963, a finely produced work by the Akademiya Nauk Azerbaydzhanskoy SSR. Two hundred maps, covering all aspects of the region at various scales up to 1 : 1,500,000, are based on statistical data mainly 1960-61. The *Atlas of the Komi Autonomous Republic* was also completed in that year as an oblong folio of forty nine pages of maps, photographs and text. Next followed the atlases of the Kustanai region and of the Virgin Lands, in 1964; sections are given to topographic maps of all the regions, maps of physical geography and economic maps, both for the USSR as a whole and for the individual regions. The *Atlas of Georgia* was published in 1965 and in 1967 the atlases of the Transbaikalia and of the Sakhalin regions. Scales, legends and cartographic symbols are standardised, to ensure comparability in study and the maps are planned always to be as evocative of ideas as possible, pointing, for example, the relevance of relief factors in the location of settlements and of communications, and including as much practical detail as possible.

Another general *Atlas of the USSR,* which should be mentioned, was published in Moscow in 1955. This small atlas, comprising seventy six pages of maps is mainly topographical, using the scale 1 : 500,000 for the Moscow, Leningrad and other main regions. It should be stressed also that Russian school atlases are of a higher standard than is usual elsewhere. The *Geographical atlas of the USSR for the seventh and eighth classes of the middle schools,* for example, published in 1953, contains sixty pages of well produced maps, with an adequate index. Layer-coloured instructional maps at 1 : 5M have been issued for middle schools by the Central Office since 1963. However, the best current large-format atlas of the whole USSR was prepared for the Central Office in 1962 under the direction of A N Baranov and others. In three sections, the first consists of general reference maps of regions, mostly at 1 : 3M or 1 : 4M; the second section contains maps of the physical features of the whole country and the third section is devoted to economic maps of each of the major regions.

Geological maps have been published mainly at two scales, 1 : 200,000 for the particularly important areas yielding economic minerals, and at 1 : 1M covering the whole country. Special maps have been compiled, called 'forecast' maps, for areas potentially valuable for coal, oil or gold. The first volume of a vast new programme of maps covering the hydrogeology of the USSR appeared in 1966. The general editor of the series is A V Sidorenko. This first volume, edited by D S Sokolov (Moscow, Publishing House Nedia, 1966), covers the central European part of the USSR, dealing with the chief natural factors affecting ground water supplies, the general features of existing ground water and the importance of ground water supplies in the national economy. Six hydrogeological maps and tables of hydrochemical data are presented separately and an introduction traces the history of hydrogeological studies. The 'Geomorphological map of the Soviet Union and neighbouring countries ', at 1 : 4M, 1959, is a particularly useful map for basic study; and a 'Tectonic map of the USSR ', 1 : 5M, edited by N S Shatskiy, with explanatory text, was published in 1956.

Russian scientists have led the world in soil studies and soil maps have been prepared in all areas where soil improvement and

261

agricultural development programmes have been planned. It would
be useful, in any study of the soil maps, to consult *Soil geographical
zoning of the USSR,* now available in an English translation by A
Gourevitch from the original Russian edition published by the
Academy of Sciences of the USSR (Jerusalem, Israel Program for
Scientific Translations, 1963). Special mapping techniques have been
implemented in the peat areas, co-ordinated by the Central Department of Peat Reserves in the Council of Ministers of the Russian
Soviet Federative Socialist Republic.

Professor A V Voznenessky made a speciality of climate studies
in the 1930s; his map of the climate of the USSR was published in
the *Transactions* of the Bureau of Agro-Meteorology, XXI 1930,
in Russian, but with English summaries. ' The climatological atlas
of the USSR ' was published first in 1933 and in a second edition in
two volumes in 1955-60 by the Central Department. The whole
structure of the varying climates is shown on maps ranging in scale
from 1 : 1·5M to 1 : 50M, based on an analysis of data covering many
years from more than three thousand Soviet meteorological stations.
The first volume comprises 286 maps of soil and vegetation and
series of maps showing air temperature, snow cover, and other
factors, and a classification of climates throughout the Soviet Union.
In the second volume, 138 maps deal with individual features such
as pressure, winds, humidity, cloud cover and sunshine.

A unique ' atlas-album ', 'Agriculture in Russia ', was published
in 1914; it included a number of diagrammatic maps showing the
distribution of crops and variations in yields, with the intention
of showing the present achievement and encouraging improvement.
A number of regional atlases of agriculture have been published in
recent years, but not until the 1950s was the available data sufficiently full to enable comprehensive treatment. The 'Agricultural atlas
of the Ukrainian SSR ' was one of the first, completed in 1958, by
the Geography Faculty of the Kiev State University. It comprises
forty seven pages of maps on a scale of 1 : 4M, with legends in
Ukrainian only. All important aspects of agriculture are mapped,
together with an interesting section of diagrams showing the development of Soviet agriculture from 1913 to 1956. The 'Atlas of agriculture in the USSR ' by A I Tulupnikov and others (Glavnoye
Upravleniye Geodezii i Kartografii SSR, 1960), is an outstanding
achievement in its field. 308 pages of maps, statistics and explanatory text present detailed information on the physical conditions
affecting agriculture, incidence of crop growing and livestock raising,
and on the agricultural character of individual major regions, not
previously available. One of the latest studies has emerged from a
seminar, ' Soviet and East European agriculture ', of which the

papers and maps have been published by the University of California Press (and OUP), 1967, edited by J F Karcs.

The mapping of forest resources has also been of great importance. Several maps of individual forest areas were published from the last years of the nineteenth century on, and surveys continued until, in 1953, the ' Map of forests of the Soviet Union ' was published in thirty two sheets at a scale of 1 : 2,500,000.

Economic mapping has been given priority over a long period. From the first, economic conditions were linked with physical features, population or transport facilities. An interesting atlas called ' Graphic tables showing the influences of railroads on the economic state of Russia ', compiled by I S Bliokj, was published in Warsaw in 1876. Several other ' commodity flow ' maps were prepared at about the same time, also a number of ' statistical ' atlases. A ' Map of industry of the European part of the Soviet Union ' at 1 : 1,500,000 was published in 1927 and the ' Map of industry of the Asian part of the Soviet Union ' at 1 : 5M in 1929. The ' Industrial atlas of the Soviet Union ', 1929-31, published by the Praesidium of the Supreme Council on National Economy, Leningrad, showed the distribution of individual industries throughout the country. In much greater detail were the maps in the 'Atlas of the Moscow region ' in 1933 and in the 'Atlas of the Leningrad region and the Karelian Autonomous Soviet Socialist Republic' in 1934. The 'Atlas of the power resources of the Soviet Union ' also appeared in 1934. More significant still was the ' Industrial atlas of the Soviet Union ' published in 1935, in which are incorporated the results of the first five year plan; heavy industry is emphasised in this atlas.

Publishers outside the Soviet Union have issued economic maps and atlases of the country. Best known, perhaps, is *The Oxford economic atlas of the USSR and eastern Europe,* 1956, the first of the Oxford regional economic atlases. A reprint with minor revision was issued in 1963. The maps and the accompanying text, prepared by the Cartographic Department of the Clarendon Press and the Economist Intelligence Unit, were based on Soviet sources not previously available to British scholars, and the atlas was well received by Russian geographers. Sixty four pages of maps and forty eight pages of notes, statistics and diagrams, with a full gazetteer, provided the first major reference work in English on this vital area. The first part contains physical maps and maps of individual regions; in the second part are the special physical, population and economic maps. Reference should be made to an interesting review by Russian geographers, printed as the first review article in *The geographical journal,* December 1956; it is a slightly abbreviated translation, by D Welsh, of the review which appeared in *Izvestiya,* Academy of Sciences of the USSR geographical series, 1956 no 4.

In 1960, Michigan University Press published an *Economic atlas of the Soviet Union,* compiled by George Kish. In this work there are sixty generalised regional distribution maps and five general maps, showing the physical background, vegetation zones, administrative divisions, population distribution and air transport, all printed in brown and black. Concise text precedes each regional map, stressing the historical evolution of the area, and there is a useful bibliography. *Soviet Union in maps: its origins and development,* edited by Harold Fullard (Philip, new edition 1960), is a valuable thirty two page brochure of maps, with notes, having an emphasis on economic aspects. ' Sowjetunion, bergbau und industrie ', a wall map on a scale of approximately 1 : 4M, was issued in 1965 by VEB Hermann Haack, Geographisch-Kartographische Anstalt. This is a striking map, using ten colours, three of which are screened to give four more tints, and the major effect is to emphasise the concentration of Russian industry and the vast underdeveloped areas. A great deal of information is included, using conventional symbols—distinguishing, for example, eighteen different types of mining. Previous small atlases published outside the USSR had been *Soviet Russia in maps,* by George Goodall (Chicago, Denoyer-Geppert, 1942); *An atlas of the USSR,* by Jasper H Stembridge (New York; OUP, 1942) and *An atlas of the USSR,* by J F Horrabin and James S Gregory (New York, Penguin, 1945).

Numerous maps and atlases have been compiled both within and outside Russia on these and other special topics and, of course, there have been increasing numbers of road maps during recent years, by many publishers. *A motorist's guide to the Soviet Union,* compiled by V E St J M Louis, 1965, is a comprehensive introduction to 'All roads open to foreigners', with information about the chief cities and towns (Yale UP, 1967; Pergamon Press, 1968). *An atlas of Russian history,* by Allen F. Chew, claims to be the first atlas of its kind in English; the maps are concerned with boundary changes, based on existing Russian historical maps, and each is accompanied by a brief textual comment. There is also *An atlas of Soviet affairs,* by R N Taaffee and R C Kingsbury (Methuen, 1965) and the *Russian history atlas,* now in final preparation. *An atlas of Russian and east European history,* by A E Adams, I M Matley and W O McCagg (Heinemann International Books 1966), presents perhaps the most balanced view, divided into four main periods, pre-1530, the Renaissance period, the nineteenth century and the twentieth century, prefaced by a short introduction,

'Basic data', giving the environmental and cultural setting. The maps are monochrome. A bibliography is included.

Two recent studies are particularly informative on aspects of Soviet thematic cartography:

Pierre Carrière: ' Quelques atlas soviétiques relatifs à la climatologie du globe terrestre ', *Annales de géographie,* September-October 1967 (416) pp 574-588.

A G Isachenko and M B Vol'f, editors : ' Thematic cartography in the USSR ', Geographical Society of the USSR, for the Academy of Sciences of the USSR (Leningrad, Nauka, 1967), in Russian, including maps.

Maps are included in the Soviet current national bibliography and in most of the current bibliographies of individual states. Other sources of information on Soviet cartography include the many scholarly geographical journals, especially *Izvestiya,* of the Academy of Sciences of the USSR, geographical series, and *Soviet geography: review and translation,* published by the American Geographical Society of New York from 1960.

Concluding this section, it may be useful to mention the *Maps of Soviet Central Asia and Kazakhstan,* published by the Central Asian Research Centre in association with St Antony's College, Oxford, Soviet Affairs Study Group, a collection of maps in a folder, with a short introduction and a gazetteer-index, 1959; the ' Road map of eastern Europe ' (Budapest, Cartographia, 1963), 1:1,850,000, which covers central and eastern Europe as far as Kiev, with an inset at 1:7,500,000 extending to Leningrad, Kazan and Sochi, with multilingual legends; the ' Eastern Soviet Union ' map, at 1:11,404,800, completed by the National Geographic Society in 1967; and the ' Soviet central Asia map ' in progress by the Central Asian Research Centre. Maps of the parts of Russia in Asia will be included in some of the items to be mentioned in the next section. The maps and the bibliography in Violet Conolly's *Beyond the Urals* (OUP, 1967) are especially relevant.

THE FAR EAST

Asian map coverage is highly complex and varied; at the larger scales, much of the area is mapped in only preliminary or incomplete series. The continent is covered by maps at 1:4M by the British Geographical Section, General Staff and most of it is covered at 1:1M. Each country has at least the well-developed regions mapped at medium or large scales; but only Israel, Japan, Korea

265

and Taiwan, India, Pakistan and Burma have been fully covered by topographic quadrangles.

The two United Nations Regional Cartographic Conferences for Asia and the Far East have been of the first importance for geodesy, topographical mapping and photo-interpretation. A series of regional geological and related maps has been prepared under the sponsorship of the United Nations Commission for Asia and the Far East in co-operation with the Commission for the Geological Map of the World of the International Geological Congress. Of those so far completed, the third, 'Mineral distribution map of Asia and the Far East', 1:5M, 1963, is a particularly useful summary of information not otherwise readily available; four sheets show the distribution of the known mineral resources in relation to their geographical background, with explanatory text. The 'Geological map of Asia and the Far East' consists of six sheets in full colour, at 1:5M, with index map and accompanying brochure; and the 'Oil and natural gas map of Asia and the Far East', at 1:5M, is complete in four sheets, showing oil and gas fields and pipelines. A 'Map of Far East oil', 1:6M, prepared and published by B Orchard Lisle, Fort Worth, Texas, in 1963, shows oil and gas fields, refineries and pipelines and, in addition, lists the owners of the concessions, leases, permits and contracts; insets show some of the oilfields in detail.

A growing number of authoritative reports and periodicals are available, frequently including original maps or sketch maps; these are the publications of numerous organisations, academic departments or special groups interested in particular aspects. The annual United Nations *Economic survey of Asia and the Far East,* 1948-, with quarterly intermediate *Economic bulletins,* cover trade, transport, agriculture, industry, monetary and fiscal problems and devolpment planning; the 'Mineral resources development' series is especially vital, also the 'Programming techniques for economic development, with special reference to Asia and the Far East' and 'Formulating industrial development programmes, with special reference to Asia and the Far East' and other series relating to water resources, irrigation and drainage, community development and the *Small industry bulletin for Asia and the Far East.* Some of this work is summarised by David Wightman: *Toward economic co-operation in Asia: the United Nations Economic Commission for Asia and the Far East* (Yale UP for the Carnegie Endowment for International Peace, 1963), which includes a map and a bibliography.

The Asian Population Conference in New Delhi in 1963 proved very important, not least for the new surveys of population necessary in the preparation of the papers involved. Similarly, the *Proceedings* of the seminar on the development of basic chemical and allied industries in Asia and the Far East, held in Bangkok in 1962, published by the United Nations in 1964, are central documents. Erudite bodies such as the School of Oriental and African Studies are invaluable both in encouraging research and in making known documents of all kinds, as they become available. A new work, *Documentation on Asia,* began in 1960, edited by Girja Kumar and V Machwe (Allied Publishers, 1963-), under the auspices of the India Council of World Affairs; arrangement of entries is first regional, then systematic. Incidental notices of maps are to be found therein, but this is only one of the innumerable sources which must be scanned for information on maps. Scholarly monographs are tending to distinguish map resources in their bibliographies and many include useful maps in the text, such as the following:

G B Cressey: *Asia's lands and peoples: a geography of one-third the earth and two-thirds its people* (McGraw-Hill, third edition, revised 1963).

E H G Dobby: *Monsoon Asia: a systematic regional geography* (ULP, 1961).

W G East and O H K Spate, editors: *The changing map of Asia: a political geography* (Methuen, fourth edition 1961).

Norton Ginsburg, editor: *The pattern of Asia* (Prentice-Hall, Constable, 1958).

Pierre Gourou: *L'Asie* (Paris, Hachette, 1953).

Erich Thiel: *The Soviet Far East: a survey of its physical and economic geography;* translated by Annélie and Ralph M Rookwood from *Sowjet-Fernost: eine landes-und wirtschaftskundliche Ueber-sicht* (Munich: Isar Verlag, 1953. Methuen, 1957).

M Tikhomirou: *Cities of central Asia* (Central Asian Research Centre, 1961).

A Zischka: *Asien: Hoffnung einer neuen Welt* (Oldenburg, Gerhard Stalling Verlag, 1950).

THE INDIAN OCEAN

The Indian Ocean has been the last of the great ocean expanses to be scientifically explored, but in recent years there has been a renewed interest in the region and a number of expeditions have been mounted by the United States of America, Australia, Japan, the USSR and the Federal Government of Germany. The most important of these expeditions was the International Indian Ocean expedition, sponsored by UNESCO and other relevant bodies under the auspices of the United Nations; with this international effort, the floor of the Indian Ocean has become one of the best surveyed areas. The results of this research have been the subject of many discussions and documents, for example, the series of papers presented at the discussion meeting of the Royal Society on November 12 1964, concerning the floor of the north-west Indian Ocean; these papers, edited by Dr M N Hill, were published in the *Philosophical Transactions* of April 1966, accompanied by maps and diagrams.

The Pan-Indian Ocean Science Association has achieved much in the stimulation of research, in organising congresses and other meetings; among its publications, the *Répertoire des principales organisations scientifiques de l'Océan Indien* (Tananarive, Impr Officielle, 1960) is most valuable and the congress *Proceedings,* 1951- contain detailed reports on work in the area. The Sixth International Symposium on Maritime History, 1962, took as the subject of two sessions ' The Indian Ocean routes '.

The Directorate of Overseas (Geodetic and Topographical) Survey has mapped the Seychelles Island Group at 1:50,000 or 1:10,000; these are excellent, clear maps, with relief shown by contours. A ' Geological map of the Seychelles Islands ' at 1:50,000, surveyed by B H Baker, was published in four sheets by the Ministry of Commerce and Industry, Mines and Geological Department, Nairobi, in 1961. Topographical maps of Mauritius are also maintained by the Directorate and an agro-climatic map at 1:100,000 was issued in 1967. Soil surveys have been carried out at 1:100,000 from 1962 by the Sugar Industry Research Institute, Réduit, Mauritius, published by the Directorate; provisional classification of soil groups has been made. *Mauritius: an economic survey,* 1964, by Barclays Bank DCO, includes a useful map. The monumental *Bibliography of Mauritius, 1502-1954,* by Auguste Toussaint and H Adolphe (Port Louis, Government Printer, 1956), includes cartographic material.

Douglas Botting: *Island of the dragon's blood* (Hodder and Stoughton, 1958) is useful, as elsewhere there is little descriptive material on the island, Socotra, and a map is included. Studies such as ' Ecology of Aldabra Atoll, Indian Ocean ', edited by D R Stoddart, produced in the *Atoll research bulletin,* no 118, 1967, well illustrated by maps, figures in the text and photographs, are gradually adding to the sum of knowledge of this vast area. Maps are included also in the following:

Burton Benedict: *People of the Seychelles* (HMSO, 1966, overseas research publication, no 14).

J D Sauer: *Coastal plant geography of Mauritius* (Louisiana State University Studies, coastal studies series, 1961).

Sir Robert Scott: *Limuria, the lesser dependencies of Mauritius* (OUP, 1961).

August Toussaint: *Histoire de l'Océan Indien* (Paris, PUF, 1961).

India: Cartographic activities in India consist of the work carried out by the Survey of India Department, Dehra Dun, in land surveying, and by the Naval Hydrographic Branch, for hydrographic surveys. The Air Survey Company of India undertakes aerial photography and surveys. With the rapid development of India, the demand for maps on various scales has greatly increased; the Survey of India Department has been expanded and contracts have been made for particular projects with organisations and firms from other countries. The Survey of India also forms the national committee representing the country at the International Cartographic Association.

The early work of the Survey does not come within the scope of this work, but it should be kept in mind, for its progress was an epic in which figures such as James Rennell, William Lambton and George Everest played leading roles, in the face of every kind of physical and human difficulty. A lecture to the Royal Geographical Society on December 11 1967, by G F Heaney, ' Rennell and the surveyors of India ', printed in *The geographical journal,* September 1968, should not be missed. It includes the discussion and a few references and a fine map by Brigadier Heaney, drawn by the Cartographic Department of the society, in shades of browns, with the river systems, boundaries and names in black; the countries bordering India and Pakistan in an arc of mountainous country from Afghanistan to Burma are magnificently portrayed as so natural a barrier and the watershed of great rivers. By the end of the nineteenth century, the Great Trigonometrical Survey had been extended to cover the whole of India and Burma and exploratory surveys were carried into the northern mountain barrier; several penetrated into unknown parts of central Asia, bringing back the first reliable geographical information of these areas. Today the Survey of India, within the Ministry of Scientific Research and Cultural Affairs, is larger than ever and is manned entirely by Indians playing an increasingly important part in the development of agricultural and irrigation projects, soil conservation and planning for the exploitation of power, industry, minerals, oil, gas, etc.

From 1905 on, cadastral surveys at sixteen inches to the mile became the responsibility of provincial administrations, based on the trigonometrical framework of the Survey. A new series of one inch to the mile maps was begun and, by the time of the second world war, more than three thousand of these sheets had been published, in addition to many others at half-inch and quarter-inch scales. Index maps recording progress are included in the *Annual reports;* reference should also be made to the authoritative paper by Brigadier G F Heaney, previously Surveyor-General of India, 'The Survey of India since the second world war', presented to the Royal Geographical Society, February 11 1952 and printed, with the discussion, in *The geographical journal,* September 1952, with a map showing the progress of topographical survey and a second map indicating special project surveys.

The metric system was introduced in 1957, one result of this change being the substitution of 1 : 50,000 and 1 : 250,000 scales as the standard series. Series are also produced at 1 : 25,000 and at 1 : 100,000. A 1 : 1M series forms the basis of numbering at all scales. The polyconic projection is now used entirely. Three separate styles are therefore in existence: 'old style', 'old maps in modern style' and 'modern style'. In the latter, all based on revised surveys, relief is shown by brown contours, grey hill-shading, red for roads and built-up areas, yellow for cultivated areas, green for vegetation and black for all other features. On the 1 : 1M sheets, contours for the greater part of the country are at fifty feet intervals. Rules for transliteration of names for about twelve of the more well-developed languages of India to Hindi (Devanagari script) and into Roman have been under active consideration for some time. Scribing has been introduced as standard practice in map drawing and experiments have been made with a special scribe coat for glass, to suit Indian climatic conditions. The introduction of plotting machines has helped to speed up the production of maps, particularly those on large scales required for special projects. A *Survey of India maps catalogue* has been issued since 1950, with quarterly addenda, but there is a time lag in production. The standard history of the survey is that by R H Phillimore, *Historical records of the survey of India,* in five volumes (Dehra Dun, Geodetic Branch, Survey of India, 1945-), in which there are illustrations and maps.

'States and districts of India' at approximately 1 : 4,245,120 and five district maps at 1 : 1,520,000, published by the Bureau of Commercial Intelligence and Statistics, Bombay, show the latest administrative boundaries. The Geological Survey of India has produced geological maps at one inch to thirty two miles, in eight sheets, 1931; a one inch to ninety six miles, in 1949; and a 1 : 2M geological map, now in its sixth edition. Quarter-inch maps of selected areas are to be found in the *Memoirs,* 1856- and the *Records,* 1868-. A 'Tectonic map of India' and a 'Metallogenic-minerogenetic map of India', both at 1:2M, are also available. The *Quarterly journal of the Geological, Mining and Metallurgical Society of India*, Calcutta, 1926- and the *Transactions* of the Mining, Geological and Metallurgical Institute of India, Calcutta, 1906- are most valuable for the maps they contain. D N Wadia: *Geology of India* (Macmillan, third edition 1953) is the standard book on rock formations, historical geology and physiography of the Indian sub-continent, based on the *Memoirs* and *Reports* of the Geological Survey of India and the author's own investigations. Several of the individual States have long-established geological surveys; for example, the Mysore Geological Department, which has published *Records* since 1894 and *Bulletins* from 1904; and the Hyderabad Geological Survey, with its *Journal, 1929-, Bulletins, 1937-* and *Annual reports, 1942-*.

Systematic land use mapping of India was begun by the National Atlas Organisation, following in general the land classification recommended by the International Geographical Union Commission on a World Land Use Survey. About two-thirds of India was mapped at a quarter-inch to the mile, showing towns, villages, mining areas, brick-fields, salt-pans and other non-agricultural areas, horticulture, orchards and plantations, crop lands, cultivable wastes, woodlands, forests, swamps, mud flats, laterite soils, etc. These maps were reduced to 1:1M for publication and a still more generalised map was made at 1:6M.

The best one sheet map is probably that published by Bartholomew, 1:4M, 1966, with insets of Delhi, Calcutta, Madras, Bombay and Karachi at 1:200,000, and the Nicobar and Andaman Islands at 1:4M.

Maps in official publications have, in general, been poor, but the Survey of India, after independence, issued a political map at 1:4,435,200, showing the internal and external boundaries. The 1941 Census publication also included a useful base map at this scale, coloured to show forest, scrub and irrigated areas and district boundaries. For the census of 1961, a nation-wide atlas project was planned, by which map series for each state were included as volume IX of the complete set of *Reports*, incorporating information from other authoritative sources as desirable. The same pattern is followed for each series: orientation, physical features, demographic data, economic data, socio-cultural topics; but scales differ according to the size and shape of each state. Substantial text accompanies most of the maps and data sources are meticulously given.

Between 1960 and 1964, the National Atlas Organization worked on a population map of India, consisting of thirteen plates at 1:1M

271

and two at 1:2M, drawn according to the recommendations of the International Geographical Union Commission on a World Population Map. Absolute distribution of population has been indicated by dots and urban population by proportional circles. Classification by occupation is also shown. Another map, in five plates, 1961, showed density of population at 1:2M, and a programme of interesting population mapping has been continuing, based on 1951 and 1961 Census material: for example, distribution of rural, urban and tribal population at 1:600,000; age and sex structure, 1:600,000, and distribution of working population, at 1:2M. No 34 of the United Nations population studies was *The Mysore population study,* a report of a field survey carried out in selected areas of the state to obtain and verify information on the trends of population, the factors which influence them and their social and economic consequences (New York, 1961); a comprehensive work, published in English.

A 'Tribal map of India' was published by the Department of Anthropology, Calcutta, in 1956. Another example of the activity in population surveying and mapping during the past few years is described in the publication *A pilot survey of fourteen villages in Uttar Pradesh and Punjab* (Asia Publishing House, 1960). During 1954-55, the Agricultural Section, Economics Research Station, Delhi School of Economics, University of Delhi, carried out a survey of fourteen villages in Meerut and Bulandshabar Districts of Uttar Pradesh and Hissar and Rohtak Districts in Punjab; this was part of a larger project known as the 'Continuous village surveys' series. The plan is to resurvey the same villages at intervals of five years, to study economic changes, etc and to apply the knowledge gained in practical suggestions relating to rural planning.

Maps and diagrams are included in *The gazetteer of India* (New Delhi, Publications Division, Ministry of Information and Broadcasting, 1965), which contains specialist articles on all aspects of the physical, economic, biogeographical and cultural features of India. The Ministry of Cultural Affairs, New Delhi, has drawn up a general plan to be followed by state governments in a completely new set of *State gazetteers. A handbook for travellers in India, Pakistan, Burma and Ceylon,* published by Murray, completely revised in the nineteenth edition 1962, contains numerous maps and plans, as also does the *Fact book on manpower,* an official publication of the Institute of Manpower Research, and *India: a reference annual,* compiled by the Research and Reference Division, Ministry of Information and Broadcasting, 1953-, widening its scope with successive issues until it now covers every major aspect of life in the country.

The great *National atlas of India* appeared in a preliminary Hindi edition in 1957, under the direction of S P Chatterjee, head of the Department of Geography in the University of Calcutta (Dehra Dun, Ministry of Education and Scientific Research, 1957). The maps are in Hindi, but a looseleaf outline map is provided, with all place names in English, and on the back is a translation of every map key into English. The text is in both Hindi and English. Scales range from the standard scale adopted for the main maps, 1:5M, to 1:80M for special topic maps. Twenty six sheets in full colour, many containing inset maps, cover India's relationship with other countries, administration and physical features, climate, geology, land use, irrigation, power resources, industry, communications and population. Not only was the atlas an up-to-date source of information at that time, but an exciting example of new methods of cartographic presentation. An English edition, edited by S P Chatterjee, has been in preparation since 1959, sponsored by the Ministry of Scientific Research and Cultural Affairs. The main scale used is 1:6M; large towns, such as Delhi and Calcutta, are mapped at 1:1M.

Several regional atlases have also been published, which are intended eventually to constitute a national atlas in greater detail. *An atlas of resources of Mysore State,* edited by A T A Learmonth and L S Bhat, in two volumes (Calcutta, Indian Statistical Institute; Asia Publishing House, 1961), is an excellent example, presenting part of a regional survey of Mysore State, undertaken by Professor Learmonth as a pilot project for the Indian Statistical Institute. The first volume is *An atlas of resources,* consisting of maps interleaved with commentary, of the physical and human resources of the state; agricultural production, transportation, industry, demography and social characteristics are portrayed against the relevant physical enviroment. Altogether, there are 105 monochrome maps at a new scale of approximately 1:2,500,000. Isopleths or pie-charts are used rather than the dot method for distribution maps and 'flow' maps are a feature. The second volume is *A regional synthesis.*

The fourth edition of the *Statistical atlas of Bombay State* was published by the Bureau of Economics and Statistics in 1950, in two parts, 'Provincial part' and 'District part'. Maps showing geographical features, agriculture, industrial development and social geography are accompanied by a wealth of tables and appendices. *The Indian agricultural atlas* was issued in 1952 by the Economic and Statistical Adviser, Ministry of Food and Agriculture, Directorate of Economics and Statistics; a second revised edition, containing fifty nine pages of maps, came out in 1958. Mainly economic and cultural geography are covered in the *Indien historisch-geograph-*

isches Kartenwerk: *entwicklung seiner Wirtschaft und Kultur,* by Professor Dr Edgar Lehmann and Dr Hildgaard Weisse (Leipzig, Verlag Enzyklopädie, 1958). Ninety maps, looseleaf in a folder, are accompanied by text and index. The *Climatological atlas of India,* edited by Sir J Eliot (Indian Meteorological Department, 1906) is said to be a splendid collection of 120 large, coloured plates; and a *Crop atlas of India* was published in Delhi in 1939. A *Historical atlas of India,* by Charles Joppen, was issued in London in a third edition 1929; and *An historical atlas of the Indian Peninsula,* by C Collin Davies (OUP, 1953, second edition 1959) covers the period from about 500 BC to 1947. Forty eight clear and detailed mono-chrome maps include thirteen which deal with contemporary geography; descriptive and analytical text is an integral part of the work.

The publications of several geographical societies are a source of useful sketch-maps—not always well produced, but presenting factual information, the results of surveys, etc. The Geographical Society of India, centred on the Geography Department, Calcutta University; the Bombay Geographical Association; the Muslim University Geographical Society, Department of Geography, Aligarh Muslim University; the National Geographical Society of India, Department of Geography, Benares Hindu University; the Indian Geographical Society, Department of Geography, Madras University and the Hyderabad Geographical Association are among the leading research bodies. The *Transactions* of the Mining, Geological and Metallurgical Insitute of India, Calcutta, are a central source of diagrammatic and survey maps; also the *Memoirs* of the Indian Meteorological Department and the *Indian weather review,* consisting of monthly reports and an annual summary. Numerous trade journals and reports, which on occasion contain valuable primary maps, are particularly important in the various aspects of agriculture and forestry.

160 informative maps are included in the authoritative work by O H K Spate and A A Learmonth: *India and Pakistan*: *a general and regional geography* (New York, Dutton; Methuen, third edition 1968). There are useful maps also in the following:

A Bopegamage: *Delhi*: *a study in urban sociology* (Bombay UP, 1957).

S B Chatterjee: *Indian climatology*: *climostatistics, climatic classification of India with special reference to the Monsoons* (Calcutta, Das Gupta, 1956); well illustrated with maps, charts and diagrams.

J Dupuis: *Madras et le nord du Coromandel*: *étude des condi-*

tions de la vie indienne dans un cadre géographique (Paris, Librairie d'Amerique et d'Orient, 1960).

A T A Learmonth: *Sample villages of Mysore State, India* (Liverpool University Department of Geography, Research papers, no 1 1962).

Jagdish Singh: *Transport geography of South Bihar* (Banaras Hindu UP, 1964).

H Uhlig *et al*: *Beiträge zur geographie tropischer und subtropischer entwicklungsländer, Indien-Westafrika-Mexico* (Giessen, Geogr Schrift 2 1962).

K Zachariah: *A historical study of internal migration in the Indian sub-continent, 1901-1931* (Asia Publishing House, 1965). Research monograph of the Bombay Demographic Training Centre.

A vast literature has accumulated on the dominant physical feature of the sub-continent, the Himalaya range of mountains stretching into Nepal and Tibet; some relevant works include the following:

P. Bordet: *Recherches géologiques dans l'Himalaya du Népal, région du Makalu* (Centre National de la Recherche Scientifique, 1961). Record of the French expedition of 1954-55.

S G Burrard and H H Hayden: *A sketch of the geography and geology of the Himalaya mountains and Tibet* (Delhi, Manager of Publications, 1908; second edition revised by Sir Sidney Burrard and A M Heron 1933); contains maps in the text and folding maps.

A Gansser: *Geology of the Himalayas* (Wiley Interscience, 1965).

Toni Hagen *et al*: *Mount Everest: formation, population and exploration of the Everest region* (OUP, 1963); translated from the original German edition of Dr Hagen's survey published by Füsseli of Zürich, 1959, with a folding map.

S L Kayastha: *The Himalayan Beas basin* (Varanasi, Banaras Hindu University, 1964).

Ulrich Schweinfurth: *Die horizontale und vertikale Verbreitung der Vegetation im Himalaya* (Bonn, Dümmler, 1957); with a vegetation map in two sheets; summaries in English, French, Spanish and Russian.

A number of studies have been made of individual, particularly significant regions. *Bengal in maps: a geographical analysis of resource distribution in West Bengal and Eastern Pakistan* (Orient Longmans, 1949), by S P Chatterjee. consisting of a folded map and text maps with commentary, is a useful record of the country before partition. A land use map of West Bengal, at 1 : 500,000, has now been completed under the auspices of the Land Utilisation Board, Government of West Bengal. *Bihar in maps,* by P Dayal (Patna, K Prakashan, 1953), is another useful study of the part of

275

this vast lowland area of the sub-continent, which has through history been a thoroughfare of peoples and disputed territories and which is now gradually showing great economic changes.

In 1955, the United Nations Cartographic Conference for Asia and the Far East was held in Mussoorie and a useful summary of the mapping of India to date was included in volume II of the *Proceedings of the conference and technical papers* (New York, 1957).

Pakistan: When Pakistan became independent in 1947, the Survey of Pakistan was established in Karachi; the Geological Survey operated separately. Records begin in 1950. Map coverage of both East and West Pakistan has been completed since 1947 by aerial surveys and the basic series of topographical maps have been revised. Pakistan is well covered by a primary triangulation of high quality and the levelling network was completed before 1947. A separate Air Survey Organisation has been established and modern mapping equipment installed. The topographic map series at 1:3,168,000 began a fifth edition in 1966. The momentous treaty signed by India and Pakistan on the division of the waters of the Indus basin enabled vast water conservation and irrigation projects to be put into operation. Through the length of Pakistan, from the Himalayas to the Arabian Sea, flows the Indus, through a land of scanty rainfall, in which for thousands of years, crop growing has depended on the annual floods; to a critical extent, the national welfare of West Pakistan depends upon irrigation agriculture. Mainly American and British engineering firms are contracted for the work and hundreds of miles in the valley have been mapped by the London and Canadian Divisions of the Hunting Survey Group. Waterlogging and salinity affect in some degree almost all the irrigated land in West Pakistan. In addition to route maps, up to 250 square miles of block mapping may be needed at the canal junctions and for judging the siting of headworks and barrages.

An *Atlas of Pakistan* (Karachi, Pakistan Publications, 1952) consists of twenty pages of political, physical and industrial maps, with indications of hydroelectric and thermoelectric production. *The Oxford economic atlas for Pakistan* is an expanded form of the Pakistan section of the 1953 *Oxford economic atlas for India and Ceylon;* prepared by the Cartographic Department of the Clarendon Press and The Economist Intelligence Unit (OUP, Pakistan Branch, 1955; USA edition 1958), it is amply illustrated with text maps and diagrams.

The publications of the Pakistan Geographical Association, Department of Geography, University of the Punjab and of the East

Pakistan Geographical Society, Department of Geography, University of Dacca are sources for original maps and sketch-maps. Maps admirably complement the text in K S Ahmad: *A geography of Pakistan* (New York, OUP, 1965), a competent basic geography of Pakistan and Kashmir.

Apart from the *Survey of India maps catalogue,* there is no central source for map bibliography. Maps are omitted from the *Indian national bibliography,* but some information can be picked up from the official *List of publications (periodical or ad hoc) issued by various ministries of the Government of India,* published by the Secretariat of the Lok Sabha since 1956 and from a *Catalogue of civil publications . . . published by the Government of India,* published by the Publications Branch of the government at irregular intervals. Maps and atlases are included in the *National bibliography of Pakistan;* there is also the *Catalogue of maps and surveys* published by the Surveyor-General, Survey Office, Muree, Pakistan.

Ceylon: In contrast to the vast land mass of India and Pakistan, principal triangulation and geodetic level networks were relatively easy to complete for the island of Ceylon. Topographical maps for civil use have been compiled by the Ceylon Survey Department since 1800. Two major series cover the country, at one inch to the mile and at a quarter-inch to the mile. Both are drawn on the Transverse Mercator Projection. The one inch series is composed of eighty two sheets, issued in both monochrome and multicoloured editions. Relief is shown by contours at 100 feet intervals and the colours used are blue for water, brown for relief, buff for cultivated land, green for woodland, red for primary and orange for secondary roads, with black for all other features. Seven special sheets have been prepared at this scale for particularly important areas, such as the Colombo district. The quarter-inch series is complete in four sheets. Contours are shown at 100, five hundred and one thousand feet intervals; relief is tinted in brown shades and the other colours used are blue, red and black. These sheets have frequently been used as bases for topical maps, such as land use. Many surveys have been carried out for irrigation and land improvement schemes to support the increasing population. Aerial survey methods were introduced in the mid 1950s. Air photographs at various scales cover the island and a new national map series at five miles to one inch has been completed, showing contours at twenty feet intervals. The department has its own printing section, which prints all government maps. A 1 : 1M scale map has been compiled. The latest group of maps to be issued has been a ten page series covering climate, mineral deposits, land utilisation, economy and population.

The *Report of the Land Commission,* 1958, includes a folding land utilisation map of the island.

Ceylon geographer, 1945-, quarterly journal of the Ceylon Geographical Society, is a source of original maps; there are twelve good maps in the text of B H Farmer's *Pioneer peasant colonization in Ceylon*: *a study in Asian agrarian problems* (New York, OUP, 1957); maps are to be found also in E K Cook: *Ceylon, its geography, its resources and its people* (Macmillan, 1931; revised edition by K Kularatnam 1951; in *The economic development of Ceylon*: *Report of a mission organised by the International Bank at the request of the Government of Ceylon* (Baltimore, Johns Hopkins P; OUP, 1953) and in Angelika Sievers: *Ceylon*: *gesellschaft und Lebensraum in den orientalischen Tropen*: *eine sozialgeographische Landeskunde* (Wiesbaden, Steiner Verlag, 1964). Ludwig Alsdorf: *Vorder, Indien, Bharat-Pakistan-Ceylon, eine landes-und kulturkunde* (Westermann, 1955) should also be noted.

Burma: The Geological Survey of India carried out a considerable amount of work in Burma until the Burma Survey Department was established in 1947. Topographical maps include four main series: 1 : 63,360, a half-inch and a quarter-inch and the 1 : 500,000. The sheet lines in all the series follow the international geographical co-ordinate lines and the sheet line system is derived from that adopted for the IMW 1 : 1M. New editions since 1947 have been based on the series completed by the Survey of India, revised by aerial photogrammetric survey and, when possible, more detailed field work. Large-scale surveys have been completed, especially for economic development or irrigation projects and new or improved railway systems. In volume II of the *Report* on the United Nations Regional Cartographic Conference for Asia and the Far East, 1955, there is a brief section, ' Recent progress in Burma '.

The government has organised a Geological Department, but progress has been slow. H L Chhibber has been the outstanding authority on Burmese geology: *The geology of Burma* (Macmillan, 1934); *The mineral resources of Burma* (Macmillan, 1934) and *Physiography of Burma* (Calcutta, 1933). Frank Kingdon-Ward specialised in the fauna and flora of Burma and the adjoining parts of Tibet and China. Most of his published works include maps in the text: *Return to the Irrawady* (Melrose, 1956); *Burma's icy mountains* (Cape, 1949); *In farthest Burma: the record of an arduous journey of exploration and research through the unknown frontier territory of Burma and Tibet* (Seeley Service, 1921). The best monograph, for scholarship of both text and maps, is probably Hugh Tinker: *The union of Burma: a study of the first years of indepen-*

dence (OUP, third edition 1961, issued under the auspices of the Royal Institute of International Affairs).

CENTRAL ASIA

NEPAL, BHUTAN, SIKKIM: The three kingdoms separating India from Tibet—Nepal, Bhutan and Sikkim—are almost entirely mountainous. Knowledge of these countries has become available from successive expeditions, frequently having special studies in mind. In 1924, the help of Indian officers of the Survey of India was asked by the Nepal Government to assist in the mapping of Nepal and by 1927 a quarter-inch reconnaissance survey map was completed. However, the triangulation covered only part of the country and the mountain chains of the Himalaya north and east of the Beri Gorge were not triangulated or even explored at that time. In co-operation with the Nepalese Government, surveyors and cartographers from the Federal Republic of Germany began work in 1962 on a map of Nepal, sponsored by the Germany Society for Promoting Research. This map, the first scientifically accurate map of the area, prepared by modern methods, was based on field work carried out between 1955 and 1962-63 and published at 1 : 50,000 by Freytag-Berndt und Artaria, Vienna. Sponsored by the International Geographical Union, a group of geographers is carrying out a survey of the Kathmandu valley, Nepal, on the basis of cadastral maps, beginning in 1968, so that this area is steadily being opened up.

Thirty five excellent maps are included in P P Karan: *Nepal: a physical and cultural geography* (University of Kentucky P, 1960); also in T Hagen: *Nepal: the kingdom in the Himalayas* (Berne, Kümmerley and Frey, 1961) and, with others, *Mount Everest: formation, population and exploration of the Everest region*, translated by E N Bowman (OUP, 1962); W Hellmich, editor; *Khumbu Himal ergebnisse des forschungsunternehmens Nepal Himalaya* (Berlin, Springer Verlag, 1964), the first published results of an expedition combining geological, geographical, meteorological, botanical, zoological and ethnographic research in the Khumbu region of Eastern Nepal; and H Kihara, editor: *Peoples of Nepal Himalaya* (Kyoto UP, 1957), scientific results of the Japanese expedition to Nepal Himalaya, 1952-53.

In 1961, the Kanjiroba Himal expedition surveyed the Karnali region of western Nepal and, in 1964, John Tyson led the West Nepal expedition. The results of these expeditions were incorporated on the map ‘ Kanjiroba Himal and adjacent areas in the Karnali region of western Nepal ', at approximately 1 : 150,000, published by the Royal Geographical Society in 1967; an inset shows western Nepal at approximately 1 : 2M. The map was controlled by a triangulation based on the positions and altitudes of hill summits above the villages of Kaigaeon and Bangthari, supplied by the Survey of India, and plotted by the single-picture method from photographs taken by the expeditions, supplemented by those of Dr H Tichy in 1953 and during the course of the Jagdula expedition, 1962. The final map in clear, clean lines, indicates contours in brown, and river systems, with spot heights, place names, woodland areas and known tracks, and was plotted by G S Holland and Captain E R Clownes and prepared for publication by the Drawing Office of the Royal Geographical Society. John Tyson's article ' West Nepal: exploring the Kenjiroba Himal ', includes a folded copy of the map, photographs and references (*The geographical journal,* September 1967). The Cartographic Department of the society issued a 1 : 100,000 map of the Everest area in 1960.

' The Kingdom of Bhutan ', at 1 : 253,440, was issued as map supplement no 5 of the *Annals* of the Association of American Geographers in 1965; the compiler was P P Karan. The topographic base was derived from Survey of India material, supplemented by field studies by Professor Karan in 1961 and 1964 and by the results of other recent research. The map is drawn on the Albers Conical Equal Area Projection. Relief is shown by hill-shading in a warm brown colour and spot heights. Roads, paths, settlements are shown and such interesting individual items as monasteries and shrines. Professor Karan's book, *Bhutan: a physical and cultural geography* (University of Kentucky P, 1967) contains illustrations, maps and a bibliography. ' Some geographical and medical observations in North Bhutan ' and ' On the highland frontier of North Bhutan ' were the titles of two lectures given to the Royal Geographical Society by Michael Ward, following his expedition with Dr F Jackson and Dr R Turner in 1964-65. The first of these lectures was printed in *The geographical journal* for December 1966, including the map compiled from observations and drawn by the Cartographic Department of the society. The map was based on the Survey of India ' Quarter-inch ' series, incorporating additional spot heights and indicating major cliff faces; further details of the river systems, marshy areas and lakes have been added. Names, other than those from the Survey of India map, have been rendered phonetically by the authors. Insets are included of ' Bhutan and adjacent countries ' and ' Villages in the Lhedi Valley and number of houses '.

TIBET: An expedition in 1913 through unsurveyed country is described by F M Bailey in *No passport to Tibet* (Hart-Davis, 1957) and the mapping of the geographic frontier between Assam and Tibet. Maps are included in the text. There are maps also in Giuseppe Tucci and E Ghersi: *Secrets of Tibet: being the chronicle of the Tucci scientific expedition to western Tibet, 1933,* translated from the Italian by Mary A Johnstone (Blackie, 1935); maps in the text and a folding map are included in the classic by F Kingdon-Ward: *The mystery rivers of Tibet: a description of the little-known land where Asia's mightiest rivers gallop in harness through the narrow gateway of Tibet, its peoples, fauna and flora* (Seeley Service, 1923).

CHINA: Important work has been done on the cartography and geography of China since the second world war; maps of China now make a most complex field of study and it is not easy to build up any complete picture of the state of topographic mapping in the country. A paper was given by Dr Tsao-Mo to the 1955 United Nations Cartographic Conference for Asia and the Far East on the 'Program of cartographic work in China'; another paper, by Tang-yueh Sun, was on the subject 'Use of the International Map of the World series in China' (both published in volume II of the *Report*). Bartholomew's 'China', in the 'World reference map' series, 1959, at 1:4,500,000, includes Mongolia, Korea, Formosa, northern Laos and Vietnam. Contours are coloured, roads are shown in red, railways in black and a wealth of other detail is included. Philip maintains a regional wall map of China at 1:5M and a 'Commercial map of China', edited by Sir Alexander Hosie at a basic scale of 1:3M, with 1:500,000 inset maps of the Peking-Tsinan and the Shanghai-Nankiang districts. Commodity distribution is indicated by names and by symbols and there is an index of place names. Hermann Haack of Gotha publishes a particularly effective wall map of China at 1:3M in four sheets. George Philip has just published *China in maps,* which traces the development of China and its peoples by twenty five pages of coloured maps, with brief text; the information covers history, climate, vegetation, agriculture, minerals and industry, population, communications and foreign trade. Also available from Philip is a coloured 'Regional wall map of China', showing physical features, boundaries, communications and other major features.

There are several reference atlases of China in Chinese, outstanding being those prepared by N K Ting, Wong Wen-hao and S T Tseng, of the Geological Survey, published by the *Shun pao,* a newspaper in Shanghai. The larger *New atlas of China* was issued

281

in 1933 and a smaller *New maps of the Chinese provinces* has appeared in several editions, all entirely in Chinese. The *New atlas of the Chinese Republic,* by V K Ting (Shanghai, 1934), includes an index which was reproduced as the *Gazetteer of Chinese placenames based on the index to V K Ting atlas,* compiled by the United States Board on Geographic Names, Washington, DC, 1944.

Some of the best atlases in English up to the 1940s, though limited to place geography, included the *Atlas of the Chinese Empire,* published by the China Inland Mission, London, 1908, and the *Postal atlas of China,* Nanking, 1936. *An atlas of Far Eastern politics,* by G F Hudson and Marthe Rajchman (New York, Institute of Pacific Relations, 1942) may be mentioned here. Another atlas compiled at this time by Marthe Rajchman was *A new atlas of China: land, air and sea routes,* with descriptive text by the staff of *Asia magazine* and an introduction by H E Yarnall (New York, John Day, for *Asia magazine,* 1941). A *Historical and commercial atlas of China,* by Adolph Herrmann, published by the Yenching Institute, Harvard, 1935-37, was issued in a second revised edition by the Aldine Publishing Company, Chicago 1966, with the title *An historical atlas of China,* edited by Norton Ginsburg, and made available in Britain in 1967 by the Edinburgh University Press. Useful historically is G T Trewartha's 'Ratio maps of China's farms and crops', published in *The geographical review,* January 1938.

A classic work, *Land utilization in China,* by John Lossing Buck, ✓ in three volumes, was a field study of 16,786 farms in 168 localities and 38,256 farm families in 22 provinces in China, 1929-33 (Shanghai, Commercial Press, 1937, reprinted in New York by the Paragon Press, 1964). Volume I analyses and summarises the data; volume II is an atlas containing 184 maps and 13 air photographs, all with descriptive comment, and volume III comprises statistical tables. The survey was suggested by Dr O E Baker, of the Division of Agricultural Economics, United States Department of Agriculture, at the Conference of the Institute of Pacific Relations at Honolulu, 1926, and the work was subsequently entrusted to the Department of Agricultural Economics, University of Nanking. The maps, in black and white, follow the outline of the first volume; sections comprise: regions, topography, climate, soils; the land; crops (distribution maps in detail); livestock and fertility maintenance; size of farm business; farm labor; prices and taxation; nutrition; population (on farms); aerial photographs of land utilization. Each section is prefaced by explanatory details and most maps carry brief annotations.

A *Provisional atlas of Communist administrative units* was published in 1959 by the United States Department of Commerce; this consists of twenty nine coloured map sheets. There have been many yearbooks of China, some of which ceased at the end of the 1930s or early 1940s; many of them included maps. The *Chinese yearbook* itself, issued by various publishers, contains useful maps, as does the *China yearbook,* formerly the *China handbook,* an annual review of political, economic, social and cultural events in China, with maps in the text and a folding map.

The National Geological Survey of China was founded in 1916; *Memoirs* and *Bulletins* were published up to the period 1937-47. A general geological map of China, 1:1M, went into a second edition in 1948. The *Bulletin* of the Geological Society of China, 1922-, includes maps and diagrams. An authoritative map is the 'Map for middle schools of China, Mongolia and Korea', 1:5M, in two sheets, published by the Chief Office for Geodesy and Cartography, Moscow, 1963; this map would be available probably only in special collections.

The National Research Institute of Geology of the Academica Sinica, Shanghai, published monographs, memoirs and other works, most of which had finished by 1947-48. Useful in this context are: J S Lee: *The geology of China* (Murby, 1939); James Thorp: *Geography of the soils of China* (Peking, the Ministry of Industries, 1939), sponsored by the National Geological Survey of China in co-operation with the Institute of Geology of the National Academy of Peiping, with the support of The China Foundation for the Promotion of Education and Culture: C P Berkey and F K Morris: *Geology of Mongolia* (New York, American Museum of Natural History, 1927).

The 'National atlas of China', edited by Chang Chi-Yun, in five volumes (Yang Ming Shan, Taiwan, National War College in co-operation with the Chinese Geographical Institute, 1950-62) is the latest in a long line of Chinese national atlases, beginning in 1718 with the 'Atlas of the empire' commissioned by the Emperor K'ang Hsi. In Chinese and English, the five volumes are arranged as follows: *1* Taiwan, 1959; *2* Tibet, Sinkiang and Mongolia, 1960; *3* North China, 1961; *4* South China, 1962; *5* General maps of China, 1962. The political maps reflect the Nationalist Chinese viewpoint.

A 'Geographical atlas of Taiwan', a work of 144 pages by Cheng-Siang Chen, was published by the Fu-Min Geographical Institute of Economic Development, 1959; and 'An atlas of land utilisation in Taiwan', compiled by Ch'en Chêng-Hsiang (Taipeh, National Taiwan University, 1950), comprises 121 pages, with accompanying text in both Chinese and English; statistics are

283

quoted from Japanese figures as well as from information forms completed by local persons. It is impossible to judge how accurate the information was, even to its date. In the small work *The development of Hong Kong and Kowloon as told in maps,* by T R Tregear and L Berry (Hong Kong UP; Macmillan, 1959), a vast amount of information has been collected and analysed; the text is interspersed with sketch-maps and plans. The *Colonial annual reports* of Hong Kong include maps, as well as other illustrations and trade graphs. In Hong Kong, the Crown Lands and Survey Office of the Public Works Department is responsible for standard topographic mapping. In 1962, the Government invited tenders for large-scale mapping of virtually the whole colony from a number of British and International Air Survey Companies, with a view to vast replanning operations to cope with increasing population problems. Hunting Surveys was given the contract to provide 1:600 plans with five feet contour intervals of most of Hong Kong island and Kowloon foothills; also 1:1,200 plans with ten feet contours of the rest of the colony. In addition, special reservoir surveys were required for calculating capacity figures. Nearly two thousand plans were required and these were prepared by the aerial survey teams working closely with Crown Lands' surveyors, who provided the necessary control points; Crown Lands and Survey Office staff use aerial photography whenever possible for the revision of existing maps.

Chinese geographical periodicals, when available, are naturally central sources of original maps and of information concerning map projects; the published work of, for example, the Chinese Geographical Society, which was founded in 1950, a merger of the former Chinese Geographical Society of Peking and the Chinese Geographical Society of Nanking. The Institute of Geography at Nanking is specially concerned with the application of geographical techniques to current economic and industrial problems; study falls into six groupings, dealing with climatology, physical geography, cartography, geomorphology, hydrography and economic geography. ' Geography of China ', bi-monthly, 1961- from the Chinese Academy of Sciences, Peking, and the *Journal* of the Geographical, Geological and Archaeological Society of the Universiy of Hong Kong, which carries English abstracts, should also be mentioned; there are numerous agricultural and trade journals, which occasionally include sketchmaps.

Many original maps, most of them prepared by Professor Allen K Philbrick, are included in *The pattern of Asia,* edited by Norton Ginsburg (Prentice-Hall, 1958); and the maps in the three volume *China proper* in the Admiralty geographical handbook series

(Naval Intelligence Division, 1944) are still useful. Other monographs including useful maps are the following:

L H D Buxton: *China, the land and the people: a human geography* (Oxford, Clarendon Press, 1929).

Sripati Chandrasekhar: *China's population: census and vital statistics* (Hong Kong UP; OUP, second edition 1960).

Wang Chung-Chi: *A geography of China* (T'aipei, Sate Translation Office, 1956); in two volumes.

G B Cressey: *China's geographic foundations: a survey of the land and its people* (McGraw-Hill, 1934).

G B Cressey: *Land of the 500 million: a geography of China* (McGraw-Hill, 1955).

S G Davis, editor: *Symposium on land use and mineral deposits in Hong Kong, southern China and south-east Asia* (Hong Kong UP; OUP, 1964). Eighteen papers on land use and twelve on mineral deposits, presented at a meeting in September 1961 as part of the Golden Jubilee Congress of the University of Hong Kong.

Albert Kolb: *Östasien: China, Japan, Korea: Geographie eines Kulturerdteiles* (Heidelberg, Quelle and Meyer Verlag, 1963); with maps in the text and separately in a pocket.

Owen Lattimore, editor: *Pivot of Asia: Sinkiang and the Inner Asian frontier of China and Russia* (Boston, Little, Brown, 1950), the work of a team of specialists taking part in Dr Lattimore's Inner Asian Seminar at the Walter Hines Page School of International Relations.

Joseph Needham: *Science and civilisation in China* (CUP, seven volumes 1954-); should be consulted for background information on all aspects; there are maps in the text.

Theodore Shabad: *China's changing map: a political and economic geography of the Chinese People's Republic* (Methuen, 1956); one of the most useful texts in clarifying the situation; includes sketch-maps.

Erich Thiel: *Die Mongolei: Land, volk und Wirtschaft der Mongolischen Volks-republik* (Isar Verlag, 1958).

T Tregear: *A survey of land use in Hong Kong and the new territories* (Hong Kong UP; Bude, Geographical Publications, 1958); monograph of the World Land Use Survey.

— *Hong Kong Gazetteer* (Hong Kong UP; Macmillan, 1958); an index to the Land Utilisation map, 1 : 80,000.

An index to the maps of China was published in 1945 by the National Geographic Society, Washington, containing 7,986 entries. Sir Aurel Stein compiled a *Memoir on maps of Chinese Turkistan and Kansu,* amounting to about 5,600 items, published in Dehra

Dun, 1923; and a 'Catalogue of maps of China proper', edited by Kô Nishimura, was published in Tokyo in 1967.

JAPAN: Japanese maps are of high technical quality and wide range. Notable work has been achieved in geomorphology, climate, land use and settlement geography. In successive field excursions, academic geographers have made exhaustive studies of the regional geography of the country, most of which remains so far in Japanese. The Geographical Survey Institute, Ministry of Construction, was founded in Tokyo in 1888 as the Intelligence Service of General Staff, which became the Imperial Land Survey in 1902 and adopted its present title in 1945. The Japan Cartographic Association, at the Geographical Survey Institute, is the national committee representing Japan at the International Cartographic Association.

The basic topographic scale is 1:50,000, which covers the whole of Japan in 1,263 sheets, at first in monochrome; the series was completely revised in 1952 and since then continuous revision has been maintained, at the rate of about sixty sheets a year. Map series at 1:10,000 and 1:25,000 have gradually increased in coverage, as air photography has come into use and modern plotting instruments, such as the Stereoplanigraph and the Multiplex, have been installed. 1:25,000 maps are available for industrial areas, most of the coast and parts of Hokkaido. Smaller scale maps are compiled, using the 1:50,000 map data. Eight sheets cover Japan at 1:500,000 and, at 1:2,500,000, Japan and an indication of its setting can be shown on one sheet. Since the early 1950s, land use mapping has become increasingly important. More than three hundred 'Land utilisation' series at 1:50,000 have been completed, also six sheets of very useful land form classification maps. Land use maps at 1:200,000 compiled by the Geographical Survey Institute, Tokyo, have been issued for the Hokkaido region. Regular soil mapping has been carried out in special areas since 1953, following individual experimental projects; an informative article on 'Reconnaissance soil survey of Japan', by C W L Swanson, is to be found in the *Proceedings* of the Soil Science Society of America, 1946.

The achievements of the institute are described in *Reports* and articles published in the *Journal of physics of the earth* and the *Journal of geodetic survey of Japan*. The *Journal* of the Japan Cartographic Association began in 1963, having a contents list in English.

The Geological Survey of Japan, Tokyo, founded in 1879, maintains series of geological maps at 1:50,000, 1:200,000 and 1:500,000. *Reports* have been issued since 1922 and *Bulletins* since 1950. A geological map of the Japanese Empire, covering Korea and Formosa, was compiled at 1:2M, with an explanatory bulletin in English. A new geological map of Japan, 1:2M, was published by the survey in 1964, drawn on a conical projection; twenty nine types of geological forms are distinguished. *The geology and mineral resources of the Japanese Empire* went into a second edition in 1926 and was completely revised again in 1956, with a second edition in 1960; in 1965, an *Outline of the geology of Japan,* including maps, was also published by the survey. The *Journal of the geological society of Japan* has been issued since 1894, until 1934 as the *Journal of the geological society of Tokyo,* and the *Japanese journal of geology and geography,* 1922-, with an emphasis on Japan, is one of the chief publications of the Science Council of Japan, Tokyo. *Geology of Japan,* by Fuyuji Takai and others (Universiy of California P, 1964), includes a historical review of geological research in Japan; each chapter has been written by an authority and is illustrated by maps. Many other journals of high quality including text maps and original work have been produced for some years, notably the *Journal* of the Faculty of Science, Hokkaido University, of which section IV is devoted to geology and minerals, 1930-; *Journal* of the Faculty of Science, University of Tokyo, section II, geology, mineralogy, geography, geophysics, 1926-; *Memoirs* of the College of Science, University of Kyoto, section M, geology and biology, 1924-; *Mining geology,* 1951- by the Society of Mining Geologists of Japan; *Science reports* of the Tohoku University, second series, geology, 1912-, third series, mineralogy, petrology, economic geology, 1921-.

The pressure of population on land resources has led to a concentration on thematic mapping; the scheme of land classification laid down in 1950 by the Commission of the International Geographical Union and advocated by the World Land Use Survey has been adopted, and aerial photography has speeded up this work. A series at 1:800,000 of three sheets each dealing with population, land use and communications has been published, also an excellent land use map in colour at 1:50,000. The Regional Forestry Office has been steadily mapping the forest areas at 1:50,000; hydrogeological maps of Japan are produced at various scales as required by the Geological Survey; and an indication of the improvement work in hand was given in ' Distribution of land improvement districts by size and work in Japan as of March 1956 ', published by the Tokyo Institute of Human Geography, University of Tokyo at 1:1,500,000, 1966. A paper, ' Present status of cartography and related surveys in Japan ', was given by Dr Naomi Miyabe to the United Nations Regional Cartographic Conference for Asia and the Far East, 1955 (published in volume II of the *Proceedings*). Maps have been issued with the 1955 and 1960 Population Censuses of Japan and an *Atlas of social studies* was published by Sanseida of Tokyo in 1957.

One of the best of the earlier physical maps of the whole Japanese Empire was that published by the Society for International Cultural Relations, Tokyo, 1937, at 1:2M. A wall map had been issued in 1931 at the same scale by the Land Survey Department. A special Olympic Games revised edition of 'Japan' at 1:2,500,000 in the Bartholomew 'World reference maps' series, 1964, includes inset maps of Tokyo at 1:500,000 and Okinawa at 1:1M; contours are indicated, roads are shown in red and railways in black. One of the latest maps of Japan is at 1:2M, published in 1967 by the Ministry of Foreign Affairs; relief is shown by layer colouring and hill-shading.

The *Standard atlas of Japan* was published in 1957 by the Teikoku Shoin Company, Tokyo; and the *Teikoku complete atlas of Japan* in 1964. The latter is a good, inexpensive general and economic atlas, in English, including large-scale maps of metropolitan areas. An *Atlas of all cities in Japan* is understood to be in progress, having reached volume IV in 1966, but no details are as yet available. A *climatological atlas of Japan* was issued in 1948 by the Industrial Meteorological Association, Tokyo; this would be helpfully used in association with T Okada's *Climate of Japan* (Central Meteorological Observatories, 1931).

Large numbers of economic, trade and industrial journals are published in Japan, many of them containing interesting sketch-maps, but difficult to come by in Britain, although there is evidence of a growing recognition of the importance of this literature; year-books and trade monographs are also numerous on individual industries, sometimes by a particular firm, many of them sump-tuously produced in English for publicity purposes. The author has been given some of these by personal contacts, but it is feared that they are not widely known outside trade circles. An *Economic atlas of Japan*, compiled by K Aki and others, was published in Tokyo in 1954 and a 'Key to the *Economic atlas of Japan*', by N Ginsburg and J D Eyre, was made available by the University of Chicago Press in 1959.

With the turn of the century, more reliable and scholarly mono-graphs on Japan began to be compiled by British and Americans studying and working in Japan. The Royal Institute of International Affairs and the Institute of Pacific Relations have done much to enable more accurate information on Japanese economic and social life to reach the west. In more recent years, the Japanese themselves have actively promoted exchange of information, especially since the organisation of the country's bibliographical services gained momentum in 1951. The monthly classified national current bibliography, *Kokumai shuppan-butsu mokuroku,* founded in 1949 by the National Diet Library, was superseded in 1955 by the weekly *Nohon shuho;* the service has included maps and annual cumulations have been issued fairly regularly since the issue for 1948-49. A bibliography of maps and atlases is contained in *Shuppen Nenkan,* a publications yearbook. *Japanese geography: a guide to Japanese reference and research materials,* by R B Hall and Toshio Noh (Michigan UP; Cumberlege, 1956) is extremely useful, in the Michigan University Center for Japanese Studies, bibliographical series, no 6.

C T Trewartha's published works on Japan are particularly notable: *A reconnaissance geography of Japan* (University of Wisconsin Studies in the Social Sciences and History, 1934) was a pioneer work, superseded in 1960 by his *Japan: a physical, cultural and regional geography* (University of Wisconsin P; Methuen, second revised edition 1965) including outstanding maps. The *Proceedings* of the International Geographical Union Regional Conference in Japan, 1957 (Tokyo, 1959) contains maps; the contributions have emphasis on Japan and Asia, with some of world-wide significance. Other relevant sources include E A Ackerman: *Japan's natural resources and their relation to Japan's economic future* (University of Chicago P; CUP, 1953); R P Dore: *Land reform in Japan* (OUP, for the Royal Institute of International Affairs, 1959); Guy-Harold Smith and Dorothy Good: *Japan: a geographical view* (New York, American Geographical Society, 1943); including a folding population map (special publication no 28). *The Japan of today,* prepared by the Public Information and Cultural Affairs Bureau, Ministry of Foreign Affairs, 1963, covers geography, economy, social conditions and cultural life. Numerous maps in the text and a folding map are included in the first of three volumes by Martin Schwind: *Das Japanische Inselreich,* and some sensible and effective sketch-maps have been compiled by Prue Dempster for her *Japan advances: a geographical study* (Methuen, 1967).

KOREA: In the Republic of Korea, the Geographical Survey Institute was set up in 1958 to develop cartographic activities. Topographic series were compiled at 1:50,000 and 1:100,000 as the basis for government planning and at 1:200,000 and 1:500,000 for public distribution. To consider geographical name standardisation, a committee was set up within the institute. Hydrographic surveying has been conducted by the Naval Hydrographical Bureau; surveys were soon completed for the sea areas surrounding the southern part of Korea.

The standard atlas of Korea, consisting of twenty nine plates of maps, with index, was compiled by Pong Su Yi and published by the Sosô Publishing Company, Seoul, 1960, in Chinese. A map is included in the *Handbook of Korea,* compiled by Chae Kyung Oh (New York, Pageant Publishing, 1958), which briefly describes the geography of Korea, agriculture, industry and commerce, trade and the people of Korea. *Korea's continuing development,* compiled by the Economic Planning Bureau and published by the Ministry of Reconstruction, Republic of Korea, 1959, presents an excellent summary to its date, illustrated and including sketch-maps in the text.

There is no current bibliography of map production in Korea, but maps and atlases have been listed in the retrospective national bibliography tentatively begun in 1956; title, name of publisher, date of publication, scale and price are given.

SOUTH-EAST ASIA

In 1929, the Tokyo Geographical Society published a ' Geological atlas of eastern Asia ' in seventeen sheets on a scale of 1 : 2M. The sheets cover Japan, Korea, Formosa, Manchuria, a large part of China and eastern Siberia, with an additional sheet for the Aleutian Islands. ' Geographia ' is maintaining a political map of the Far East, showing railways, canals and oil pipelines; an inset map shows the spread of populations. ' South-east Asia ', at 1 : 5,800,000, in the Bartholomew ' World reference map ' series, revised in 1964, gives a good idea of the extent and variety of this fascinating part of the world. It covers an area from Hong Kong to Timor and from the Andaman Islands to the eastern frontier of Indonesia. An inset shows northern Burma, Papua and most of the Trust Territory of New Guinea at the same scale; another inset shows the Malayan peninsula at 1 : 3M. The Transverse Mercator Projection was used. Land relief is shown by layer tints and spot heights, and submarine topography, particularly important in this area, is clearly shown by blue layers. Roads are graded, in red, railways in black and two kinds of airfields are indicated. Political boundaries are clear and a great deal of other information is included. A political map of south-east Asia, 1 : 1M was issued in two sheets, north and south, in 1965, by the Department of Lands and Surveys, Wellington.

In 1964, Macmillan published the *Atlas of South-east Asia,* the second in the series, in the same format as the *Atlas of the Arab world and the Middle East*. Sixty eight coloured maps were compiled with the help of geographers in south-east Asia. Detail on individual maps varies according to the quality of the available data. A number of maps of the physical features, climate, vegetation, population, minerals and communications of the area are followed by more detailed maps of the Philippines, Indonesia, Singapore, Malaya, Thailand, Indochina and Burma, showing seasonal climatic distributions, irrigation, industries and local communications. Eight historical maps figure as end-papers. The city plans clearly show land use and particular features of the sites. The text section, by D G E Hall, is illustrated with photographs. St Martin's Press of New York published the atlas for the United States and it was issued at the same time by Djambatan of Amsterdam.

Maps accompany the text of the *Bibliography of the peoples and cultures of mainland South-east Asia*, edited by John F Embree and Lillian O Dotson (Yale UP, 1950, in the South-east Asia Studies series); this work does not include Malaya. *Studies in the geography of South-east Asia: a selection of papers presented at the Regional Conference of South-east Asian Geographers, Kuala Lumpur* (Philip, 1964), also contains useful maps; there are thirty nine papers, originally published in volumes XVII and XVIII of *The journal of tropical geography*. The following is a selection of works including notable maps, providing reliable guidance on the understanding of this very complex area:

J F Cady: *Southeast Asia: its historical development* (McGraw-Hill, 1964); with useful sections on geographical setting and economic development.

C D Cowan, editor: *The economic development of south-east Asia: studies in economic history and political economy* (Allen and Unwin, 1964); Studies on modern Asia and Africa series).

C A Fisher: *South-east Asia: a social, economic and political geography* (Methuen; New York, Dutton, 1964).

H H Heinisch: *Südostasien: Menschen, Wirtschaft und Kultur der Staaten und Einzelräume* (Berlin, Safari Verlag, 1954). Industry, culture and people in Burma, Thailand, Malaya, Indochina, South China, Formosa, Singapore and Hong Kong, Macao and Timor, British Borneo, Indonesia, Philippines and New Guinea.

K J Pelzer: *Pioneer settlement in the Asiatic tropics: studies in land utilization and agricultural colonization in south-eastern Asia* (New York, American Geographical Society, 1948).

K Lê Thank: *Histoire de l'Asie du Sud-Est* (Paris, PUF, 1959; Que sais-je? series, on the thesis of a region and people united by climate and its influence on life.

THAILAND (SIAM): The Royal Thai Survey Department, Bangkok, was established in 1908; it now comprises a Control Survey Division, Air Photo Division and an Educational Division. Military Survey was transferred to the army in 1909, cadastral survey became a new service administered by the Ministry of Agriculture and the Royal Survey Department has since confined its work to topographical mapping. Plane-table methods were used at first to create four series, at 1:50,000, 1:100,000, 1:200,000 and 1:500,000. Their sheet lines, as well as their reference numbers, are derived from the IMW 1:1M. The sheets of the 1:100,000 series are mostly monochromes; in each of the other series, there are some monochrome and some in colour. In 1952, a co-operative mapping programme with the United States resulted in a revised survey for almost the whole country at 1:50,000, based on aerial mapping. With the assistance of the United States Army Map Service, thirteen first-order base lines were measured by 1955; all geodetic data were adjusted and tied into the Indian datum by the United States Coast and Geodetic Survey. The Universal Transverse Mercator Projection was used.

Since 1955, the Royal Survey Department has set up modern equipment and training courses. Maps showing the stages of cartography in Thailand at the time of the Second United Nations Regional Cartographic Conference for Asia and the Far East are included in volume II of the *Proceedings, 1958*. The Thai Ministry of Community Development, in co-operation with Shell, has been carrying out a detailed land use survey. Maps are included in the official report *Land utilization of Thailand* (Ministry of Agriculture, 1963), printed in Thai and English, and in 1964, eighteen maps, mainly at 1:2,500,000, were published on such themes as climate, forests, soils, minerals, land use, population distribution, communications and power resources.

Excellent maps are in the comprehensive study by Wilhelm Credner: *Siam: das land der Tai, eine landeskunde auf grundigener reisen und forschungen* (Stuttgart, J Englehorns Nachf, 1935). At a later date, maps also figure in J E de Young: *Village life in modern Thailand* (University of California P; CUP, 1955) and in *A public development program for Thailand: report of a mission organised by the International Bank at the request of the Government of Thailand* (Baltimore, Johns Hopkins P, 1959).

LAOS, CAMBODIA, VIETNAM (INDOCHINA): An *Atlas de l'Indochine* was issued in 1920 and 1928 by the Service Géographique de l'Indochine, Hanoi; an *Atlas météorologique de l'Indochine*, prepared by the Service Météorologique, appeared in 1930 and the *Atlas van Tropisch Nederland* (The Hague, Martinus Nijhoff, 1938) also remains a useful introduction to this whole area. An outstanding achievement was the *Atlas des colonies françaises, protectorats et territoires sous mandat de la France,* edited by Guillaume Grandidier (Paris, Société d'Editions Géographiques, Maritimes et Coloniales, 1934), issued in parts, with supplementary maps, illustrations and diagrams. A pictorial map of French Indochina was prepared about 1946-47 on a scale of 1:2M, by Lucien Boucher, published by the Association Nationale pour l'Indochine Française, Paris; and a ' Carte générale de l'Indochine Française ' at 1:1M was produced under the direction of the Service de Documentation Cartographique. The Institut Géographique National, Paris, issued a ' Carte routière de l'Indochine ', at 1:400,000 in fifteen sheets, in 1945, on the Bonne Projection, with legends in French and English; and the Ecole Française d'Extreme-Orient published a ' Carte ethnolinguistique ' at 1:2M, prepared by the Service Géographique de l'Indochine in 1949.

A good deal of mapping in the area had therefore been accomplished when, between 1952 and 1954, an aerial survey of the greater part of the territories of Cambodia, Laos and Vietnam was undertaken by the Institut Géographique National, at the scale of 1:40,000. In 1955, the governments of Cambodia, Laos and Vietnam took over the responsibility for the cartographic work relating to their respective countries. At that time, maps existed at scales ranging from 1:25,000 to 1:2M, according to the terrain, and most of the three territories had been mapped at 1:100,000. The Service Géographique National du Viet-Nam was founded in 1955. Ensuing events have not encouraged continued progress.

In 1951, the Service Géologique de l'Indochine, Hanoi, was transferred to the Centre de Recherches Scientifiques et Techniques de l'Indochine. A *Bulletin* had been issued from 1913 to 1950, which was succeeded by *Archives géologiques du Cambodge, du Laos et du Viet-Nam,* 1952-. *Mémoires* were published between 1912 and 1927 and a ' Carte géologiques de l'Indochine ', at 1:2M, was issued in a second edition, 1952. Maps were included in the annual *France d'outre-mer,* from 1946 (Paris, Didot-Bottin, Annuaire de Commerce); and the maps in the Admiralty handbook *Indo-china* (Naval Intelligence Division, 1943) are still useful. A folding map accompanies Donald Lancaster's *The emancipation of French Indo-China* (Royal Institute of International Affairs, 1961). Charles Robequain's fundamental works on this region are, unfortunately, not well furnished with maps, but there are useful maps in Eugène Teston and Maurice Percheron: *L'Indochine moderne: encyclopédie administrative, touristique, artistique et économique* (Paris, Librairie de France, 1932).

293

A map 'Royaume du Cambodge Carte Routière et Administrative' at 1 : 1M was published by the Service Géographique in 1961. *Indo-China: a geographic appreciation,* prepared by the Canada Department of Mines and Technical Surveys, Geographical Branch (Foreign geography information series, no 6, 1953), in eighty eight pages presents maps covering physical and economic topics, alternating with pages of analytical discussion, an effective introduction to the whole area. The National Geographic Society has issued two map supplements to the *National geographic,* in 1965 and 1967 respectively: 'Vietnam, Cambodia, Laos and eastern Thailand', at 1 : 4,707,648, drawn on the Lambert Conformal Conic Projection, showing relief by hill-shading and spot heights; and 'Vietnam, Cambodia, Laos and Thailand', at 1 : 1,900,800, with an inset of south Thailand. Maps are included in the United Nations Report of the Economic Survey Mission to the Republic of Vietnam, *Toward the economic development of the Republic of Viet-Nam* (New York, 1959). Evocative sketch-maps are included in G C Hickey: *Village in Vietnam* (Yale UP, 1964), an account of Khanh Hau, in the Mekong river delta, taken as an example; and in D J Steinberg and others, editors: *Cambodia: its people, its society, its culture* (New Haven, HRAF P, 1957).

In Cambodia, there is no current national bibliography. Publications in French issued in Cambodia are included in the *Bibliographie de la France* and government publications are listed in the *Journal officiel du Cambodge,* weekly, 1946-, in French and Cambodian. With regard to French publications, the same applies to Laos; there is a system of compulsory legal deposit, but, so far as is known, no lists are published. The Directorate of Archives and National Libraries, established in 1959, made plans for the issue of a monthly current bibliography and various individual bibliographies were published in the following year. The library of the National Institute of Administration published a classified catalogue of the library in 1957, followed by a second edition in 1959, and a monthly bulletin began in 1957; the Vietnamese-American Association at Saigon published a catalogue of its collection in 1958.

MALAYSIA: In dealing with 'Malaysia', an area of some 130,000 square miles is under consideration, composed of nine sovereign Malay states, the former British Straits Settlements Colonies of Penang Island, Malacca and Singapore and the former British Crown Colonies of Sarawak and North Borneo, now known as Sabah. It occupies the Malay peninsula in South-east Asia and the northern and north-western regions of the island of Borneo. Land frontiers with Thailand lie on the Kra isthmus and touch the

Republic of Indonesia in the island of Borneo. For most of this complex and multiracial area, the Directorate of Overseas (Geodetic and Topographical) Survey has produced maps by field and air survey at 1.25,000 and/or 1:50,000, according to the nature of the area. The Federation of Malaya Survey Department was established in 1909 at Kuala Lumpur and it also forms the national committee for representation at the International Cartographic Association. In the 1960s, the Survey Department has made great progress, at both large and small scales, as required. The 1:2M map of Malaysia, 1964, issued by the department, in two sheets, was the first official map of the Federation; a reduction, at 1:4,350,000 is available as a brochure for tourists, giving information on international and state boundaries, residency or divisional boundaries, railways, main roads and chief towns; at the base of the map and on the reverse are brief notes on places and facilities in the country, with small inset photographs of particularly important settlements or buildings.

A series at 1:63,360, of individual areas, is in progress, for example 'Malacca', 1964, showing predominant land use and communications. The Survey Department is also carrying out a land use survey at 1:25,000, based on aerial photographic interpretation, with the aim of calculating the statistics of agricultural land on a state-wide basis, with the assistance of the Commission on World Land Use Survey.

In 1958, the Singapore Improvement Trust and Department of Geography, University of Malaya, began to issue a series of over ninety detailed land use maps at 1:6,366, of rural Singapore, for urgent planning. Forty three categories of land use are clearly defined on the maps, using nine basic conventional colours, in combination with symbols, which distinguish individual local crops, such as pineapples or pepper. The Malaya Geological Survey, founded in 1903, has issued *Memoirs* since 1937, *Annual reports* since 1946 and a 'Geological map of Malaya' at a scale of one inch to twelve miles, in 1948. In 1963, a sixth edition of the geological map was published and a fourth edition of the 'Mineral distribution map' in 1966, both at 1:500,000. A map is included in *The geology of Malaya,* by J B Scrivenor (Macmillan, 1931).

The publications of the British Association of Malaysia, 1920-, are important for information on developments of all kinds within the area; and a number of other reliable periodicals on special aspects of the economy of the country include notices of specific surveying projects. Sketch-maps are included in the publication *Malaysia in brief,* produced by the Department of Information, Kuala Lumpur, in 1963, covering physical, social and political aspects of the territories, the economy, industries, trade and commerce and international relations.

In the Nelson series of overseas atlases is included the *Primary school atlas for Malaysia and Singapore,* 1965, consisting of twenty six pages of maps, with special emphasis on Malaysia, and explanatory text and index, in English, distributed in Singapore by the Tien Wah Press. Standford's ' General map of Malaysia ' is at the scale of 1 : 633,600 (Philip, revised edition 1966). Excellent maps are included in Norton S Ginsburg and C F Roberts, junior, editors : *Malaya* (University of Washington Press, 1958) in the series of publications of the American Ethnology Society; environment and the economic, political and social characteristics of Malaya are considered, including Singapore. Useful maps are to be found also in the following :

The economic development of Malaya: report of the International Bank for Reconstruction and Development (Baltimore, Johns Hopkins P, 1955).

Pierre Fistré : *Singapour et la Malaisie* (Paris, PUF, 1960).

Ooi Jin-Bee : *Land, people and economy in Malaya* (Longmans, 1963).

Charles Robequain : *Malaya, Indonesia, Borneo and the Philippines : a geographical, economic and political description of Malaya, the East Indies and the Philippines* (Longmans, second edition 1958, in co-operation with the International Secretariat, Institute of Pacific Relations, in the Geographies for advanced study series); translated from the French by E D Laborde.

State of Singapore Development Plan, 1961-1964 (Ministry of Finance, 1961).

There is no systematic current national bibliography in the Federation of Malaya and Singapore, but legal deposit of publications is compulsory and they are recorded in the official *Gazette* of the Government and Colony of Singapore and the Federation of Malaya respectively.

INDONESIA : With the proclamation of independence in 1945, a new era began for Indonesia. The Geographical Institute, Directorate of Topography, Ministry of Defence, at Djakarta, was established in 1947. For many years, Dutch scientists and geographers had prepared maps, research monographs and reports, mostly only in Dutch. The institute has continued this work, but maps and reports are now usually published in English as well. The primary responsibility of the institute is to produce maps for the army, but it prepares other topographical maps as required by the government departments. It includes an Institute of Geodesy, an Institute of Photogrammetry, an Institute of Geography, an Institute of Topography and a Printing and Reproduction Office. Aerial photography was introduced in the 1950s and has been used increasingly, not only for military maps, but also in the preparation of forestry, land use, agriculture and soil survey maps. From 1955, a new topographic series for Bali has been in progress at 1 : 25,000; a series for Sumatra at 1 : 50,000 and of eastern Sumatra at 1 : 100,000 also; and a revision of the existing 1 : 50,000 for Java was begun, supplemented by land surveying.

A Geological Survey had been established by the Dutch authorities in 1850 and this has been continued since independence as the Bureau of Mines and Geological Survey, in Bandung. The latest geological map of Indonesia, at 1 : 2M, was prepared by the Bureau and published by the United States Geological Survey, under the auspices of the Agency for International Development, 1965. R W Van Bemmelen: *The geology of Indonesia,* in two volumes, is a classic study (The Hague, Government Printing Office, 1949), with a portfolio of forty two plates of maps. The Philippine Islands, the Andaman and Nicobar Islands, Christmas Island, New Guinea and Timor are included.

Geographers, agricultural economists and other scientists have been much concerned with land use mapping, but there have been obvious hindrances to progress. For the map so far produced a base map at 1 : 25,000 has been used, reduced to 1 : 50,000; the recommendations of the International Geographical Union Commission on World Land Use Survey have been adopted, with suitable modifications. A national atlas of Indonesia began publication in 1960 (Ganaco NV, Djakarta and Bandung).

The *Indonesian journal of geography,* bi-annual 1960-, from the Gadjah Mada University, Faculty of Letters and Culture, Department of Geography, is a central source of original maps. The text maps and folding map in the Admiralty Handbook *Netherlands East Indies* (Naval Intelligence Division, 1944), in two volumes, are still useful. *The Timor problem, a geographical interpretation of an underdeveloped island,* by Ferdinand J Ormeling (J B Wolters, 1955), discusses the physical environment vis à vis the human geography and is well documented with maps and charts. Maps are also included in *The Indonesian town: studies in urban sociology, selected studies on Indonesia by Dutch scholars,* 1958-, one of the series in progress, published by W van Hoeve for the Royal Tropical Institute, Amsterdam; in H T Verstappen and others: *Drie geografische Studien Java,* published in 1963-, which includes ' Geomorphological observations on Indonesian volcanoes' (in English): ' Java: population growth and demographic structure' (in Dutch, with English summaries); ' Population densities in Besuki ' (Java) (in Dutch, with English summaries); R T McVey, editor: *Indonesia* (Yale University and HRAF Press, 1963; Yale University Southeast Asia Studies, Survey of world culture Series); Leslie Palmier: *Indonesia and the Dutch* (OUP, 1962, issued under the auspices of the Institute of Race Relations).

BORNEO: In 1949, the Geological Survey Department was established at Kuching in the British Territories in Borneo; *Annual reports* have been issued since 1949 and *Bulletins* since 1951. A folding map is included in the *Colonial reports* for Borneo and Brunei. Also government sponsored was the reconnaissance soil survey of the Semporna Peninsula, North Borneo, directed by T R Paton for the Department of Technical Co-operation; text maps and maps separately in a pocket accompany the report (HMSO, 1963, Colonial research studies, no 36).

Useful maps are included in:

E C J Mohr: *The soils of equatorial regions, with special reference to the Netherlands East Indies* (Michigan, Ann Arbor; Edwards, 1944; originally published by the Colonial Institute of Amsterdam, 1933 and 1938); translated from the Dutch by R L Pendleton.

P Pfeffer: *Bivouacs à Borneo* (Paris, Flammarion; Collections l'Aventure Vécue, 1963). The account of an expedition by a natural historian, describes the inhabitants and the flora and fauna.

K G Tregonning: *North Borneo* (HMSO, 1960, Corona Library series); includes a folding map.

J F H Umbgrove: *Structural history of the East Indies* (CUP, 1949).

Information on Sarawak was prepared by the Sarawak Information Service (Kuching, Borneo Literature Bureau, 1961).

THE PHILIPPINES: The Bureau of Coast and Geodetic Survey of the Philippines, founded in 1901 at Manila, includes, besides the Administration Division, the Divisions for Operations and Technical Services, Cartographic Services, Geodesy and Geophysics and Photogrammetry. The ' Geological map of the Philippines ', 1:1M was produced by the Bureau of Mines, Manila, in 1963; and a ' Mineral distribution map of the Philippines ', 1:2,500,000, in 1964. A map accompanied the study, *Geology and mineral resources of the Philippine Islands,* by W D Smith (Manila, Bureau of Science of the Department of Agriculture and Natural Resources, 1924). An ' Urban and rural population distribution map of the Philippines ' was published in 1960, at 1:2M, by the Bureau of the Census and Statistics, which is compiling a *Census atlas of the Philippines.* A source of original maps is the *Philippine geographical journal* quarterly in English, 1953-, the journal of the Philippine Geographical Society at Manila, which also gives notice of new cartographical projects.

NORTH AMERICA

UNITED STATES: *A list of maps of America in the Library of Congress, preceded by a list of works relating to cartography,* compiled by P L Phillips (Washington, 1901; reprinted in Holland 1967), comes only partially within the scope of this work; ' Cartography in the Americas ', in *World cartography,* I 1951, provides a good introduction.

Mapping in both American continents is encouraged by the activities of the Pan-American Institute of Geography and History, created formally in 1928 at the Sixth International Conference of American States, although intermittent work had been in operation since about 1898. In 1941, the institute established a Commission on Cartography, the earliest of the three Commissions, those for Geography and History being formed in 1946. The central mapping organisation, the United States Geological Survey, issued a new geological map of North America at 1:5M in 1965; the Survey compiled a ' Glacial map of North America ' at 1:4,555,000 (special paper, no 60), which, by the excellence of the draftsmanship, included considerable detail clearly even at this small scale.

Most of the commercial map publishers have issued maps of the North American continent, sometimes including also parts of Mexico and central America. The Bartholomew regional map, at 1:10M, was revised in 1967; it includes the Aleutian Islands, Greenland, the West Indies and central America. Geographia issue a general purpose map of the Americas and a political map of North America, showing the principal roads, railways, state and province boundaries and airports, with relief indicated by hill-shading. The National Geographic Society has published maps of the whole continent from time to time; a particularly interesting one was ' Vacation lands of the United States and Canada ' on one sheet at 1:5,132,160 in July 1966. The latest edition of the *Rand McNally road atlas,* 1967, includes 108 maps at various scales, covering the United States, Canada and Mexico; information is given on turnpikes and toll roads and a useful section of 170 city maps is included. Stanford's ' General map of the United States ' includes the southern provinces of Canada and Newfoundland and part of Mexico, at 1:5,332,240.

The United States and Canada is the fourth volume in the Oxford regional economic atlases series, 1967, comprising 128 map plates printed in up to seven colours, with a gazetteer of some 10,000 names. Topographic maps are at a scale of 100 miles to one inch, with selected areas in the east, south, midwest and west at scales of thirty miles to one inch or fifty miles to one inch. Puerto Rico and the Hawaiian Islands are shown at fifty miles to one inch. Hill-shading is used on the larger scale maps. An important sequence of double-spread thematic maps deals with topics such as relief, solid geology, pleistocene glaciation, river flow and run-off at a uniform scale of 250 miles to one inch, followed by a section devoted to demography—population growth during the period 1950 to 1960, the distribution of important minorities, etc. Some fifty pages of maps, mostly at the 250 mile scale, then deal comprehensively with agriculture, industry, mining and transport; much of the material incorporated into these maps was derived from recently compiled statistics and investigations. Another useful feature is the seven page section of city plans at a scale of six miles to the inch, showing major commercial and industrial zones, transport facilities and planning projects of the principal cities in the United States and Canada. Much thought has obviously been given to the diagrammatic mapping of specific factors, such as ' river flow ' and ' ethnical groups '; and the documentation of the atlas is most helpful, including not only acknowledgements, but source maps and books, with their dates, and an explanatory introductory text. The work is available in both hard cover and paperback editions.

The special *Atlas of paleographic maps of North America,* published by Wiley in 1955, made available in Britain by Chapman and Hall in 1956, is a work of great scholarship compiled by C Schuchert, with an introduction by C O Dunbar.

Numerous periodicals and monographs including original maps or interesting sketch-maps are available covering the whole or part of the North American continent. The publications of the American Geographical Society of New York should be particularly noted; for example, *Readings in the geography of North America: a selection of articles from the Geographical Review 1916-1950* (1952). The classic *The physiographic provinces of North America,* by W W Atwood (Boston, Ginn, 1940), includes maps and a selected list of topographic maps follows the references for each regional chapter.

The national committee representing the United States at the International Cartographic Association is the American Congress on Surveying and Mapping, Cartography Division, Washington.

In addition to the United States Geological Survey, four other departments are responsible for national mapping; the United States Department of the Interior, Bureau of Land Management; the United States Department of Commerce, Coast and Geodetic Survey; the United States Department of the Army, Corps of Engineers, Army Map Service, all based on Washington; and the Headquarters Aeronautical Chart and Information Centre (USAF) in Missouri.

The United States Geological Survey was founded in 1879 to establish base maps and to make a systematic survey of geology and natural resources. Having taken over the functions and records of earlier surveys, the Survey created five branches, administrative, geological, topographic, conservation and water resources. The Survey also publishes and distributes original topographic maps prepared by the Department of the Army and the Coastal Geodetic Survey. Nine series of maps were planned, on scales ranging from 1 : 20,000 to 1 : 1M. The earliest maps, at 1 : 250,000, 1 : 125,000 and 1 : 62,500 were derived from the 1 : 1M. Later, to meet the requirements of regional development plans, maps at 1 : 24,000 were introduced for relevant areas. The whole country is covered by the small-scale series, with the exception of Alaska. Most of the maps are now prepared or revised by means of aerial photography, in conjunction with ground survey in suitable areas. In general, blue is used for water features, brown for relief, green for vegetation, red for roads and built-up areas and black for all other features. The Survey issues descriptive brochures and progress reports. A feature of both United States and Canadian maps that seems to be on the increase is the use of the reverse of the map sheets for printing information about the areas concerned or for photographs or diagrams.

The United States Geological Survey works closely with the Surveys and Mapping Branch of the Canada Department of Mines and Technical Surveys in the preparation and publication of maps along the international boundary, while mapping along the Mexican border has been co-ordinated through the Office of the Chief of Engineers, United States Army, the Joint United States-Mexican Defense Commission and the Inter-American Geodetic Survey. Within the whole programme of the Survey, both new maps and revisions are planned on a state basis, cartographic improvements being made as occasion allows. For example, the new map of the State of Washington, 1 : 500,000, compiled in 1961, published in 1964, showed an attractive new form of representing shaded relief; and improved detail of National Forests and Parks was introduced in the new 1965 edition of the ' State of Texas map ' on the same scale, first published in 1962. For Alaska, a geological map was published in 1964, at 1 : 1,584,000, from the survey undertaken by NV Karlstrom and others in 1960 and a series of mineral investigation resource maps, at 1 : 2,500,000, is in progress. ' Tectonic map of Alaska : peninsula and adjacent areas ', 1 : 1M, was completed in 1965. Some of the maps published by the Geological Survey are based on originals prepared by other agencies, namely, the Department of the Army, the Coast and Geodetic Survey or special authorities such as the United States Forest Service or The Tennessee Valley Authority.

The collection of information about maps and aerial photographic coverage of the United States is one of the functions of the Map Information Office of the Geological Survey. Index maps are published from time to time showing the state of topographic mapping at all scales, horizontal control, vertical control, aerial mosaics and aerial photography. Index sheets showing the coverage by aerial photography are available also from the various governmental and commercial agencies, also ozalid prints. A series of geological map indexes, compiled by Leona Boordman, has been published by the Survey, also a similar index for Alaska, by E H Cobb, including a description of the geology of Alaska. The topographical maps available are delineated on a Geological Survey index map for each of the states. Details of production and work in progress are given in the United States *National cartographic report*, presented to the International Cartographic Association Technical Symposium at Edinburgh, 1964. Reference should be made also, not only to the *Annual reports*, 1880-, but also to the *Special reports* contained in the Professional papers series, 1902-; *Geological Survey research, 1966*, for example, by W T Pecora or the *Water supply papers*, 1896-, prepared by the Department of the Interior and the Geological Survey. One useful point to note is that sets of twenty five or 100 specially selected maps, together with indexes of topographical features, are available for educational purposes at very reasonable prices.

Three groups of standard topographical maps intended to cover the whole country are in current use : at 1 : 24,000 and 1 : 31,680; at 1 : 62,500; and at 1 : 125,000 and 1 : 250,000. The sheet lines adopted for the first three are based on the international longitude and latitude degree lines and the maps are all on the Conical Projection. Each sheet at 1 : 250,000 is divided into four sheets at 1 : 125,000, which in turn divides into four sheets at 1 : 62,500 and again into four. The sheets at 1 : 24,000 and 1 : 31,680 are the largest scale standard maps available, based upon surveys designed for use in areas in which major public projects are being undertaken. Contour intervals vary from one foot to three feet, depending on the topography. Colours used are blue for drainage and swampy areas, green for vegetation, brown for relief, red in-fill for primary roads, broken red in-fill for secondary roads and black for all other features. The published sheets at these two scales cover most of the main cities and their vicinities and other densely populated areas. The 1:31,680 series is to be replaced in time by the 1:24,000 series. Surveys for the maps at 1:62,500 were primarily designed for areas in which the public works are of average importance; contour intervals vary from five to 100 feet, depending on the terrain, and the colour scheme is the same as for the 1:24,000 series.

The series at 1:125,000 was designed chiefly for areas such as the deserts and semi-deserts in the west of the country. Contour intervals vary from twenty feet to 250 feet. Some of the 1:125,000 sheets are based on old surveys, but most of them are being replaced by 1:62,500 sheets and no new sheets are being compiled at 1:125,000. In the western part of the United States, there are several sheets covering sparsely populated areas at 1:250,000.

Each state is covered by a series at 1:500,000. Sheet sizes and contour intervals vary according to topography and the shape of the state. Colours used are blue for drainage, brown for relief, purple for roads, an orange tint for large cities and towns and black for all other features.

A new Army map series is to cover the whole of the United States in 468 sheets. The sheet divisions and numbering system are based on those of the International Map of the World 1:1M, but the projection used is the Universal Transverse Mercator. Contour intervals vary from fifty to 100 feet. Blue is used for drainage and swampy land, green for vegetation, brown for relief, an orange tint for large conurbations, red and broken red in-fill for roads. Other federal agencies and some state agencies also issue topographical maps to meet special needs. It is estimated that rather more than two-thirds of the country are adequately mapped; map maintenance is a major problem, which is carried out mainly by aerial photography.

To produce a satisfactory national atlas of so vast and complex an area as the United States presents great difficulties. The task was begun some years ago under the direction of the National Research Council, National Academy of Sciences, and by 1957 the first sheets had been completed, covering agriculture and land use and the map of general types of farming. The standard scale adopted was 1:10M. In 1962, the *National atlas* project was transferred to the Geological Survey for replanning and co-ordination; this was achieved by 1964, the published work consisting of some four hundred pages, including historic, physical, economic and cultural features of the United States, incorporating the latest census figures for population, business, industry and agriculture. Details of planning and production are given by A C Gerlach: ' The national atlas of the United States of America ', in *The cartographer,* May 1965. The National Atlas of the United States section of the US Geological Survey is working also on a marketing study based on eight thousand questionaires using electronic computer methods of indexing.

An *Atlas of the United States* is also in progress, 1964-, by the National Geographic Society. The complete atlas will comprise at least sixty four plates, each state, areas of special interest and major cities being represented by individual maps. Thirteen double-spread maps, with index-gazetteer, had been issued in 1960 by the society under the title *National Geographic atlas of the fifty United States;* it included Alaska and Hawaii.

The Coast and Geodetic Survey was established as the Coast Survey in 1807, the extension of geodetic work being provided for by an Act of 1871. The duties of the Survey are to prepare charts and related information for the safe navigation of marine and air commerce and to provide the basic data required in development projects in agriculture, commerce and industry. A primary network of horizontal and vertical control along the coasts and in the interior was established, following which hydrographic, oceanic and topographical surveys have been undertaken and the coastal surveys are co-ordinated to provide a framework for mapping and engineering work. Geomorphological research of all kinds is part of the programme, also field surveys to supplement aeronautical charting and the compilation of air charts. The data for all these activities are analysed in the Washington Office and publications include nautical and aeronautical charts, geodetic control data, annual lists and charts of the United States, and other earthquakes, coast pilots and annual tables of tide and current predictions. C H Deetz gave a summary of the work to date in his *United States Coast and Geodetic Survey,* published in a special publication in 1943. The Survey is also engaged on the production of a series of base maps of the continental shelf of the United States. The USAF Aeronautical Chart and Information Center has developed an automated management system for the planning, analysis and control of production programmes. The centre is responsible for providing the Air Force with aeronautical and astronautical charts and all the other data and documents required.

Numbers of special topic maps and atlases are concerned with

the whole country or with individual states, sponsored or compiled by both government agencies and by academic geography
departments, the American Geographical Society of New York
and the National Geographic Society. A recent atlas which has
created much interest is *Lithofacies maps: an atlas of the United
States and Southern Canada,* compiled by the graduate students at
Northwestern University engaged in regional stratigraphic studies
under the editorship of L L Sloss, E C Dapples and W C Krumbein
(New York, Wiley, 1960). This is a unique source of information
on the thickness and lithology of the sedimentary rocks of the
United States. 153 maps, printed in black and red, show the thickness of specific geological formations by ispachytes superimposed
on maps showing lithological type. A map of the whole area is
included for each system, with a more detailed section of maps for
smaller areas. There is an explanatory introduction and a table of
formations at the end. In studying the geology of the United States,
a valuable reference is G P Merrill: *The first one hundred years
of American geology,* published in 1924 by Yale UP, reprinted by
Hafner in 1964.

The climatic map of the United States, compiled by Dr S S
Visher (Harvard UP, 1954), includes 1,031 maps and diagrams
dealing clearly and simply with every imaginable aspect of American climate for which usable data were available, interspersed with
explanatory text. Typical phenomena of the United States weather
are presented in series of maps and graphs—temperature, winds
and storms, sunshine, humidity and evaporation, precipitation, the
effect of climate on topography and agriculture, climatic regions
and changes. The work summarises the results of decades of observations by the United States Weather Bureau, to whom the volume
is dedicated. A relief map from the United States Geological Survey
is included.

Compiled by the Water Information Center (Fort Washington,
1963), the *Water atlas of the United States: basic facts about the
nation's water resources* is a collation of maps, with introduction,
presenting the occurrence of water and other related climatic
phenomena and the various uses of water. The *Atlas of American
agriculture: physical basis, including land relief, climate, soils and
natural vegetation of the United States,* planned in six volumes
by the United States Department of Agriculture, was unfortunately
never completed. Under the direction of O E Baker, six folios were
issued between 1918 and 1936, dealing only with the physical geography of the country; some of these plates were detailed and carefully executed and the explanatory text was authoritative. Even
with its limitations of date and scope, it is therefore a useful source.

An atlas, with tabular material and brief text entitled *Agriculture, 1954, graphic summary: land utilization, farm machinery and facilities, farm tenure,* was issued in 1957 by the United States Department of Agriculture (special report, no 4). Based on the census of 1954, the work brings out the main characteristics of American farming at this date. Other cartographic source documents for the natural features of the United States economy include *A tectonic map of the United States* in two sheets, prepared as a joint project of the Geological Survey and the American Association of Petroleum Geologists in 1964, at 1:2,500,000, based on 1963 data; the map also shows the presence of oil shale, areas of basement rock and selected sedimentary areas. A map of the 'Potential natural vegetation of the conterminous United States', compiled by A W Küchler, at 1:3,168,000, was published by the American Geographical Society in 1964, together with an explanatory text. The geographical distribution of 116 types of vegetation is shown by colours and numbers and, in the text, each type is illustrated by photographs.

The official map of 'Forest regions in the United States' was revised in 1960; the text issued with it lists the principal trees of each region and gives a description of the distinctive physical character of each. With the classic study, *Deciduous forests of eastern North America,* by E L Brown, is a linen-backed map issued separately. A map of 'Land resource regions and major land resources areas of the United States (exclusive of Alaska and Hawaii)' at 1:10M was published by the Soil Conservation Services, United States Department of Agriculture, 1965.

The *Atlas of the Pacific northwest: resources and development,* edited by Richard M Highsmith (Oregon State UP, third edition, 1962), covers the physical features and economic resources of Washington, Oregon, Idaho and Western Montana. J M Leverenz was responsible for the cartography and there are maps, diagrams and black and white photographs also illustrating the text, which was contributed by several specialists. The evolution of the economy is dealt with, also the physical basis of the country, land resources, agriculture, fisheries, mineral resources, manufactures, communications and trade. Rand McNally's *Pioneer atlas of the American West* (Chicago, 1956) contains facsimile maps and indexes from the 1876 first edition of *Rand McNally's business atlas of the Great Mississippi valley and Pacific slope,* together with contemporary railway maps and travel literature, with text by D L Morgan. The United States Geological Survey 'Permafrost map of Alaska', compiled by Oscar J Ferrians, jr, the 'Tectonic map of the Alaska Peninsula and adjacent areas', 1:1M, published by the Geological Society of America and the 'Mineral resources and industry map of Missouri' published by the Division of Geological Survey and Water Resources, all appeared in 1965.

The production of a series of regional atlases was a much more viable proposition than a comprehensive national atlas. The *Atlas of Illinois resources,* for example, published from 1958- by the State of Illinois, Department of Registration and Education and the Division of Industrial Planning and Development, Springfield, is a fine collection of maps, well documented by text, indexes and bibliographies. Looseleaf sections to date include water resources and climate, mineral resources, forest, wild life and recreational resources, transportation, manpower resources and agriculture in the Illinois economy. Relevant authorities have been consulted, in addition to the Department of Geography, University of Illinois.

The *Atlas of Texas,* by Stanley A Arbingast and Lorrin Kennamer for the Bureau of Business Research, University of Texas, 1963, was a third revision of *Texas resources and industries: selected maps of distribution,* issued by the Bureau in 1955 and 1958. The present volume gives good coverage of locational maps, physical agriculture, mining, population and manufacture maps and there are plans for a complete atlas of the state in full colour. ' The national survey map of New Hampshire ', at a scale of approximately 1 : 300,000, published by the National Survey, Chester, Vermont, is based on the latest official government and state data; it is a well-designed map, printed in eight colours, showing relief by layer colouring and hill-shading.

The University of Missouri, College of Agriculture, Agricultural Experiment Station sponsored the *Agricultural atlas of Missouri* in 1955. General maps of the state, the soils, climate, forests, use of land, types and sizes of farms, distribution of crops and of livestock, agricultural values and selected agricultural industries are all mapped, presenting a comprehensive picture of the agricultural resources and their actual and potential utilisation, with brief text for each map. The *Atlas of Oregon agriculture* was prepared by Richard M Highsmith for the Oregon State College, Corvallis, Agricultural Experiment Station. Published in 1958, the atlas collected together some maps already in existence and included new maps based on the 1954 Census of Agriculture. The forty three pages of maps and statistics, with brief textual notes, are intended as a reference handbook on the location of the basic resources of agriculture, farmlands, practical farming and the marketing and processing facilities of the state.

Compiled especially for school use is the interesting work *Patterns on the land: geographical, historical and political maps of California,* by R W Durrenberger (Northridge, California, Roberts

307

Publishing Company, second edition 1960). It is illustrated by photographs, text maps and diagrams. A commentary on regional mapping is presented by R E Stipe: *Officials' views on the uses of and need for topographic maps in North Carolina* (University of North Carolina, Institute of Government, 1962). One of the best of the regional atlases to date is the *Atlas of North Carolina,* compiled under the direction of Richard E Lonsdale (University of North Carolina P; OUP, 1967). Arrangement of the maps is by topics, each prepared by a specialist, with text, diagrams and photographs. Physical features and resources are considered first, followed by historical maps, population changes, urbanisation and cultural aspects; the final section deals with economic factors of all kinds. Three scales are used, 1 : 6,336,000, 1 : 3,960,000 and 1 : 3,113,000, and plastic overlays are inserted in a back pocket for the identification of counties with any map in the atlas. Outlines of the state and counties are also drawn on many of the maps, in blue, symbols and other information being plotted in black.

The *Atlas of Washington agriculture* was produced by the Washington State Department of Agriculture in co-operation with the Washington Crop and Livestock Reporting Service, Seattle, in 1963. Under the direction of E C Wilcox, the maps and charts, prepared by Fred Nammacher, the text by Cecil Ouellette and W R Clevinger and the introductory notes by Joe Dwyer and Don Olson, all combine to create a most useful reference source on all the chief aspects of the agricultural resources of the State. *Richards atlas of New York State* is a looseleaf atlas issued in a binder, edited by R J Tayback and others (Phoenix, New York; F E Richards, 1957-59) and illustrated with text maps and diagrams. The *Atlas of forestry in New York,* an illustrated collection of maps, edited by N J Stort, was published by the State University College of Forestry at Syracuse University in 1958. Effective bibliographies are included.

A colourful and attractively produced atlas is the *Atlas of Florida,* prepared by Erwin Raisz and his colleagues in the Department of Geography of the University of Florida, with text by John R Dunkle (Gainsville, 1964). A variety of methods is used to present the development of the state from 1845; it has had, it seems, the greatest increase in population of any state in the Union. Time charts, bar graphs, divided circles and flow diagrams, with maps of all kinds and notes of past and present activities, make an easily understood whole, significant of the current trend in experimentation of the graphic presentation of geographical factors. More scholarly, perhaps, in impression, but still demonstrating the same trend, is *Kansas in maps,* prepared by R W Baughman for the Kansas State Historical Society (McCormick-Armstrong, 1960).

The statistical cartographic work of the United States Bureau of the Census began in about 1870. In 1874, the *Statistical atlas of the United States* was published under the direction of F A Walker. A series of maps traced population density at each census from 1790 to 1870, based on county and smaller area figures. Other maps show types of population by percentages, physical geography, agriculture, including distribution of individual crops, and vital statistics. Another atlas, edited by Henry Gannett, then Geographer of the Census, was based on the 1890 census; here, techniques of mapping were introduced and, significantly enough, a section of maps of manufacturing and mining areas was necessarily added. The dot method was introduced into the statistical atlas for 1910, for the agricultural maps. Another Geographer of the Census, C S Sloane, was responsible for the statistical atlas based on the fourteenth census (Government Printing Office, 1925) and yet another, C E Batschelet, began the production of large-scale maps of each state, based on the 1930 census; these maps, most useful as base maps for further work, have remained a permanent feature. In the series of maps illustrating the 1960 census, more than four hundred city block maps were included, also 180 census tract maps. It is illuminating to examine these successive atlases, each increasing in complexity and clarity of execution. For further details, Robert C Klove's article ' Statistical cartography at the US Bureau of the Census ' should be consulted (*International yearbook of cartography*, VII 1967). Typical of the variety of maps issued by the bureau are ' Families with incomes under $3,000 in 1959 by counties of the United States—1960 ', 1:5M, 1965; ' US showing employment in professional, technical and kindred occupations ', 1:5M, 1966; ' US showing employment in manufacturing ', 1:5M, 1966; 'American Indians in the United States, 1960 ', 1:5M, 1967; ' Japanese and Chinese in the United States, 1960 ', 1:5M, 1967; ' Negro population as percentage of total population, by counties of the United States, 1960 ', 1:5M, 1967; ' Standard metropolitan and statistical areas of the United States and Puerto Rico, 1967 ', 1:5M, 1967.

Professor Wilbur Zelinsky has directed the compilation of a population map at 1:1M in the United States and Canada, in collaboration with geographers throughout the two countries. At the meetings of the Association of American Geographers held in Denver in 1963, a formal committee was established, the Subcommittee for a population map of Anglo-America, under the auspices of the association. Since that time, Professor Norman Thrower has taken over the major responsibility for this work, as Professor Zelinsky was committed to population mapping in Latin America. Anglo-America has been divided into forty three zones, having grid boundaries related to those of the International Map of the World, 1 : 1M. Work has been based on nineteen centres in the United States and six in Canada. Preliminary pilot plotting was carried out for areas with varied conditions—southern British Columbia, northern Washington, southern California, Hawaii, Iowa and New England. Typical was the map ' California population: distribution, 1960 ', prepared by Professor Thrower for issue as map supplement no 7, by the Association of American Geographers, 1966. The IMW 1 : 1M sheet lines, projection, grid, etc, have been used as a base; dot symbols, one dot representing fifty persons, have been used to portray rural population; proportional circles for urban centres about one thousand people; and the five largest cities on any sheet have been distinguished by an initial letter and listed in the legend, together with the population figure. Urban areas having a density above 350 persons per square mile are shown by screened tonal patterns, which show up very effectively. Insets of particularly significant areas may be included. Aerial photography and ground survey are used to assist in the accurate placing of symbols in conjunction with the census figures.

The first major historical atlas of the United States, using modern methods of compilation, was the *Atlas of the historical geography of the United States,* by Charles O Paullin, edited by J K Wright, published jointly by the Carnegie Institution of Washington and the American Geographical Society, 1932. 688 maps, compiled from a multitude of sources, endeavoured to show ' the essential facts of geography and history that condition and explain the development of the United States '. Some contemporary drawings were reproduced, but most of the plates, dealing with physical, historical, social, political and economic topics, were specially prepared. The Clarendon Press at Oxford produced the *Atlas of American history* in 1967, which consists of 147 map plates in monochrome. Sixty four historians co-operated in the work and, in the foreword, the editors give their opinion that ' the need was for maps that would interpret our history through the location of places as they actually existed . . . at a given time '. The maps were prepared under the direction of LeRoy H Appleton. Arrangement of the map plates is chronological and the maps themselves are drawn mainly by hand, incorporating, as an additional aid to interpretation and interest, pictures of trees, ships on trade routes, etc, rather reminiscent of early cartographic design, but now combined with accuracy of map content. Place symbols vary with the scale of the map, from conventional town stamps to the shapes of individual buildings. Different type founts are used uniformly to indicate different features.

Between these two atlases, there have appeared a number of

other historical atlases. The *Atlas of American history,* edited by J T Adams and R V Coleman, was published in New York by Scribner in 1943. This work was designed to be used in conjunction with the *Scribner dictionary of American history* and covers discovery, exploration, frontier posts, different kinds of settlement, territorial organisation and the extension of communications, with brief explanatory text. The *Historical atlas of the United States,* edited by C L and E H Lord (New York, Holt, 1944, with a revised edition in 1953), is especially useful for the maps showing economic development and the distribution of population. It is intended as an aid to students. *An atlas of economic development,* by Norton Ginsburg, is the most comprehensive and scholarly work on this aspect of American development, produced by the University of Chicago, Department of Geography, in 1960, research paper no 68. An *Economic development atlas* was prepared by the United States Office of Domestic Commerce and published by the United States Government Printing Office in 1949. Perhaps better known in Britain is *Wesley's historical atlas of the United States,* produced in the States for sale in the British Commonwealth, except Canada, by Denoyer-Geppert, one of several historical atlases by this publisher; George Philip produced a second edition in 1961 with the title *Our United States—its history in maps.* A new *Atlas of American history,* edited by James Truslow Adams, was issued by Scribner in 1967.

Several regional atlases with a historical approach have been issued. *The Southeast in early maps,* for example, is a monumental work published by Princeton University Press, 1958 (OUP, 1958), with a new edition in progress. An annotated check list of printed and manuscript regional and local maps of south-eastern North America during the colonial period is included.

Popular mapping has been greatly stimulated in America during recent years. The map supplements as well as the maps in the text of the *Magazine* of the National Geographic Society have played a large part in this. The publication of *Hammond's pictorial travel atlas of scenic America* in 1955 recognised the requirements of the developing tourism. Edited by E L Jordan, the atlas was designed to provide practical information and to demonstrate the variety of American scenery. Private and official agencies specialising in travel information assisted in the preparation of the small, but useful, sectional maps, the numerous coloured photographs, informative text and the 'sightseeing gazetteer'. In 1963, Rand McNally produced a special road map of the United States, prepared by the National Automobile Club, San Francisco at a scale of approximately one inch to eighty five miles. A 'Communications' series for the individual states is at 1 : 1,250,000. Other publishers also produce individual maps of a state or particular area, for example, *California: communications, roads,* at 1 : 1,250,000, and *San Francisco and vicinity,* 1 : 300,000, by the H M Goushá Company of Chicago.

The journals and other publications of the American Geographical Society of New York, the Association of American Geographers, the National Geographic Society and the academic geography departments abound in original maps and evocative sketch maps.

No central current bibliography covers the entire United States, although bibliographical source material exists to a higher degree than in almost any other country. Maps are included in the relevant section of the *Catalog of copyright entries.* The individual map publishers issue catalogues and the bibliographical work of the Map Division of the Library of Congress is invaluable. For many years, the Map Division has maintained a *Bibliography of cartography,* of which microfilm copies are available and entries for maps and atlases form a special section of the published catalogue. Clara Le Gear's *United States atlases* and *A list of maps of America* are essential sources, also *Marketing maps of the USA: an annotated bibliography,* compiled by M C Goodman and W W Ristow (second revised edition 1952), which includes not only all the maps and atlases on the theme in the Library of Congress collection, but also statistical studies illustrated by maps.

Many of the individual states have issued their own map bibliographies, notably *A selected bibliography of southern Californian maps,* compiled by Edward L Chapin (University of California Press, 1953).

Of central importance are the files of *Surveying and mapping* and of the *Bulletin* of the Geography and Map Division of the Special Libraries Association of New York. The *Annual reports* of such specialised organisations as the Bureau of Mining, Washington and the United States Soil Survey are primary sources of information. The published papers of the United Nations Regional Cartographic Conference for Asia and the Far East, 1955, include a report on ' Recent progress in cartography and interesting technical developments in the United States ' (volume II, pp 106-112). *States and trends of geography in the United States, 1957-1960* should also be consulted; this is a report prepared for the Commission on Geography, Pan-American Institute of Geography and History, by the Committee of the Association of American Geographers in collaboration with the National Academy of Sciences, National Research Council, 1961.

CANADA: Not only the vast area of Canada, but the contrasts in physical conditions and population density made the mapping of the whole country at even medium scales an impossible task until aerial photography and advanced photogrammetric equipment came into use. Much of the country still remains to be mapped at large scales and for special purposes, only preliminary maps or advanced prints are available. The chief mapping agencies are the Surveys and Mapping Branch, Department of Technical Surveys, and the Directorate of Military Survey, Department of National Defense. The Canadian Institute of Surveying, Cartographic Committee, Ottawa, forms the committee to the International Cartographic Association, but the national mapping organisation is the Department of Mines and Technical Surveys, which includes the Surveys and Mapping Branch; Marine Sciences Branch; Geological Survey, Geographical Branch; Dominion Observatories and Mineral Resources Division. The department had its origin in 1943, when J L Robinson was appointed Geographer to the Northwest Territories Administration of the Federal Department of Mines and Resources 'to compile, organise and analyse information about Northern Canada for wartime purposes and peacetime development'. As a result, in 1947, a Geographical Bureau was established 'to collect and organise data on the physical, economic and social geography of Canada' and in 1951 this bureau became the Geographical Branch of the main department, having two divisions dealing with Canadian work—regional geography and systematic geography—and a Foreign Geography Division, which works on studies of foreign areas of interest to Canada. The *Geographical bulletin* is published, also *Memoirs* and *Papers*.

Under the general title 'National topographic series', the following are the main series by scale: 1:50,000, 1:63,360, 1:126,720, 1:250,000, 1:253,440 and 1:506,880. All sheets are based on the Universal Transverse Mercator Projection and are drawn along the international geographical co-ordinates, on a systematic principle, so that four sheets of the 1:50,000 or 1:63,360 are included in one sheet of the 1:126,720 and four sheets of the latter are covered by one sheet of the 1:250,000 and 1:253,440. About 1,500 sheets of the 1:50,000 series cover areas having great population density and those parts of the country in which considerable economic development has taken place or is envisaged. The contour interval varies from twenty five, fifty or 100 feet, according to the topographical features on individual sheets. Five colours are used: blue for water features, browns for relief, green for areas permanently covered by vegetation, red for primary and secondary roads and black for all cultural symbols.

The 1:63,360 series may eventually replace the 1:50,000; seven colours are used, including orange to distinguish secondary roads and grey for built-up areas, in addition to those used for the 1:50,000. On the 1:126,720 series, contour intervals again vary according to the nature of the relief and the colouring system is the same as that for the 1:63,360. The 1:250,000 series does not cover the country. In some areas, five hundred feet contours are included in addition to the others and six colours are used: blue, browns, green and black as for the larger scale series, red for primary roads, broken red for secondary roads and for air and navigation symbols and yellow for outlined towns and cities. The 1:253,440 series is being replaced by the 1:250,000; the contour interval on existing sheets is the same as on the 1:250,000, while the colour scheme is similar to that for the 1:63,360. The 1:506,880 series, which covers the whole of Canada, is intended primarily for aeronautical purposes, but a topographical edition is also available. It has been compiled mainly from trimetrogon aerial photography, controlled by astronomical fixes, and four colours are used, blue, brown, red and black, with the addition of town circles to indicate areas of dense population. A new series at 1:25,000, drawn on a Transverse Mercator Projection, was begun in 1965 by the Surveys and Mapping Branch of the Department of Mines and Technical Surveys. Relief is indicated by contours at ten or twenty five feet. Series are maintained also at 1:125,000, 1:500,000, in a third edition, and at 1:15,840,000, which is available in French and English. A series at 1:25,000 is prepared by the Army Survey Establishment.

A map of Canada in two sheets, 1:4,055,040, was published by the Department of Energy, Mines and Resources in 1966. Insert is a map of Canada at 1:25,344,000, showing the main Canadian airways, international routes and time zones. The National Geographic Society has produced numerous maps of all parts of the country; a particularly useful one is 'Eastern Canada', at 1:2,851,000, issued in 1967, for it shows effectively this whole area so vital to the Canadian economy, with insets of the St Lawrence seaway, Cape Breton Island, Montreal, Quebec, Halifax and St John. The Royal Canadian Geographical Society has recently completed a useful reference map of the less-known 'Northeastern Region, Canada', at 1:4,500,000.

The Geological Survey of Canada was established in 1841 and, in 1886, photographic survey work was begun. Canada has been a pioneer in oblique aerial surveying, by which thousands of square miles of the interior have been mapped. In 1904, the Canadian government sponsored an expedition for the exploration of the northern fringe of Canadian territory, the mainland to the north of Hudson Bay and the islands of the Arctic Ocean. Between 1913 and 1918, further surveying work was carried out, of which full reports, the corrected maps and all the scientific data have been published. *Memoirs* have been published from 1910, dealing with specific map areas, usually accompanied by coloured lithographed geological maps; an ' Economic geology series ' was begun in 1926, presenting topical reports on specific metals and minerals; a series of *Papers* in 1935, the *Bulletin* in 1945 and in 1948 a further series, ' Geophysics papers ', reporting on the results of geophysical surveys. William E Logan, the first director, achieved a great deal of valuable work. Progress is shown in the *Annual reports* and in Logan's own *Geology of Canada,* 1863. Much of the exploration of Canadian territory, especially in the west, was carried out by the Survey and, by the turn of the century, detailed mapping was being systematically planned, together with studies of mineral resources. F J Alcock: *A century in the history of the Geological Survey of Canada* (Ottawa, 1952) should be consulted for further details.

The accurate mapping of Canada is thus progressing rapidly and, since the second world war, the whole country has, in addition, been photographed by the Royal Canadian Air Force. Large-scale ground survey, however, will continue for a very long time, owing to the vast size of the country. One special survey must be mentioned particularly, the ' Glacial map of Canada ', 1 : 3,801,600, published by the Geological Association of Canada in co-operation with the Geological Survey, the Defense Research Board and the National Research Council. The project originated in the Geophysics Laboratory of the University of Toronto and a great number of experts have worked together on this map, which was based on the maps compiled from air photographs by the Map Compilation and Reproduction Division, Department of Mines and Technical Survey. The glacial features of Canada are shown also on the ' Glacial map of North America '. A fine piece of mapping was produced by the American Geographical Society as part of the International Geophysical Year programme for mapping glaciers throughout the northern hemisphere; entitled *Nine glacier maps, northwestern North America,* in nine sheets at 1 : 10,000 (special publication, no 34), the method of cartography and skilful production were acclaimed throughout the world. The construction processes are described in an accompanying thirty eight page brochure. A ' Glacier map of southern British Columbia and Alberta ', 1 : 1M was produced by the Department of Mines and Technical Surveys in 1966.

The Geological Surveys of the individual Canadian provinces

now also produce their own regional geological maps, reports, bulletins and information circulars. A particularly fine piece of geological mapping was the ' Geological map of the province of Nova Scotia ', 1 : 506, 880, published by the Department of Mines, Nova Scotia, 1965; ' Ontario geology and principal minerals ' was produced at 1 : 4,224,000 by the Ontario Department of Mines, 1965. Surficial geology maps are published by the Geological Survey at 1 : 63,360 and at 1 : 253,440. A fine ' Tectonic map of the Canadian Shield ' was completed by the Survey in 1965 at 1 : 5M and a similar map was produced in the same year covering Saskatchewan and western Manitoba at 1 : 1,267,200. *An index to the publications of the Geological Survey of Canada 1845-1958,* compiled by A G Johnson, was published in 1961. Supplements for 1959-60, 1961-62 were compiled by H M A Rice, published in two volumes, 1961, 1963, with an author index, 1963. The *Annual reports* are also well indexed; a special publication covered maps without reports, 1927-50.

Other special maps have been prepared, usually either for tourist attraction, as in the case of the attractive 1 : 50,000 sheet for ' Mount Revelstoke National Park, Kootenay district, British Columbia ', published by the Surveys and Mapping Branch, Department of Mines and Technical Surveys, using a combination of hill-shading and contours for relief representation; or for economic development purposes, such as the ' Predominant types of commercial farms in Canada, 1961 ', 1 : 1M, published by the Economics Branch, Department of Canada, from information supplied by the Census Division, Dominion Bureau of Statistics. Similarly, the Canadian Government Travel Bureau brought out in 1960 a new ' Highways of Canada and the northern United States ', at a scale of one inch to approximately fifty one miles, the ' Terra Nova National Park, Newfoundland ', at 1 : 50,000, and the ' Centennial Range, Yukon territory ', 1 : 125,000, published by the Department of Energy, Mines and Resources, 1967. A ' Hydroelectric power in Canada ' map at 1 : 2M was completed in five sheets in 1964, by the Department of Mines and Technical Surveys; and ' Principal mineral areas of Canada ' at 1 : 7,603,200 in 1965. One of the earliest regional air surveys was *Northernmost Labrador mapped from the air,* comprising six maps and navigational notes on the Labrador coast, by Alexander Forbes and others, published by the American Geographical Society of New York, 1938. ' La province de Québec (physique) ', compiled by B Brouillette and P Dagenais, was published as a wall map by Beauchemin of Montreal, at approximately sixteen miles to the inch, in ten colours. Most attractive and informative brochures are prepared by individual state departments, mainly for the use of visitors: ' Ontario: province of opportunity ', by the Department of Mines, Ontario in two maps, north and south, one on each side, showing provincial and district boundaries; roads, classified; settlements, distinguished according to size; town plans for all the largest settlements, an index of cities, towns and villages, an index of lakes and a great deal of other useful information. Official guide maps are also excellently produced, such as the ' Map of the Municipality of Metropolitan Toronto ', issued by the Convention and Tourist Bureau of Metropolitan Toronto, with a larger scale plan of the central area and information on many points of interest, on the back.

The Geographical Branch, Canada Department of Mines and Technical Surveys is working on a new ' Canadian land use ' series, of which the first to be completed were those for Nova Scotia and the Niagara province of Ontario. The maps are basically following the recommendations of the World Land Use Commission of the International Geographical Union. Nearly all are based on the existing topographic series at 1 : 50,000, 1 : 250,000 and 1 : 500,000, the latter for sparsely populated areas such as the prairies of western Canada. The exception is Prince Edward Island, for which maps are being drawn at two miles to the inch. ' Predominant types of commercial farms in Canada, 1961 ', at 1 : 5M, was compiled by the Economics Branch, Department of Agriculture, from information supplied by the Census Division, Dominion Bureau of Statistics. In addition, in the individual provinces, much progress has been made, largely by geographers, in the making and use of regional land use, population and resources surveys, especially by the Ontario Department of Planning and Development, the Saskatchewan Department of Natural Resources, the Alberta Department of Municipal Affairs and the British Columbia Department of Lands and Forests. The co-ordinating body for all this work is the Service Géographique, established in 1952 in Quebec, which developed from the Economic Research Bureau set up in 1937 as part of the Department of Commerce and Industry.

Soil survey and economic survey maps are prepared by the Department of Agriculture and by the Dominion Bureau of Statistics; fisheries maps by the Information and Educational Service, Department of Fisheries; forestry maps by the Information and Technical Services, Department of Forestry; meteorological maps by the Meteorological Branch of the Department of Transport; planning maps by the National Capital Commission and, in the case of maps which need to incorporate census material, by the Dominion Bureau of Statistics. The soil surveys and *Reports* of the Canada Department of Agriculture are particularly important. A number of contributions have been made to the World Population Map, by provinces,

317

as mentioned above, beginning with Alberta and British Columbia. A 'Petroleum and natural gas map of Canada', 1:2,500,000, was published by the Toronto Dominion Bank, Calgary, 1964; it shows the oil and gas fields and the location of refineries and pipelines. The Geological Survey completed 'Alberta and northeastern British Columbia showing oil and gas fields and oil and gas discoveries', at 1:1,267,200, in 1966, with insets of 'Oil and gas' and 'Pipelines of Canada' at approximately 1:50,000. Finally, a number of studies have been made of the great cities by Rolph C Stone for BP of Toronto: for example, 'Montreal and vicinity', at 1:126,720, 'Ottawa and vicinity', at 1:36,000, and 'Province of Quebec' at 1:672,000.

The first official *Atlas of Canada* was published in 1906 by the Department of the Interior of the Canadian government under the direction of the geographer John Johnson; this atlas was revised and enlarged in 1915. The third edition, dated 1957, was issued in 1959, by the Geographical Branch of the Department of Mines and Technical Surveys, and is a superb achievement covering all aspects, with a special emphasis on economic factors, 'an outline of the physical background and the economic development of the nation at mid-century', showing how these factors contributed to the characteristics of Canadian life. Five thousand copies of each map were printed, of which two thousand were bound immediately in atlas form and the remaining sheets were bound as required. French and English editions were prepared; unfortunately, the English edition is now out of print. Most of the maps were drawn on the Lambert Conformal Conic Projection, with seventy seven degrees north and forty nine degrees north as the standard parallels. The basic scale is 1:10M, on which the whole of Canada can be shown on one double spread. Four maps on the scale of 1:20M can be shown on a double page; and for special sheets, the 1:50M scale is used. Economic factors are stressed, but maps cover also all aspects of physical features, climate, agriculture, population, industry and communications. One of the interesting new maps is that showing the routes taken by the early explorers of Canada. Awareness of world co-operation is shown in various ways, notably in the series of urban land use maps at 1:100,000, which link with the World Land Use Survey; these have been compiled in conjunction with development maps of Quebec, Montreal, Toronto, Ottawa, Winnipeg, Edmonton, Vancouver and Victoria. There is also a final plate showing the connections between Canada and the Commonwealth, the Colombo Plan, NATO and UNO. This series of national atlases, designed to portray all aspects of the life of a country by cartographic means, has been most influential.

Inspired by the first two editions of the national atlas and also by the British Columbia Natural Resources Conference, the *British Columbia atlas of resources*, edited by J D Chapman and D B Turner, was issued in 1956 to mark the centenary of British Columbia. The atlas is in looseleaf form, arranged in three parts, on a basic scale of 1 : 3,500,000; geology, land forms, climate, soils and other geographical features are followed by the section depicting the utilisation of natural resources; the third part comprises the indexes and gazetteer. Facing each map is an explanatory text, illustrated by diagrams, tables and photographs. Informative maps and diagrams appear also in the *Annual reports* of the British Columbia Natural Resources Conference and a valuable collection of prints of air photographs is available for loan or purchase from the Library of the Department of Lands and Forests, Victoria.

A second splendid regional atlas is the *Economic atlas of Manitoba*, edited by Thomas R Weir and published by the Department of Industry and Commerce, Province of Manitoba, 1960. The work is scholarly and practical, in three divisions, dealing with the basic resources, population and settlement and the use of resources, the third section being the most comprehensive, especially with regard to agriculture. Scales vary according to subject content, from 1 : 72,000 to 1 : 5M. Winnipeg is drawn at 6,000 feet to the inch. The map plates are supplemented by explanatory text, graphs, diagrams, statistics and photographs. An *Atlas of Alberta* has been sponsored by the government of Alberta and the University of Alberta, Department of Geography. Two members of the editorial committee are government officials and three are members of the department. An account of the atlas by J J Klawe is included in *The cartographer*, May 1965. The Department of Geography has assumed the responsibility of producing complete map coverage of land use in Alberta for the Agricultural Rehabilitation and Development project, Department of Agriculture, under the direction of A H Laycock. Regional atlases, with emphasis on resources use, are in preparation for other provinces also.

The *Historical atlas of Canada*, edited by Donald G G Kerr for the Historical Association of Canada (Nelson, 1959, 1963) with a second edition in 1966), is a fine atlas, presenting clearly all aspects of the development of Canadian national life against the physical background. The 154 map reproductions, original maps and diagrams maintain a very high standard of cartography. The reproductions include Champlain's maps of 1612 and 1632, and a series of maps and diagrams traces the exploration of Canada and the evolution of the economy. The present agricultural distribution receives careful treatment and the St Lawrence Seaway, so vital to the economy of Canada, is clearly shown. A bibliography and explanatory notes complete this scholarly work. *An historical atlas of Canada,* edited with introduction, notes and chronological tables, by Lawrence J Burpee, with maps by Bartholomew (Toronto, Nelson, 1927) and *An atlas of Canadian history,* edited by W J Eccles and J W Chalmers (Philip, 1964), are not so relevant to the work in hand.

The *Atlas historique du Canada français des débuts à 1867,* edited by Marcel Trudel (Quebec, Les Presses de l'Université Laval, 1961), comprises ninety three maps of New France and Quebec; a pictorial history of this part of Canada is presented, some maps showing also the relationship with neighbouring regions. It is a well-documented work and the town plans are particularly useful.

Two special subject atlases must be mentioned: the *Climatological atlas of Canada,* edited by M K Thomas (National Research Council, Division of Building Research, 1953), and the *Ice atlas of Arctic Canada,* compiled by Charles W M Swithinbank (Canada Defense Research Board, 1960). The latter includes sixty seven pages of maps expertly prepared at the Scott Polar Research Institute, Cambridge, and the Department of Mines and Technical Surveys, Ottawa, Surveys and Mapping Branch. Data collected since 1900 from 325 stations between northern Alaska and western Greenland are shown by symbols, including the number of months of five types of ice cover and of four degrees of difficulty of navigation. Additional sketch-maps, diagrams and a bibliography all contribute to the scholarship of this atlas, which includes also the west coast of Greenland and the north coast of Alaska.

As noted earlier, it is an increasing trend among publishers to issue world atlases designed for a particular regional market. *Nelson's Canadian atlas,* edited by J Wreford Watson, *Nelson's Canadian junior atlas* and *Nelson's Canadian school atlas* all give more comprehensive treatment to Canada; the *Canadian Oxford atlas of the world* (second edition 1957) is divided into two parts: *1* Canada, distribution maps, regional maps and gazetteer of Canada; *2* the rest of the world. There are also the *Canadian Oxford school atlas* and *Nelson's historical atlas of Canada.*

Canadian cartography has been issued by the Canadian Institute of Surveying since 1962; the first volume included the *Proceedings* of the Symposium on Cartography held in Ottawa in the February of that year. Other high level geographical journals frequently including original maps are *The Canadian geographical journal,* monthly journal of the Royal Canadian Geographical Society, Ottawa, since 1930 and the journal of the Institut de Géographie, Université Laval, *Cahiers de géographie de Québec.* The Department of Mines and Technical Surveys, Geographical Branch, issues a *Geographical bulletin* twice a year and irregular *Geographical papers. Canadian maps 1947-1954* was published in 1956. Reference should also be made to the first issue of the special papers of the Centre de Documentation Cartographique et Géographique, *Le Canada: documentation cartographique, bibliographique et photographique,* 1949, and to ' Surveyors of Canada ', which was the theme of a special number of *The Canadian surveyor,* volume 20, no 5 1966, with maps in a separate pocket. Other relevant references include :

W A Barnard: ' The evolution of cartography in Canada ', *Canadian cartography,* Volume I 1962.

T E Layng: ' The first line in the cartography of Canada ', *The Canadian surveyor,* March 1964.

L C Murray, J B McClellan and L E Philpotts: 'Air photo interpretation and rural land use mapping in Canada ', *Photogrammetria,* June 1966.

L J O'Brien, editor: *Canadian cartography, 1962.* Proceedings of the symposium on cartography, held at the Canadian Institute of Surveying, February 1962 (Canadian Institute of Surveying, 1962); ill, maps.

D J Packman and L E Philpotts: *Elementary agricultural air photo interpretation, with particular reference to Eastern Canada* (Ottawa, Canadian Department of Agriculture, 1955). Annotated air photographs, showing their use in the interpretation of soils, crops, etc.

D W Thomson: ' The history of surveying and mapping in Canada ', *Surveying and Mapping,* March 1967.

D W Thomson: *Men and meridians: the history of surveying and mapping in Canada.* The work is to be completed in three volumes, of which the first, dealing with the period prior to 1867 was published by the Queen's Printer in 1967.

J W Watson: ' Mapping a hundred years of change in the Niagara peninsula ', *Canadian geography,* 32, 1946.

MEXICO, CENTRAL AND SOUTH AMERICA

MEXICO: The central topographical mapping body in Mexico is the Dirección General de Geografía, Meteorologia, founded in 1899 as the Commissión Geodésica Mexicana, changing to the present name in 1942. The national committee is the Mexican National Committee of the International Cartographic Association, at the Facultad de Filosofía y Letras, Colegio de Geografia, Universitad Nacional de México.

A number of thematic maps have been published by the institute, such as 'Places of more than 10,000 inhabitants in the economic zones of Mexico' in two sheets, based on the 1940 and 1960 Census figures, at approximately 1:6M; 'Places of more than 100,000 inhabitants in the economic zones of Mexico', at 1:4M, based on the 1960 figures; 'Mapa de vegetacion natural de Mexico', at 1:4M, 1962; 'Mapa de suelos de Mexico', at approximately 1:6M; 'Zonas y regiones geoeconomices de Mexico', at 1:4M, 1961, which distinguishes seven zones and nearly ninety economic regions, etc. The main topographic series is at 1:100,000, of which a new edition was begun recently covering central Mexico. Topographic maps at 1:500,000 have been compiled, especially for southern Mexico, by the Comité Coordinador del Levantamiento de la Carta Geografica de la Republica Mexicana.

The Geological Survey of Mexico has issued *Boletines* since 1895 and *Anales* since 1917. The 'Carta geológica de la Republica Mexico 1:5M' was completed in 1942. A new 'Carta geólogica' was prepared at 1:2M for the 1956 International Geological Congress. A new land form map was compiled in 1958, sponsored by the Office of Naval Research; the work was constructed on existing base maps, particularly on the 1:1M World Aeronautical Chart for rivers, roads and cities, on monographs and articles recording individual research and, especially, on a set of trimetrogon aerial photographs. The shape of Mexico left plenty of room for insets on the map sheets, and text and glossary were added. An article by Erwin Raisz, 'A new landform map of Mexico', in the *International yearbook of cartography,* 1961, gives further details, illustrated with maps and diagrams. In 1961, the Geological Society of America issued a 'Tectonic map of Mexico', at 1:2,500,000; and in 1966, a major work, 'Carta general de la Republica Mexicana', at approximately 1:3,500,000, was published by the Coleccion Geografica Patria, with a series of district maps at various scales compiled by Jorge L Tamayo.

As in many other countries, the past decade has produced much research and survey work designed for planning economic and social studies, also showing the tendency to bring mapping activities into line with those of other countries. Such was the 1:1M 'Population map' drawn in the Department of Geography of the Faculty of Philosophy and Letters of the University of Mexico in collaboration with the Military Cartographic Department in accordance with the recommendations of the International Geographical Commission; the Lambert Conformal Projection was used and data was taken from the 1960 Census. Aerial photography has been used for some years; in 1966, a most useful work was published by the Organization of American States, Washington—the *Annotated index of aerial photographic coverage and mapping of topographical and natural resources,* 1 : 6,022,000, consisting of thirty three maps and text, forming a comprehensive record of Central and South American mapping. The Instituto Nacional para la Investigacion de Recursus Minerales has been responsible for the mapping and description of mineral deposits; *Boletines* have been published since 1945.

Various tourist maps have been issued by the Departamento de Turismo, Petroleos Mexicanos; Geographia have a general reference map of Mexico, showing roads and railways; Rand McNally completed their ' Rand McNally Imperial map of Mexico 1 : 3M ' in 1959, with an inset of Mexico City; there is also the Rand McNally *Guide to Mexico,* by Herman and Jaunita Liebes, containing quick reference information and a folding map; and a plan of Mexico City itself at 1 : 200,000, was brought out by the Instituto de Geografia in 1961. ' Mexiko : Land der Olympiade 1968 ', by H Bleckert (VEB Hermann Haack, Geographische-Kartographische Anstalt, Gotha, 1968) is a coloured fold-out map, with good photographs.

From the time of publication of the *Atlas of Antonio García Cubas,* Mexico, 1858, the first comprehensive cartographical publication of the nineteenth century, to 1962, with the issue of J L Tamayo's *Atlas* to accompany his four volume *Geografía general de México,* there has been a constant production of atlases in Mexico, totalling at least some 120, the majority being geographical in scope; they contain important cartographic, statistical and graphical material of vital significance in the study of geographic ideas and methodology in Mexico. The 1962 *Atlas geografico general de México* was the second edition of the work, published by the Instituto Mexicano de Investigaciones Económicas; Jorge Tamayo himself prepared a number of the maps. Small sketch-maps illustrate economic factors in or about 1960. A one volume *Atlas universal y de México,* edited by C Reyes Orozco (Librería Británica, 1966), was made available in Britain by Nelson; and a more specific work, *Estudio climatico de la Cuenca Hidrografica Lerma Santiago* became available in its latest edition (Guadalajara, 1966) with English summaries. This atlas, compiled by E Jauregui, has been issued by various publishers in Spanish; it consists of 105 sheets at a basic scale of 1 : 2M, with explanatory text.

The papers presented to the International Geographical Union Latin American Regional Conference, Mexico City, 1966, include many original and interesting maps and sketch-maps. The publications of the Mexican Society of Geography and Statistics, of the Mexican Association of Professional Geographers and the *Boletin de la Sociedad Geologica Mexicana,* 1905, all include maps and information about map production.

So far as is known, there is no special current bibliography for maps, but the National Bibliographic Institute began publication in 1958 of the *Anuarios bibliográficos,* and from 1950, the national library has issued the *Boletin de la Biblioteca Nacional,* which includes classified lists of recent accessions. There is also the *Bibliografía geográfica de México,* compiled by B A Bassols (Dirección General de Géografía y Meteorologia, 1955), and the *Bibliografía economica de Mexico,* quarterly from 1955 from the library of the Department of Economic Studies, Bank of Mexico, Mexico City.

More detailed studies of individual regions have lately been undertaken: for example, B A Arnold: *Late Pleistocene and recent changes in land forms, climate and archaeology in Central Baja* (University of California, publications in geography, X (4) 1957), which includes illustrations, maps and diagrams, on the Chapala basin and the surrounding area; Q D Hill: *The changing landscape of a Mexico municipio, Villa las Rosas, Chiapas* (University of Chicago, Department of Geography, research paper no 91, 1964); T T Poleman: *The Papaloapan project: agricultural development in the Mexican tropics* (Stanford UP for the Food Research Institute: studies in tropical development, 1964); and J Walter Thompson: *The Mexican market,* edited by the Marketing Department of J Walter Thompson de Mexico, second edition 1963; an excellent geographical summary, with maps in the text.

Maps of Mexico, Central America and the West Indies are included in the first volume of *A catalogue of Latin American flat maps 1926-1964,* mentioned below.

CENTRAL AMERICA: A comprehensive record of Central and South American mapping is given in the series of twelve sheets at approximately 1:11M, ' Status of topographic, geologic and soil mapping and vertical aerial photography of the OAS members of northern and southern South America and Middle America ', prepared and published by the Natural Resources Unit of the Department of Economic Affairs of the Pan-American Union, 1965-. These annotated index maps bring together much information otherwise difficult to obtain. The Pan-American Union has sponsored a number of regional maps, maps of agriculture, ecology and soil maps of Central America, usually at 1:250,000 or 1:1M.

British Honduras has been mapped by the Directorate of Overseas (Geodetic and Topographic) Survey at 1:50,000, and in 1959, the Colonial Office issued the report *Land in British Honduras: a report of the British Honduras Land Use Survey team,* edited by D H

Romney (Colonial Research publication, no 24), which makes a thorough analysis of all aspects of the physical and economic geography, with suggestions for development. There are many maps in the text and, issued separately, in a box, seven maps concerned with land use, vegetation, rainfall and soils. Previous Colonial Office reports should also be consulted in studying the development of the economy, notably the *Report of the British Guiana and British Honduras Settlement Commission.*

The General Directorate of Cartography, a department of the Ministry of Communications and Public Works, is responsible for geodetic topographic work, in co-operation with the Inter-American Geodetic Survey; other governmental agencies prepare soil, vegetation, population and economic maps, based on data provided by the General Directorate of Cartography. The Directorate's first concern is to complete the base map of the Republic, which is being plotted from aerial photographs. The government of the Federal Republic of Germany has contributed valuable photogrammetric equipment and the services of technical staff to co-operate with the Directorate in the compilation of a national cadastral survey. The annual reports of the Forest, Agricultural and Labour Departments and the *Trade report* frequently allude to surveys in progress and an *Atlas of British Honduras* was issued by the Survey Department in 1939.

The national mapping agency for Guatamala is the Instituto Geográfico Nacional, Dirección General de Cartografía. The basic scale for these Central American countries is 1 : 50,000, published for Guatamala by the Dirección General de Cartografía. The Department of Mapping and Cartography had been set up in 1945, completely reorganised as the Department of Cartography in 1954 and again expanded and reformed as the National Geographic Institute in 1964. The Inter-American Geodetic Survey of the United States of America, which has been assisting in the mapping of Guatamala since 1946, continues to support the institute, and the German Technical Mission has, since 1964, provided valuable equipment and technical assistance with the aim of establishing in due time a cadastral programme to cover the country.

In the southern part of Guatamala, traditional methods of triangulation have been used from the beginning until the recent introduction of modern methods of distance measurement. North of the sixteenth parallel, the densely wooded areas were surveyed by photogrammetric methods. The basic map of the republic is on the scale 1 : 250,000, printed in seven colours, with contour lines at 100 metre intervals and fifty metre auxiliary contours. These sheets have also been published as plastic relief maps. The second important series is the 'Topographic map of the Republic' at 1:50,000, comprising five-colour sheets, with twenty metre contour intervals and ten metre auxiliary contour lines. Four preliminary maps of the republic were completed on scales of 1:200,000, 1:500,000, 1:750,000 and 1:800,000. Maps of the principal towns are available at 1:50,000, with five metre contour lines; eighty one map sheets cover Guatamala City on scales ranging from 1:1,700 to 1:5,000. There are also 125 'Area maps of the Republic' on scales from 1:5,000 to 1:500,000.

Geological mapping is in progress on two scales, 1:250,000 and 1:50,000. Other special topic maps include tourist maps of Antigua and of Guatamala City and a gravimetric map of the republic at 1:1M. A *Preliminary atlas of the Republic of Guatamala* brings together all studies so far completed on the economic and social developments, accompanied by an explanatory booklet 'Preliminary draft plan for the economic and social development of the country in twenty one years'. There is also a *Departmental atlas of the Republic,* containing separate sheets for each department and an index map. Detailed maps showing the state of first-order triangulation, levelling and index maps of the 1:250,000 and 1:50,000 topographical series are included in volume VII 1967 of *World cartography*. One of the most useful maps of Guatamala, 'Mapa de la Republica de Guatamala', at approximately 1:905,000, was produced in 1953 by the General Drafting Company of New York for Esso Standard Oil (Central America), with explanatory text in Spanish. The issues of the *Anales de la Sociedad de Geografía e Historia de Guatamala* are a central source of information, also the government department publications, such as the *Revista de la Economia Nacional,* monthly since 1946 by the Ministerio de Economía, the monthly *Boletin* of the Instituto de Fomento de la Producción and other trade journals. The economic surveys by Barclays Bank DCO are most valuable and up-to-date; particularly useful was the issue for 1965, comprising twenty two pages of close analysis and a map.

The situation and terrain of Nicaragua, having both an Atlantic and a Pacific coast, have affected development and cartographic reproduction of the area. The Atlantic zone is thickly wooded, hot, humid and unhealthy; the north central zone enjoys a fresh, pleasant climate and the hot, dry Pacific zone is the most heavily cultivated and most densely populated part. The Geodesy Office was established in 1946 by agreement with the United States government; with expanding activities, the name was changed in 1960 to the General Directorate of Cartography, within the Ministry of Development and Public Works, and co-operation with the Inter-American

Geodetic Survey takes the form of technical assistance and equipment. The Directorate is concentrating on basic maps of the republic at scales of 1:50,000, 1:100,000 and 1:250,000. Planimetric maps at 100,000 and 1:250,000 are also produced and topographic maps of the towns are being steadily completed on scales of 1:5,000 and 1:10,000. Preparations are in hand for the production of soil maps. The Pacific zone had been covered by aerial photography by the United States Air Force in 1954, at 1:64,000, and this area has been completed at 1:50,000, showing contour lines at intervals of twenty metres. Work on maps of the central and south Atlantic zones is well advanced.

The National Geological Service uses the main topographic maps as the bases of its surveys. 'Nicaragua' by the Ministerio de Fomento, Oficina de Geodesia, compiled in collaboration with the Inter-American Geodetic Survey, on the Transverse Mercator Projection, with relief shown at twenty metre intervals, is another useful general purpose map.

The El Salvador mapping service is the Dirección General de Cartografía, founded in 1946, which began the 1:50,000 series covering the country in 1955, the coastal areas being completed first. The Lambert Conformal Conic Projection was chosen and contours are at fifty metre intervals. The *Atlas censal de El Salvador* (and the *Atlas estadístico de Costa Rica*) was compiled under the direction of Alford Archer, seconded for the purpose by the United States Bureau of the Census; both are clear, well-balanced atlases, with explanatory text, which is also illustrated. Close analysis of population is a feature, both throughout the countries and in the chief cities; climatic and physical conditions are shown as conditioning agriculture and industry.

Costa Rica, one of the smallest countries of central America, presents great diversity of physical features, from the mountain masses and volcanic peaks to the broad plains and more developed area of the central valley. The body responsible for mapping is the Instituto Geográfico de Costa Rica, established in 1945, which maintains a topographic series at 1:50,000 and 1:25,000. Aeronautical charts, 1:1M, have been prepared by the United States Hydrographic Office and the United States Aeronautical Chart and Information Service. In common with nearly all the cartographic institutions of Latin America, work is carried out in conjunction with the Inter-American Geodetic Survey, a branch of the United States Army Map Service. This co-operation began in Costa Rica in 1945 and has gradually extended, tending to an increased standardisation of format. The Pan-American Institute of Geography and History also attempts to encourage uniformity of production in accordance with agreed resolutions adopted at ' meetings of consultation '; and the Central American Cartographic Weeks have been particularly successful in achieving co-operation in cartographic projects. The contour interval on the 1:25,000 series is ten metres and relief is shown by five or six colours. The series began in 1952; the 1:50,000 maps developed later, having a contour interval of twenty metres. Full details were included in *Detailed work plan for the initiation and completion of the project for acceleration of basic topographic mapping,* published by the institute in 1964. Photogrammetric plotting of the basic map on the scale 1:50,000 was completed in 1968. At the same time, a general map of the whole country has been prepared at 1:500,000; this is a physical and political map, revised to include the latest available information. A wall map edition at 1:350,000 and a relief map are available for educational purposes. An *Atlas estadístico de Costa Rica* was published in San José in 1953.

Maps of Costa Rica: an annotated bibliography, by Albert E Palmerlee (Lawrence: University of Kansas Libraries, 1965; University of Kansas Publications, library series, no 19), attempts to be comprehensive, including also maps found in books and other publications. Five major divisions are distinguished: General maps covering the whole of Costa Rica; subject maps, of climate, agriculture, industry, population, etc; regional maps; provincial, cantonal and district maps; city plans. The work is all the more valuable as the national current bibliography does not contain maps. The author refers also to the *Lista de mapas parciales o totales de Costa Rica,* by Luis Dobles Segreda (San José, 1928) and the *Bibliografía aborigen de Costa Rica,* by Jorge A Lines (San José, 1944). The *Informe semestral,* published twice a year, 1954-, by the Instituto geográfico de Costa Rica, is a source of maps and of information concerning geographical and cartographical projects.

Many fine maps are included in *Middle America: its lands and peoples,* by Robert C West and John P Augelli (Prentice-Hall, 1966); also in Karl Helbig: *Die Landschaften von Nordost-Honduras: auf Grund einer geographischen Studienreise im Jahre* 1953 (Gotha, VEB Hermann Haack, 1959); *The economic development of Nicaragua:* report of a mission organised by the International Bank at the request of the Government of Nicaragua (Baltimore, Johns Hopkins P, for the International Bank, 1953); B W Taylor: *Ecological land use surveys in Nicaragua* (FAO, 1959) prepared by the Ministerio de Economía, Instituto de Fomento Nacional, with text in Spanish and English.

D A G Weddell: *British Honduras: a historical and contempo-*

rary survey (OUP, 1961), was issued under the auspices of the Royal Institute of International Affairs.

In the Caribbean area, the Directorate of Overseas (Geodetic and Topographical) Surveys is working particularly on large-scale maps at between 1:2,500 and 1:10,000 for planning and development purposes and at smaller scale maps suitable for tourists. Throughout the West Indian islands, the Directorate maintains maps at 1:25,000; Jamaica at 1:50,000 and at 1:250,000; the Cayman Islands at 1:125,000; Trinidad and Tobago at 1:10,000 and 1:25,000. In 1964, a new series at 1:10,000 was begun for the latter, using field surveys by the Trinidad and Tobago Lands and Surveys Department. The Cassini Projection was used and relief was shown by contours at twenty five feet intervals and spot heights. The Windward Islands are mapped at 1:50,000 and individual islands, such as the island of Montserrat and Dominica at 1:25,000, drawn on the Transverse Mercator Projection, showing relief by contours and hill-shading, settlements and communications. Saint Lucia island is similarly mapped, with a small inset map of Castries at 1:10,000. Important towns, such as Kingston, Jamaica, have been mapped at 1:10,000. The Directorate has completed much new mapping throughout the islands during the past few years, such as the new maps of the British Virgin Islands, issued in 1963, all of very high standard both in design and printing. New mapping of the Bahama islands has been completed at 1:25,000. The Netherlands Antilles Cadastral Survey Department maintains topographical maps at two main series, 1:10,000 and 1:25,000; relief is shown by contours at ten metre intervals.

Barclay's Bank DCO issues excellent concise economic surveys, frequently revised and including maps—Jamaica . . ., The Windward Islands . . ., The Leeward Islands . . ., Barbados . . ., The Bahamas . . . and Trinidad and Tobago . . . The Colonial Office *Annual reports* for Barbados and Trinidad must be remembered; also the Barbados Development Board issues a very useful handbook, *Barbados: a social, political, economic, commercial, industrial, agricultural, labour survey for businessmen,* including statistics and maps; and a *Handbook of the Leeward Islands,* compiled by F H Watkins (The West India Committee, 1924), contains a map. *Fodor's Guide to the Caribbean, Bahamas and Bermuda,* 1960-, is particularly informative, including a map of the whole West Indian area and twenty five island and regional maps. Skinner's *West Indies and Caribbean yearbook* contains useful reference maps.

The *Atlas de Cuba,* compiled by Gerardo Canet and Erwin Raisz (Harvard University Press, 1945), comprises a large folding map and sixty pages of diagrams and maps, with text in Spanish and

English. Prepared in the Institut de Exploraciones Geográficas at Harvard University, in co-operation with the Ministry of Agriculture of the Republic of Cuba, this is an evocative atlas covering all aspects, with pictorial and block diagrams used to great effect. The Commission del Mapa Geologica del Ministerio de Agricultura, Havana, completed a geological map of Cuba, 1 : 1M, in 1946. A ' Map of the Republic of Cuba ' is maintained by the United States War Department, Washington, at 1 : 500,000. Other maps of Cuba include the ' Carta militar de la República de Cuba, Cuartel General del Ejército, Socción de Ingeniería', 1 : 100,000; a ' Soil map of Cuba ' by the Tropical Plant Research Foundation, Washington, DC, 1928 and a ' Mapa de la Isla de Cuba Mostrando los Ferrocarriles de Servicio Público ', 1948.

The economic development of Jamaica; report of a mission organised by the International Bank at the request of the government (Baltimore, Johns Hopkins Press for the International Bank, 1952) includes useful maps; other relevant monographs include J Butterlin : *La constitution géologique et la structure des Antilles* (Centre national de la Recherche Scientifique, 1956); C F Jones and R Pico, editors : *Symposium on the geography of Puerto Rico* (University of Puerto Rico Press, 1955) under the auspices of the Social Science Research Centre of the University of Puerto Rico in co-operation with the Puerto Rico Department of Agriculture and Commerce, the Puerto Rican Planning Board and the Department of Geography of Northwestern University; D L Niddrie : *Land Use and population in Tobago, with map* (Bude, Geographical Publications, 1961, monograph no 3, World Land Use Survey).

Current systematic bibliography throughout the area is still in the early stages. Maps are included in the *Bibliografía Cubana,* 1937-; and the *Current Caribbean bibliography,* 1951- is a record of all items printed in the Caribbean countries served by the Commission (French, British, Dutch and American members of the Commission). *A descriptive list of maps of Barbados,* by E M Shilstone, is included in the *Journal* of the Barbados Museum and Historical Society, V (2) February 1938, comprising sixty four titles. *West Indies: a catalogue of books, maps etc* by Edward Bros was published in 1929.

SOUTH AMERICA : A useful introduction to South American cartography is *A catalogue of Latin American flat maps, 1926-1964,* published 1967- by the University of Texas Institute of Latin American Studies (guides and bibliographical series, no 2), under the editorship of Palmyra V Monteiro. This catalogue is envisaged as a continuation of the work in the *Catalogue of maps of Hispanic*

America, published by the American Geographical Society, 1930-32. Maps issued by government agencies are given priority, but many privately produced maps are included; arrangement is by category. Useful appendices give the names and addresses of organisations from which maps and map information in the various countries may be obtained. An excellent summary is provided by P H Freeman and others in their article 'An inventory of Latin American mapping' in the *Bulletin* of the Geography and Map Division, Special Libraries Association, September 1963.

The most important series of maps covering the whole of South America is the *Map of Hispanic America,* completed in 107 sheets in 1947 by the American Geographical Society, conforming in essentials to the specifications of the International Map of the World 1:1M. This impressive work involved the co-ordination of several thousand original regional surveys. The United States Bureau of the Census, in co-operation with the International Co-operation Administration, United States Department of Commerce, has compiled some census maps of Central and South America; and Denoyer-Geppert of Chicago are publishing a Hispanic America series in Spanish and English, edited by J F King and E Herbert, in two main sections: *A* relief and culture maps; *B* population maps. A new edition of the Bartholomew 'Reference map of South America' was issued in 1964, at 1:10M; contour colouring is shown, roads are in red and railways in black. 'Geographia' issue a political map of South America, showing main roads, railways and shipping routes, canals, pipelines, airports, etc. A physical map in two sheets shows also roads, railways and airports. All the great map publishing firms mentioned in chapter 2 have produced useful general maps of South America and the dramatic nature of the country's relief has made it the subject of many experiments in relief form mapping. *Latin America in maps,* by A G Wilgus (New York, Barnes and Noble, 1951), and *Ferriday's map book of South America* (Macmillan, 1967) are two handy sources of general maps. The latter consists of twenty seven sheets at various scales, dealing with each state, with explanatory text. Before mentioning the individual countries' mapping services, one other thematic map of great interest should be mentioned: the 'Ethno-linguistic distribution of South American Indians', 1:8,500,000, compiled by Čestmír Loukotka and issued as map supplement no 8 of the *Annals* of the Association of American Geographers, June 1967. *The ethnographic bibliography of South America,* by T J O'Leary, 1963, contains maps.

The Dirección de Cartografía Nacional del Ministerio de Obras Publicas, Caracas, founded in 1935 and reorganised in 1939, has established three topographical map series for Venezuela, at

1:25,000, at 1:100,000 and at 1:250,000, the latter being started in 1948, beginning with the northern part of the country. There is also a geographical map in twenty four sheets at 1:500,000 and larger scale plans for the major cities, for example, Valencia at 1:10,000 and Caracas at 1:20,000, showing streets and principal buildings against the background of vegetation. The sheet lines of the topographical maps are based on international longitudes and latitudes and a systemic division between the series has been adopted, so that a uniform numbering system can be used. All the maps use the same colour scheme: blue for water, brown for relief, red for roads, green for vegetation and black for all other features. The geographical map served as the base for the International Map of the World 1:1M sheets of Venezuela. A new series of the 1:100,000 was begun in 1962, drawn on the Transverse Mercator Projection; in this series, design and printing are of high order; relief is shown by contours at twenty or forty metre intervals.

The Ministry's cartographic department published a political and physical map of the country at 1:2M in 1965. International and state boundaries are clearly drawn and relief is shown by hillshading; railways and four categories of roads are included. Index maps of aerial photography coverage and of the areas covered by 1:100,000 maps to date are included in *World cartography,* VIII 1967.

In 1960, an *Atlas agrícola de Venezuela* was published by the Ministerio de Agricultura y Cria, Dirección de Planificación Agropecuaria, Consejo de Bienestar Rural, Caracas. The Geological Survey, Dirección de Geologia, Ministerio de Minas e Hidrocarborus, Caracas, has issued maps and a *Boletin de geologia,* 1951-, and a ' Geologico-tectonic map of Venezuela ', 1:1M, was compiled in 1958 by W H Bucher and published by the Geological Society of America, New York.

There is no current map bibliography issued, so that information on map production must be gleaned from individual catalogues and from such publications as the *Boletín bibliográfico,* 1953-, by the library of the Agricultural Faculty of the Central University; the *Boletín informativo,* issued by the Ministry of Agriculture Library and a similar *Bulletin* issued by the central library of the Central University since 1950.

British and French mapping agencies maintained the basic topographical maps of Guyana until independence. The Directorate of Overseas (Geodetic and Topographic) Surveys has maintained a map series at 1:500,000. The Institut Géographique National, Paris, has maintained photogrammetric surveys at 1:100,000 and the national organisations have prepared thematic maps. The Geological

Survey of British Guiana has issued *Annual reports* since 1933 and *Bulletins* since 1936; and the Department of Lands and Surveys, Georgetown, has prepared a 'Map of British Guiana showing vegetation, minerals and communications', 1 : 1M, 1959. A very interesting series of more than 100 large-scale estate plans is maintained at various scales, showing the location of sugar estates, by Bookers Sugar Estates, Georgetown. The government issued a useful thirty five page booklet on the resources of the country, with a map, *British Guiana: industry and development* (Georgetown, 1958).

In Brazil, the Serviço Geográfico do Exército is the national mapping agency; maps are also produced by the Conselho Nacional de Geografía, the Instituto Geológico e Geográfico de Sao Paulo, the Departmento Geográfico de Estado Minas Gerais and the Prefeîtura do Distrito Federal. The national committee representing the country at the International Cartographic Association is the Sociedade Brasileira de Cartografia, Rio de Janiero. Photogrammetry was first used in Brazil in 1913, and in 1922 the first photogrammetric map of the Federal District was completed by the Brazilian Army Service. Other government departments, the Navy Hydrographic Department, the Bureau of Reclamation and the Mineral Department have since installed their own photogrammetric equipment and great progress has been made in the production of maps for economic and regional planning projects. In 1937, the Conselho Nacional de Geografía was created to prepare a new edition of the 1 : 1M map, to initiate a general census of the republic, which was done in 1940, and to set up a national network of first-order triangulation and levelling on which systematic mapping of the country could be based.

The main publications of the Serviço Geográfico do Exército include the 1 : 25,000 series, the 1 : 50,000 'Carta Normal' series, and the 1 : 100,000 and 1 : 250,000, on which special maps are constructed. The regular series are based on the Gauss Conformal Projection. Many of the sheets are in monochrome, with contour intervals of ten metres. The basic 1 : 50,000 maps are printed either in five colours, black, blue, brown, orange and green; or in three, black, blue and brown, depending on the importance of the region and the completeness of the detail shown. Several sheets are planimetric and a few provisional sheets are in monochrome. The primary importance of the 1 : 50,000 series is to assist in the redeployment of resources and the nationalisation of planning social and economic development; they represent a combined effort by all the mapping agencies of the country, based on aerial photographic surveys, with ground field work to complete the detail. A complete revision of the 1 : 100,000 was sponsored by the Conselho Nacional de Geografía in 1964, to complete the series and replace individual sheets previously compiled by the institute and other geographical bodies. The 1 : 250,000 series is in a provisional edition, printed in three colours, blue for drainage, red for roads and black for other features. Some of these sheets include contours in brown at fifty metre intervals.

The Conselho Nacional de Geografía, Rio de Janeiro, published the three ' Carta do Brasil ' series at 1 : 250,000, 1 : 500,000, the special map of the Distrito Federal, 1 : 100,000, and the 1 : 1M, which conforms closely to the International Map of the World 1 : 1M specifications. The conselho also publishes general maps for particular states, as occasion arises. The first edition of the 1 : 1M map was edited by the Brazilian Engineering Club; reprinting of the forty six sheets covering the country was completed by the Conselho Nacional de Geografía in 1946 and new editions of individual sheets have since been issued. The format of the 1 : 1M map sheet is now six degrees in longitude and four degrees in latitude, in conformity with the Bonn specifications of 1962.

The Geological Survey of Brasil, working through the Divisão de Geologia e Mineralogia, the Departmento Nacional da Produção Mineral and the Ministerio de Agricultura, Rio de Janeiro, has produced monographs since 1913, *Relatorios* annually from 1919 and *Boletins* from 1920. A geological map of Brazil at 1 : 5M was completed in 1942. A Centre of Applied Hydrology is being established at the Institute of Hydraulic Research in the Federal University of Rio Grande do Sol at Porto Alegre. Basic surveying and mapping will occupy the centre for some time, together with the training of water resources planners. Eventually, it is anticipated that programmes will be extended to meet the needs of other Latin American countries.

The Instituto Brasileiro de Geografía e Estatistica issued a second edition of the *Atlas do Brasil* in 1959. Map plates, accompanied by brief text, cover relief, climate, vegetation, population, economic activity, transport and communications, for Brazil as a whole and for a number of the major regions, at scales ranging from 1 : 30,000 to 1 : 14M. A topographical map at 1 : 1M is included and the work is admirably illustrated by photographs, sketch-maps and diagrams. *A statistical atlas of Brazil,* compiled by C A Ribeiro Campos and published in Rio de Janeiro in 1941, is useful for purposes of historical geography; in the same year a special *Atlas corografico de la cultura cafeeira* was issued by the Departamento Nacional do Café, Secçáo de Estatística. The *Atlas pluviométrico do Brasil* was compiled by the Hydrological Bureau of the Divisão de Aguas and prepared by the Serviço Gráfico do Instituto Brasileiro de Geografia a Estatística, at a basic scale of about 1 : 1M. Covering such an immense territory, this is a major work, showing the results of monthly observations from 387 rain-gauge stations over a twenty five year period. The text is in Portuguese and Spanish. The Serviço de Meteorologica, Rio de Janeiro, has produced an *Atlas climatológico do Brasil* in three volumes, 1960, with the maps at a basic scale of 1 : 7,500,000, under the direction of Adalberto Serra.

Much planning and development is in progress in Brazil: for example, a new surveying project is in progress on the River Paraguay, which is described in the article 'A large-scale hydrological study in Latin America: the upper basin of the Rio Paraguay in Brazil', by Newton V Cordeiro, *Natural resources,* III (2) June 1967. The Instituto Geológico e Geográfico de Sao Paulo has published maps of the state of Sao Paulo at 1 : 100,000 and the Departamento Geográfico do Estado de Minas Gerais has issued a sixty sheet series covering the state. The Federal District has also been well surveyed and maps at 1 : 1,000, 1 : 2,000, 1 : 5,000, 1 : 15,000 and 1 : 20,000 have been published by the Prefeîtura do Distrito Federal.

A comprehensive summary of the state of mapping in Brazil, with index maps showing the basic geodetic network, aerial photographic coverage, including trimetrogon photography and indications of mapping coverage at various scales, is included in *World cartography,* VII 1967. Maps are listed in the annual *Bibliografia cartográfia do Brasil;* there is also the annual *Bibliografia geográfia do Brasil* and, for the north-eastern region, a systematic bibliography was published in 1965, the *Bibliografia cartográfica do Nordeste,* by the Superintendencia do Desenvolvimento do Nordeste, Divisão de Cartografia. More than one thousand cartographic publications relating to the Brazilian north-east available in the map collection of the SUDENE Division of Cartography are arranged according to the Universal Decimal Classification. *La cartografia da região Amazonica: catalogo descritivo (1500-1961)* by Isa Adonias, was published in Rio de Janeiro by the Conselho Nacional de Pesquisas, Instituto Nacional de Pesquisas da Amazonia, 1963, in two large volumes and is undoubtedly the most important contribution of recent years to the historical cartography of South America. Maps are listed chronologically within regional groups; each map is given a detailed entry and reference, but there is no index.

The mapping of Colombia is not yet entirely integrated due mainly to the extremes of relief in the west and east of the country; but although the mapping of relief, climate and soils has been difficult, it has also been rewarding in the planning of mineral, agricultural and industrial potentials. A topographic series at

1 : 25,000 was begun by the Institut Geografico 'Agustin Cadazzi' in 1944. The Gauss Conformal Projection was used and relief shown by contours at fifty or 100 metre intervals. Bogota and the south-western regions were completed first and more than two hundred sheets are now available. In 1948, the institute began a second series at 1 : 100,000 for coverage of the northern and north-eastern parts of the country, using aerial photography; relief on these sheets is shown by contours at various intervals, depending on the terrain and with spot heights indicated in metres. An *Annotated index of aerial photographic coverage and mapping of topography and natural resources in Colombia* was published by the Pan-American Union, Washington, DC, in 1966.

An 'Administrative map of the Republic of Colombia', 1 : 1,500,000 was issued in 1958 by the Instituto Geografico de Colombia. The geology of a country like Colombia is obviously of extreme interest. The Geological Survey of Colombia, Servicio Geologico Nacional, Ministerio de Minas, Petroleros, Bogota, has published an irregular series ' Compilacion de los estudios geologicos oficiales en Colombia ', from 1933 and a *Boletin geologico* from 1952. A geological map of the country at 1 : 2M was issued in 1944 and a new map, in a preliminary edition, at 1 : 1,500,000, came out in 1962, compiled in co-operation with the petrol companies. The publications of the Sociedad Colombiana de Ingenieros, founded in 1887, are a central source of information and reference, also those of the Sociedad Geográfica de Colombia from 1903, especially the *Boletín,* 1934-. In *World cartography,* 1967, details are given of the state of geodetic first-order triangulation, first- and second-order levelling and aerial photographic coverage carried out by the Instituto Geografico 'Agustin Cadazzi', with an index map; also the topographic map coverage on the scale 1 : 25,000 and less extensive coverage at 1 : 100,000.

Only one institution is responsible for the topographical mapping of Ecuador, the Institut Geográfico Militar, established in 1928. A topographical series at 1 : 50,000 is maintained. A geological map of Ecuador at 1 : 1,500,000 was compiled by Walter Sauer and published by the University of Quito in 1950. No regular current map bibliography is as yet available.

On the other hand, cartographical work in Peru is considerably advanced, carried out there by three governmental agencies, the Military Geographic Institute, responsible for basic geodetic control and the national mapping service; the National Aerial Photography Service, which primarily prepares photographic and topographic special purpose maps; and the Hydrographic and Lighthouse Service, responsible for marigraphs, nautical charts and navigational aids. A number of official bodies collaborate with the government in preliminary mapping for development projects; the National Land Reform Office, for example, the National Office of Natural Resources Evaluation, the National Town Planning and Town Development Office and the National Planning Institute.

The chief photographic coverage of Peru is at the scale 1 : 60,000; other areas are mapped at 1 : 30,000 and 1 : 10,000. Maps at 1 : 1M and 1 : 500,000 cover the entire country; especially important parts are covered at 1 : 200,000 and 1 : 100,000, with a few at 1 : 20,000, in addition to large-scale surveys for special projects. The institute published a 'Mapa fisico-politico 1 : 1M, Republica del Peru' in 1963; this is an attractive map in eight sheets, clearly marking the departmental and provincial boundaries and showing relief by hill-shading and spot heights. An impressive topographic map of Panta, Cordillera Vilcabamba, Peru, was prepared at 1 : 25,000 by the Swiss Foundation for Alpine Research in 1965, as an appendix to *Mountain world,* 1964-65. The Geological Survey is part of the Instituto Nacional de Investigacion y Fomento Mineros, Lima; the *Boletin* was published between 1945 and 1949 and the *Boletin del Instituto Nacional de Investigacion y Fomento Mineros* from 1950. The *Boletin* of the Geological Society of Peru dates from 1925 and is a valuable source of information on the growth of geological knowledge of the area. Two other main sources of original maps and of information concerning new projects are the publications of the Asociación de Geógrafos del Perú, Lima, and of the Sociedad Geográfica de Lima, 1888-. Current map production is included in the *Anuario bibliográfico peruano,* 1943-, founded by the National Library in 1945. The quarterly *Boletín,* 1891-, of the Sociedad Geográfica de Lima, has provided a long record of cartographical information.

Bolivia is in the heart of Latin America, a vast area of high mountains, deep valleys and tropical plains crossed by great navigable rivers. Up to about the past two decades, the country was mapped only by general political, orographic and physical maps at scales of 1 : 200,000 and 1 : 1M. Since then, the Agrarian Reform Act created a new economic structure and a National Planning Secretariat was established (including a Cartographic Department), which has co-ordinated the maps prepared by the Military Geographic Institute and maps, based on aerial photogrammetry, produced by such companies as Gulf and Shell, having interests in the petroleum areas. All such maps available were checked and incorporated into a series suitable as a basis for planning. Other areas needing to be covered for immediate planning have been mapped at varying scales according to immediate requirements. The early 1970s will see

337

the results of the long-term planning now in operation. The Military Geographic Institute has been provided with all the equipment necessary for the completion of modern air surveys, including electronic equipment; the basic network already covers the country. In 1966, censuses of population, agriculture and livestock were carried out, providing new comparative statistics for the basis of thematic maps. At the same time, general, physical, political and ecological maps at 1:1,500,000 are in hand, the whole series comprising the sheets of an *Atlas,* of which a limited edition is planned for publication. There is also in existence a 'Mapa de la Republica de Bolivia', at 1:1,875,000, by the Libreria La Universitaria, La Paz. An *Atlas de Bolivia* was published in 1958, prepared by René R Camacho Lara at La Paz and in Novara, at the Istituto Geografico de Agostini; eighteen physical and political maps of the departments of Bolivia are included; but there is no index. The *Bibliografía boliviane* was established as an annual in 1962 (Los Amigos del Libro, 1963-); official publications are included, but it is not reliable for the inclusion of maps. A similar cartographical organisation exists for Paraguay, where topographical mapping is the responsibility of the Instituto Geográfico Militar, Asuncion.

The Instituto Geográfia Militar, Santiago, founded in 1891, faced a difficult task in the topographical mapping of Chile, a long, narrow strip of territory, extending from north to south for a distance of about 4,270 kilometres, with an average width of 190 kilometres. Gravimetric surveys are particularly important in Chile, owing to the physical characteristics of the country and they are used also for mineral prospecting. Considerable progress has been made in triangulation work and levelling; re-triangulation and re-levelling became necessary in the areas affected by the earth tremors of May 1960 and March 1965. Geodetic and aerial photogrammetry are being increasingly used.

The national series of topographical maps prepared by the institute are at four scales, 1:25,000, 1:100,000, 1:250,000 and 1:500,000. The sheet lines of the 1:100,000 series are based on a kilometric grid system, but those of the other three follow the international longitude and latitude lines. International geographical co-ordinates are indicated on all the maps. The 1:25,000 map sheets are mainly in monochrome, showing contours at twenty metre intervals. The 1:100,000 series is in two types: one is printed in five colours, blue for water, brown for the contours, red for primary roads, green for vegetation and black for all other features. The second type shows a uniform yellow tint for land areas, blue for water, red for roads and black for the contour lines and other planimetric features. The 1:250,000 preliminary map was compiled from trimetrogon photography and other available information between 1950 and 1961. The Lambert Conical Conformal Projection was used. Sheet coverage varies; contours are at five hundred, then one thousand feet, but spot heights are indicated in metres. Five colours are used; blue for water features, grey for contour lines and sandy areas, red for roads, airfields and seaplane bases, green for vegetation and black for all other features, and sometimes relief shading is added in an olive-brown tint. The 1 : 500,000 sheets are considered as atlas size, each covering normally three by two degrees. Blue is used for water features, brown for relief shading and indications of land forms, red for roads, green for woodland and other cultivated areas, purple for airfields, and black for all other features. A red over-print is also used for shading major boundaries and important names.

In recent years, much mapping and charting has been carried out by the Naval Institute of Hydrography and the Aerial Photo-grammetry Service of the Chilean Air Force; since 1947, valuable co-operation has been given also by the Inter-American Geodetic Survey. The coastal islands and the Chilean Antarctic territory have been mapped on the scale of 1 : 250,000 or smaller scales. The maps prepared by the Army Geographical Institute are printed on offset presses within the organisation; since 1960, plastic scribing has been adopted for colour separation.

A geological map of Chile was published by the institute at 1 : 3,100,000 in 1950; and a 'Mapa geomorfologico de Chile', at 1 : 3,500,000, was completed in 1964 by the University of Chile at Santiago, also a climatic map at approximately 1 : 3M. The Insti-tute of Geological Research, founded in 1957, is responsible for maintaining the basic geological maps and, with the assistance of the United States Geological Survey, has produced a geological map of Chile, 1960 edition, at 1 : 1M and a 1962 edition of the 'Mineral resources map of Chile', at 1 : 1,500,000, both based on the physical map of Chile 1 : 1M. Transport maps are produced by the Ministry of Public Works and an 'Electrical resources map of Chile' was issued by the National Electricity Board in 1957. A national agro-logical map, published by the Soil and Water Conservation Depart-ment at 1 : 2,500,000, includes the chief groups of soils and agri-cultural conditions. The department has published several other special maps in connection with resources development projects, as has also the Irrigation Office in its own field. Much preparatory work has been accomplished by the Institute for Natural Resources Research, under the aegis of the Development Corporation, while land use maps have also been prepared by the Ministry of Agricul-ture and the Aerial Photogrammetry Project, the Department of Agrarian Geography of the Institute of Geography of the University

339

of Chile and the Institute of Forestry. Large-scale mapping has so far been concentrated in the area of large centres of population, notably Santiago and other parts of the Santiago province. Other urban surveys for planning purposes have been made by the Aerial Photogrammetry Service of the Chilean Air Force and various private organisations.

The Instituto Geografico Militar published an *Atlas de la Republica de Chile* in 1966, consisting of thirty three plates at various scales. Physical maps at 1 : 1M cover the country, each accompanied by photographs illustrating the landscape, the towns, land use and local industries. The atlas is intended to present a popular view of this unique and varied country; it includes also Easter Island and the Juan Fernandez archipelago and the Chilean Antarctic Territory. Text is in Spanish only, but the photographs have been given captions in Spanish, English and French.

There is no current national bibliography in Chile, so that information on map publication is scattered. Large numbers of geographical periodicals or periodicals including specialised geographical interest have been launched in Chile, as in the other major South American countries, but it is not always easy to come by copies of these in Britain, or to find precise information concerning their state of publication, as they frequently change their titles or become absorbed in other series. The *Publicación* of the Instituto de Geografía, Santiago, has appeared irregularly since 1948, with occasional English summaries; *Informaciones geográficas,* annually 1951- from the Institute of Geography, Santiago University and the *Revista geográfica de Chile,* irregularly from 1948, from the Instituto Geográfico Militar are valuable sources of information.

The official topographical maps of Argentina are maintained by the Instituto Geografico Militar at five main scales; 1 : 25,000, 1 : 50,000, 1 : 100,000, 1 : 250,000 and 1 : 500,000. They are based on surveys carried out at the three larger scales, supplemented by photogrammetric plotting. Early editions were drawn on polyhedric, polyconic and some other projections, but a change to the Gauss-Krüger Projection was made in 1927, when a new series was begun. The contour interval on the 1 : 25,000 maps is 1·25 metres; on the 1 : 50,000 maps, 1·25 or 2·50 metres according to the topography. Both series are printed in four colours. The 1 : 100,000 series uses four contour intervals, at five, ten, twenty five and/or fifty metres, according to the nature of the country, bearing in mind the high mountain chain in the west and the broad river valleys and plains of the centre and east. In both the 1 : 250,000 and the 1 : 500,000 series, hypsometric tints are used to show relief. The new map at

1 : 500,000 covers the whole country in seventy nine sheets. In 1966, the institute published a new map of the republic at 1 : 10M.

The Geological Survey of Argentina, Dirección Nacional de Geologia, Minéra, established in 1904, has made great progress in mapping and analysing the structure of this extremely complex country. *Boletines* have been issued in six sections : geology, mining, hydrology, mineral chemistry, bibliographical and reports; *Publicaciones* have dated from 1924 and *Anales* from 1947. A new geological map at 1 : 5M was issued by the department in 1964. Other serials containing maps or information about them include *Revista minera, geologia y mineralogia*, 1926-45, irregularly by the Sociedad Argentina Mineria y Geologia; *Revista del Instituto Nacional de Investigacion de las Ciencias Naturales, Ciencias Geologicas*, 1949-, published annually by the Museo Argentino de Ciencias Naturales, Buenos Aires; and the *Revista de la Asociacion Geologica Argentina*, 1948-.

Numbers of other thematic maps have been compiled as parts of Argentina have begun to develop their vast natural resources. A most useful bibliographical source is the *Annotated index of aerial photographic coverage and mapping of topographical and natural resources*, 1 : 6,300,000, a volume of twenty seven pages of maps and text, published by the Organization of American States, Washington, 1966. An *Atlas de la Republica Argentina* began publication by the Instituto Geografico Militar in 1965, with a section of twenty seven plates dealing with political and provincial boundaries. Physical and population maps are in preparation; of particular interest is the development of the Welsh colony in Patagonia. In this connection, it is appropriate to mention the informative talk given to the Royal Geographical Society on October 25 1965, by Professor E G Bowen, 'The Welsh colony in Patagonia 1865-1885: a study in historical geography'; in the course of his talk, Professor Bowen draws attention to a number of atlases and bibliographical records and the printed article in *The geographical journal*, March 1966, includes a fold-out plate of thirteen maps compiled by Professor Bowen on the natural features of the area and stages in its colonisation. Other journals including original maps are the *Boletín* del Instituto Geografico Argentino, founded in 1879, annually 1881-1901, since then irregularly; *Atlas*, the official organ of the Argentina Instituto Geográfico Militar, Buenos Aires, 1954-; and *Gaea*, published irregularly by the Sociedad Argentino de Estudios Geográficos, Buenos Aires, 1924- . The publications of the Academia Argentina de Geografía and of the Instituto Geográfico Militar are also central sources for new maps and of information about them.

The 'Carta del Uruguay', 1 : 50,000, has been maintained by

the Instituto Geográfico Militar, Montevideo, from 1926, beginning with the southern border areas. As in Argentina, the Gauss-Krüger Projection has been used and relief is shown by contours and spot heights in metres. Field work on the basic geodetic network for the whole country has been completed; work continues on second- and third-order networks. The national cartographic plan suggested for map coverage of Uruguay has been hampered by economic reasons and by shortage of trained personnel. A basic series at 1:50,000 was planned to cover the country in 292 sheets; about ten per cent of the total area has so far been completed. Special surveys on various scales have been made independently, for both civil and military purposes. Aerial photogrammetric coverage of the country is planned on scales of 1:100,000 and 1:50,000. Maps of soil, vegetation, ecology, forest cover, etc are in the first stages of production. Hydrographic surveys are the responsibility of the Hydrographic Service, which operates within the Ministry of National Defence, and the Hydrography Board, under the Ministry of Public Works. Town plans are included in the future programme, but so far only that for Montevideo has been completed.

The Instituto Geologico del Uruguay, Montevideo, founded in 1912, has carried out a systematic study of the structure of the country; a ' Geological map of Uruguay ', at 1:750,000, was completed in 1946. *Boletines* have been published since 1914.

The National Bibliographical Group, constituted in 1950, faces a gigantic task in organising current or retrospective bibliographies. Individual compilations of reference sources, accessions lists and catalogues have been attempted, but, so far as is known, no comprehensive list of maps exists.

The *British bulletin of publications on Latin America, the West Indies, Portugal and Spain,* published twice yearly by the Hispanic and Luso-Brazilian Councils, London, includes in each issue some 150 monographs and articles. The publications of a number of other London organisations interested in Latin America frequently include maps or sketch-maps or draw attention to surveys in progress; in addition to the two above mentioned, the Argentine Chamber of Commerce in Great Britain and the Argentine Information Bureau, the Brazilian Chamber of Commerce and Economic Affairs in Great Britain and the Brazilian Embassy Commercial and Information Service, the British/Mexican Society, etc. An inventory of mapping and photography in Latin America has been made by the Inter-American Committee for Agricultural Development, forming part of an inventory of information preparing for the Alliance for Progress programme. A Natural Resources Unit was set up by the Organization of American States within the Department of Economic Affairs of the Pan-American Union. In 1963, the unit began publication of a series of index atlases for each country, with text in Spanish and English; at the time of writing, the volumes for Bolivia, Chile, Colombia, the Dominican Republic, Ecuador, Haiti, Paraguay, Peru, Uruguay and Venezuela have been completed, each showing the available map coverage at various scales by aerial photographs, topographical, geological, soil and land use and vegetation maps. An analysis such as this is of the utmost importance in all studies of land potential and the development of natural resources.

Numerous conferences, symposia and surveys have been carried out on the whole or parts of South America during recent years, from those of the United Nations to local planning organisations. All produce factual reports and most, but not all, of the published works include maps. The *South American handbook* (London, Trade and Travel Publications) contains maps and town plans. There are also the invaluable economic studies undertaken by the International Bank for Reconstruction and Development, which have covered many of the Central and South American countries. G J Butland is one of the English-language authorities on South American geography; most of his published works contain valuable maps in the text; for example, *Chile: an outline of its geography, economics and politics* (OUP, for the Royal Institute of International Affairs, third revised edition 1956); *The human geography of Southern Chile* (Philip, for the Royal Institute of British Geographers, 1957); *Latin America: a regional geography* (Longmans, 1960). George Pendle is the other English authority notably his *Argentina* (OUP, for the Royal Institute of International Affairs, third edition 1963); *Paraguay: a riverside nation* (OUP, second edition 1956); *Uruguay: South America's first welfare state* (OUP, third edition 1963) and several works on the other states in Black's 'Land and people' series. Other relevant monographs include:

F Ahlfeld: *Geologia de Bolivia* (Instituto del Museo, Universidad Nacional de la Plata, 1946).

R L Carmin: *Anépolis, Brazil: regional capital of an agricultural frontier* (University of Chicago, Department of Geography, research paper no. 35 1953).

L F Chaves: *Geografía agraria de Venezuela* (University of Central Venezuela, 1963).

J P Cole: *Latin America: an economic and social geography* (Butterworth, 1965).

L P Cummings: *Geography of Guyana* (Collins, 1965).

P J V Delaney: *Quaternary geologic history of the coastal plain of Rio Grande do Sol, Brazil* (Louisiana State UP, 1964).

343

Marvin Harris: *Town and country in Brazil* (Columbia UP; OUP, 1956).

A E Krause: *Mennonite settlement in the Paraguayan Chaco* (University of Chicago, Department of Geography, research paper no 25 1952); with maps separately in a pocket.

Edwin Lieuwen: *Venezuela* (OUP, for the Royal Institute of International Affairs, 1961).

A G Ogilvie: *Geography of the Central Andes* (New York, American Geographical Society, 1922); the handbook which was prepared to accompany the La Paz sheet of the Map of Hispanic America 1:1M.

Irmgard Pohl and Joseph Zepp: *Latin America: a geographical commentary*, edited by K E Webb (Murray, 1964); translated and adapted by John Paterson from the *Harms Erdkunde*.

Emilio Romero: *Geografía económicà del Perú* (Lima, Universidad Nacional Mayor de San Marcus, 1961).

D H Romney, *editor*: *Land in British Honduras*. Report of the British Honduras Land Use Survey Team (Colonial Office, Colonial Research Publication, no 24; HMSO, 1959, second impression 1963); with maps in the text and in a separate folder.

J Walter Thompson Company: *The Latin American markets: a descriptive and statistical survey of thirty markets made up of almost 173 million people* (New York, McGraw-Hill, 1956).

Pablo Vila: *Geográfia de Venezuela* (Caracas, Ministerio de Educación, Departamento de Publicaciones, 1960); in three volumes.

Kempton Webb and John Paterson, *editors*: *Latin America: a geographical commentary* (Murray; New York, Rand McNally, 1963), translated and adapted from *Harms Erdkunde*.

R C West: *The Pacific Lowlands of Colombia: a negroid area of the American tropics* (Louisiana State UP, 1957).

Herbert Wilheny and Wilhelm Rohmeder: *Die La Plata-länder: Argentinien, Paraguay, Uruguay* (Braunschweig, Westermann Verlag, 1963).

AFRICA

The size of the African continent and its uneven development have led to a great variety of maps. A few areas had already been mapped in some detail while vast areas in the interior were completely unknown. The explorations of Heinrich Barth and other notable nineteenth century scientists and geographers gradually filled in the empty spaces on the map of central Africa and a vast literature contains published editions of their own notes as well as commentaries on their exploits; in 1965, for example, a centenary edition, *Travels and discoveries in north and central Africa,* by Heinrich Barth, was published by Frank Cass, with a biographical note by A H Kirke-Green. Scientific and systematic mapping was introduced by British, French, German and Portuguese cartographers.

1 : 1M and 1 : 2M maps of the whole of Africa have been issued by the Geographical Section, General Staff of the War Office; 'Afrique' at 1 : 5M has been compiled by the Service Géographique de l'Armée, Paris. Bartholomew's 'Reference map of Africa', 1 : 10M, revised in 1966, is printed in full contour colouring; principal main roads, other roads and railways are marked, and international and provincial boundaries are shown. 'Africa, central and east' at 1 : 4M, was revised, showing contour colouring, main roads and secondary routes, also caravan routes and tracks, railways and international boundaries; in extent, it covers the area from the southern Sudan to the territory which was formerly Northern Rhodesia. 'Africa South and Madagascar', 1 : 4M, shows Madagascar as an inset. Stanford's 'General map of Africa', 1 : 10M, includes railways, principal roads, airports and oil pipelines. Five later maps include a wall map of the whole continent, 1 : 6M, issued by George Philip, 1966, showing principal roads, tracks, railways, airports, rivers, canals and oil pipelines with large numbers of place names marked; 'North-east Africa'; 'West Africa'; 'Central and southern Africa', all at 1 : 5,500,000, show population by size of towns, and these, with another general map at 1 : 10M are bordered with African scenes. *Afrique: carte politique: situation 1er Janvier, 1967,* at 1 : 10M, is one of the most up-to-date maps of Africa, published by the Institut Géographique National.

Practical geography in Africa, by T E Hilton (Longmans, 1961), draws on African material in illustration of his exposition on general cartography and includes thirteen specimens of coloured maps from African countries. A very useful reference map, published by Philip in 1966 is 'Relief of land, political and communications', at 1 : 9M. Philip's 'Library map of Africa' is a magnificently produced political map showing clearly the many African nations, on a set of sheets at 1 : 6M. Some sixteen thousand place names are marked. An earlier 'Tribal map of Negro Africa', by C B Hunter, was published in 1956 by the American Museum of Natural History, New York. Also in 1966, Hermann Haack issued a wall map of Africa, 1 : 6M, in four sheets, and Westermann published 'Population density in Africa', 1 : 10M, based on statistics in or near 1960. 'Major mineral resources', 1 : 10M, formed one of the 'Map studio productions', Johannesburg, 1966. The volumes in the series *The ethnographic survey of Africa,* edited by Daryll Forde, have been prepared and published by the International African Institute with the aid of grants from the British government and the governments of the territories concerned and with the collaboration of the Musée Royal de l'Afrique Centralê. Each volume presents a concise summary of available information on an African people, or group of related peoples, based on original field studies, as well as on published sources; each contains a bibliography and a specially drawn map. The standard work edited by K M Barbour and R M Prothero: *Essays on African population* (Routledge and Kegan Paul, 1961; New York, Praeger, 1962) should be remembered; it contains several maps.

Several interesting atlases are available for the general study of Africa. *The Oxford regional economic atlas—Africa,* the third in the series of regional economic atlases published by the Cartographic Department of the Clarendon Press in 1965, is a well-designed reference atlas, prepared under the direction of P H Ady. The maps were based on 1956 data, but in the introduction, some data are quoted to 1960. There are textual commentaries and the format and layout follow the style of the parent atlas. This, the first economic atlas to deal exclusively with the entire African continent, consists of forty three pages of regional topographic maps, followed by sixty nine pages of maps of the whole continent dealing with themes ranging from solid geology to the distribution of locusts. The gazetteer contains some eighteen thousand entries. An attractive series of twelve maps at one inch to 100 miles shows relief by layer colouring and hill-shading. *The shorter Oxford atlas of Africa,* consisting of forty eight pages of topographic maps and a fifty page name-gazetteer, containing some eighteen thousand entries, was based on the regional economic atlas. The whole continent, except for a small part of the Sahara, is mapped at a scale of one inch to 100 miles, showing relief by layer colouring and hill-shading. The most densely populated areas are also mapped at scales of one inch to fifty, thirty three or twenty five miles.

An atlas of Africa, prepared by J F Horrabin (Gollancz, 1960), is a small atlas containing fifty diagrammatic maps on the history, politics and economics of the continent from prehistoric times. The clear black and white maps do not present a comprehensive coverage, but a great deal of information is effectively introduced. *An atlas of African affairs,* by Andrew Boyd and Patrick van Rensburg (Methuen, 1962; also in a paperback edition), is in octavo book form. The chief topics of current interest are covered, presenting economic and historic facts and commentary, with monochrome maps by W H Bromage on facing pages. Aspects of population, social problems, transport and the development of research are

346

followed by sections on individual regions, each also illustrated by a map. *A map book of Africa,* by A Ferriday (Macmillan, 1966), consists of eight pages of maps and descriptive text; and *An atlas of African history,* compiled by J D Fage (Arnold, 1958), presents in sixty four pages the historical background against which present-day African affairs can be understood. R D Hodgson and E A Stoneham: *The changing map of Africa,* in a second edition 1968, includes eight particularly interesting maps.

So fast have political and economic conditions changed in many parts of Africa, however, that much of the factual information contained in these atlases is outdated. In 1961, at Addis Ababa, representatives of the states and territories of Africa undertook, with the help of UNESCO and the Economic Commission for Africa, the long-range planning of its human resources, and in 1964 a further fifteen year plan was agreed for the development of scientific research centred on natural resources. The International Conference on the Organisation of Research and Training in Africa in relation to the Study, Conservation and Utilization of Natural Resources, published an *Outline of a plan for scientific research and training in Africa* (UNESCO, 1964), known as the Lagos plan, calling for considered action by each country to establish a national research body and a national research budget, a national research manpower register and other steps necessary to improve national resources and national science education, at continental level; all-African programmes and inter-African co-operation was necessary, co-ordinated by the Organization of African Unity, and maps of all kinds were required as bases for all aspects of the work. A Scientific and Technical Committee on Natural Resources was established by the organisation, with the co-operation of UNESCO, as well as the Economic Commission for Africa and other appropriate United Nations agencies. At the same time, the International Union for Conservation of Nature and Natural Resources, with the help of UNESCO and FAO revised the 1933 convention on the conservation of the flora and fauna of Africa. A rationalisation of research institutions was planned to include cartography, hydrology, energy resources, arid zone, savannah zone, humid tropic zone, geophysics and seismology, mining and economic geology, vulcanology, soil sciences, irrigation and drainage, oceanography and marine biology, plant pests and diseases, forestry, etc. The two reports issued from the conference were *Final report of the Lagos Committee,* 1964; and *Lagos Committee: selected documents,* 1965. A further valuable document, *Scientific research in Africa: national policies, research institutions* was published in 1966.

The fifteen year plan aimed at a large degree of fulfilment by

1980; unfortunately, unforeseen domestic events have hampered development in some areas. Internationally, much co-operation has ensued; participation in international scientific programmes, for example, the International Hydrological Decade, the Upper Mantle Project and the International Oceanographical Research Programme. On the suggestion of the Economic Commission for Africa, UNESCO compiled a *Review of the natural resources of the African continent and Madagascar*, in 1963, an excellent document, including comprehensive bibliographies and fifty seven maps, also a folding map in a separate pocket. The investigation was carried out in such major fields as topography and maps, geology, applied geology, especially with a view to location of mineral resources, climate, water resources and soils. Many individual research projects have come to fruition, of which the reports are vital. *African development planning*, 1964, edited by Ronald Robinson, includes thirteen papers and comments, with an introduction, from the conference organised in 1963 by the Overseas Studies Committee, Cambridge. *Economic transition in Africa*, 1964, edited by M J Herskovits and Mitchell Harwitz, comprises papers from the conference on indigenous and induced elements in the economies of sub-Saharan Africa, held by the Commission on Economic Growth, Social Science Research Council, Northwestern University, 1961; and *Economic development for Africa south of the Sahara*, 1964. A Conference on Economic Research in Africa was held at Bellagio, Italy, towards the end of 1964 and, in the same year, an International Seminar on African Primary Products and international trade was held in Edinburgh, of which the papers were edited by I G Stewart and H W Ord (Edinburgh UP, 1965).

There has thus been great activity in the thematic mapping of Africa during recent years. The Commission for Technical Co-operation in Africa South of the Sahara produced a *Climatological atlas of Africa* in 1961 (Lagos, CTCA and the Scientific Council for Africa South of the Sahara, 1961; New York, International Publications Service, 1963). Fifty five sheets, in looseleaf format, cover mean annual rainfall, mean monthly rainfall, mean daily temperature, mean humidity mixing ratio and the contours of the 850, 700 and 500 millibar surfaces. A *Climatological atlas of Africa* was compiled and edited in the Africa Climatological Unit, Johannesburg, in 1964, with the collaboration of the African Regional branch of the Association of World Meteorological Vegetation and the cartographic services of member governments of the Commission for Technical Co-operation in Africa South of the Sahara. The work consists of fifty five coloured folded maps, with text in English, French and Portuguese, edited by S P Jackson. The following year, B W Thompson prepared an *Atlas of the climate of Africa* (OUP, 1965), consisting of 131 charts at scales ranging from 1:22M to 1:32M. This presents a valuable collection of material not easily available elsewhere, based on 1956-60 data, with an introduction analysing the methods used in the preparation of the atlas; the whole work is related to the physical conditions prevailing in Africa, and tropical meteorology generally. The monochrome maps are arranged in two groups: first, monthly or annual charts of solar radiation, sunshine, rainfall, temperature and relative humidity; and monthly analyses of such features as winds, pressure and dewpoints for January, April, July and October. Land over one thousand feet in height is shaded. There is an emphasis on east Africa, and one of the aims is stated to be the 'training of the new generation of African meteorological personnel'. An interesting article 'An agroclimatic mapping of Africa', by M M Bennett, was included in *Stanford Food Research Institute Studies*, III (3) 1962. B W Thompson's *The climate of Africa* (Nairobi: OUP, 1965) is a comprehensive source book, containing 132 pages of maps.

The International Africa Seminar held in Leopoldville in January 1960, under the auspices of the International African Institute, produced some basic material published under the title *African agrarian systems: studies presented and discussed at the Second International African Seminar* (OUP, for the Institute, 1963), edited by Daniel Biebuyek, with an introduction by Daryll Forde; maps accompany the papers, which, in English or French, with summaries in the other language, analyse actual area studies illustrating the basic factors and problems connected with the land in tropical Africa today. *Africa, maps and statistics,* prepared by the Africa Institute, is a valuable statistical atlas, published in twelve quarterly issues, 1962, by the Cape and Transvaal Printers, Johannesburg, to enable current information of the significant political, social and economic evolution of African countries to be disseminated. Each folio highlights one or two special topics; all aspects are excellently portrayed, covering populations, vital and medical statistics, culture and education, transport and communications, energy resources, production and consumption.

The Association of African Geological Surveys has been working on the analysis and co-ordination of a mass of hitherto unpublished material to prepare a complete revision of the geological map of Africa published in nine sheets between 1938 and 1962. The *Geological map of Africa*, 1:5M, published under the auspices of UNESCO in 1964, is in effect a completely new map; the topographical base remains the same, but only in a few areas where no more recent research has been carried out are the maps the same as in the first edition. It has been drawn in accordance with the specifications of the Commission for the Geological Map of the World. Each sheet represents the first scientific post-war survey in English of the sub-surface resources of Africa, particularly rich in minerals. The divisions run as follows : *1* Western Mediterranean region : Morocco, Algeria, Sahara, Rio de Oro, Mali, Mauritania, Nigeria, Upper Volta, Senegal, Gambia; *2* Eastern Mediterranean region : Algeria, Tunisia, Libya, Sahara, Sudan, Nigeria, Tchad; *3* Red Sea region: Egypt, Abyssinia, Arab peninsula, Middle East; *4* Western Atlantic region : Portuguese Guinea, Guinea, Ivory Coast, Mali, Upper Volta, Togo, Dahomey, Ghana, Nigeria, Sierra Leone; *5* Central Africa region : Nigeria, Cameroon, Gabon, Congo, Central African Republic (as it then was), Tchad, Angola, Uganda, Northern Rhodesia, Tanganyika, Nyasaland, Sudan; *6* Eastern Africa: Somalia, Abyssinia, Kenya, Tanganyika, Nyasaland, Northern Rhodesia, Mozambique; *7* South Atlantic Region : territories of continental Africa; *8* South and South-west Africa : African territories, Republic of South Africa, South-west Africa, Swaziland, Bechuanaland, Southern Rhodesia, Angola, Mozambique, Northern Rhodesia; *9* Indian Ocean and Madagascar region : Mozambique, Madagascar. A brochure containing explanatory notes on the map, *African geological bibliography* and notes on basic documents was published by UNESCO in 1964. The map sheets carry English and French captions. A detailed commentary on the map is given by Georges Choubert and Anne Faure-Muret in *Nature and resources,* June 1968; it would also be advantageous to study the map sheets in connection with Raymond Furon's classical *Géologie de l'Afrique* (Paris, Payot, 1950), translated by A Hallam and L A Stevens from the second edition 1960, as *The geology of Africa* (Edinburgh, Oliver and Boyd, 1963), containing sketch-maps and a bibliography. Two maps, ' Geologic background maps of Africa ', intended to serve as a basis for various syntheses and further thematic maps, were published in three colours by UNESCO, 1967, with the Association of African Geological Surveys, Paris; the first map is at 1 : 10M, the second at 1 : 20M, and captions are in English and French. In preparation, 1968, are nine sheets of a ' Tectonic map of Africa ', 1 : 5M, by UNESCO and the Association of African Geological Surveys, Paris.

The ' Soils map of Africa ', 1 : 5M, published by the Commission for Technical Co-operation in Africa (Lagos, 1964), covers the whole continent. The generalised soil map is the work of the Inter-African Pedological Service, with explanatory monographs by J L d'Hoore. A general introduction is followed by a review of the sixty two elements of the legend making up the 275 cartographic units of the map. The third part is devoted to characteristic profiles and analytical data and part IV presents additional or complementary data in the form of tables or lists. The Inter-African Pedological Service, housed in the Institut National pour l'Etude Agronomique de Congo Belge, 1953, is a regional inter-governmental organisation specialising in the cartography and classification of African soils, working closely with the Inter-African Bureau for Soils and Rural Economy in Paris. The Office de la Recherche Scientifique et Technique Outre-Mer, Paris, published a ' Carte de la vegetation de l'Afrique Tropicale Occidentale, 1 : 1M ', in 1962-, prepared by Guy Roberty. On nine sheets, two hundred types of vegetation are differentiated; a comprehensive volume of text accompanies each sheet, as well as a separate introduction and glossary.

A ' Coal map of Africa ', 1 : 10M, was published jointly by the Economic Commission for Africa and the Association of African Geological Surveys, 1965. Explanatory notes were completed by Robert Feys and Jean Fabre, in 1966. The *Munger map book,* compiled by A H Munger and E Placid (Los Angeles, Munger Oil Information Service, 1960), covers petroleum developments and generalised geology over a wide field, including Africa, the Middle East, India and Pakistan. The reports of the United Nations Regional Cartographic Conference for Africa, held in Nairobi in 1963, are essential reading; as also is *A review of the natural resources of the African continent* (UNESCO; natural resources research, 1; New York, International Document Service, Columbia UP, 1963). This comprehensive work contains a series of essays on work completed or in progress in the field of topographic mapping and in the study of geology, meteorology, climatology, hydrology and soils. Useful also is the article by A S Chapman, ' Some aspects of recent mapping in Africa ', in *Surveying and mapping,* September 1966. An Inter-African Conference on Hydrology was held in 1961, of which the *Proceedings,* published in English and French, are most valuable, illustrated with photographs, maps and diagrams.

Of the numerous organisations specialising in African studies and surveys, the following are among the foremost: the Institut français d'Afrique Noire, Le Centre d'Analyse et de Recherche Documentaires pour l'Afrique Noire, Paris, the International African Institute, London and the Centre de Documentation, Economique et Sociale Africaine, with headquarters in Brussels, in collaboration with specialists throughout the world. Many academic institutions have concentrated on African research, for example, the School of Oriental and African Studies, London, the Centre of West African Studies, University of Birmingham, the Centre of African Studies, University of Edinburgh, the African Studies Group, University of Leeds, and the University of Cambridge Group for Afro-Asia Social Studies. In America, the African Studies Association was founded in 1957 and there are centres of African studies at Northwestern University, Boston, Howard, California, Columbia, Indiana, Johns Hopkins, Michigan, New York, Stanford, Syracuse, Wisconsin and Yale. The University of Leiden specialises in African regional studies and some of the most erudite studies on Africa have been published by Brill of Leiden. The Academy of Sciences of the Soviet Union sponsors an African Institute in Moscow and there is a strong interest in the University of Leningrad. The Italian work, concentrated in Ethiopia and Somaliland, has been predominantly in anthropological and linguistic studies. Recently, the increase in research and advanced level studies by Africans themselves has been of the greatest importance, especially in the University of Ibadan, where six international conferences have been held to discuss many aspects of emergent Africa. The first International Congress of Africanists was held in Accra in December 1962; the *Proceedings* were edited by Lalage Bown and Michael Crowder (Longmans, 1964).

In 1964, the CIBA Foundation for the Promotion of International Co-operation in Medical and Chemical Research, together with the Haile Selassie I Prize Trust arranged an international conference at Addis Ababa. A number of eminent specialists prepared papers in scientific, social and economic aspects of African development, published, with the discussions, as *Man and Africa: a CIBA Foundation Symposium jointly with the Haile Selassie I Prize Trust under the patronage of His Imperial Majesty Haile I, Emperor of Ethiopia,* edited by Gordon Wolstenholme and Maeve O'Connor (Churchill, 1965), illustrated with many maps and including a bibliography. Themes range over the whole environmental development, including industrial and commercial aspects.

Maps are included in the three following works, each an outstanding co-ordinating analysis of the volume of research on Africa: *Science in Africa: a review of scientific research relating to tropical and southern Africa,* by E B Worthington (OUP, for the Committee of the African Research Survey under the auspices of the Royal Institute of International Affairs, 1938); maps and surveys are covered, also geology, meteorology, soil science, forestry, botany and zoology, agriculture and fisheries and various aspects of population. The same author produced *Science in the development of Africa,* 1958. The second work is *The African world: a survey of social research,* edited by Robert A Lysted (Pall Mall Press, for the African Studies Association, 1965), which attempts for the humanities what E B Worthington did in the previously mentioned work;

covering a wide field, including history, anthropology and economic development. Thirdly, *The study of Africa,* edited by Peter J M McEwan and Robert B Sutcliffe (Methuen, 1965), which comprises essays on the main social, economic and political issues in present-day Africa, by leading authorities in the English-speaking world. *The geographical journal,* June 1968 contains an article by Anthony Young, 'Mapping Africa's natural resources', which will usefully supplement the preceding notes.

In many African countries, English, French, German or Italian surveyors have carried out the control surveys, with local help in filling in the detail. In some cases, a survey handbook has been compiled, including instruction on traverses, levelling and all methods for routine maintenance. Skilled staff is still inadequate in some areas, but the framework for locally manned survey departments is in most cases well established and modern equipment installed.

NORTHERN AFRICA: Apart from such general maps as the 1:5M 'Map of north-west Africa' by Bartholomew, latest revision 1966, and the Michelin 'Map of Africa, north and west', 1965, at approximately 1:4M, the majority of the vast literature available to the student of north and north-west Africa is in French or Arabic. The French carried out systematic mapping of their colonial territories, one of the largest works being the *Atlas des colonies françaises, protectorats, et territoires sous mandat de la France,* 1932.

The Service Géographique du Maroc, established in 1907, became attached to the Institut Géographique National in 1941 and, since 1963, has taken the name Section Géographique du Service Topographique. The national committee is the Service Topographique, within the Ministère d'Agriculture, Rabat. Topographical maps of Morocco are maintained at scales of 1:50,000, with main centres, such as Rabat, at 1:10,000. The 'Carte générale du Maroc' 1:1M was published by the Service Géologique and the Service Topographique du Maroc, in two sheets, 1966, showing relief by 250-metre contours and layer colouring. A 'Carte géologique du Maroc', at 1:500,000, was prepared in four sheets, 1952-53. The Service Géologique has been issuing *Notes et Mémoirs,* illustrated with maps since 1922. Research and mapping of specific physical characteristics are carried out at the Station de Recherches Présahariennes, at Aouinet-Torkez, and at the Station Météorologique de Montagne, at Ifrane, both attached to the Institute Scientifique Chérifien, Rabat.

Special maps are available for the Casablanca area; 'Population activities', 1965; and 'Croissance et types d'habitat' in three parts, 'Plan de la ville', 'Evolution de la surface construite' and 'Types

353

d'habitat', 1965, all at the scale 1 : 50,000. The Institute Géographique National issued a 'Carte des tribes', at 1 : 500,000, in 1958. Original maps are to be found in the twice-yearly *Revue de géographie du Maroc*. Two interesting special atlases are the *Atlas historique, géographique et économique*, produced by E Levi-Provençal, Paris, in 1935; and the *Atlas et commentaire des cartes sur les genres de vie de montagne dans le massif de Grand Atlas Marocain*, compiled by J Dresch and published in Tours in 1941. The first sheets of the current *Atlas du Maroc* appeared in 1954-55, prepared by the Laboratoire de Géographie Physique de l'Institut Scientifique Chérifien, with the co-operation of many experts. The complete atlas is planned in fifty four chapters, grouped in eleven sections, with further sub-divisions as necessary. The whole of Morocco is mapped at 1 : 2M or 1 : 4M, with 1 : 1M as the basic scale for individual regions. Every aspect of the geography is to be covered, physical, agricultural, economic and social.

A 'Carte géologique du nord-ouest de l'Afrique' was prepared for the Nineteenth International Geological Congress at Algiers in 1952; in two sheets, at 1 : 2M. As early as 1923, the Institut Géographique National sponsored an *Atlas d'Algérie et de Tunisie*, jointly compiled by Augustin Bernard and R de Flotte de Roqueraire, produced by the Service cartographique de la Direction de l'Agriculture, du Commerce et de la Colonization. Sixteen parts were planned but up to 1933 only nine had been issued, covering geology, climate, relief, vegetation, population, colonisation and agriculture, with explanatory text; no further issues have been traced. The basic scale was 1 : 1,500,000. Topographical maps are maintained at 1 : 25,000, of which a new series began in 1960. The 'Carte d'Algérie', 1 : 200,000, has insets of Algiers and suburbs, 1 : 50,000, of Tunis, 1 : 35,000, environs of Oran, 1 : 300,000, Algiers and Kabylie region, 1 : 300,000 and the Tunis region at 1 : 500,000. Two companion maps were published in Algiers; the 'Carte phytogéographique de l'Algérie et de la Tunisie', in 1926 and a 'Carte Forestière de l'Algérie et de la Tunisie', compiled by R Maire and P de Peyerimhoff, in 1941, both including bibliographies.

Two atlases were published in Paris by Horizons de France, *Algérie: atlas historique, géographique et économique*, 1934, and *Tunisie: atlas historique, géographique, économique et touristique*, 1936, by S Leconte, both with descriptive text. The Institut Géographique National now maintains topographic maps at 1 : 50,000 and 1 : 100,000 for all Algeria and Tunisia, and some of the interior areas. A 'Carte de reconnaissance des sols d'Algérie', at 1 : 200,000, was published by the Inspection Générale de l'Agriculture in 1955; three sheets have accompanying descriptive booklets. Two most interesting and useful maps were compiled in 1958 by Henri Gaussen and A Vernard, a 'Carte des pluies en Algérie au 1 : 500,000' and 'Carte des pluies en Tunisie en 1 : 500,000'. The *Service de la Carte Géologique de l'Algérie* has been published in six series, of which the first began in 1885; a second edition of the 'Carte géologique at 1 : 500,000' began in 1952.

The University of Algiers was founded in 1829; to it a number of relevant institutions are attached, such as the Institut de Recherches Sahariennes, the Institut de Météorologie et Physique du globe de l'Algérie and the Institut de Géographie, whose published works include many original surveys and maps. Valuable also is the *Bulletin* of the Société du Géographie d'Alger et de l'Afrique du Nord, founded in 1896. The Tunisian Topographical Service was established in 1886 to carry out the cadastral surveys required by the Land Act of July 1885. Work on the geodetic plan was begun with the co-operation of the Institut Géographique National; survey work, levelling and revision went on for many years with constant interruptions. The basic topographical service, at 1 : 40,000, carried out by plane-tabling methods, is available in two forms: the 'Algeria-Tunisia' type for sheets published before 1922, and the '1922' type for newly plotted or revised sheets. A 1 : 100,000 series began in 1895; sheets in seven colours with relief shown by twenty five metre contours, cover the central part of the country. Thirdly, a 1 : 200,000 series, compiled between 1896 and 1929, covers the whole country in forty six sheets. Aerial photography was introduced in 1935 and coverage was completed by 1954, revised again by the latest techniques in 1962. In 1959, the Topographical Service completely revised the precision levelling network. Tunisia is therefore now very adequately mapped. Three-colour sheets have been compiled for some particularly important areas, and town plans have been made of most of the towns by the Société Tunisienne de Topographie at 1 : 10,000, 1 : 5,000 or 1 : 2,000. In *World cartography*, VIII 1967, may be seen an index map showing map coverage of Tunisia at scales between 1 : 25,000 and 1 : 50,000. A 'Carte administrative de la Tunisie', 1 : 750,000, showing the latest boundaries, was issued in 1964 by the Topographical Service. A revised 'Carte géologiques de la Tunisie', at 1 : 500,000, in two sheets, was issued in 1962.

Maps are included in Nevill Barbour's *A survey of North West Africa (The Maghrib)* (OUP, second edition 1962), issued under the auspices of the Royal Institute of International Affairs. This work covers the whole area from Morocco to west Egypt and the Sudan. Some of the maps in the Admiralty geographical handbooks *Algeria*, in two volumes, 1944, and *Tunisia*, 1945, are still useful. Nearly all the works by Jean Despois on the area contain very interesting and evocative maps in the texts, especially *L'Afrique blanche française*, Volume I, *L'Afrique du Nord*, of which the second volume was written by Robert Capot-Ray, *L'Afrique blanche française* (Paris, PUF, 1949, 1953). *Journées d'information* should be mentioned for its maps—*Médico-sociales Sahariennes*, by the Comité de Coordination Scientifique du Sahara (Paris, Arts et Métiers Graphiques, 1960), also Elio Migliorini: L'esplorazione del Sahara (Turin, Unione Tipografico, Editrice Torinese, 1961); Roger Coque: *La Tunisie présaharienne: étude géomorphologique* (Paris, Colin, 1962), which includes excellent maps; also René Raynal: *Plaines et piedmonts de la Mouloya* (*Maroc oriental*): *étude géomorphologique* (Rabat Bibliothèque de la Faculté des Lettres, 1961).

The United States Geological Survey has been responsible for recent mapping in Libya. A ' Topographic map of the United Kingdom of Libya ', 1 : 2M, was published by the Survey, prepared under the joint sponsorship of the ministries of national economy, petroleum affairs and industry of the United Kingdom of Libya and the United States Department of State. It was compiled from the latest maps, field exploration and ground traverse data and it served as the base for a ' Geologic map of Libya ' (miscellaneous geologic investigations map), 1 : 2M, also published by the Survey. A map at 1 : 2M, showing oilfields and pipelines, with indications of the concessions granted to various oil companies, was prepared in 1964 by BP. Particularly useful official surveys, with maps, include the *Second annual report of the United Nations Commission in Libya*, prepared in consultation with the Council for Libya, Paris, 1951; the *Supplementary report to the second annual report . . .*, 1952 and the *Economic and social development of Libya*, 1953.

The Sahara is mapped at 1 : 100,000 (Type Algérie) by the Institut Géographique National; there is also a special map at 1 : 500,000, ' Carte géologique du Sahara, Massif du Hoggar ', comprising nine sheets and text, published in 1962 by the Bureau de Recherches Géologiques et Minières. The *Travaux* of the Institut de Recherches Sahariennes, University of Algiers, frequently include reports of new surveys, and sketch-maps on agricultural and vegetation aspects are to be found occasionally in *Sahara*, 1957-   (Paris, Agence France-Presse). In the enormous literature on the Sahara, in many languages, original or interesting sketch-maps are frequently to be found. Particularly numerous and informative maps are included in B E Thomas: *Trade routes of Algeria and the Sahara* (University of California, publications in geography, 1957). The Free University of Berlin recently opened a new research station at Bardai, a large oasis in the Sahara in the northern part of the Republic of

Chad, sponsored by the German Research Association and the Berlin Senate. Its main function is to develop research in geomorphology and vegetation conditions. The results of this research, both text and maps, will be valuable.

The Institut Géographique National has also been responsible for the basic cartographic work in Mauritania and Mali. The Mauritanian Survey Department has begun to carry out topographical and cadastral surveys. An astronomic network and first-order levelling provided the basis for standard maps at 1:200,000, 1:50,000, beginning with the most important areas, and general maps at 1:500,000 and 1:1M. Special purpose maps are at various scales. The Senegal river area has so far been the most extensively mapped. The institute has now completed a series of vertical aerial photographs at 1:50,000. To assist in development projects, detailed topographical mapping has been carried out in the major urban centres; in most cases, direct ground surveys have been made, or ground surveys in conjunction with photogrammetric methods, at a basic scale of 1:2,000, with larger scales, 1:1,000, 1:500 and 1:200, as desirable. Such organisations as the Société des Mines de Fer de Mauritanie and the Société des Mines de Cuivre du Mauritanie have made detailed surveys in their special fields.

The Institut Géographique National has in general taken over the basic cartographic responsiblities for French West African territories. A map of the République du Mali at 1:2,500,000 was published in 1962 by the Dakar Branch of the IGN, showing administrative divisions and communications. A ten year plan was begun in 1964 in Mali, to complete accurate astronomic control, levelling and a topographic map series at 1:200,000. Thereafter, topographic map series at 1:50,000 and local surveys are to be maintained by the Institut National de Topographie de Mali or by private companies. The whole territory has already been covered by photographic surveys at 1:50,000. ICAO charts at 1:1M cover the whole country on the Lambert Conformal Conical Projection; otherwise, the projection used in all the French-speaking countries of west Africa is the Universal Transverse Mercator.

In the Niger territory, an accurate levelling network has been completed, linking up with the Upper Volta, Mali, Algeria and Chad. Astronomical points have been established sufficiently to compile base maps at 1:200,000, using vertical aerial photographs at 1:50,000, but the 1:50,000 map sheets so far completed have been for the more economically important areas. Other map series maintained by the IGN are at 1:500,000, 1:1M and 1:2·5M derived from the 1:200,000 surveys. A series of 1:50,000 special subject maps cover French-speaking west Africa, for geology, economy,

357

agriculture and ethnography. Many specific maps of individual areas have been compiled by the Office Française de la Recherche Scientifique et Technique d'Outre-mer. The Niger Topographic Service has so far prepared 1 : 2,000 topographic surveys for all the main centres of population, and has completed special surveys for hydro-agricultural development at 1 : 5,000 or 1 : 10,000. Plane-table surveying is usually the method used, with photographic surveys, if possible, as a check. The office du Niger, Ségou, established in 1932, took over from the French Government in 1958 an extensive programme of irrigation and agricultural improvement in the central Niger Valley, based on large-scale surveys. Maps of Chad are also published by the Brazzaville Geographical Department.

No current bibliography of maps is available for French-speaking west Africa, but legal deposit is compulsory with the library of the Institut Française d'Afrique Noire, so that maps are recorded in the card catalogue. Some find their way also into the *Bibliographie de la France*.

UNITED ARAB REPUBLIC (EGYPT REGION), SUDAN AND ETHIOPIA : Geodetic triangulation and precise levelling of Egypt (United Arab Republic) were carried out between about 1906-7 and 1914 by the Survey of Egypt. The main series of topographical maps were at first the 1 : 500,000, 1 : 250,000, 1 : 100,000 and 1 : 25,000. Their sheet lines were all based on a kilometric co-ordination system. The 1 : 500,000 series covered the whole of Egypt in twelve sheets; relief was shown by 100-metre contours and by hypsometric tints; blue was used to show the drainage system, red for roads and black for other features. The 1 : 25,000 series covered individual areas, such as the four sheets which comprised the ' Northern Sinai ' series, and other areas, such as the Bhariya, Farafra and Dakhlas oases, which are each mapped on an individual sheet.

The 1 : 100,000 series consisted of several groups of maps, each covering a specific area. On some sheets, topographic detail is shown only for the populated and cultivated areas, leaving the desert blank. The series termed the ' Normal' series covered the lower Egypt, Farijûm, the Nile valley and the Kharga and Dakhlas oases, in about seventy sheets, published in English and Arabic editions. The contour interval was one metre, while on some sheets hachures were also used to indicate relief details. A grey screen indicated built-up areas. Very individualistic mapping was demanded by the nature of the terrain; three shades of green were used to show cultivated areas and vegetation cover, pale green indicating cultivated areas not dependent upon irrigation, a darker green showing cultivated areas depending on irrigation and a very dark green showing wooded

358

areas, overprinted with black vegetation symbols. Roads were indicated in red, drainage in blue and all other features in black. The 1:100,000 'Red Sea coast' series ranged from Suez to the Egyptian-Sudan boundary, the 'Eastern desert' (now Arabian desert) series ran roughly parallel. Then there were the eighteen sheets of the 'Northern Sinai' series and the twelve sheets of the 'South Sinai' series. The 'North-west' series covered the north-west part of the country, their southern sheet lines meeting the sheets of the 'Normal' series. Finally, a special set of three sheets was compiled to show the Qena-Quseir Road and its adjacent areas, overlapping with the other series.

The 'Normal' series at 1:25,000 covers the same area as the 'Normal' series at 1:100,000 and is available in two editions, English and Arabic. The contour interval was one metre and five colours were used, blue to show drainage, brown for relief, red for woods, green for all cultivated areas and black for cultural features. The Survey also made a beginning on topographical series, in Arabic and English, at scales of 1:10,000 and 1:50,000; some cadastral mapping was done at 1:2,500 and town plans at 1:500, 1:1,000 and 1:5,000. A second edition of the 1:25,000 series was completed in 1956; the 1:50,000 series for some desert areas of economic importance and the 1:100,000 covering the whole country were maintained, both production and revision being greatly facilitated by the introduction of aerial photographic methods. Much topographic and soil mapping was required for the planning of the Aswan High Dam: 1:25,000 topographic maps, with contours at five metre intervals, were prepared from wide-angle photography. Similar mapping was involved in the whole New Valley project.

The Geological Survey of Egypt, Cairo, has issued reports and memoirs since 1900, with which should be associated two major works: the study, in four volumes, including maps and diagrams, by W F Hume: *The surface features of Egypt: their determining causes and relation to geological structure; The fundamental Pre-Cambrian rocks of Egypt and the Sudan; The later plutonic and minor intrusive rocks;* and *The minerals of economic value associated with the intrusive Pre-Cambrian igneous rocks and ancient sediments* (Cairo, Geological Survey of Egypt, 1927-37); and, secondly, *The Pre-Cambrian along the Gulf of Suez and the northern part of the Red Sea,* by H M E Schürmann and others (Leiden, Brill, 1967). This volume is a compilation and summary of work done in the field and in laboratories during the past fifty years. Three folded maps in nine colours and three in two colours, as well as sketch-maps, are an essential part of the work. The large maps may be obtained also unfolded: I 'A geological reconnaissance map at 1 : 250,000 of the northern Red Sea Hills'; II 'Sinai'; III 'Southern part of northern Red Sea Hills'; IA 'Locality map of samples on sheet I showing chemical analyses in green and age determinations in red'; IV 'Tectonic sketch map of northern Red Sea area, the Gulf of Suez and the Gulf of Akaba'; 1 : 2M; V 'Sketch map of field trip made with Glen Brown in northwest Saudi Arabia and tentative correlation with Pre-Cambrian of the Gulf of Suez and northern Red Sea area', 1 : 1M. A 'Geological map of Egypt', 1 : 5M, was completed in 1939. A 'road map of the United Arab Republic, Egypt region', 1 : 500,000, was revised by the Cairo Drafting Office, in 1963; and a special 'Guide plan of Cairo', 1 : 15,000, is maintained, prepared by the Survey of Egypt.

The atlas of Egypt was prepared by the Survey for the International Geographical Congress at Cambridge, 1928. Eleven pages of text and thirty one plates of maps and diagrams give an account of the orography, geology, meteorology and economic conditions of the country. The general map at 1 : 2M shows motor routes across the desert; the orographic map follows the 1 : 1M specifications, except in the portrayal of arable land in green, to distinguish it from the desert land above sea level, which is shown in brown and from the land below sea level, shown in grey. The maps of weather types and the relative humidity diagrams are particularly effective. Population and economic data are based on the latest census figures, the text for the economic maps and diagrams being contributed by the Controller-General of the Census. Material was included also on the Anglo-Egyptian Sudan, Kenya, Uganda and Tanganyika. For the International Geographical Congress in Paris, 1931, a meteorological atlas was prepared, consisting of forty one sheets, which illustrate all aspects of the climate of the Nile basin.

Stanford's 'General map of North-east Africa', at 1 : 5,500,000. includes the whole of Egypt, the Sudan and Ethiopia, overlapping into Kenya and Tanganyika. Main roads, railways, canals and oil pipelines are shown, with heights indicated in feet above sea level. Population is indicated by the size of the town markings. Michelin Service de Tourisme produced a map of 'North-east Africa showing main roads', 1 : 4M, in 1964. Original maps frequently accompany the articles in the annual *Journal* of the Egyptian Geographical Society, published in French and English.

Sudan: The Sudan Survey Department, established in 1902, began work on three main series of topographical maps, at 1 : 100,000, 1 : 250,000 and 1 : 500,000, all based on the specifications of the International Map of the World 1 : 1M. In addition, a number of large-scale surveys were made for irrigation projects in the Blue

Nile and White Nile river areas. Publication of the series at 1:100,000 began in 1949, with the addition of several colours from 1955. The 1:250,000 covered the country in 167 sheets, about half of which are coloured, based on the former GSGS African series; and the 1:500,000 series was completed in forty five sheets. After the second world war, when large areas needed to be surveyed for development projects, additional sections were established, responsible for air survey, town planning and training. The extremely varied and difficult nature of the terrain must, however, be borne in mind and as, in addition, a record flood in 1946 had washed away the boundary marks, in 1952 the department began virtually new control surveys for new topographical series. First-order and second-order triangulations were begun and, at the International Union of Geodesy and Geophysics 1951 Conference, the United States Coast and Geodetic Survey agreed to help in this work, which was completed in 1954.

The Sudan became independent in 1955 and the department was taken over by the first Sudanese Director of Surveys. The whole organisation was overhauled and the air photography section was amalgamated with the topographic section. In 1958, the United States agreed an aid programme for extensive survey work, especially in areas where agriculture, forestry and mineral resources needed to be developed. Maps were produced for the whole area at 1:100,000 and for some special areas at 1:50,000 and 1:25,000, which replaced the earlier outdated topographic series. The Sudan Survey Department, now one of the best in Africa, consisting of Cadastral, Topographic, Geodetic, Computation, Development and Printing Departments, under the Ministry of Mineral Resources, demarcates and levels annually in the region of 1,300,000 acres at fifty centimetre contour intervals, also some thirty five thousand new plots in town lands. Twenty one provincial survey offices are now fully working. For special survey projects, contracts are made with overseas firms, as necessary. The regional study involved in the Jebel Marra Investigations, carried out by Hunting Technical Services, is a case in point. The aim was to devise a scheme whereby the copious rainfall caused by the 10,000 foot mountain range of Jebel Marra could be used for irrigated agriculture during the dry season. In 1957, twelve thousand miles of country were surveyed and mapped by a team comprising geologists, an agricultural ecologist, a forester and surveyors, from the Sudan Ministry of Irrigation and Hydroelectric Power. The maps produced in the course of the Equatorial Nile Project and its effects on the Anglo-Egyptian Sudan should also be mentioned; the *Report* of the Jonglei Investigation Team was published in five volumes, with maps and diagrams, in 1955.

Geological Survey *Bulletins* have been issued since 1911 and a

361

' Geological map' at 1:4M has been several times revised. Many new maps have appeared as a result of the first population census of Sudan 1955-56. An Arid Zone Research Unit, attached to the University of Khartoum in 1961, carries out scientific surveys of fauna, flora and geology of the arid regions of the Sudan. Notable for the maps included are also J H Schultze: *Der Öst-Sudan: entwicklungstand zwischen wüste und regenwald* (Berlin, Dietrich Reiner Verlag, 1963); H E Hurst et al: *The Nile basin,* in five volumes (Cairo, Ministry of Public Works, 1931-55; K J Krolki, editor: *21 facts about the Sudanese* (Population Census Office, Ministry for Social Affairs, Republic of Sudan, 1959), which includes five maps showing population centres and two density and distribution maps; J D Tothill, editor: *Agriculture in the Sudan: being a handbook of agriculture as practised in the Anglo-Egyptian Sudan* (OUP, 1949), well illustrated with maps in the text and with a separate map in a pocket. *Land use in Sudan,* by J H G Lebon, is the fourth of the World Land Use Surveys regional monographs (Bude: Geographical Publications, 1965), a most comprehensive work, including not only the 1:1M, but other maps of the Sudan. From time to time, the Survey Department issues a *Catalogue of topographical maps.*

Ethiopia: The *Report on cartographic activity in Ethiopia,* presented in 1963 by the Imperial Ethiopian Mapping and Geography Institute to the United Nations Regional Cartographic Conference for Africa, stated ' Maps of Ethiopia or parts thereof with a desirable standard of accuracy and incorporating a sufficient amount of information are a rarity to this day '. In the Second Five Year Development Plan, therefore, 1963-67, mapping activities of all kinds formed a major provision, particularly in connection with the economic development of the country. The cartographic resources of the imperial Ethiopian Air Force, Photography and Aerial Survey Services and of the government departments concerned with water resources surveys, highways, power and mines are at present nominally coordinated by the Imperial Ethiopian Mapping and Geography Institute and, with the completion of a new plan, will in fact be centralised within the Imperial Institute, which was founded in 1954. The Mapping Section includes three divisions: the Photogrammetry Division, responsible for aerial photography, photo-interpretation, photomapping, photomosaics and the planning of planimetric and topographic maps; the Cartographic Section, which is mainly concerned with the preparation of maps and charts; and the Geography Section, which sponsors original research on the geography of Ethiopia, some of which is published in the *Ethiopian geographical journal,* semi-annual, 1963-, and compiles atlases, topical maps and gazetteers. The Cartography Section is planning expansion to include cadastral, geodetic and geological survey in the near future (and may now have indeed done so).

Aerial photography was used first to map especially important areas, such as the Blue Nile river basin and the Awash river valley. In 1963, the Ethiopia-United States Mapping Mission was created, to complete aerial surveys for the whole country at approximately 1 : 50,000, also some areas at 1 : 25,000, to establish geodetic control data, to assist the Mapping and Geography Institute in producing a map of Ethiopia at 1 : 50,000 for initial planning purposes and to prepare topographic maps at 1 : 250,000. Training programmes were also begun and modern equipment installed at the institute to enable it to function efficiently. A team of about 150 members of the United States Corps of Engineers began ground operations in 1964 and the basic surveys are expected to be completed by 1969. A map of Ethiopia at 1 : 50,000 was published by the Deutsche Forschungs-gemeinschaft in 1966 and issued as a supplement to *Zeitschrift für Vermessungswesen,* April 1967. The Mapping and Geography Institute completed a transportation and administration map at 1 : 2,800,000 in 1966; and in the same year Mobil produced two useful maps, a ' Road map of Ethiopia ' at 1 : 3M, indicating road distances in miles and kilometres; and an 'Addis Ababa City map '.

AFRICA SOUTH OF THE SAHARA

The Scientific Council for Africa South of the Sahara, Commission for Technical Co-operation in Africa South of the Sahara published *Mapping and Surveying of Africa south of the Sahara* in 1956, a key document, in French and English, for the study of the tremendous tasks involved in the mapping of this vast area. The commission sponsors research of all kinds, in collaboration, when possible, with other bodies, or, for example, with the Inter-African Research Fund in the preparation of a climatological atlas for Africa; the commission arranges conferences and symposia, of which it usually publishes the results. UNESCO recently established a Regional Centre for Science and Technology for Africa, centred in Nairobi, to serve the following countries: Basutoland, Bechuanaland, Barundi, Cameroun, Central Africa Republic, Chad, Congo (Brazzaville), Congo (Leopoldville), Dahomey, Ethiopia, Gabon, Gambia, Ghana, Guinea, Ivory Coast, Kenya, Liberia, Madagascar, Malawi, Mali, Mauritania, Mauritius, Niger, Nigeria, Rwanda, Senegal, Sierra Leone, Somalia, Swaziland, Tanzania, Togo, Uganda, Upper Volta and Zambia. The centre is responsible for the planning, execution and supervision of the UNESCO regional activities in science and technology, for putting into effect the aims of the Lagos Plan and for taking practical steps in any specific tasks and missions as they are agreed. The *Centre d'analyse documentaire pour l'Afrique noire* should be mentioned in passing; its work is vital, but its relevance here is its role in making known the research, including any form of cartographic research, being done in this part of Africa.

Malcolm Hailey's *An African survey—revised 1956: a study of problems arising in Africa south of the Sahara* (OUP, 1957), issued under the auspices of the Royal Institute of International Affairs, must be mentioned in any review of information, cartographical or otherwise, on this area; also E B Worthington: *Science in the development of Africa: a review of the contribution of physical and biological knowledge south of the Sahara,* includes many maps in the text (Commission for Technical Co-operation in Africa South of the Sahara and the Scientific Council for Africa South of the Sahara, 1958).

'A vegetation map of Africa south of the Tropic of Cancer ' was sponsored by l'Association pour l'Etude Taxonomique de la Flore d'Afrique Tropicale, published by Oxford University P with the assistance of UNESCO. The work, edited by A Aubreville and others, consists of twenty four pages of explanatory notes, in English and French, by R W J Keay, descriptive of a large map folded into the back of the text. The map is also available unfolded. The FAO *Crop ecology survey in West Africa* was completed in 1962. The second volume consists of twenty four maps at 1:5,500,000. Invaluable also is *Tropical soils and vegetation,* the *Proceedings* of the Abidjan Symposium arranged by Unesco and the Commission for Technical Co-operation in Africa South of the Sahara, October 20-24 1959 (UNESCO, 1961), which includes textmaps. In *Geography,* July 1968, is a relevant article by Anthony Young, ' Natural resource surveys for land development in the Tropics ', *West Africa: the French-speaking nations, yesterday and today,* by R Adloff, 1965, includes some useful maps; the geography, culture, history and economy of Senegal, Mauritania, Sudan, Niger, Guinea, Ivory Coast, Upper Volta and Dahomey are discussed; another standard work including maps is S H Houghton: *The stratigraphic history of Africa south of the Sahara* (Edinburgh, Oliver and Boyd, 1963).

The international atlas of west Africa is in progress in Senegal. It is understood that the complete work, in English and French, will consist of forty eight maps, at a basic scale of 1:5M, with extensive notes, graphs, diagrams and statistics, dealing with history, languages, economic factors, industrialisation and general environmental features. *Atlas des cartes ethnographiques et administratives des differentes colonies du gouvernment général* was prepared by the Service Géographique de Gouvernment Général de l'Afrique Occidentale Française and published in Dakar in 1922; the *Atlas des cercles de l'Afrique occidentale française,* by Forest of Paris in 1926. Nelson publishes a special *West African secondary school atlas.*

A number of organisations sponsor original surveys, depending on the pattern of administration in previous years. In the tropical countries of west Africa, the latitude and longitude lines are frequently used as sheet boundaries and this has accounted for a degree of uniformity in an otherwise complex cartographic status. In Senegal, the Institut Géographique National is maintaining topographical maps at 1:50,000. Gambia has been mapped by the Directorate of Overseas (Geodetic and Topographical) Surveys at 1:50,000 and 1:125,000. The especially important sheets covering the River Gambia area began a third edition in 1966. In 1958, the Directorate completed a Gambia Land Use map at 1:25,000; this is a practical map, mainly related to rice cultivation and with it are included very effective air photographs. Small-scale topographical and geological maps of Guinea have been issued by the Spanish and Portuguese authorities.

By 1951, topographic maps of Sierra Leone at 1:62,500 by the Directorate of Overseas Surveys covered the country in 104 sheets, showing contours at fifty feet intervals. A second edition of the topographic series at 1:50,000, produced photogrammetrically, is nearing completion; an uncontoured edition has already been made available and has proved of special importance in the development of transport and communications. Large-scale maps for town planning are being made at scales ranging from 1:1,200 to 1:4,800. The Survey and Lands Department is responsible for field work and the provision of place names and other local information. There is also a topographical series at 1:10,000. The Directorate of Overseas Surveys 1:1M map of Sierra Leone went into a sixth edition in 1965. The Geological Department is steadily producing geological maps, having begun with the areas in the centre and east, where mineral resources exist. Bulletins and annual reports have been issued since 1928.

In 1953, the Survey and Lands Department at Freetown published a sixteen page *Atlas of Sierra Leone,* at scales ranging from 1:15,000 to 1:3M. The quality of the maps varies both in accuracy and in methods of presentation, but here, for the first time, authoritative maps of the Colony and Protectorate were brought together; climate, vegetation, soil, tribes, agriculture, population, minerals and town

365

plans are included. *Sierra Leone in maps,* edited by John I Clarke, was published in 1966 by the University of London Press, for Fourah Bay College, Freetown; the work comprises fifty one maps, with commentaries by thirteen past and present members of the Fourah Bay College staff, which present comprehensively the nature and resources of the country. Nelson's special *Sierra Leone atlas,* edited by M I Cran, 1966, is intended for school use; twelve plates are devoted to Sierra Leone, its characteristics and relation to the rest of Africa, with eight plates representing other continents, a double-page spread of the countries of the world and an index of place names.

Sierra Leone: a modern portrait, by Roy Lewis (HMSO, 1954), in the Corona Library series, includes useful text maps and an informative folding map; also useful is A R Stobbs: *The soils and geography of the Boliland region of Sierra Leone* (Government of Sierra Leone and Crown Agents, 1964).

Legal deposit of publications in Sierra Leone is compulsory and these are recorded at the Colonial Secretary's Office; government publications are listed from time to time in the *Sierra Leone royal gazette;* otherwise there is no central check on map publication.

The Government of Liberia, in co-operation with the German Federal Republic, has published sheet maps, compiled from aerial photographs, region by region, at 1:20,000. Prior to 1960, when the Ivory Coast became independent, the Bakar branch of the French Institut Géographique Nationale was responsible for preparing topographical maps on the scales of 1:200,000 and 1:50,000. Through the Fonds d'aide et de Coopération, this work continues, linking up with similar plans for all the states of west Africa, based on isolated astronomical points. The institute publishes an *Annual report.* The Topographical Office, attached to the Ministry of Buildings, has gradually enlarged the scope of its work, from local land surveys to road and population projects. The Land Registration Office, established in 1962, is the specialist agency for large-scale maps on scales of 1:500 to 1:10,000. The cadastral maps published by this Office form the basis of all development plans; the office is responsible for field surveying and topographical work relating to land registration. The great area of equatorial forest makes geodetic triangulation impossible, but experiments have been undertaken in the use of apparatus using electromagnetic waves for traversing. 'Carte géologique de la Côte d'Ivoire', 1:1M, was published by the Direction des Mines et de la Géologie, Abidjan, 1965.

The Survey of Ghana was established in 1919, but before that time some ground surveys had been carried out for particular areas.

The varying physical characteristics of different parts of the country have affected the speed of working. Scientific surveying did not really begin until about 1930 and by 1948 primary and secondary triangulations and levelling had made considerable progress and topographical maps at 1:62,500 covered the southern parts, drawn on the Transverse Mercator Projection. In the northern less-developed and populated areas, series at 1:125,000 or 1:250,000 were compiled and most of the towns and larger villages were mapped at 1:1,250 or 1:2,500. Maps at 1:250,000 of almost the whole country were made by photographic reproduction from the 1:125,000 series. For the Volta Dam Project, a new topographic survey was required; using photogrammetric methods, this was completed at 1:50,000 between 1948 and 1953. Many pamphlets, usually including maps, were published by various government departments, describing the project, such as 'The geology of the Volta River Project, with geological plans and sections', by W B Tevendale (Accra, Government Printer, 1957).

The United Kingdom-Ghana Technical Assistance Scheme made a grant of over £200,000 available for the period 1963-68, towards the cost of air photography, field survey and mapping. The Ghana government also made corresponding increases in expenditure for the completion of the basic maps needed for development projects. Photogrammetric mapping was undertaken by the Directorate of Overseas Surveys, with the assistance of local ground surveyors. Much of the plotting is still done in Britain, but local trained staff are steadily taking over more of the work. The national Survey is rapidly increasing its efficiency, concentrating primarily on the 1:50,000 topographic series to replace the former sheets at 1:62,500 and revising the large-scale plans of the main towns and cities.

The Department of the Geological Survey, established in 1913, made considerable contributions to the expansion of the industrial economy, planned when independence was achieved. Systematic geological mapping and prospecting were based on the 1:62,500 scale maps in the south and on the 1:125,000 scale maps in the north. Annual reports have been issued since 1913, bulletins since 1925 and memoirs since 1929.

An *Atlas of the Gold Coast* was compiled by the then Gold Coast Survey Department, Accra, in 1950. The *Ghana population atlas*: *the distribution and density of population in the Gold Coast and Togoland under the United Kingdom trusteeship* was published by Nelson, 1960, on behalf of the University College of Ghana. Forty pages of maps, compiled by T E Hilton, were based on the 1948 population census, on scales ranging from 1:1,500,000 to 1:3M. Some existing maps were included, such as that of the closed forest areas, 1908 to 1953. The commentary is informative and stimulating. The Ghana Survey Department published a portfolio of twelve Ghana maps in 1962. Various maps at 1 : 1M were completed, also a road map in two sheets, north and south, at 1 : 500,000 in 1959. Especially important was the *Accra Survey: a social survey of the capital of Ghana, formerly called the Gold Coast, undertaken for the West African Institute of Social and Economic Research, 1953-1956,* which includes maps (edited by Ione Acquah, for the University of London Press, 1958).

A growing number of monographs contain maps of Ghana, such as the following: P M Ahn: *Soils of the lower Tamo basin, southwestern Ghana* (Accra, Government Printing Department, Crown Agents for Overseas Government and Administration, 1961; Ghana Ministry of Food and Agriculture, Scientific Service Division, Soil and Land Use Survey Branch, memoir no 2); P R Gould: *The development of the transportation pattern in Ghana* (Northwestern University Studies in Geography, no 5 1960) and J B Wills, editor: *Agriculture and land use in Ghana* (OUP, for the Ghana Ministry of Food and Agriculture, 1962; with maps folded in a pocket). The flow of textbooks from Britain to newly independent countries began in the mid 1950s, followed by atlases specially designed for the individual country. The *Ghana junior atlas,* for example (Nelson, 1965), edited by E A Boateng, devotes ten pages to Ghana; on five of these are twelve thematic maps at various scales, with brief explanatory text. Four town plans are included, each on a different scale, and twelve coloured photographs illustrate vegetation and economy. A further eight pages are given to the rest of Africa, including thematic maps showing aspects of climate, vegetation, population and economy. The other continents have one general map apiece and there are two world maps, one political, the other showing population.

Maps published by the Survey of Ghana are listed by the Director of Surveys in Accra; otherwise, information about map publishing is difficult to come by.

In Nigeria, the Directorate of Overseas (Geodetic and Topographical) Surveys continues mapping programmes, in association with the Federal Survey of Nigeria, which was founded in 1900 within the Ministry of Works and Surveys, Lagos. The largest scale topographic series covering the whole country is the 1 : 500,000. Parts of the country are mapped at 1 : 50,000, 1 : 62,500, 1 : 100,000 or 1 : 125,000. A number of large-scale town plans, showing contours, have been issued, as of Enugu and Port Harcourt, at 1 : 4,800. Lagos and outskirts have been mapped at 1 : 1,200 and a special study was made of the City of Ibadan, for a symposium, of which the papers were presented by P C Lloyd and others. A comprehensive survey of the history, geography, sociology and political structure of the city was made, illustrated by maps. The Federal Survey maintains maps at 1:250,000 and 1:750,000. Special topic maps have been issued by the Survey at 1:3M, showing climate, vegetation, minerals, agriculture, communications, tribes and languages, and there is also a 1:1M population map, 1960.

The Nigeria Geological Survey has issued a 1:2M map since 1943, of which the latest revision was in 1964; fully coloured maps are maintained at 1:100,000 and 1:250,000. The Survey has issued bulletins from 1921, 'Occasional papers' from 1925 and annual reports from 1930, in addition to some valuable monographs. The Survey has been assisted by technical missions from the United States, which has undertaken numerous geological and hydrological studies in the country. Others studies have been made by the Directorate of Overseas Surveys of Great Britain. So far as is known, considerable areas of northern Nigeria have still to be mapped at medium and large scales. *A sketch-map of Nigeria*, compiled by H O N Oboli (Harrap, 1960; second edition 1962), consists of some twenty maps, with explanatory notes, intended mainly for school use. Nelson publishes a general atlas with a large section devoted to Nigeria and to the rest of Africa.

Maps are to be found in the *Nigerian geographical journal*, semi-annual from 1957, in English, from the Department of Geography, University College, Ibadan, and either maps or reports of surveys are included in the monthly *Journal of Nigerian studies* (African Book Company, Lagos), in the *Directory of the Federation of Nigeria* and in the numerous reports of individual bodies, such as those of the regional marketing boards, the Nigerian Timber Association, etc, and the HMSO Overseas Economic Survey—*Nigeria*. Maps illustrate the work of Daryll Forde and his colleagues in Nigeria: *Efik traders of old Calabar*, for example (OUP, for the International African Institute, 1956) and *The native economies of Nigeria*, written with Richenda Scott, as the first volume of *The economics of a tropical dependency*, edited by Margery Perham (Faber, 1946) with folding maps. The publication *The industrial potentialities of northern Nigeria*, issued by the Ministry of Trade and Industry, Kaduna, in 1963, is excellent, with maps and a bibliography.

Two cartographical commentaries make constructive reading: O M Dixon: 'The selection of towns and other features on atlas maps of Nigeria', *Cartographical journal*, June 1967, which includes references and an appendix showing the various means of 'ranking' towns; and J M Zarzycki: 'Experience with a new mapping system

Atlas of Nigeria (loose leaf)?

369

employed on a topographical survey in Nigeria ', *Report of Proceedings,* Conference of Commonwealth Survey Officers, 1963 (1964). Current map production is included in the national bibliography, *Nigerian publications,* 1950-, annually from 1953, supplemented by quarterly *Lists,* compiled at the University College Library, Ibadan.

The Service Géographique du Cameroun, Yaounde, maintains local mapping and an *Atlas de Cameroun* was compiled by the Haut Commissaire de la République Française au Cameroun in 1948. The work comprises a straightforward, factual text and seven folded coloured maps, showing administrative regions, geology, races, density of population, communications, the hydroelectric works in special regions and the industrial activity in the capital city, Douala. The Républic Fédéral du Cameroun prepared a map ' Organisation administrative ' at 1:3,666,666, which was published in 1965 by the Annexe en Afrique Equatoriale de l'Institut Géographique National; the revised administrative boundaries are shown.

Congo .

Turning now to the vast area forming the drainage basin of the River Congo—Equatorial Africa proper—its tangled political history must be borne in mind, also the realisation of its potential resources, particularly after the second world war. Great progress has been made by the Paris Branch of the Institut Géographique National, with the Topographic and Cadastral Service of the Congo and the Société Africaine de Travaux et d'Etudes Topographiques. The Brazzaville branch of the institute was established in 1945. From 1946, work on astro-geodetic points progressed and first- and second-order levelling were completed. A basic map for the whole country was established at 1:200,000 and many of these originals are now in process of revision. Special maps at 1:50,000 have been made for government departments in connection with urban planning or with agricultural and industrial development; the speed and accuracy of mapping is increasing, with the introduction of modern equipment. In 1950, the institute undertook the publication of the 1:200,000 map, the whole of the office work being financed from the French public works budget. A new 1:50,000 series, begun in 1958, financed by the Aid and Co-operation Fund in Paris, is growing steadily, having begun with the most important areas.

A map of the Brazzaville region is available at 1:500,000, being a generalisation of the 1:200,000; and plans of the towns of Brazzaville and Pointe-Noire have been issued at 1:10,000, maps of each town in its region at 1:20,000 and a new edition of the Brazzaville plan in seven colours in 1965. A 1:1M map shows the distribution of villages, using data received from the Central Statistical Office. Road maps have been published at 1:500,000 and, at various scales, maps of the economy, agriculture, forestry and stock, mining and industry, also a population density map and a monochrome map of administrative regions, all based on aerial photographic plotting.

The Cadastral Service, set up in 1950, immediately began to prepare surveys of the main towns at 1 : 2,000, revising them as necessary. The Société Africaine des Travaux et Etudes Topographiques carries out surveys for the Topographique and Cadastral Services, for the Directorate of Construction, Town Planning and Housing and for firms and private individuals, as required. The Service Géologique du Congo Belge maintained a map at 1 : 500,000 in seven sheets; this service terminated with the independence of the states of the former French Equatorial Africa and a national service was established in 1962, the Mining Bureau of Geological and Mining Research. A new general map is being prepared at 1 : 100,000. Bulletins have been issued since 1945. A geological map at 1 : 2M, with an explanatory bulletin, prepared in four sheets by the Ministère des Colonies, Service Cartographique, Brussels, went into a fourth edition in 1951.

The work of the Direction des Mines et de la Géologie de l'Afrique Equatoriale Française, Brazzaville, frequently overlaps between French west African areas and equatorial Africa. Since the early 1930s, a great quantity of useful work has been accomplished; Bulletins have been issued irregularly since 1938, annual reports since 1946; *Cartes et notices,* comprising maps, mainly at 1 : 500,000, with explanatory pamphlets, have been issued as completed and a ' Carte géologique de l'Afrique Occidentale française ' at 1 : 6M was issued in 1952. Also published in 1952 was a ' Carte géologique de l'Afrique Equatoriale française et du Cameroun ' in three sheets, at 1 : 2M, compiled by M Nickles.

Previous mapping of Katanga and other parts of the vast area now the Democratic Republic of the Congo was carried out by Belgian cartographers at the Belgian Military Geographical Institute. In 1919, the Special Committee of Katanga set up a Geographic and Geological Service, including a triangulation and topography service. The basic chains were completed by 1951, forming a grid covering the whole country. An *Atlas du Katanga,* compiled by H Droogmans and others, was published under the auspices of the Comité Spécial du Katanga, in three volumes, 1928-29; it is understood that a new edition is in preparation. In 1938, the *Atlas géographique et historique du Congo belge et des territoires sous mandat du Ruandi Urundi,* prepared by René de Rouch, was published in Brussels.

Since independence, the functions of the Special Committee of Katanga have been transferred to the Geographic Institute of the Congo, established in Leopoldville, 1949-50. The institute, which

371

combines geodetic surveying, field surveys, plotting, aerial surveys, drawing and printing, has systematically completed aerial photography of the country at scales mainly of 1:33,000 or 1:40,000. The mapping of the tropical forest area has been possible only since the introduction of photogrammetric survey techniques and, even so, atmospheric conditions frequently make photography difficult. Topographic map sheets are published at 1:50,000; these surveys have greatly helped the advance of economic planning in the more favourable areas; for special projects, mapping at 1:10,000 or 1:25,000 is frequently undertaken. Maps at 1:200,000 have been prepared for some years by plane-table surveying, supplemented by local information; road and administrative maps have been derived from these; also the 1:1M map, all of which have now established revision programmes. The more populated areas, having greater economic development, are mapped at scales between 1:5,000 and 1:30,000.

The Geological Service of the Democratic Republic of the Congo also uses aerial photography for exploratory and geological surveys. A first edition of the ' Carte géologique du Congo ', 1:200,000, was published in 1964. A diagrammatic map showing the official map coverage of the republic accompanies a report on current mapping included in *World cartography,* VIII 1967. In addition, ' Les provinces de la République démocratique du Congo ', at approximately 1:11,500,000, was published in 1966 by Mwana Shaba of Kinshasa. *Cartes de la densité et de la localisation de la population de l'Ancienne Province de Léopoldville,* 1:1M, consists of three maps compiled by R E de Smet and published by the Centre Scientifique et Médical de l'Université Libre de Bruxelles en Afrique Centrale, 1966; forty six pages of explanatory text in French make an extremely useful contribution to this subject. The ' Carte routière de Congo belge et du Ruanda Urundi ' is in two sheets, at 1:2M, published in Brussels by Établissements Cartographiques, 1953. A ' Carte géologique de Rwanda ', 1:100,000, is maintained by the Institut Géographique Militaire de Belgique, Brussels.

The *Atlas générale du Congo,* published by the Centre Scientifique et Médical de l'Université Libre de Bruxelles en Afrique Centrale, completed between 1962 and 1966, has been designed on an interesting principle, endeavouring to show knowledge of the Congo at different periods of history, in an effort to relate the problems of the present to those of the past. Some individual map sheets were revised in 1951; at a basic scale of 1:5M, they are arranged under the general headings of history, geodesy, physical environment, biogeography, anthropology and culture, political, administrative, social and economic features.

372

No record of current mapping is maintained, so far as is known, but important official maps may be found in the *Bibliographie de Belgique*.

A vast periodical and monograph literature covers this fascinating part of the world, much of it including interesting maps, so that it is difficult to single out individual items for comment; a detailed study of a region otherwise little known is by Henri Nicolai: *The Kwilu: étude géographique d'une région congolaine* (Bruxelles, Edition Cemobac; Centre Scientifique et Médical de l'Université Libre de Bruxelles en Afrique Centrale, 1963). This work contains maps throughout the text and a folding map and is excellently documented. The maps in the text and in the pocket of *Belgian Congo*, in the Admiralty geographical handbook series (Naval Intelligence Division, 1944) are still useful; the monograph by G Sys and others: *La cartographie des sols au Congo, ses principes et ses méthodes* (Brussels, INEAC, 1961) is an important study, including illustrations, maps and diagrams; also Maurice Robert: *Géologie et géographie du Katanga* (Brussels, Union Minière de Hart-Katanga, 1956).

To the southward of the Congo area, Angola has not been mapped in any detail. The latest administrative map, the ' Carta de Angola ', 1:2M, was issued by the Ministério do Ultramar in 1966. A geological map at 1:2M was completed by the Servicos de Geologia e Minas, Luanda, in 1951. Bartholomew issues a 1:4M ' Map of central and east Africa ', revised in 1963; and the Institut Géographique National maintains a ' Carte de l'Afrique central ' at 1:500,000. The special *Oxford atlas for east Africa* (Clarendon Press, 1966), is intended mainly for school use. The layout and presentation are therefore straightforward and explanations are included of the meaning of such terms as ' scale ' and ' projection '. Of the sixty five pages of maps, thirteen are devoted to east Africa; topographic maps at fifty or 100 miles to the inch, smaller scale simplified maps of the climate and economy of the country and very useful sketch-maps of Dar es Salaam, Kampala, Mombasa and Nairobi follow; topographic maps of the rest of the world are given at various scales, also good maps of the Atlantic, Pacific and Indian Oceans and some generalised world maps form the final section of the atlas, at 1:100M or 1:150M. A useful table sets out the population, area and land use of each country.

A pioneer in population mapping was the work ' Tribal maps of east Africa and Zanzibar ' by J E Goldthorpe and F B Wilson (Kampala, East African Institute of Social Research, 1960). Eight maps show population based on the 1948 Census, four maps show tribal distribution in Uganda and Kenya; two rural population maps of Zanzibar and Pemba, based on data obtained in 1931—invaluable maps drawn in the Geography Department of Makerere College. Since this time, population mapping has developed by more advanced cartographical techniques, as reliable figures became available. In the former ' Northern Rhodesia ', the first attempt at an accurate survey was the 1950 sample Census, followed by the 1956 and other estimates, which have provided data for several population maps. 'East Africa: population distribution as at August 1962 ', by P W Porter, was derived from the series of east African censuses; Tanganyika, 1957, Zanzibar, 1958, Uganda, 1959 and Kenya, 1962. The difference in the dates of information has been adjusted to match that of the Kenya figures; while admitting the limitations of this compromise, due to the political exercises in the area, the map provides a basic document for the study of this part of Africa, published as map supplement no 6, at 1 : 2M, in the *Annals* of the Association of American Geographers in 1966.

The article by A M O'Connor: ' East African topographic mapping' in the *East African geographical review,* April 1966, is a most useful source of information and the maps in the same author's *An economic geography of East Africa* (Bell, 1966), drawn by J C Sebunnya, draughtsman in the Makerere Department of Geography, should also be consulted on this aspect of the country's development. The *East Africa Royal Commission report, 1953-55* (HMSO, 1955) is of the greatest value to geographers, including the maps; also G Dainelli: *Geologia dell'Africa orientale,* in four volumes (Rome, Reale Accademia d'Italia, 1943), includes a geological map 1 : 2M. Frank Debenham's works on central and east Africa include many informative maps: *Kalahari sand* (Bell, 1953), *Report on the water resources of the Bechuanaland protectorate, Northern Rhodesia, the Nyasaland protectorate, Tanganyika territory, Kenya and the Uganda protectorate* (HMSO, 1948) and *Study of an African swamp: report of the Cambridge University expedition to the Bangweulu Swamp, Northern Rhodesia, 1949* (HMSO, for the Government of Northern Rhodesia, 1952).

The Directorate of Overseas (Geodetic and Topographic) Surveys has covered Uganda, Kenya and Tanganyika, mainly at 1 : 500,000, but the basic topographical series is to be the 1 : 50,000, which has still to be completed in some areas. In Uganda, the Department of Lands and Surveys at Entebbe is maintaining the 1 : 50,000 topographic map, also the series at 1 : 2,500, and is working on a series of district maps at 1 : 125,000. These are attractive maps, showing relief by contours at 100 feet intervals and layer colouring. The Geological Survey has issued maps at 1 : 125,000 and 1 : 250,000, to which its progress reports are a guide. Annual reports have been issued since 1920, memoirs since 1926, bulletins between 1933 and 1939 and ' records ' since 1950. The Department of Agriculture is mapping soils at 1 : 250,000; and 'A subsistence crop geography of Uganda ', by D N McMaster, in the world land use survey series (Bude: Geographical Publications, 1962), is an excellent study, designed to serve as a model for other monographs in the series. A ' Map of vegetation ', 1 : 500,000, in four sheets was completed by the Department of Lands and Surveys in 1964, to accompany *The vegetation of Uganda*, published by the Uganda Government. Eighty six types of vegetation are distinguished by colours and numbers.

The *National atlas of Uganda* was prepared by the Department of Lands and Surveys to mark the country's independence (Kampala, 1962), under the direction of Bruce Whittaker, with the co-operation of other government departments and the staff of Makerere University College Geography Department. Thirty eight sheets of maps cover the main aspects of Uganda's historical, physical and human geography, including ethnic and tribal divisions of the people. The economic maps and trade diagrams are most useful, as are the special maps depicting tsetse, malaria, and medical and educational facilities. Most of the maps are at 1 : 1,500,000. Commentaries and photographs, an adequate gazetteer and a valuable bibliography complete the atlas. Nelson published a *Uganda junior atlas*, edited by Yunia Mugahya, in 1966; the work comprises twenty pages, in the usual style of this Nelson series, with emphasis on Uganda. In *Geography,* July 1968, is a relevant article ' Tsetse control and livestock development: a case study for Uganda ', by Brenda J Turner and P Randall Baker, which includes four useful maps and several references.

An index of maps of Uganda is maintained by the Department of Lands and Surveys.

The Directorate of Overseas (Geodetic and Topographical) Surveys has mapped Kenya at 1 : 50,000 and 1 : 100,000; these are maintained by the Survey of Kenya, Nairobi, which has also prepared a map at 1 : 250,000 and a 1 : 1M ' Route map of Kenya ', 1965, with the northern part on one side and the southern on the other. A special set of twenty three sheets covering Nairobi and district at 1 : 2,500 was issued in 1961; ' Mombasa Island and environs ' at 1 : 14,000 was completed in 1963, compiled from large municipality plans, air photographs and field surveys; in addition, the Survey is steadily mapping all the smaller towns at 1 : 5,000. *Population of Kenya: density and distribution: a geographical introduction to the Kenya Population Census of 1962,* by W T W Morgan and N M Schaffer, was prepared at the Department of Geography, University College, Nairobi (Nairobi: OUP, 1966) as an aid in assessing the carrying capacity of the land, labour supply,

375

potential markets and the placing of social services; four map sheets of Kenya, 1:1M, and thirty six pages of text, including tables, graphs and charts, are issued in a plastic container. Distribution of population, by superimposed dots and density by locations in colour are shown on a standard base map. The country has been divided into natural regions and a concise description is given of each. The department also compiled an ' Ethnic map of Kenya ', based on the 1962 Census, published in 1965; five groups are distinguished within each province. A map of the Nairobi region, 1:250,000, was also prepared by the department for *Nairobi: city and region,* edited by W T W Morgan (OUP, 1967), which is a collection of essays considering aspects of the city and its region from a series of specialised standpoints, with the map in a folder.

A 1:1M map of Kenya has been compiled by the Survey in two sheets, using five colours and altitude tints, also a 1:3M map in similar style, but using three colours, both of which have been recently made available. A special relief map of Mount Kenya has been prepared at 1:25,000. In 1961, the Survey published an interesting map ' 20% probability map of annual rainfall of east Africa ' at 1:2M. The activities of the Kenya Geological Survey are recorded in the reports and annual reports issued since 1933. The *Catalogue of maps,* published by the Survey of Kenya, gives additional information and prices.

The *National atlas of Kenya: a comprehensive series of new and authentic maps prepared from the National Survey and other governmental sources, with gazetteer and notes on pronunciation and spelling,* was completed in 1959 by the Survey of Kenya. The finely produced plates, showing historical aspects, physical features, natural resources, population, social services, industry, communications, power and urban centres, provides an indispensable guide to the country and its resources. Some early maps of the country are reproduced. Nelson has published a *Kenya junior atlas,* conformal with the rest of this excellent series for individual countries.

In Tanzania, the mapping service is the Survey Division, which is responsible for geodetic, topographic and cadastral surveys. The basic topographic series is the 1:50,000, in which the Survey Division is assisted by the Directorate of Overseas (Geodetic and Topographical) Surveys. A number of sheets of a 1:125,000 series exists, but they are being replaced by a new series at 1:250,000, usually compiled from the 1:50,000; this is a multicolour series, using layer colouring. The whole country is covered by 1:500,000 sheets, now out of print, prepared by the British Army Engineers in the 1940s. Six sheets at 1:1M cover Tanzania, of which the Directorate of Overseas Surveys began a second series in 1959. A one sheet map at 1:2M is regularly revised and is available with or without layer colouring; 1:3M maps are also prepared, using normally as base maps for atlases or for the illustration of reports and other documents. Tanzania is a rapidly developing country, so that a careful revision plan has been undertaken, by which at least 100 of the 1:50,000 series are revised each year. A series of district maps is in preparation for the Ministry of Lands and Surveys; individual districts, mostly at 1:250,000, are prepared on one sheet, showing towns, villages, communications and the major topographical details, printed in three colours, but without contours of layer colouring. All the major towns are covered by plans at 1:2,500, printed in four or five colours, with contours at five feet intervals. Larger-scale plans are available for Dar es Salaam and Tanga. Tanzania is completely covered by air photography at 1:30,000 and 1:40,000, carried out by the Air Survey Division, and an Air Photo Library is maintained.

The Geological Survey of Tanzania, attached to the Mineral Resources Department, maintains a geological series at 1:125,000. J F Harris: *Summary of the geology of Tanganyika* (Dar es Salaam, 1961) should be consulted in conjunction with the maps. Bulletins have been issued since 1927, papers and annual reports since 1928, and pamphlets since 1939.

A special map of Kilimanjaro at 1:100,000 should be mentioned; this was prepared by the Directorate of Overseas Surveys for the United Republic of Tanzania, in 1965, using the 1:50,000 base map, supplemented by aerial photographs. Printing was done by the Ordnance Survey. Relief is shown by contours at 100 feet intervals, together with hill-shading, in clear, striking colours, and spot heights. The settlement pattern is shown and roads are classified. The map is decorated with a cartouche composed of a sketch of Kilimanjaro and with grasses, flora and fauna.

The first edition of the *National atlas of Tanganyika* was published by the Survey Division, Department of Lands and Mines, Dar es Salaam, in 1942, followed by a second edition in 1948. The maps were at a basic scale of 1:4M. In 1956, a third edition was completed, composed of twenty nine pages of maps at scales ranging from 1:2,500 to 1:3M. Physical features, geology, climate, vegetation, sociological factors, details of the economy, communications and plans of major towns are included, also, separately, a 1:2M sheet of the whole country. The *Tanzania junior atlas* is available from Nelson in series with the other atlases specially compiled for individual countries.

Two studies are particularly relevant: D W Malcolm: *Sukumaland: an African people and their country: a study of land use in*

Tanganyika (OUP for the International African Institute, 1953) contains informative maps and appendices on native administration,
surface water catchment works, soils, vegetation and planned group
farming; and J F R Hill: *Tanganyika: a review of its resources and
their development* (Dar es Salaam, Government of Tanganyika,
1955), a massive work, edited by J F Moffett, with maps and excellent documentation.

The mapping of Malawi is being maintained at 1:50,000 and
1:100,000 by the Directorate of Overseas (Geodetic and Topographical) Surveys. Special projects for the development of agriculture, irrigation or swamp reclamation are undertaken for the
government by firms such as Hunting Technical Services, which
recently carried out the Shire Valley Project, involving control of
the Shire river, stabilisation of the water level in Lake Nyasa and
the development of a fishing industry, hydroelectric power, swamp
reclamation and irrigation. The Geological Survey has issued bulletins and annual reports since 1923 and several editions have been
issued of a geological map, 1:4,200,000.

An official *Atlas of Rhodesia and Nyasaland* (now Zambia and
Malawi) was published in nineteen sheets between 1960 and 1962 by
the Federal Department of Trigonometrical and Topographical
Surveys. Emphasis was on the natural farming areas of the Federation, shown on maps at 1:2,500,000. When considering the mapping
of Zambia, Rhodesia, Mozambique and the southern African states,
the vast areas involved must be borne in mind; areas of economic
importance have taken precedence in successive mapping programmes. Frank Debenham: *Nyasaland: the land of the Lake,* in
the Corona Library series (HMSO, 1955), includes maps in the text
and an informative folding map; George Kay: *Changing patterns
of settlement and land use in the eastern province of Northern
Rhodesia* (Hull UP, 1965) is no 2 in the Occasional papers in geography series, excellently documented and including maps; and there
are useful sketch-maps in J G Pike and G T Rimmington: *Malawi:
a geographical study* (OUP, 1965), which includes a reproduction of
a sheet from the Geological Survey of Malawi.

The Directorate of Overseas Surveys maintains maps of Zambia
at 1:50,000; the Ministry of Defence, in London, sponsored a map
at 1:1,500,000 in 1966; and the Department of Geography, University of Hull, in 1964 published a map of the urban population of
Zambia, compiled by George Kay at 1:633,000, based on the 1963
Census of Africans. A vegetation and soil map at 1:1M was published in two sheets by the Directorate of Overseas Surveys, compiled by the members of the Department of Agriculture and the
Forestry Branch, (then) Northern Rhodesia. A *Reference atlas of*

378

Rhodesia, edited by M O Collins, was published by him in Johannesburg, 1965; and in the same year, he compiled twenty one maps, accompanied by twenty two pages of text, *Rhodesia: its natural resources and economic development* (Salisbury, Mardon Printers, 1965). Geological bulletins, short reports and annual reports have been issued by the Geological Survey since 1917, and a 1 : 1M 'Geological map' has been issued in several revised editions. The Federal Department of Information issues from time to time a *Catalogue of official publications,* and the Surveyor-General published a list of maps of the Federation in 1955.

Little is known about the exact state of mapping in Mozambique. A 'Carta de Moçambique' at 1 : 2M has been published by the Instituto Geografico e Cadastral, Lisbon; and an *Atlas de Mozambique* was published by the Empresa Moderna, Lourenço Marques, in 1962. This is a balanced and informative atlas, consisting of ten maps at 1 : 1M and twenty physical, administrative and thematic maps at 1 : 6M. The Serviços de Geologia e Minas, at Lourenço Marques, has issued a *Boletin* since 1937. In 1949, the Serviços de Industria e Geologià completed a geological map at 1 : 2M.

The Directorate of Overseas Surveys maintains a 1 : 50,000 topographical series of Swaziland. Particular attention has been given recently to the geological surveying of the country; the Geological Survey of the Directorate is working on a geological map at 1 : 50,000 and the Geological Survey and Mines Department at Mbabane produced a 'Geological map of Swaziland', 1 : 250,000, in 1967. A map of the land systems and soils of the new Lesotho, at 1 : 250,000, has recently been completed by the ministry of Overseas Development and the Special Commonwealth African Assistance Plan (Department of Overseas Surveys, 1967). The Directorate also mapped Bechuanaland at 1 : 500,000 and 1 : 125,000, and the new Botswana at 1 : 3M in 1967. Basutoland is covered at 1 : 50,000 by Directorate maps. There is also a 'Shell road map of Basutoland', 1 : 1,267,000 by Map Studio Production for Shell, Johannesburg, 1966. In the Corona Library series, *Swaziland,* by Dudley Barker (HMSO, 1965), contains an informative folding map. *Experiment in Swaziland: report of the Swaziland sample survey, 1960* (OUP, 1964) includes maps and a comprehensive analysis of the aims and methods of the survey by a research team from the University of Natal on behalf of the Swaziland Administration. Useful also in this context are G M Stockley: *Report on the geology of Basutoland* (Basutoland Government, 1947) and D M Doveton: *The human geography of Swaziland* (Philip, for the Institute of British Geographers, 1937).

The Trigonometrical Survey Office, founded in 1919, is the

official mapping agency for the Union of South Africa and the South African national committee for representation to the International Cartographic Association and is based at the Science Co-operation Division, Pretoria. The Trigonometrical Survey prepares all official maps except those for geology. Primary, secondary and tertiary orders of triangulation are established over the greater part of the country. Aerial photography now forms the basis of compilation of the larger scale series of topographic maps, from 1:12,000 to 1:70,000. The basic series is 1:50,000. Other series are at 1:250,000, 1:500,000, 1:1M and 1:2,500,000 (second edition 1965-), the state of production being summarised in a *Catalogue of maps* issued every two years by the Government Printer, Pretoria. Most of the maps are printed in English and Afrikaans.

Influential in the development of aerial photographic surveying in southern Africa have been the Field Aircraft Services Africa, based at Germiston, near Johannesburg, and Field Aircraft Services Central Africa of Salisbury. The first photographic survey was undertaken in 1929 by the then Aircraft Operating Company of London; two years later, the forerunner of the present South African Company was formed—the Aircraft Operating Company (Aerial Surveys) of Johannesburg. Continuous changes during the war and the succeeding peace have led to the establishment of the two organisations above mentioned, which are aiding considerably in industrial expansion by the production of accurate and up-to-date survey projects.

The Geological Survey, based at Pretoria, Department of Mines, uses a basic scale of 1:250,000, the sheets of which are issued with explanatory notes. In connection with geological studies, the classic work *The geology of South Africa,* by A L Du Toit (Edinburgh, Oliver and Boyd, third edition, 1954), should be remembered, also his 'Geological map of South Africa, revised by S H Haughton (Oliver and Boyd, 1950). The modern study, *South Africa today: its riches and geography* (Johannesburg, Map Studio Productions, 1966) includes a map at approximately 1:3M, with insets of Africa showing dates of independence, agriculture, forestry and fishing, mineral resources, geology, rainfall and weather areas and the disposition of industry. The Geological Survey, Pretoria, has issued memoirs since 1906 and bulletins since 1934. The *Transactions* and *Proceedings* of the Geological Society of South Africa, 1896- are also invaluable sources of studies and maps. Most governmental departments provide training in cartography. In addition, the Trigonometrical Survey Office functions as a central mapping organisation, undertaking map projects as required. The Survey Office contributed also to the *Climatological atlas for Africa south of the Sahara.*

'Population distribution of the Union of South Africa', a map in two sheets, was published by the Natural Resources Development Council in 1951, with accompanying explanation by Dr J H Moolman (Pretoria, Government Printer, 1954). Under the direction of the planning staff of the council, the distribution of non-Europeans and other nationalities was plotted at 1:250,000. From this map and from the 1951 Census, a further map at 1:1,500,000 was compiled. The studies and large-scale maps involved in this work proved most useful in economic development and planning.

Numerous maps of roads and other communications have been recently produced for the Union by a number of organisations. The Bartholomew reference map series, 'South Africa and Madagascar', at 1:4M, was revised in 1967; roads and railways are shown. The Pretoria Publicity and Travel Department, South African Railways, produced a 'Railway map of South Africa', showing air routes also, at 1:2,534,400, 1954. Michelin, Service de Tourisme, Paris, issued in 1964, 'Central and South Africa and Madagascar', showing main roads, 1:4M, with insets of the Comores Islands, 1:4M, and of the Mauritius and Reunion at 1:2M. The oil companies, however, have published the greater number of road maps in the region. Shell has issued 'South Africa, Shell road map', at 1:2,400,000, with a list of caravan parks on the reverse, 1966; 'Cape Peninsula and Western Cape Province, Shell road map', 1:125,000 and 1:530,000, 1966, with a plan of Cape Town on the reverse; 'South west Africa, Shell road map', 1:2,500,000, 1966; 'Witwatersrand Shell road map', 1:407,000, 1966, with a street map of Johannesburg on the reverse; 'Natal, Shell road map', 1:1M, 1966, with a map of Drakensburg on the reverse, at 1:785,000; all these have been published for Shell by Map Studio Productions, Johannesburg. 'Durban, arterial road map', 1:50,000, has been published by the same firm for Caltex, 1966; also, 'East London/Port Elizabeth, arterial route map' at various scales, 1966-. *Road maps of Southern Africa*, published by BP Southern Africa and Map Studio Productions, 1965, comprises sheet maps, unfolded, in booklet form; they are printed in black, blue, yellow and dark green. The greater part of the region is at 1:2M, with smaller scales for the neighbouring areas of south-west Africa, Zambia, Rhodesia and Malawi, and 1:500,000 for the densely populated area round Cape Town and the coast south of Durban. A 'Total road map of southern Africa', at 1:2,700,000, was issued in its seventh edition in 1967 by Map Studio Productions for Total; on the reverse is a 1:5,500,000 road map of Rhodesia, Malawi, Moçambique and Zambia.

The national *Atlas of the Union of South Africa*, edited by A M and W J Talbot (Pretoria, Government Printer, 1960), presents 'a mid-century survey of natural resources and of changing demographic and economic conditions through the first four decades since Union'. Work began in 1947 by government departments, private organisations and individual experts; sources and dates are quoted and discussed in the accompanying text. The 598 maps, monochrome except for four in colour, are mainly the work of the Trigonometrical Survey Office and were compiled with the information needs of the Social and Planning Council largely in mind. The seven main sections deal with relief, geology and water resources; population, mining, soils, fisheries; climate; agriculture; industries and occupations; transportation and external trade. The exploration of the country is included, also plans of modern cities. Text and captions are in English and Afrikaans. J H Wellington: *Southern Africa: a geographical study* in two volumes (CUP, 1955) forms a valuable introduction to the whole area and includes text maps.

There is as yet no system for recording current maps in detail. *Maps and statistics* began publication in Pretoria in July, 1962.

The most useful existing map of the Malagasy Republic is the ' Carte de Madagascar ', 1 : 500,000, published by the Institut Géographique National, Annexe de Tananarive, 1964. This is a new map in twelve sheets, which replaces the previous 1947 road map in twenty four sheets. The Laborde Conformal Projection has been used and relief is shown by contours at 100 feet intervals and hillshading. Inset maps show magnetic declination, natural regions, ethnology and administrative divisions. The institute issued a ' Carte administrative et de densité de population : situation au 1 Janvier 1965 ', in 1 : 2M, 1965, and Pierre Gourou compiled ' Madagascar : centres de densité et de localisation de la population ' in three sheets, two at 1 : 1M and the third at 1 : 2M (Brussels and Paris, Cemubac and Ormstom, 1967). Twenty eight pages of text, in French, comment on the maps.

The Service Géographique de Madagascar, Tananarive, produced a ' Carte géologique ' in three sheets edited by Henri Besairie, 1 : 1M, in 1965. The *Annales géologiques du Service des Mines de Madagascar* have appeared since 1931; ' Carte géologique de reconnaissance ' was issued in sheets at 1 : 200,000, with explanatory booklets, and a ' Carte géologique et minière ', 1 : 1M, was completed in 1952. Vast numbers of articles and monographs have been sponsored by organisations of many kinds and reports of surveys are legion; the quality of production is normally high, including good illustrations and text maps. The Institute of African Affairs, the School of Oriental and African Studies and publishers such as Methuen and Longmans have been influential in maintaining the high standards of textbooks.

There is no centralised current bibliography in the Malagasy Republic, but in 1954 the Geological Service of Madagascar issued a *Catalogue des archives et périodiques*, followed by annual supplements.

PACIFIC OCEAN AND ISLANDS, AUSTRALASIA

In the Pacific area, the Directorate of Overseas (Geodetic and Topographical) Surveys has for some time been using both ground and air survey to obtain the best possible coverage of the island groups as quickly as possible. Physical conditions hamper survey work in many of these regions, especially in parts like the New Hebrides, where the hazards include heavy rain, hurricanes and volcanic earth tremors. A framework of control has been established in the Solomon Islands on which the local survey department is working. The basic series is at 1:50,000. New Georgia and the Shortlands group have been well mapped at 1:50,000. The first geological map of the area, ' Geological map of the British Solomon Islands protectorate ', at approximately 1:1,030,000, was published by the Directorate in 1963, based on investigations between 1950 and 1962.

In the Fiji Islands, the Directorate is mapping at 1:50,000 and at 1:250,000; a new series of the latter was begun in 1963—a general map with relief shown by contours and layer colouring. A geological map of Fiji, compiled by the Geological Survey at Suvo, was published by the Directorate of Overseas Surveys in 1965. A geological map at 1:50,000 is maintained by the Directorate, also a ' Provisional geological map of Fiji ' at 1:500,000, with an inset of the Lau group at 1:1M. A land use map of Fiji, at 1:250,000, has been compiled by R G Ward, of University College, London, from aerial photography and field reconnaissance, published by the Directorate. A map showing electoral constituencies and polling stations was completed in eight sheets by the Department of Lands, Mines and Surveys at Suva. The delimitation of satisfactory census boundaries for urban areas has proved difficult; *The boundaries of towns and urban areas* in Fiji was prepared for the 1966 Census by G T Bloomfield, in which a number of anomalies were removed. The full text, mimeographed by the University of Auckland in 1967, summarises the bases on which the boundaries were suggested and the limits are described clearly in text and maps; appendices include previous definitions and maps of former town boundaries.

Maps of the French New Hebrides are maintained by the Institut Géographique National at 1:100,000; and an aerial photographic coverage at 1:40,000 was carried out by the institute in 1955 for

383

the entire area of the islands of Tahiti, Mopélia, Maupiti, Bor-Bora, Bellingshaussen, Seilly, Tetiaroa, Tubuai, Manu, Raiatea-Tahaa, Tupai, Mooréa and Huahiné.

The Institute of Pacific Relations, an independent international organisation founded for the study of the social, economic and political relations of the area, frequently sponsors maps in the course of research and publication. The South Pacific Commission, also, is influential in co-ordinating cartographic, as well as other aspects of research in the economic and social development of the non-self-governing territories administered by the six member governments in the South Pacific region—Australia, France, Great Britain, New Zealand, the Netherlands and the United States of America. The Charles Darwin Foundation for the Galapagos Islands includes a sub-section devoted to cartography.

A great deal of high level research goes on in the Pacific area; the results of such research, and the papers given to congresses, for example, those given to the Congress of the Oceanographic Institute of the USSR Academy of Sciences, usually contain original maps, as does the *Encyclopaedia of Papua and New Guinea* (Melbourne UP, 1968, in association with the University of Papua and New Guinea) and other reference works. Maps are also featured in *Pacific viewpoint,* semi-annual from 1960, published by the Geography Department of the Victoria University of Wellington, New Zealand; in the government annual reports on individual territories and in association with the various development plans by such organisations as the Land Development Authority, Natural Resources Office, Suva; the Pacific Science Association; the Polynesian Society; the Société des Etudes Océaniennes and the Australian National Research Council.

The *Atlas van tropisch Nederland,* published by Martinus Nijhoff, The Hague, 1938, has already been mentioned; *Meteorologie Nederlands Nieuw Guinea : a provisional atlas, pending a more detailed survey, with additional maps and diagrams* was published by the Ministerie van Marine, Afdeling Hydrografie, The Hague, 1959, with preface and explanatory text in Dutch and English. Maps of New Guinea at 1:250,000 are maintained by the Division of National Mapping, Canberra; a reconnaissance map at 1:100,000 was prepared by the Department of National Development, Canberra, in 1962. Most of New Guinea is particularly difficult surveying country.

Monographs containing useful maps of parts of the vast area include :

Jacques Barrow, *editor*: *Plants and the migrations of Pacific peoples* (Hawaii, Bishop Museum P, 1963); report on the symposium.

Sir Alan Burns: *Fiji* (HMSO, Corona Library series, 1963).

François Doumenge: *L'homme dans de Pacifique Sud*: *étude géographique* (Paris, Musée de l'Homme, 1966; Société des Océanistes, no 19).

The economic development of the territory of Papua and New Guinea (Baltimore, Johns Hopkins P; OUP, 1965). Report of a Mission organised by the International Bank for Reconstruction and Development.

Brian Essai: *Papua and New Guinea* (OUP, 1961).

H I Hogbin: *Transformation scene*: *the changing culture of a New Guinea village* (Routledge and Kegan Paul, 1951); in the International Library of Sociology and Social Reconstruction.

Gavin Souter: *New Guinea*: *the last unknown* (Angus and Robertson, 1964). Summarises eighty years of exploration and discovery, with maps and a bibliography.

R G Ward: *Land use and population in Fiji*: *a geographical study* (HMSO, 1965, Department of Technical Co-operation, Overseas Research Publication, no 9); maps in a pocket, maps throughout the text and an appendix on field mapping.

H A Wood: *Northern Haiti*: *land use and settlement*: *a geographical investigation of the Département du Nord* (Toronto UP, 1963); includes four field maps.

The maps in the four volumes of the Admiralty geographical handbook series, *Pacific Islands* (Naval Intelligence Division, 1944-45) are still useful; and the mapping work of the Department of Geography, Research School of Pacific Studies of the Australian National UP, especially with regard to climatic conditions, should be mentioned, also *Memoirs* nos. 1 and 2, 1955 and 1958, of the Solomon Islands Geological Exploration and Research expedition 1955-56. Original maps are also to be found in *Australian geographical studies,* published biennially since 1963 by the Institute of Australian Geographers, in the geographical studies of the Royal Geographical Society of Australasia, the Australia and New Zealand Association for the Advancement of Science, the Australian Institute of Agricultural Science, CSIRO, the Commonwealth Bureau of Mineral Resources and the Commonwealth Solar Observatory and many other research institutions and societies.

A descriptive atlas of the Pacific Islands was compiled by T F Kennedy and published by A H and A W Reed; Bailey Brothers and Swinfen, 1966. The atlas, which covers Australia, New Zealand, Polynesia, Melanesia, Micronesia and the Philippines, is printed in black and grey tones. The concise text is particularly strong on the development of population in the area.

A bibliography of Pacific area maps, a report in the international research series of the Institute of Pacific Relations, by C H Mac-Fadden, was published in 1941 by the American Council Institute of Pacific Relations (studies of the Pacific, no 6). The Department of National Development of Australia, Division of Mapping, has issued annually from 1963 a record of the air photography, field survey and mapping activities carried out by the various agencies covering Papua and New Guinea. Publications available in the year of issue are illustrated, previous and forthcoming publications mentioned.

AUSTRALASIA

Australia: The increasing interest in cartography in Australia, until recently almost the worst mapped of continents, was reflected in the holding of the Fifth (1967) United Nations Regional Cartographic Conference for Asia and the Far East in Canberra; forty eight papers were presented by Australian geographers and cartographers. By virtue of the vast area of the continent and its complex terrain, it is not difficult to understand why complete mapping of the country was slow, until aerial photography could be used in cartographic work. Now, the latest electronic measuring equipment and plotting machines are in use. G C Ingleton: *Charting a continent,* published in Sydney in 1944, gives some idea of the initial difficulties.

The Department of National Development, through the National Mapping Council, is responsible for geodetic surveys, topographical surveying and mapping and aerial photogrammetry for the continent as a whole and also for those territories without local mapping branches. There are state mapping agencies in Victoria, Tasmania, New South Wales, South Australia, Queensland and Western Australia; these, in general, confine their activities to the large-scale maps needed for development projects within their own territories. The national cartographic committee for Australia is the Australian Institute of Cartographers, Melbourne.

The Royal Australian Survey Corps, Division of National Planning, having made accurate planimetric maps, continued with standard contoured maps, all being broadly classified as topographical, planimetric and photomaps. Three chief map series are published, for which the grid system and sheet lines are all based on international longitude and latitude degree lines. The 1:63,360 scale 'Photo map' series sheets are prepared from unrectified aerial photography, using the Transverse Mercator Projection. The sheet lines divide Australia into eight longitudinal zones five degrees wide and, in each zone, the sheets are numbered from west to east along the parallel lines. The purpose of these photomaps is to provide a workable base for various types of special topic maps, including soil, timber, geological, geographical and aeronautical maps. A second 'Photomap' series is at 1:253,440; each sheet at this scale covers 1½ degrees by one degree and the sheet line and numbering systems are based on the system used by the International Map of the World 1:1M. The maps are either prepared from unrectified aerial photographs, controlled by slotted template plotting, usually based on astronomical fixations or they may be a collection of annotated rectified or unrectified aerial photographs. Contours are sometimes combined with layer tints and features such as drainage and vegetation, railways, roads and towns are shown as well as relief. The third series, the 'Planimetric' series, at the scale of 1:253,440, is similar to the 'Photomap' series at the same scale, but the sheets are available for certain areas only; they show all types of natural and cultural features, with relief indicated by hill-shading, hachures and spot heights. Since 1960, the Commonwealth agencies have abandoned mapping at 1:253,440 and 1:63,360 in favour of 1:250,000 and 1:50,000.

The series of topographical maps at 1:250,000 will gradually replace the previous four mile maps. These are compiled from aerial photography checked when possible by ground survey. They are drawn on the Transverse Mercator Projection and relief is indicated by contours at 250 feet intervals, or by hill-shading combined with spot heights. The northern and central areas are now mostly covered by these sheets. The Survey Corps is maintaining also a series at 1:100,000, which is still in its first edition.

The 'Australian geographical' series at the scale 1:1M is planned to cover the whole of Australia and New Guinea. These sheets are also used as a base for the World Aeronautical Chart, 1:1M, drawn on the Lambert Conformal Conic Projection. Relief is indicated by contours with hypsometric tints and spot heights, or by hachuring. Five colours are used: blue for drainage, brown for relief, yellow for land areas where relief details are incomplete, red for roads and black for cultural features and names. A geographical map at 1:2,500,000 covering Australia in four sheets was published in 1967 by the Division of National Mapping, Department of National Development.

Geological information is contained chiefly in the publications of the State Geological Survey, the State Royal Societies and the Linnean Society of New South Wales. The Bureau of Mineral Resources, Geology and Geophysics, maintains series at 1:250,000 and 1:63,360, with special sheets at 1:500,000. A new 'Geological map of the Commonwealth of Australia', based on the maps already published by the Geological Surveys of the various states, was completed in 1932, with a *Memoir,* itself illustrated by text maps and diagrams, by Sir T W Edgeworth David (Commonwealth Council for Scientific and Industrial Research, 1932). David's classic work, *Geology of the Commonwealth of Australia,* edited and enlarged by S R Browne in three volumes (Arnold, 1950), should also be consulted. The Resources Information and Development Branch of the Department of National Development produced a special geology map at 1:1M in 1966 in the 'Resources' series, with accompanying text. Various topical and regional maps are being planned in this series, all at 1:1M and all with descriptive text.

Maps frequently portray the results of research carried out by the divisions and sections of the Commonwealth Scientific and Industrial Research Organization. Research on the plant industry is carried out mainly at Canberra, on soils at Adelaide, on forestry products at Melbourne and other centres, on fisheries and oceanography at Cronulla, on dairy products at Melbourne and on wheat farming at Sydney. The 'Land Research' series is particularly important.

The Commonwealth of Australia Bureau of Meteorology, Melbourne, founded in 1907, publishes daily 'Weather charts', also monthly and annual climatic maps of all kinds. The *Australian meteorological magazine* and *Meteorological studies* also contain original maps. The parts of the National Resources Survey compiled by the Town and Country Planning Branch, Ministry of Works, such as 'The West Coast region', 'The Bay of Plenty region' and 'The Northland region' contain excellent coloured folded maps. The Division of National Development has completed a 'Surface water' series at 1:1M, with descriptive text.

Details of the series at various scales, also of geodetic surveys and levelling are shown in *Indexes* issued by the National Mapping Council. A *Map Catalogue* was first published in 1955, to which amendments have been issued from time to time since 1956. The Department of National Resources, Canberra, has published a *Classified index* to resources maps of Australia, in English, published between 1940 and 1959. The index includes maps bound up with or accompanying publications which appeared during this period as well as loose map sheets. Classification is by subject, subdivided by states.

In the Bartholomew 'World reference map' series, Australia is on one spread, at 1:5M, revised in 1966. Contour colouring is used, with roads in red and railways in black. Stanford's 'General map of Australia', also at 1:5M, printed in full colour by George Philip, shows railways, roads and air and shipping routes. The Great Barrier Reef is clearly portrayed and artesian basins are marked. Insets show railway line gauges and the geographical relationship of Australia and New Zealand. A map of the roads of Australia, with insets of the main towns, at various scales, was completed in 1966, in seventy six sheets, by BP Australia, Melbourne. In 1963, the Division of National Mapping, Department of National Development, Canberra, issued a small-scale map of Australia; the browns of the hill-shading give an excellent impression of the relief forms and the river systems in blue show up very clearly. Tasmania is included and, in the north, the southernmost tips of Indonesia and Papua show their relation to Australia.

The *Atlas of Australian resources,* the first modern comprehensive atlas of Australia, was prepared by the Department of Natural Development, Canberra, and edited by Dr Konrad Frenzel (Angus and Robertson, 1952-60). Thirty plates, with accompanying text, were issued in loose sheets, mounted or unmounted, and with a binder or box containers, as required. The base map was on a polyconic projection, at 1:6M. The sheets cover structure, geology, physical features, drainage and climate; land use and agricultural production; transport, all aspects of population, industries and manufactures. A special feature is the chart of 'Major developmental projects'. The maps were clearly printed by offset and careful consideration was given to the range of colours used. Commentary by a specialist accompanies each large map, illustrated by informative diagrams and special maps. In 1961, work began on a second series, to replace the original sheets. Most sheets are revisions, but some deal with entirely new topics. A new geology sheet at 1:6M, for example, was contributed by the Department of Natural Development, 1966. Projections have been changed where necessary and drawing and reproduction techniques have been improved where possible. The basic scale is the same, 1:6M. The reverse of some sheets is used for additional maps and lists of references; a special booklet accompanies each map sheet. Further details may be found in the articles 'Progress on Australia's national atlas' by T W Plumb, in *Cartography,* 5 1964; and 'Planning the second series of *The Atlas of Australian resources*', by T W Plumb and L K Hazlewood, in *The Australian geographer,* March 1965.

On similar principles, a *Regional atlas of New South Wales* is being issued by the State Government.

An *Atlas of Australian soils* is nearing completion by the Melbourne University Press in association with the Commonwealth Scientific and Industrial Research Organization. It has been planned in ten sheets for presentation to the Ninth International Congress of Soil Science at Adelaide, August 1968; the first sheets to appear have been those covering the Port Augusta-Adelaide-Hamilton area, with an explanatory fifty page handbook, 1960; the Melbourne-Tasmania area, 1962; Canberra-Sydney-Bourke-Armidale area, 1966. The handbooks are by K H Northcote. Other special subject atlases of Australia include the *Atlas of Western Australian agriculture,* by J Gentilli (University of Western Australia Text Books Board, 1941); *An Atlas of population and production for New South Wales,* by J Macdonald Holmes (Angus and Robertson, 1931); and the *Atlas of Australian rural industries,* published by the Rural Bank of South Wales in 1948. *A map book of Australia,* by A Ferriday, consisting of twenty three plates, was published by Macmillan in 1966.

' The population structure of Australian cities ' was the subject of a talk by Professor Peter Scott to the Royal Geographical Society on March 8 1965, printed in *The geographical journal,* December 1965. The maps accompanying the talk, one reproduced as a pull-out, the others in the text, are good examples of the specificity achieved in such a detailed and expert survey of a particular topic and the variety of skilled cartographic method used to interpret the statistical information on which the conclusions are based.

The file of *Cartography* is the best source for record and commentary on the progress of mapping in Australia; ' The use of maps in Australia ' was a particularly interesting article by G C Irving, V 1964. P J Smailes wrote on ' The large-scale cadastral map coverage of Australia and the parish maps of New South Wales ' in *The Australian geographer,* September 1966.

Tasmania: The island is usually included in any comprehensive cartographical publication concerning Australia. Dealing only with Tasmania is the *Regional planning atlas of Tasmania,* first published in 1945 by the state government. The maps, prepared by the Mapping Branch of the Department of Lands and Surveys, included physical features, land utilisation, crop and animal distributions, mineral and power resources, forestry, industries and communications. Now in a fourth revised edition 1965, edited by J L Doris, for the Lands and Surveys Department, the topographic maps are at 1 : 500,000 and the thematic maps at smaller scales, drawn on the Transverse Mercator Projection. Relief is shown by hill-shading.

The Royal Society of Tasmania published a fine *Glacial map of Tasmania,* 1 : 250,000, in 1965 (special publication no 2), showing such features as ice flow and erosion.

New Zealand: Early surveys of New Zealand were prepared on a regional basis, usually in connection with land tenure and development projects, and it was not until 1906 that a geodetic triangulation of the whole country was begun, completed in 1946. The Lands and Survey Department is the national cartographic organisation. A topographical series at one inch to the mile was begun in 1939; with the development of aerial surveying, mapping proceeded much more rapidly and the North Island was covered by 1962, followed soon after by the complete series for the South Island. In an article contributed to *The geographical journal* for December 1956, N E Odell describes the aerial surveying of the New Zealand Alps; some magnificent photographs are reproduced and a map.

Topographical map series are now at four scales: 1:25,000, 1:63,360, 1:500,000 and 1:250,000. The 1:500,000 map is drawn on the Lambert Orthomorphic Conical Projection and the others on the National Transverse Mercator Projection. A 'Strip map' series is published at 1:506,880 and this is drawn on the Polyconical Projection. The basic aerial coverage of New Zealand has been made at the scale of four inches to the mile, but some areas have been photographed at three or six inches to the mile.

Each of the 1:25,000 map sheets covers about forty eight square miles. The contour interval is fifty feet and six colours are used, blue for drainage features, browns for relief, green for vegetation, orange for roads, purple for grid lines and black for cultural features and names. Mosaic maps were produced at 1:25,000 and 1:15,680 for important areas, before topographic maps were available. The sheets at 1:63,360 each cover an area of about 435 square miles. The sheet lines of some of the maps in the coastal regions have been modified to include small land areas of adjacent sheets. The contour interval is 100 feet. The colour scheme follows that for the 1:25,000, except that drainage is in black. 161 sheets cover the North Island, 191 sheets the South Island.

Thirty one sheets at the scale of 1:250,000 cover the whole country. In these, the contour interval is five hundred or one thousand feet. Blue is used for hydrographic features, except for the shorelines, which are in black; brown is used for the contour lines, with brown and sepia shading for hypsometric tints; green is used for vegetation and black for cultural features and names. At 1:500,000, seven sheets cover the country, based on the international geographic co-ordinate lines. Contours are at one thousand feet intervals. There are minor differences only in colouring: blue is used for drainage features and names, brown for the contour lines and buff, brown and sepia for the hypsometric tints; roads are red and cultural features black.

Six sheets make up the 'Strip map' series at 1 : 506,880, adjacent sheets overlapping. Contours are at one thousand feet. Blue is again used for drainage, but roads and contour lines are shown in red, purple shading is used for hypsometric tints and the graticules are in green.

From 1949, the Lands and Survey Department *Map catalogue* has been issued in looseleaf form, kept up to date, as necessary. Maps are included in the General Assembly Library's *Copyright publications,* which has been issued annually since 1934 and since 1949 in monthly lists, cumulated annually.

A map at 1 : 4M is issued by the department, with an inset of New Zealand territory. Four editions have been published of the 1 : 2M map of New Zealand, the fourth in 1964 being in two sheets, in two colours, showing the 1964 administrative divisions. Relief is shown by contours at two thousand feet intervals, with layer colouring and spot heights. The 1 : 1M standard map was issued in a second edition in 1963, also in two sheets, for the North and the South Island. The department is now issuing maps of special regions at 1 : 10,000 and 1 : 100,000, also plans of towns, set within their environments, at various scales. A 'Map of Blenheim and environs' at 1 : 10,000, issued in 1964, includes an inset of the town centre at 1 : 5,000, with a list of streets and public buildings. Similarly, in the same year, 'Christchurch and environs' was published, at 1 : 20,000, with the town centre inset at 1 : 10,000. Second editions of the 'Map of Auckland south' and of 'Auckland west' at 1 : 20,000 were issued in 1966. The complete 'Map of Auckland and environs', at 1 : 20,000, went into a fourth edition in 1965, with an index of streets and public buildings and an inset of Auckland City at 1 : 10,000. To the same specifications, the 'Map of Dunedin and environs', at 1 : 20,000, became available in 1965. A particularly interesting and well-designed special map is the 'Map of Bay of Islands', issued by the department in 1965, at 1 : 80,000. Other special maps include a map of the statistical area boundaries of New Zealand, at 1 : 2M, 1966, with an inset showing New Zealand territory at 1 : 42,750,000; New Zealand communications, at 1 : 1M in two sheets, 1963; and a physical map of New Zealand at 1 : 1,637,000, 1966.

The Geological Survey is based on the Department of Scientific and Industrial Research, Wellington. Maps have been prepared at various scales. In 1948, a sixteen miles to one inch map was completed, with an explanatory memoir, *Outline of the geology of New Zealand.* The first geological map at 1 : 2M was completed by G W Grindley and others in 1959. There are also series at 1 : 250,000

392

and at 1:63,360. Climate maps are issued by the Climatological Section, New Zealand Meteorological Service. Weather records and maps have been rationalised and co-ordinated by the DSIR and the Air Department. Dr E Kidson and other geographers have been influential in this, as in other aspects of New Zealand geography. Land utilisation and soil maps are the responsibility of the Soil Bureau, DSIR and the Department of Agriculture. A 'Soil map of the North and South Islands of New Zealand', 1:1M, was published by the DSIR, Wellington, in 1963, compiled from soil surveys by officers of the Soil Bureau.

Hydrographic surveys, as such, are beyond the scope of this work, but the series of charts made by HMNZS Lachlan of the coastal waters of New Zealand during recent years must be mentioned. Dr J W Brodie, director of the New Zealand Oceanographic Institute, initiated the surveys under the general title of 'Coastal chart series 1:200,000 at latitude 41 degrees'. These charts are printed at the Government Printing Office in Wellington and they show by close contouring and colour grading considerable detail of the New Zealand continental shelf. Together with other research emanating from the New Zealand Oceanographic Institute, significant pioneer contributions are being made for use by the various scientists currently interested in the characteristics and potentialities of the sea floor.

'NAC route systems' at a scale of about 1:1,584,000, by the New Zealand National Airways Corporation, 1964, is issued in the form of a poster, based on a map, using the full range of cartographic techniques. As seen from the air, the land forms are portrayed in detail and in natural colours; the air routes of the corporation are shown by fine red lines. The varied landscapes of New Zealand show excellently, from the volcanic peaks to the sandy beaches. Towns are shown by their generalised shapes by black filled in with yellow.

R P Hargreaves has recently compiled two most helpful map bibliographies: *Maps of New Zealand appearing in the British parliamentary papers* (Dunedin, University of Otago P, 1962) and *Maps in New Zealand provincial council papers*, published by the same press in 1966. The latter bibliography lists, with brief annotations giving the essential details, 110 maps printed for the council, 1853-75. Another map bibliography, *Maps of Canterbury and the West Coast: a selected bibliography* was prepared and mimeographed by the Canterbury Branch of the New Zealand Geographical Society, 1958.

A descriptive atlas of New Zealand, edited by A H McLintock (Wellington, Government Printer, 1959-), was the first systematic

13*

atlas of New Zealand and so much in demand was it that a second edition was put in hand immediately and published in 1960. An official publication of high standard, in handy format, it provides a wealth of information about New Zealand—' an analysis and assessment' of the country's resources, on maps of varying scale; the topographical map is at 1:1M, while maps of geology, soils, population and other factors are mainly at 1:3,200,000. Descriptions accompany each map sheet and a comprehensive text, illustrated by diagrams, graphs, sketch maps and photographs, was contributed by members of various government departments. The atlas is now out of print and a new issue is planned. There are excellent articles on the discovery, colonisation and settlement of New Zealand, on the landscape, geology, climate, farming, soils, land classification, resources, trade and commerce; and a full gazetteer has been included.

The first edition of the *Atlas of New Zealand geography* was compiled by G J R Linge and R M Frazer and published by A H and A W Reed of Wellington, in 1964, with a second edition in 1966. Aimed at secondary school level, thirty maps and descriptive text present the physical and economic geography of New Zealand. The second, enlarged edition is still a small atlas, of some sixty four pages, but excellently balanced and packed with information. There are thirty major maps in black and white, with many insets and diagrams; geology, relief, climate, soils, agriculture, population and industry are all treated fully, but with clarity. A series of topographic maps illustrates particular regions and land use and industrial maps of Auckland are most useful. Each map is accompanied by a page of text; short bibliographies are appended for each topic and a bibliography of articles from journals also encourages the scholarly interest of this atlas.

A W Reed, in addition to his contribution to publications concerning New Zealand, is also himself a notable scholar on the subject, especially on the Maori people. *Reed's atlas of New Zealand,* which he compiled (Wellington, A H and A W Reed, 1952), comprises large-scale maps of the regions, followed by distribution maps of vegetation, farming, livestock, etc. Air routes are shown and much statistical information incorporated.

The Oxford atlas for New Zealand (OUP, 1966) is a second edition under the direction of R G Lister. Seventeen maps of New Zealand and seventeen of Australia and the Pacific Ocean are followed by less comprehensive coverage of the rest of the world; the last section comprises world maps on such topics as physical features, rainfall, soils, vegetation and distribution of commodities. Maps of world population, population area and land use, world sea communications and time zones and a world air communications map are most useful in orientating New Zealand with the rest of the world. There is a gazetteer of New Zealand, Australia and the South Pacific Islands and a select gazetteer of the remaining countries.

New Zealand, then, is well mapped, and cartobibliographical work is also making progress; this is in the main due to the active influence of New Zealand geographers and scholars in allied disciplines, notably those working through the New Zealand Geographical Society and the Department of Geography, University of Canterbury, the Royal Society of New Zealand and the Australian and New Zealand Association for the Advancement of Science. Original maps and information about maps frequently appear in the pages of *The New Zealand geographer,* the *Transactions* of the special sections on geology, zoology, botany, etc of the Royal Society of New Zealand and the proceedings of conferences and congresses. Excellent maps demonstrate the themes in B H Farrell: *Power in New Zealand: a geography of energy resources* (Wellington, A H and A W Reed, 1962) and in the article by N E O'Dell: 'Air survey of the New Zealand Alps ', *The geographical journal,* December, 1956.

POLAR REGIONS

ARCTIC: The development of mapping and charting in both polar regions follows the history of successive exploratory expeditions, which mapped the new features as they were discovered. The style of mapping, especially in the Antarctic, depended to a great extent on the nationality of the exploring team. A relevant monograph here is W L G Joerg: *Brief history of polar exploration since the introduction of flying,* produced as special publication no 11 of the American Geographical Society, 1930, to accompany a physical map of the Arctic and a bathymetric map of the Antarctic, in a slip case.

The Arctic Institute in Leningrad was founded in 1920, with the primary object of research on the hydrographical, meteorological and geographical conditions along the northern sea route, or North East Passage. Since 1933, the institute has been the scientific centre for the Chief Administration of the Northern Sea Route, in Moscow, and has worked in close collaboration with the Central Hydrological and Meteorological Offices of the Soviet Union in Moscow, the Department of Navigation and the State Oceanographical Institute. The Central Geophysical Observatory in Leningrad has, through the past thirty or forty years, made comprehensive snow and ice studies; nearly all of these have been published in Russian, with increasingly frequent summaries in English, German or French. An important summary of Russian Arctic expeditions was made by V F Burkhanov, leader of the 1954-55 expedition—' The latest geographical discoveries in the Soviet Arctic '—in his address to the 1955 Congress of the USSR Geographical Society.

The Scott Polar Research Institute and the Norsk Polarinstitutt have also been particularly active since their foundation in 1920 and 1928 respectively; also the Istituto Geografico Polare, 1944, the Arctic Institute of North America, 1945, the Centre d'Etudes Arctiques et Antartiques, Paris, 1958, the DSIR Ross Dependency Research Committee, Wellington, 1958 and the Mawson Institute for Antarctic Research, University of Adelaide, 1961.

The *Oceanographic atlas of the polar seas* was published in two parts by the United States Navy Hydrographic Office: *1* 'Antarctic ', 1957 and *2* ' Arctic ', 1958; topics treated include tides and currents, physical properties of the oceans, ice, wind, marine geology and biology. 'Arctic Island region ', at 1 : 4,500,000, was issued by the Royal Canadian Geographical Society, Ottawa, in 1966. Meteorological maps of the Arctic are frequently published by the Meteorological Office, such as the series ' Ice at the end of May, June, August, September 1965 ', at 1 : 22M, 1965 in four sheets; ' Ice at the end of October, November, December, 1965 and January, 1966 ' at 1 : 22M, 1966; ' Ice at the end of February, March, April, May, June, July, August, 1966 ' at 1 : 13,100,000, 1966, with sea temperatures charted on the verso, 1966, etc.

The latest map of the Arctic Ocean, with adjacent land areas, has been published by The Times Newspapers in association with *The times atlas of the world,* tracing the route currently being followed by the British Trans-Arctic expedition 1968-69 and giving details of the venture. The map is drawn on the Polar Azimuthal Equidistant Projection and printed by Bartholomew at 1 : 10,000. The expedition ' will practise and perfect survival techniques in one of the most hostile environments, and bring back observations covering a wide range of studies '. It is the first expedition to attempt the crossing entirely on foot and it will be a test of Stefansson's theory that food for men and dogs can be found anywhere in the Arctic Ocean area. The map plots the route of the expedition vividly in red, crossing the North Pole and the Lomonsov Ridge; the route of the *Fram* and of Nansen on foot is shown in black and supply drops are shown by broad grey arrows. Six colours are used, the deepest blue being hatched in black to indicate 5,000 metres depth. Land areas are shown in yellow for low-lying areas up to two hundred metres; and a great deal of information is added to the basic features

—the main direction of the drift, the caps and glaciers being superbly drawn, and a full representation of the river system draining into the Arctic.

Maps or information regarding new mapping projects are included in the semi-annual *Polar times,* 1935-, journal of the American Polar Society, New York; the *Polar-forschung, leransgegeben von der deutschen gesellschaft für polarforschung und dem deutschen archiv für polarforschung;* in the *Research papers* of the McGill Sub-Arctic Research Laboratory; *Acta Arctica,* Copenhagen, 1943-; *Bulletin d'information du Centre national de recherches polaires,* Brussels; in the *Arctic circular,* irregularly from 1948, from the Arctic Circle Club, Ottawa; and the *Arctic summary,* of the Department of Transport, Meteorological Branch, Toronto. Monographs such as *Arctic frontiers: United States explorations: the Far North,* by J E Caswell (University of Oklahoma P, 1956) and Pierre George: *Les regions polaires* (Paris, Colin, 1950) include useful maps and an interesting map accompanies the study by L M Gould: *The polar regions in their relation to human affairs* (American Geographical Society of New York, 1958, Bowman Memorial Lectures, series 4). John Grierson: *Challenge to the poles: highlights of Arctic and Antarctic aviation* (Foulis, 1964) should be mentioned, also L P Kirwan: *The white road: a survey of polar exploration* (Hollis and Carter, 1959) as authoritative and interesting introductions to the gradual discovery of the whole area. Most of the works of Vilhjalmur Stefansson contain maps and other evocative illustrations.

The Arctic and Antarctic areas were selected for special study during the International Geophysical Year, 1957-58; the results of the research amount already to thousands of items, including maps, articles, manuscripts, monographs, reports which are collected as a special collection of Arctic and Antarctic research within the National Academy of Sciences. The *Annals* of the International Geophysical Year, 1957- are being published by Pergamon Press, Oxford. More than a thousand maps have been collected in the Arctic Health Research Center, in Anchorage, Alaska, in the broad subject fields of sanitation, epidemiology, physiology, nutrition, animal-borne diseases and entomology in the Arctic regions.

ANTARCTICA: Antarctica has been the last area to be explored in outline. Small-scale maps of the continent, at 1:10,000, are now fairly good, and major discoveries of large-scale topographic features are unlikely. Detailed mapping, however, is far from complete and is limited to areas of specially planned field work. Only a few of the exploratory expeditions can be mentioned here. With Britain and Australia, New Zealand joined in the BANZARE expedition, 1922-

397

31, which charted much of the remaining coastline south of
Australia; after this—the first of New Zealand and Australia expeditions—New Zealand interest, especially in the Ross Dependency
area, was greatly heightened, leading to the establishment of a
permanent New Zealand Antarctic Research Programme. A most
stimulating address on ' New Zealand's field work in the Antarctic ',
given to the Royal Society of Arts on April 4, 1968, by Malcolm
Ford, formerly of the Research Programme, was printed in the
society's *Journal* for September 1968, with maps of surveyed areas.

The Norwegian-British-Swedish Antarctic expedition 1949-52,
was of historic importance, as the first truly international project.
The *General report of the expedition,* by J Giaever and V Schytt
(Oslo University P for the Norsk Polarinstitutt, 1963) is comprehensive, with maps separately in a pocket and John Giaever's *The white
desert: the official account of the Norwegian-British-Swedish Antarctic expedition* (Chatto and Windus, 1954) is illustrated with
photographs and maps; all the reports of expeditions contain maps,
but unfortunately not all are readily available; for example, the
excellent reports of the New Zealand Geological Survey Antarctic
expedition 1957-58 and for 1958-59 are in mimeographed form
and at present therefore available only in the great libraries.

The first edition of the Australian government map of Antarctica
was published by the Property and Survey Branch of the Department of the Interior in 1939 and this remained for some years the
best map of the continent. A revised second edition, 1:10M,
appeared in 1956, from the National Mapping Office, Australian
Department of the Interior, for the Antarctic Division, Australian
Department of External Affairs, incorporating the results of research
from further exploration. A number of map series of Australian
Antarctic territory are now published by the Division of National
Mapping, Department of National Development, Canberra. A series
at 1:100,000 is maintained, based on air photographs and the results
of expeditions between 1954 and 1962; another series at 1:250,000,
was also based on aerial photography and the results of expeditions
between 1956 and 1958—'Antarctica ', at 1:2M, at 71 degrees south,
for example, published in 1965. A single sheet map at 1:20M has
been constructed on the Polar Sterographic Projection.

A series at 1:250,000 is maintained also by the Department of
Lands and Survey, Wellington, New Zealand, and a particularly
useful map of the ' Facilities in the region of Scott Base ', by the
department, at 1:6,250, was reissued in a second edition 1965;
inset is Scott Base at 1:2,000. ' Dronning Maud Land ' at 1:250,000
was prepared by the Norsk Polarinstitutt, Oslo, in 1964, revised
in 1966. Other maps of Antarctica have been published at various

398

scales by the Geographical Survey Institute, Tokyo, particularly at
1 : 5,000 and 1 : 25,000.

Two major and epic projects stimulated scientific work in the
Antarctic, the programme of the International Geophysical Year,
1957-58 and the Commonwealth Trans-Antarctic expedition. The
IGY scientific programme in Antarctica involved thirteen nations
and about 750 individual explorers and research workers in all
branches. Since that period, twelve nations continue to work in the
Antarctic area: Argentina, Australia, Belgium, Chile, France,
Japan, New Zealand, Norway, South Africa, the United Kingdom,
the United States of America and the USSR. Air survey methods
have proved particularly rewarding in this vast region. Research
was also carried out by different countries at this time for the compilation of tectonic and geomorphological maps of the southern
ocean, using echo-sounding and seismo-acoustic apparatus. A useful
summary of Antarctic geography was presented in the *Geophysical
monograph* no 1 of the American Geophysical Union of the National
Academy of Sciences, 1956—*Antarctica in the IGY,* based on a
symposium on the Antarctic, co-sponsored by the United States
national committee for the International Geophysical Year. It is
illustrated by maps, with a 1 : 6M map by the American Geographical Society in a separate pocket. *Antarctic research: a review of
British scientific achievement in Antarctica,* edited by Sir Raymond
Priestley and others (Butterworths, 1964), covers the period 1944
to the end of 1963 and includes maps in a separate folder.
British Antarctic survey is the collective title for the scientific
reports on all branches of British scientific work, except hydrography; the reports are illustrated by plates and folded maps.
The publications, with the general title *Information bulletins,*
of the Soviet Antarctic expedition, bi-monthly since 1964, are invaluable; an English translation is available, well illustrated with
maps and charts. They are issued by Dr M M Somov of the Russian
Arctic and Antarctic Scientific Research Institute, sponsored by the
Geophysical and Polar Research Center of the University of Wisconsin, with support from the National Science Foundation, having
the aim of bringing together all the data obtained from the USSR
IGY expedition, especially in glaciology and meteorology.

The systematic mapping of the continent has proceeded steadily
by the various nations interested. The *Antarctica Reconnaissance*
series, 1 : 250,000, is being published by the United States Geological
Survey, in co-operation with the National Science Foundation; contours are shown and names follow the forms approved by the United
States Board of Geographic Names. The American Geographical
Society in 1965 issued a 1 : 5M map of Antarctica, revised in 1967,

with insets of McMurdo Sound and Victoria Land at 1 : 40M, also one of the Antarctica and adjacent seas at 1 : 40M and 'Antarctica in relation to the world ocean ' at 1 : 100M. Antarctic strip charts are being made at various scales by the United States Naval Oceanographic Office.

The Directorate of Overseas Surveys has produced a general map at 1 : 15M, showing contours and traverse routes, with spot heights indicated. The Directorate's sheets of British Antarctic territory are at 1 : 500,000, 1 : 200,000, of which a new series began in 1951, and at 1 : 25,000. A good general map of British Antarctic territory north of 75 degrees south is available at 1 : 3M. *Antarctica,* edited by Trevor Hatherton (Methuen, 1965), includes a folded map of the Antarctic regions at 1 : 16M, which incorporates the latest research in the physical and biological sciences. Incidentally, the second edition of the *Antarctic pilot* (Admiralty Hydrographic Department, 1948) was the first to apply conformity in the spelling of place names; it is pleasant to note during the past twenty years or so the co-operation between the the Australian Committee on Antarctic Names, the United Kingdom Antarctic Place-names Committee and the United States Board on Geographic Names.

Phillip Law, in an article included in *The geographical journal,* June 1962, describes ' New ANARE landings in Australian Antarctic territory 1960 '; of the coast of Wilkes Land, he says, 'Areas of rock are so rare that each becomes important as a permanent reference point for mapping and scientific work '. A map illustrates the text. In the March 1963 issue, Captain E A McDonald discusses the problems involved in mapping Antarctica. ' Probably no more than 10 per cent of the Antarctic continent has been covered cartographically ', he says, and shows a map of Marie Byrd Land, Ellsworth Land and round by the peninsula of Graham Land (Palmer peninsula) to the coast of the Weddell Sea. In the same issue of the *Journal* is a detailed description of ' Recent exploration of Victoria Land north of Terra Nova Bay ', by H J Harrington, with fine photographs, maps and references. More maps, photographs and relief profiles illustrate the printing of a lecture by D L Linton to a joint meeting of the Geographical Association, the Institute of British Geographers and the Royal Geographical Society on January 2 1962 (*The geographical journal,* September 1963). In the *Journal* for June 1963 are some remarkable articles on Antarctic surveying —D J H Searle: ' The evolution of the map of Alexander and Charcot Islands, Antarctica ', includes photographs and maps in the text, also a fold-out series of maps showing ' The changing map of Alexander and Charcot Islands ', and a particularly fine map at 1 : 1M of ' British Antarctic territory, Alexander Island and Charcot

Island ', compiled by D H Searle in 1959 and 1960 from trimetrogon air photography taken by the Ronne Antarctic research expedition 1947-48 and by the United States Navy Operation Highjump, published by the Royal Geographical Society. The flight lines are indicated on the map. Relief is shown by form lines, definite or indefinite, in blue continuous or dotted lines; areas of exposed rock, crevassed areas, etc are finely drawn and particularly complete details are plotted, based on information obtained by ground survey by many explorers. The article contains references. Also in this issue of the *Journal* are ' The Antarctic landfalls of John Bisere, 1831 ' by John Cumpston, also including maps in the text, references and a note, containing many references to successive expeditions, by Griffith Taylor, ' Probable distintegration of Antarctica '. In the December 1963 issue is the text of W W Herbert's personal experiences as leader of a New Zealand expedition, ' In Amundsen's tracks on the Axel Heiberg glacier ', published with some of the magnificent photographs he showed and with a particularly fine and beautiful pair of maps based on information held by the New Zealand Department of Lands and Surveys, Wellington and compiled by The Royal Geographical Society. The larger map is of the Axel Heiberg glacier region, the other of Queen Maud range, at 1:316,800 and 1:1,267,200 respectively, showing contours in blue and hill-shading in a bluish brown. In the caption, it is stated that the ground survey was based on a field triangulation by P M Otway, of the New Zealand Geological and Survey expedition, 1961-62, controlled by astronomical fixes and adjusted to the US Geological Survey Tellurometer Traverse 1961-62; additional control and the relief shading was added as a result of the observations of Professor Herbert.

The *Antarctic map folio series,* begun by the American Geographical Society in 1964, sponsored by the National Science Foundation, is planned to comprise about twenty folios, each containing several sheets of maps and text on a specialised subject, under the general editorship of V C Bushnell. The aim is to summarise existing knowledge of the Antarctic continent and adjacent waters. Particularly relevant here is folio 3 1965, 'Antarctic maps and surveys 1900-1964, showing map coverage at various scales and aerial photographic surveys 1900-1953 and 1954-1964 ', and with text by G D Whitmore, who has been working on Antarctic mapping since the time of the International Geophysical Year research.

Most ambitious of the Antarctic cartographical works to date is the *Atlas Antarktiki,* a magnificent atlas of 224 plates, recording the results of research during the Soviet Antarctic expedition and incorporating all knowledge of the Antarctic to about 1964. The atlas was sponsored by the Soviet Academy of Sciences (Moscow,

GUGK, 1966-67). Sections deal with historical topics, topography, geophysics, geology, climate, glaciology, oceanography and biogeography. Commentary on the maps comprises the second volume published in 1967.

An *Acquisitions list of Antarctic cartographic materials . . .: a source of information on maps and associated cartographic data of the Antarctic region* was first issued through the Office of the Co-ordinator for Maps, US Department of State, 1961-; the Office continues to prepare the list from contributions sent in by participating organisations and agencies, arrangement being under four headings: gazetteers, geodetic data, maps and charts, and photography. Since 1964, it has formed part of the *Antarctic report*.

The *Antarctic record,* published by the Ministry of Education, Tokyo, and *Antarctic: a news bulletin,* issued by the New Zealand Antarctic Society are important sources of information on cartographic publications; also the new *Antarctic journal of the United States,* bi-monthly from 1966, issued jointly by the Office of Antarctic Programs of the National Science Foundation and the United States Naval Support Force, Antarctica, of the Department of Defense. The ten page report of the Special Committee on Antarctic Research, working group on Geodesy and Cartography, on 'Standard symbols for use on topographic maps of Antarctica' (Canberra, Department of National Development, Division of National Mapping, 1962) is important, as is Hans-Peter Kosack: *Die polarforschung* (Braunschweig, Friedr, Vieweg, 1967), a condensation into one volume of all the essential data relating to both polar regions, from geologic structure to logistic problems, summarised in thirty three sections, each complemented by a set of tables. There are lists of polar organisations, of scientific stations and a chronology of Arctic and Antarctic expeditions; a third sequence presents basic data cartographically, with short bibliographies and a comprehensive index.

Another central document was the special issue no 1, *Proceedings* of the symposium on Pacific-Antarctic Sciences; Japanese Antarctic research expedition *Scientific reports;* papers presented to the Eleventh Pacific Science Congress held at the University of Tokyo, August 1966, edited by Takesi Nagata (Tokyo, Department of Polar Research, National Science Museum 1967); illustrated with maps. Roald Amundsen's classic *The South Pole: an account of the Norwegian Antarctic expedition in the 'Fram' 1910-1912* (in English translation, Murray, 1912) contains maps; also the two works by Frank Debenham *Antarctica: the story of a continent* (Jenkins, 1959) and *British (Terra Nova) Antarctic Expedition 1910-1913: report on the maps and surveys* (Harrison for the Committee of the

Captain Scott Antarctic Fund, 1923), which contains folding maps and a useful chapter on polar survey instruments. Most important also is the work edited by R J Adie: *Antarctic geology: proceedings of the First International Symposium on Antarctic geology*, Cape Town, September 16-21 1963 (1964), sponsored by the Scientific Committee on Antarctic Research (SCAR) of the International Council of Scientific Unions, with the support of the International Union of Geological Sciences. Illustrated by figures and maps, seventy six papers cover the whole continent, together with the sub-Antarctic islands.

L B Quartermain: *South from New Zealand: an introduction to Antarctica* (New Zealand, Antarctic Division, DSIR, 1964) includes informative maps, and further excellent basic information on mapping in these inhospitable regions is included in Phillip Law: 'Antarctic cartography', *Cartography*, March 1957; E A McDonald: ' Charting the coast of Antarctica ', *The geographical journal*, March 1963; M B McHugo: ' Mapping of British Antarctic territory ', *British Antarctic survey bulletin*, no 4 1964; and G D Whitmore and R B Southward, jr: ' Topographic maps in Antarctica by the US Geological Survey ', *The Antarctic journal of the United States*, March-April, 1966.

FALKLAND ISLANDS: The Falkland Islands region includes about two hundred islands, amounting to an area of some ten thousand square miles. Many of the islands are difficult of access and frequently the only means of transport is on horseback. The Falkland Islands Dependencies Survey functioned between 1945 and 1962, after which it became known as the British Antarctic Survey. The Falkland Islands and Dependencies Aerial Survey expedition 1955-57, sponsored by the Colonial Office, was carried out by a civil operating company, Hunting Aerosurveys, and by the end of 1956, photographic coverage of the Falkland Islands was completed. During the next few years, the Directorate of Overseas Surveys surveyors had carried out the trigonometrical control of the colony and in 1961 and 1962 twenty nine sheets appeared, comprising the topographical map series at 1:50,000. A paper on the methods and results was given by the leader of the expedition, P G Mott, to the Royal Geographical Society on October 21 1957, printed, with the discussion, in *The geographical journal*, March 1958. *The Falkland Islands Dependencies Survey: scientific reports* (HMSO, 1953-) include both maps in the text and comment upon cartographic activity. M B McHugo wrote on ' Mapping the Falkland Islands Dependencies and British Antarctic territory, 1956-64 ' in *The Polar record*, 12 1965.

4
THEMATIC
MAPS AND ATLASES

' I am told there are people who do not care for maps, and
I find it hard to believe.' R L STEVENSON.

RECENT YEARS HAVE seen a growing realisation that cartographic
representation of a wide variety of topics stimulates ideas, promotes
fresh deductions by the close comparison of a number of factors and
provides the most satisfactory basis for planning. The potentialities
of special subject mapping are currently among the most pressing
topics under discussion at both national and international level.
Original thematic maps are compiled from primary sources; they
may present the results of research by one scholar, in which case
they may remain hidden in theses or appear in a relevant journal;
or they may be part of a government-sponsored survey or be prepared for a national atlas, congress, conference or symposium. One
map may represent the fruits of years of patient collection of factual
data, or, on the contrary, a series of maps on the same base may
be used to depict immediately some element changing daily or at
some other regular interval of time. Their format varies from line
maps or folding maps in a text to sheet maps or maps reduced from
large-scale originals to atlas or smaller plate size. Much of the work
being done in thematic mapping is still in the experimental stages;
new combinations of colouring and symbols are being tried out in
economic and population mapping. In the mapping of climate and
oceanographic data, the work of numerous scientists is only now
coming into maximum usage through international co-ordination. In
the use and interpretation of all such maps also much remains to be
learned and methods of reproduction are constantly under review, to
avoid as much as possible any risk of misleading collocations.

As thematic maps are not merely descriptive, the responsibility of
the cartographer is correspondingly greater, to ensure that the
correct deductions are made clear. The new technical developments
have been particularly helpful in the preparation of these maps;
greater precision of photographic and copying equipment, the

404

employment of more refined graduated screens, screen foils, etc can produce fine gradation of tonal values and improved methods of colour separation enable much greater detail to be included without loss of clarity. Internationally agreed ranges of colours and types of symbols are gradually facilitating the interpretation of comparable maps. Geological and meteorological maps already conform to a great degree to internationally accepted rules and certain conventions have become universally acceptable in soil, vegetation and population mapping during recent years. Standardisation of road classifications is extending. The international exhibitions of topographic and special subject maps arranged in connection with the Technical Conferences of the International Cartographic Association and the frequency of displays at other widely attended meetings have enabled the mapping styles of other countries to become more familiar and the most effective ideas to be exchanged.

Some general references to monographs and articles on thematic mapping follow before dealing with individual topics in more detail:

Erik Arnberger: *Handbuch der thematischen Kartographie* (Vienna, Deuticke, 1966). Ill, maps, dia, bibliog.

— *Thematische Kartographie: Theorie, Entwicklung, Entwurf und Darstellungsmoeglichkeiten: ein Handbuch in Zusammenfassender Darstellung* (Vienna, Deuticke, 1965). Ill, maps, dia, bibliog.

E D Baldock: 'Considerations in producing special maps ', *The cartographer,* May 1965.

J Barbier: *Thematic cartography: problems particular to illustration,* ICA Technical Symposium, Edinburgh, 1964.

T W Birch: *Maps, topographical and statistical* (Oxford, Clarendon Press, second edition 1964).

H Bosse: *Kartentechnik* (Lahr, Schwarzwald: Astra; Bailey and Swinfen, third edition 1961).

J Dejeumont: 'Improved colour chart applicable to thematic overprints printed by offset ', ICA Technical Symposium, Edinburgh, 1964.

G C Dickinson: *Statistical mapping and the presentation of statistics* (Arnold, 1963). Techniques used in construction and design; types of statistical maps; sources of statistics and an appendix, ' List of Census Reports for Great Britain, 1801-1931 '.

S Gregory: *Statistical methods and the geographer* (Longmans, 1963).

W R Heath: ' Technical problems in thematic mapping ', *Cartographic journal,* December 1964. Methods used in the Department of Geography, University of Washington.

— ' Technical problems of thematic mapping ', ICA Technical Symposium, Edinburgh, 1964.

405

F Joly and S de Brommer: 'Proposal for the standardisation of symbols on thematic maps', ICA Technical Symposium, Edinburgh, 1964.

G Krauss: 'Difficulties of maintaining topographic maps and possibilities of overcoming them', ICA Technical Symposium, Edinburgh, 1964.

J R MacKay: 'Some cartographic problems in the field of special (thematic) maps', *Canadian cartography,* vol 1 1962.

F V Maure: 'The uniformity of the thematic maps of a nation', ICA Technical Symposium, Edinburgh, 1964.

F J Monkhouse and H R Wilkinson: *Maps and diagrams* (Methuen, University paperback, second edition 1964).

H Stump and E Spiess: 'Experience in three-colour printing for coloured areas on thematic maps', ICA Technical Symposium, Edinburgh, 1964.

W R Tobler: 'Automation in the preparation of thematic maps', ICA Technical Symposium, Edinburgh, 1964; *Cartographic journal,* June 1965. Bibliog.

E D C Wiggins et al: 'Specialist maps', *Report of proceedings,* II, Conference of Commonweath Survey Officers, 1963 (1964).

E C Willatts: 'Designing thematic maps', *Penrose annual,* 1964. —'Some principles and problems of preparing thematic maps', *Report of proceedings,* II, Conference of Commonwealth Survey Officers, 1963 (1964).

GEOMORPHOLOGY, CLIMATOLOGY, OCEANOGRAPHY

Sketch-maps in illustration of theories regarding physical geography began to appear throughout the eighteenth century, but the starting-point of this investigation undoubtedly lies in the work of Henri de la Beche. His experiments in mapping in the West Country attracted much interest and the government commissioned from him a stratigraphical map of south-west England. In 1835, de la Beche was awarded a Treasury grant to enlarge the scope of his work and, by 1845, the British Geological Survey was in being as a separate organisation, with de la Beche as the first Director-General, a development of great significance, for it established the science of geology in its own right on a scientific basis in Britain, together with systematic geological mapping. The International Geological Congress in Bologna, 1881, had reached agreement on proposed international standards for the compilation of base geological maps.

Detailed geomorphological maps must be based on field studies, supported where possible by air photographs and plotted on contoured base sheets. Scales must be chosen according to the terrain, but, ordinarily, should not be smaller than 1 : 100,000. In countries such as Britain, in which the state of topographical mapping is advanced, attention can be concentrated on improvements in techniques and in more complete thematic coverage; the British Geomorphological Research Group, for example, has been working on the problems involved in mapping land forms in detail. All large countries in the world today and many of the smaller ones have their own official Geological Surveys. Information is gathered impartially by a Geological Survey for the benefit of the whole nation, for whom it is codified and made available; both government-sponsored and commercial interests make specific surveys in areas containing minerals, oil, etc. The French Geological Survey was in fact the first state geological survey, for which plans were formulated in 1822; the Canadian Geological Survey followed the British, in 1842, and the Geological Survey of India was established in 1846. Some individual states in the United States of America started surveys from the 1820s, but the Federal United States Geological Survey did not begin until 1879. Geological mapping differs obviously from topographical mapping in being concerned with the strata beneath the surface, frequently of the greatest complexity.

Because of the third dimension involved, geological mapping is much slower and there is also the obvious inability to survey all areas in equal detail. On the other hand, changes are not constantly taking place, in normal conditions, as on the surface and revised editions of these maps either incorporate additional information that has come to hand by some kind of excavations or reproduce the original information in improved form. A greater measure of agreement exists in the use of colour or symbols to distinguish different rock strata, fault lines, mineral deposits or outcrops. Vertical sections, core diagrams or block diagrams are usually provided in conjunction with geological maps, as aids to correlation with surface features, and memoirs or other textual matter, usually also illustrated by maps and diagrams, are provided to aid interpretation.

M S Bishop's *Subsurface mapping* is most helpful in this context, also *Problems of geomorphological mapping,* being data from the International Conference of the Sub-commission on Geomorphological Mapping, Poland, May 1962 (Polish Academy of Sciences, in English, 1962).

With the increasing importance of mineral resources throughout the world and the realisation of the value of geological mapping to miners and civil engineers, scientific surveying developed rapidly;

407

information on water supplies and underground conditions rapidly assumed a more vital importance. A special Water Department of the Geological Survey of Great Britain was formed in 1934 and similar developments occurred in many countries faced with increasing size of cities or inadequate rainfall conditions. Holland, to whom such matters are of central significance, has its Government Office for Water Supply and the Federal Government Geological Survey has built up a comprehensive hydrological network of departments in the United States. Geological surveys and maps differ from one country to another. Everywhere, a primary survey is the first stage; in Britain, this was completed at a scale of one inch to the mile, followed by detailed mapping at six inches to the mile. The *memoirs* and *annual reports* of national surveys form the central source of information for every country.

The Overseas Geological Surveys, founded in 1947 by the Colonial Office, became in 1961 part of the Department of Technical Co-operation, now the Ministry of Overseas Development. Their new responsibility is to assess the mineral resources of a country by geological mapping and other scientific techniques, maintaining close liaison with the government Geological Survey concerned. The quarterly bulletin, *Overseas geology and mineral resources,* includes maps and information on the publication of geological and mineral maps.

The *Physikalischer atlas of Heinrich Berghaus* (Justus Perthes, 1845; second edition 1852) is generally thought to be the first thematic world atlas. In any case, it was a unique achievement and of particular importance because of its influence on atlas production and the use of techniques. Heinrich Berghaus was one of the leading German cartographers of the nineteenth century; he worked with Alexander von Humboldt and with Carl Ritter and was for a long time with the Cartographic Institute Justus Perthes at Gotha. Berghaus considered that the ' theme ' of a map and, particularly, of an atlas, was of paramount importance, the topographic basis serving the purpose of orientation. In his atlas, he gave the most careful thought to the choice of symbols; colour gradations were used to show changing relief, the colours being added by hand, as the newly introduced lithographic method was considered too expensive. Maps of ethnographic interest were also included, showing population density, distribution of races and such aspects as education and government, thus revealing an original concept of geography at that time, to which there are signs of a returning trend. A section of maps showing the geographical distribution of diseases was published in 1847. For further details concerning this atlas, reference should be made to the article by Gerhard Engelmann,

'Der Physikalische Atlas des Heinrich Berghaus: die kartograph-ische technik der ältesten thematischen kartensammlung', *International yearbook of cartography,* 1964.

Another physical atlas of about the same date should be mentioned, for the interest of its content and its associations: *The physical atlas: a series of maps and illustrations of the geographical distribution of natural phenomena,* by A K Johnston 'with the co-operation of men eminent in the different departments of science' (Edinburgh and London, Blackwood, 1849), dedicated to and with acknowledgement to Alexander von Humboldt, by whom the work was suggested. It is a large folio production in four parts: geology, hydrography, meteorology, natural history, each part being preceded by explanatory text. The cartographic technique is interesting and there are many finely prepared inset diagrams and figures. The geological section contains maps of the mountain system of Europe, geological structure of the globe, mountain chains of Europe and Asia, of North America and of South America, glacier regions, phenomena of volcanic action, comparative views of geological phenomena and palaeontology of the British Isles. In the hydrographical section are physical charts of the Atlantic, Indian and Pacific oceans, a tidal chart of the British seas and river systems of Europe, Asia and America. The meteorological section includes Humboldt's system of isothermal lines, the geographical distribution of air currents, hyetographic or rain maps of the world and of Europe and the projection of lines of equal polarisation. The contents of the fourth section, now usually constituting completely separate topics of geographical studies, include the geographical distribution of plants and of the cultivation of food plants, the distribution, etc of mammalia, of carnivora, of rodentia and rumminantia, of birds and reptiles and, finally, an ethnographic map of Europe and of Britain. In the second edition, 1856, a world map showing the geographical distribution of health and disease was added, with text by A K Johnston.

If, now, the latest in the long line of physical atlases is mentioned, the changes that have taken place in geographical thinking and method will be obvious. The *Physical-geographical atlas of the world,* published by the Academy of Sciences of the USSR, in conjunction with the Department of Geodesy and Cartography GGK, SSSR, in 1964, is the most detailed atlas of physical geography yet produced. It is an entirely new atlas, compiled by the leading scientists of the USSR, incorporating much new research, especially regarding the atmosphere, hydrosphere and upper layers of the lithosphere and including the latest earth sciences theories. In the section dealing with general geography, some maps from the 1954

Atlas mira have been used in revised forms, but all other maps are original and previously unpublished. They are arranged in three main parts. Seventy maps are devoted to the natural features of the world, including the Arctic and Antarctic; geological structure, tectonics, deposits of useful rocks, earthquakes and volcanoes, quaternary deposits of the Pacific Ocean; sea and ocean beds; geomorphology; climate; hydrology; soil; biogeography; all on relatively small scales, 1:60M to 1:150M. A summary world map at 1:80M, of natural landscape zones, shows the world's major physical zones and sub-divides each into constituent types of natural landscape. The second section contains maps of the individual continents, 1:10M to 1:40M, each series comprising about twenty pages of maps covering land morphology, tectonics, minerals, geology, quaternary deposits, soils, climate, vegetation, fauna; and a final zone map for each continent. An extensive study of the USSR forms the third section, consisting of over eighty maps at 1:15M, 1:20M and 1:35M. Specialised maps are on such topics as climatological indicators, water régimes, hydrochemistry of surface and ground waters, depth of snow cover, and, again, a summary map of natural zones and provinces. Forty pages of explanatory text summarise the basic conclusions derived from the content of the maps and discuss the methods used in data selection and map compilation. The whole atlas notably demonstrates new cartographical techniques. A complete translation of legend matter, contents and explanatory text is presented in a combined two month issue of *Soviet geography: review and translation* (American Geographical Society, May-June, 1965).

Between these two atlases, many others could be mentioned, differing both in date and in standard. The *Palaeogeographical atlas,* compiled by L J Wills (Blackie, 1951), is a small atlas containing twenty two map plates, with text and bibliography; the *Atlas de palaeogéographie,* by H and G Termier (Paris, Masson, 1960), includes thirty six maps and explanatory text. *The atlas and glossary of primary sedimentary structures,* by F J Pettijohn and P E Potter, published in Germany in 1964, has been translated into English, French and Spanish. A superb example is the *Relief form atlas,* published by the Institut Géographique National in 1956; a wide range of typical land forms in French territory is depicted by shaded relief maps, air photographs and anaglyphs. There are now several examples of map sheets or atlases devoted to the close analysis of particular land forms. The *Atlas of landforms,* compiled by Major James L Scovel and others (Wiley, 1966), is a classified collection of material illustrating a great variety of land forms, with examples taken mainly from the United States, with some from other countries. The 155 maps, at various scales, include the collection used in the teaching of land forms at the United States Military Academy. The plates are double spreads, each being devoted to a particular land form type, grouped under broad headings, such as 'Shore lines'. Concise explanatory texts, photographs and diagrams supplement the maps. The index could be improved and the bibliography is rather general. *Lithofacies maps: an atlas of the United States and southern Canada* was the work of graduate students at Northwestern University engaged on regional stratigraphical studies, under the editorship of L L Sloss, E C Dapples and W C Krumbein (Wiley, 1960). Together with the accompanying text, it presents a unique source of information concerning the thickness and lithology of the sedimentary rocks of the United States. The first volume of *Mayer's grosser physischer Weltatlas* to be published was the *Atlas zur Bodenkunde* (Mannheim, Bibliographisches Institut, 1965), to be mentioned below.

The Sub-commission of the International Geographical Union on geomorphological maps, set up in 1962, aims to encourage geomorphological surveying and the preparation of detailed geomorphological maps, to establish a standard basis and uniform principles and to encourage the use of the geomorphological map in economic planning. Scientific papers read to the commission at the Twenty Second International Geological Congress, 1964 (Paris, 1966), consist of sixteen papers, eleven in English and five in French, illustrated with maps. The Commission on Coastal Geomorphology, established under the name Commission on Coastal Sedimentation in 1952 at the International Geographical Congress in Washington, DC, was voted to continue at the next congress in Rio de Janeiro. The purpose of the commission is primarily to act as a centre for the co-ordination of scientific coastal research, through field observation and measurements, aerial photography and mapping. Since the inclusion of cliff shorelines was also intended, the Stockholm congress in 1960 decided that the commission should continue, with the changed name.

An international map of the world at 1:5M has been sponsored by the International Geological Congress. Individual national surveys are undertaking the mapping of relevant areas on behalf of the Commission for the Geological Map of the World; for example, the Canberra Bureau of Mineral Resources has carried out the mapping of Australia and Oceania (1965). Other relevant special commissions of the International Geological Congress are the Commission for the Study of the Earth's Crust; the Commission on the Geological Map of Europe, the Commission on the Geological Map of Africa and the Study of Arid and Sub-Arid Regions.

411

Numerous other organisations throughout the world issue geological and geomorphological maps; the International Association for Quaternary Research, for example, at Copenhagen, the International Association of Scientific Hydrology, Gentbrugge, and the International Association of Sedimentology, based on the Geological Institute at Wageningen. In Britain, the Geological Society of London, founded in 1807, the Geological Societies of Edinburgh and Glasgow, the Geologists' Association and many other local societies have done much to advance the knowledge of geological structure in relation to the general configuration of the country. Their *Proceedings* and *Transactions* and other publications are of central importance. The Institute of Geological Sciences was set up in 1966 by the Natural Environment Research Council; the new body incorporates the Geological Survey of Great Britain, the Museum of Practical Geology and the Overseas Geological Surveys, with a wide term of reference—' to include the range of scientific disciplines which are employed in modern techniques of geological survey and in geological and geophysical research generally '.

The Centre for International Topographical Research is based on the Federal Institute for Topographical Research at Hanover. About four hundred geologists, mineralogists, geophysicists and general scientists are permanently at work abroad, especially in Asia and Africa, where they advise on prospecting for minerals, the construction of railways, dams and reservoirs, the location of artesian wells and the development of all kinds of agriculture. The first task is usually the mapping of the local strata.

The world-wide activities of the International Geophysical Year programme, co-ordinated by the International Union of Geodesy and Geophysics, have given rise to a vast literature, including numerous maps. The *Annals of the IGY* have been published, 1957 —by the Pergamon Press, Oxford. UNESCO's present programme with regard to natural resources research, concerned essentially with the study of the various components of the natural environment capable of providing man with resources, and with the interplay of these components, is having far-reaching results, of which notice is given in *Nature and resources*: *newsletter about scientific research on environment, resources and conservation of nature,* the quarterly bulletin of the International Hydrological Decade, 1965-.

The improvement of techniques used in gathering data is also beginning to be felt in physical mapping. An outstanding example of this is to be seen in the mapping of continental shelves, using data based on sonic soundings. The mapping of ice bound coastlines is difficult, owing to the changing thickness of the ice; two methods of graphic presentation have been chiefly adopted, namely by the use of isopleths showing the mean and extreme limits of the ice, or by means of sector diagrams, indicating accurate conditions at selected points at certain times. Considerable advance is currently being made in the mapping of all aspects of snow and ice formations. A symposium of glacier mapping was held in Ottawa in 1965; the papers were published in a special issue of the *Canadian journal of earth sciences,* edited by H C Gunning (Ottawa National Research Council, 1966), with plates and maps in a separate container. The maps are at varying scales and many of them demonstrate new methods of photogrammetric plotting. The symposium stressed the need for standardisation of symbols to be used in glacier mapping, the preparation of a technical manual on glacier mapping, research into improved methods and techniques and the establishment of scales suitable for different types of study. Large-scale topographic mapping is necessary in any survey of avalanche slopes, ice dams, etc and the mapping must be repeated at frequent intervals to record the behaviour patterns of the formations.

The mapping of snow, ice and permafrost demands special techniques. The Commission of Snow and Ice of the International Association of Scientific Hydrology and the Associate Committee on Geology and Geophysics of the National Research Council in Canada are engaged on such research. *Fluctuations of glaciers, 1959-1965 : a contribution to the International Hydrological Decade,* by Peter Kasser (UNESCO; the International Association of Scientific Hydrology, Belgium, 1967), which includes maps, is a work of fundamental importance. The *Ice atlas of the northern hemisphere,* compiled by the Hydrographic Office, United States Navy (Washington, DC, 1946), gave a highly accurate and comprehensive picture of ice conditions in the northern hemisphere at that time, both at sea and in the rivers: it is now being revised. *The southern hemisphere glacier atlas* (United States Army Earth Sciences Laboratory, Natick, Mass, 1967) is not really an atlas, for only fifteen pages of maps appear in more than three hundred pages of text by John H Mercer of the American Geographical Society of New York. Six regions are treated: the Andes, New Guinea, East Africa, sub-Antarctic islands, New Zealand and Antarctica. Regional accounts are accompanied by detailed bibliographies and lists of photographs and map sources.

In addition to their use in pure research, geological maps are used by civil engineers and industrialists in the planning of water supply for settlements and other practical problems. In Britain, the Ordnance Survey base maps are normally used. In many areas of the world the control of water supply is becoming an increasing problem and special organisations have been set up to deal with it,

413

one of the most recent being a new department of the American Geophysical Union, established to co-ordinate activities in water science; their work is reported in *Water resources research,* quarterly from 1965. A special case in water research has been the coastal and continental shelf surveying carried out in connection with the massive reclamation programmes in the Netherlands. The thematic map, the 'Waterstaatkaart', was begun in 1863 by the Royal Dutch Institution of Civil Engineers, with the co-operation of the hundreds of governing institutions, the Polderauthorities, and the first two sheets were published in 1865. It now covers the whole of the territory involved at the scale 1:50,000 in sixty two sheets, showing the heights of water level, the courses of water in the discharging systems, the pumping stations and dykes. By the nature of the terrain, this map is probably unique; it serves also as a base for a multicoloured overprint. Each colour represents polders connected in one discharging system, but different shades show specific water levels. The pumping stations and other parts of the control systems are shown by red symbols.

A vast periodical literature exists concerning the earth sciences; original maps and charts are frequently included in the issues of the *Proceedings* of the earth science section of the Academy of Sciences of the USSR; the geological series of *Izvestiya; Acta geologica* of the Hungarian Academy of Sciences; *Acta geologica Polonica; Erde;* the *Bulletin* of the American Association of Petroleum Geologists; *Zentralblatt für geologie und paläontologie; Geologisches jahrbuch; Earth science* from the Midwest Federation of Mineralogical Societies, Chicago; the *Bulletin* of the Société Géologique de France; *Revue de géologie et de géographie* from the Rumanian Academy; the *Annales de géophysique* of the Centre National de la Recherche Scientifique and many more. *A select list of British scientific periodicals*: *geology* was issued by the Ray Society, with the British Council, National Book League and the Nuffield Foundation, in 1961.

Relevant yearbooks and handbooks exist in many countries. Germany has produced several such source books of which one unique example is the *Berge der welt,* now in its sixteenth issue, 1966-67 (Munich, Nymphenburger Verlagshandlung, 1967; London, Allen and Unwin), a beautifully produced record of mountain and mountaineering containing fine pictures and maps. New physical maps and revisions are noticed in the relevant sections of the general map bibliographies, in current lists of additions to the specialist libraries, such as the *Monthly list of additions* to the Geological Survey Library, London, and in the periodical literature. *Reports,* such as those of the Institute of Geological Sciences, under the Natural

Environment Research Council, mentioned above, are naturally of central importance.

A definitive work by Professor Eduard Imhof, *Kartographische gelandedarstellung*, was published by Walter de Gruyter of Berlin in 1965, the latest in a long tradition of German studies in this subject. Although the text is in German only, the finely produced plates and illustrations are readily understandable. A thorough examination is made of base materials, methods of survey, height measurement and contouring, landscape sketching, photogrammetry and air-photo interpretation; colour problems are analysed, contours, hachures, shading and layer tinting, spot heights and depths; three-dimensional graphic representations, with clear explanations of practical methods of draughtsmanship; fine rock drawing and the presentation of special features such as potholes, with examples mainly drawn from central Europe. The bibliography includes more than three hundred items. Another standard production is *Intérêt des stéréogrammes aériens dans l'enseignement de la géographie physique*, compiled by M Horlaville, published by the Institut Géographique National in 1951 for the International Geographical Union Commission ' pour l'utilisation des photographies aériennes dans les études géographiques'; fifteen plates show types of relief features, with introduction and notes. The following monographs are also particularly relevant:

G M Bennison: *Introduction to geological structures and maps* (Arnold, 1964).

M S Bishop: *Subsurface mapping* (New York, Wiley, 1960). Maps, dia, bibliog. The use of maps particularly in assisting to solve oilfield problems.

W A Black: ' Cartographic techniques in mapping sea ice ', *The cartographer*, May 1965.

F G H Blyth: *Geological maps and their interpretation* (Arnold, 1965). Notes on the back of each map describe the main features shown.

C C Carter: *Landforms and life: short studies on topographical maps* (Christophers, 1959, revised by M O Walter).

R M Chalmers: *Geological maps: the determination of structural detail* (OUP, 1926).

K J Ewing and M G Marens: ' Cartographic representation and symbolisation in glacier mapping', *Canadian journal of earth sciences*, November 1966.

E H Hammond: 'Analysis of properties in land form geography: an application to broad scale land form mapping ', *Annals of the Association of American Geographers*, March 1964.

Louis Herbert: *Geomorphologische Studien*: *Machatschek-Festschrift* (Gotha, Haack, 1957).

F E Koltlowski: *Measuring stratigraphic sections* (Holt, Rinehart and Winston, 1965, Geologic field techniques series). Ill, facsims, dia, bibliog.

E Lehmann: 'Physical atlas maps', *International yearbook of cartography*, IV 1964.

A I Levorsen: *Paleogeologic maps* (San Francisco, Freeman, 1960). World examples, especially from the United States and Canada.

Peveril Meigs: *Geography of coastal deserts*: *distinctiveness, limits, classification and potential uses of coastal deserts* (UNESCO, 1966). Ill by fourteen maps.

R L Moravetz: 'Cartographic preparation and printing of geologic and hydrologic maps', *Report of proceedings*, Conference of Commonwealth Survey Officers, 1963 (1964).

F J Pettijohn and P E Potter: *Atlas and glossary of primary sedimentary structures* (Springer Verlag, 1964).

W E Powers and C F Kohn: *Aerial photo-interpretation of landforms and rural cultural features in glaciated and coastal regions* (Northwestern University, Department of Geography, 1959; Northwestern University studies in geography).

C Rathjens: *Geomorphologie für Kartographen und Vermessungingenieure* (Lahr, Schwarzwald: Astra, 1958).

M M Ridd: 'The proportional relief landform map', *Annals of the Association of American Geographers*, December 1963.

J-P Rothé: 'Tableau de la séismicité du globe pendant l'année 1965; chronique séismologique (Geneva: *Revue*, Union International Secours, 1966, 5 3-61).

R A G Savigear: 'A technique of morphological mapping', *Annals of the Association of American Geographers*, 55 1965.

A I Spiridonou: *Geomorphologische Kartographie* (Berlin, Verlag d Wissenschaften, 1956).

J Tricart: *Principles et méthodes de la géomorphologie* (Paris, Masson, 1965).

O Wagenbreth: *Geologisches Kartenlesen und Profilzeichnen* (Leipzig, Teubner, 1958).

Werner Witt: *Thematische kartographie*: *methoden und probleme, tendenzen und aufgaben* (Hanover, Gebrüder Jäneche Verlag, 1967).

MAPPING CLIMATE AND WEATHER FEATURES

The earth's atmosphere being the one feature influencing the entire earth's surface, first individually, then in co-operation, the nations of the world have appreciated the importance of studying climatic conditions and applying the knowledge thus gained to the various activities of man in all parts of the world. Climatological maps represent primary data, which may not become meaningful until they are charted. Scientific accuracy can be acquired in these studies, however, only by the interpretation of quantitative observations. Köppen, Thornthwaite, de Martonne and Miller have made outstanding contributions to knowledge of the major climatic regions of the world; much detail still remains to be filled in, particularly about conditions in the Arctic, the Antarctic and the lesser-known areas of South America, Asia and the Pacific. Greater attention now is being given to micro-climatology, the microclimates of plants and soils and local variations in climate, due to a variety of factors, which frequently affect very considerably the livelihood and health in both urban and rural environments.

Following a series of conferences at Leipzig, Vienna, Utrecht and London, in 1878 at Utrecht, the International Meteorological Organisation was founded. Air travel increased the need for improved weather services and technical developments, such as universal radio communications, made possible a complex network for the co-ordination and dissemination of information on world weather conditions. In 1947, in Washington, a World Meteorological Convention approved the merging of the International Meteorological Organisation into a new World Meteorological Organisation, which came into effect in December, 1951, as a Specialised Agency of the United Nations, created ' to co-ordinate, standardise and improve the services rendered by Meteorology throughout the world to various human activities . . .' This is not the occasion for a full account of the work of the WMO; suffice it to say that six regional associations, one for each continent, and eight technical commissions have been established for agricultural, maritime, aeronautical, hydrological and synoptic meteorology, aerology, climatology and instruments and methods of observation. Working groups and panels of experts carry out detailed technical work, co-ordinated by the Secretariat. The central document, among many publications, is the series known as publication no 9 *Weather reports—stations, codes, transmissions*—a vast handbook representing the work of some 8,500 observing stations, all data being transmitted by radio or teletype machines in a universally understood code. The *International cloud atlas,* issued by WMO, is a beautiful and vitally informative collection of 224 illustrations of cloud formations, classified by genera and species.

One of the most far-reaching developments of recent years was the launching of the first artificial satellite designed solely to obtain meteorological data. Tiros I, in 1960, proved that the new information obtained in this way could contribute to the detection of storm centres and the associated cloud systems; Tiros II, launched late in 1961, enabled regular transmission over radio facsimile circuits of weather charts based on meteorological data from this satellite. The Americans put up satellites of two kinds, the Command and Data Acquisition System, which records pictures from around the world on tape for subsequent transmission to special ground stations in Alaska and Virginia; the other, the Automatic Picture Transmission System, which sends out pictures continuously for reception by any country which has set up receiving stations, one of the first of which was at Bracknell in Berkshire, where the British Meteorological Office has its headquarters. In its *First report* to the United Nations, June 8 1962, the World Meteorological Organisation recommended the creation of a World Weather Watch, combining satellite weather information with an expanded network of conventional observations, an international undertaking aimed at improving weather services throughout the world, having three major world centres at Washington, Moscow and Melbourne. In some instances, two or more nations combine for a particular task; for example, the United States and some western European countries share the maintenance of the ocean vessels that gather meteorological data for the North Atlantic.

The eleventh Tiros satellite became operational in February 1966 and the United States Tiros Operational Satellite System was then co-ordinated to make a further major advance in global weather observation; it provides, for example, information on cloud cover over the entire earth every day, automatic picture transmission equipment permitting direct readout of cloud photographs at local stations as the satellites pass over.

The improvement in computers has added the final development in the processing of weather data. In Britain, the daily forecasts are based on predictions made by the Comet computer at Bracknell. Even more powerful machines are required to make the fullest use of the volume of complex data that comprises world weather. To compile weather forecasts and maps, then, there now exist three main systems, the Global Observing System, which collects meteorological information from all around the planet, the Global Processing System, which collates this information in a form in which it can be used and the Global Telecommunication System, which enables instantaneous pooling of readings and forecasts. Increasingly, therefore, accurate daily weather maps and charts are being compiled on the pattern of those existing in Britain, the United States and many other countries. A useful article on this topic was contributed by J

Brian Bird and A Morrison, ' Space photography and its geographical applications ', to *The geographical review,* October 1964.

An independent body of research resulted from the meteorological programme of the International Geophysical Year, outlined in *International Geophysical Year: meteorological programme, general survey,* published by WMO in 1957. In the *Annals of the IGY* (Oxford, Pergamon Press), volume XXXII is devoted to meteorology and many other volumes are concerned with specific aspects; maps and charts naturally form an integral part. Numerous individual research projects are constantly being undertaken on special aspects of the subject. Meteorologists from fifty one countries, for example, met in Wiesbaden in March, 1966, at which the new form of numerical weather forecasting was one of the many important topics under discussion.

The German Weather Service Centre at Offenbach-am-Main is one of the most important information centres of its kind in the northern hemisphere; and the Ionosphere Institute near Breisach-am-Rhein was the first research institute in Europe to receive consistent data about conditions in the ionosphere.

In reading climatic maps, the common practice of reduction to sea level of pressure and temperature figures must be allowed for. If insufficient readings can be made to provide a reasonably good picture, spot readings should be indicated and, in any case, the fullest documentation should accompany all such maps, to avoid any misinterpretation. Frequently, additional charts, graphs and/or statistics are printed at the side of or accompanying the maps. Series of comparable climatic maps prepared over a long period are of the greatest usefulness in determining trends and revealing the interrelationship of various factors. The Meteorological Office has been experimenting with automatic recording stations, so that any operator's errors can be eliminated; the standard weather maps of the United States Weather Bureau are prepared by automatic methods. The Jodrell Bank radio telescope is also being used to compile radio maps. The second edition of *Meteorology for glider pilots,* by C E Wallington (Murray, 1966), includes a chapter on the construction of weather maps from forecasts. Computer-drawn charts are already being made and the next innovation to be expected is the construction of animated charts.

Numerous publications of the Meteorological Office explain and comment upon the various series of maps and charts; full details of these publications are given in the *HMSO Sectional List* of government publications, no 37. The *Weather map* itself is an introduction to weather forecasting. Such maps are drawn several times a day, based on observations of pressure, temperature, wind, rain and other elements near the ground or in the upper air. The complete METMAP form includes sections for taking down the general synopsis, coastal weather reports, sea area forecasts and gale warnings. A list of useful abbreviations and geostrophic scales for gauging wind forces and moving fronts are also included. The actual METMAP form is printed in green at a scale of 1 : 10M. Maps accompany almost all meteorological publications; *Weather in Home Fleet waters*, for example, in two parts; *The northern seas and weather in the Mediterranean*, 1964, and the companion volume, *Weather in the Black Sea*.

It would really be invidious to mention individual climatic maps, so many have there been in recent years, published by almost every nation. A map that made an outstanding contribution to rainfall studies, however, was 'La carte des precipitations de l'Afrique du Nord', compiled by Henri Gaussen and A Vernet at 1 : 500,000. 'The world maps of climatology', in five sheets at 1 : 45M must also be mentioned, prepared by H E Landsberg and others in 1963 (second edition, Springer Verlag 1965). H E Landsberg is also the editor of a forthcoming world survey of climatology in twelve volumes which will summarise current knowledge both in a causal and geographical context. Five maps on an equal-area projection, at 1 : 45M, include 'Mean January sunshine' and 'Mean July sunshine', both by Landsberg; 'Total hours of sunshine', 'Generalised isolines of global radiation' and a map of the earth's seasonal climates by Professor C Troll are among the first to be completed. The maps are produced by the Geomedical Research Unit of the Heidelberg Academy of Sciences and are in series with those in the *World atlas of epidemic diseases*. Explanatory notes are in German and English.

One of the outstanding pioneer attempts in systematic climatological mapping was the *Atlas of meteorology: a series of over four hundred maps . . .*, compiled by J G Bartholomew and A J Herbertson, edited by A Buchan and issued under the patronage of the Royal Geographical Society, 1899, as volume III of the complete physical atlas planned by Bartholomew.

A modern landmark in the approach to climatic studies was the symposium in Rome organised by UNESCO and the WMO on 'Changes in climate', of which the *Proceedings*, published in English and French, 1963, include papers on the changes during the period of meteorological records; changes during the late geological and early historical record; theories of changes of climate; the significance of changes of climate; and conclusions. Another example of the major works which result from the growing number of scientific congresses and symposia was *World climate from 8000 to 0 BC:*

Proceedings of the international symposium held at Imperial College, London, April 18-19 1966, prepared under the direction of the council of the Royal Meteorological Society, edited by H S Sawyer (the Society, 1966), containing illustrations, maps, diagrams and a bibliography.

The *Klimadiagramm weltatlas,* compiled by Heinrich Walter, Helmut Lieth and Elizabeth Harnickell (VEB Gustav Fischer Verlag, in three parts, 1960-67), aims to present climatic regions as accurately as possible, both as a factual analysis, making use of the mass of climatic data now available throughout the world, and as a basis for further studies on vegetation, soils and agriculture. Each diagrammatic map represents the data compiled for an individual station, stating the altitude of the station and the period of time during which data have been accurately noted. The authors have distinguished ten broad climatic regions, with relevant sub-divisions, which, in their judgement, are an improvement upon the scheme devised by V P Köppen. In any case, the atlas is valuable for its careful plotting of comparable data throughout the world. The introductory text is in German, English, French, Russian and Spanish; there are glossaries, explanatory notes and a sound bibliography.

Die Weltmeere: taschenatlas mit den wichtigsten tatsachen aus meteorologie und nautik (fifteenth edition, Gotha: Haack 1956) contains twenty three maps of the seas and oceans and twenty port plans. Text covers all aspects of weather, oceanography, navigation and navigational equipment, and shipping, showing the close relationship between climatic and oceanographic features. The United States Navy *Marine climatic atlas of the world,* in four volumes; *1* North Atlantic Ocean, 1955; *2* North Pacific Ocean, 1956; *3* Indian Ocean, 1957; *4* Atlantic Ocean, 1958, also emphasises this conjuction; and thirdly, the *Climatological and oceanographic atlas for mariners* (Washington, DC, Weather Bureau, 1959)—in two volumes). A mass of data is presented for the North Atlantic and the Pacific oceans; coverage includes ice features, tides, cloud cover, winds, precipitation, visibility and storms. A *Cloud and weather atlas,* by Hugh Duncan Grant (New York, Coward McCann, 1944; Harrap, 1946), contained a fine collection of nearly five hundred photographs illustrating clouds of all types, as well as rainbows, tornadoes, water spouts, dust storms and other phenomena. Many departments and organisations contributed to the work, which includes also an explanatory text. A very specialised *Atlas of 500 mb wind characteristics for the northern hemisphere,* containing eighty four maps, was produced by the University of Wisconsin in 1958.

Most of the specialist periodicals contain maps and charts; among these are *Météorologie: revue de météorologie et de physique du globe,* 1826-, of the Société Météorologique de France; the *Meteorologische zeitschrift,* 1884-; *Local climatological data,* 1910-; *Papers in meteorology and geophysics,* published 1950-, in English, by the Meteorological Research Institute, Tokyo; the *Meteorologische rundschau,* of the Deutsche Meteorologische Gesellschaft, 1947- the looseleaf *International journal of bioclimatology and biometeorology,* in English or German, from the International Society of Bioclimatology and Biometeorology, from 1957; and the *Miscellaneous reports* series of the Chicago University Institute of Meteorology and the *Climatological research* series issued by the Geography Department of McGill University.

Particularly relevant examples from the vast literature on all aspects of the subject include:

Joachim Blüthgen: *Allgemeine Klimageographie* (Berlin, Walther de Gruyter, 1964).

R A Craig: *The upper atmosphere: meteorology and physics* (Academic Press, 1965).

E S Gates: *Meteorology, and climatology, for sixth forms and beyond* (Harrap, third revised and enlarged edition 1965).

Bernard Haurwitz and J M Austin: *Climatology* (McGraw-Hill, 1944). With maps in a pocket.

W G Kendraw: *The climates of the continents* (Oxford, Clarendon Press, fifth edition 1960).

W G Kendrew: *Climatology, treated mainly in relation to distribution in time and place* (Oxford, Clarendon Press, second edition 1957).

C E Koeppe and G C De Long: *Weather and climate* (McGraw-Hill, 1958).

H H Lamb: *The changing climate: selected papers* (Methuen, 1966).

Sverre Petterssen: *Introduction to meteorology* (McGraw-Hill, second edition 1958).

Sverre Petterssen: *Weather analysis and forecasting* in two volumes (McGraw-Hill, second edition 1956). With additional charts in a pocket.

W D Sellers: *Physical climatology* (University of Chicago P, 1965).

G T Trewartha: *The earth's problem climates* (University of Wisconsin P; Methuen, 1961, second printing 1962).

G T Trewartha: *An introduction to climate* (McGraw-Hill, third edition 1954).

The modern period of oceanographical research began with the work of Matthew Fontaine Maury (1806-73); during previous centuries, there had been no want of interest by individual scientists, but the necessary equipment had been lacking. Maury helped to prepare the way for the intensive study of physical oceanography and, during his lifetime, charts of a fair accuracy became available for the major currents, prevailing winds and storm frequencies. In *Matthew Fontaine Maury, scientist of the sea* (Rutgers UP, 1963), F L Williams makes a close analysis of this work. Not only was Maury's work fundamental to ocean research, but it inspired one of the first international scientific conferences, the Brussels conference of 1853, which considered the establishment of a system of marine weather observation which might be operative throughout the world. Meanwhile, the Scottish universities were producing a generation of students studying marine biology, foremost among whom was Edward Forbes. Contemporary with him were Michael Sars and Henri Milne-Edwards, and by their inspiration and research, they together laid the foundations of modern oceanography.

A student of Professor Forbes, Charles Wyville Thomson, later Sir Wyville, continued his pioneer work. During the years 1872 to 1876, he led the famous expedition in HMS *Challenger;* the *Challenger reports* and charts, in fifty volumes, became the nucleus of a rapidly accumulating body of material gathered by expeditions inspired by the *Challenger.* Institutions began work in many countries; Alexander Agassiz charted more than one hundred thousand miles of the Caribbean, the Indian Ocean and the tropical Pacific and made some studies of the Gulf Stream. Prince Albert of Monaco carried out numerous research expeditions into the water conditions and submarine topography of the Atlantic and the Mediterranean. The first international conference for the discussion of the exploration of the sea was held in 1899, with a second following in 1901. In 1949-50, negotiations were completed for a Royal Charter for the National Oceanographic Council, so that the United Kingdom, in co-operation with the Commonwealth countries, should play its part in marine science research. International co-operation could now come fully into operation, which, together with technical improvements in the precision echo-sounder, deep sea photography, the technique of using horizontal sound beams for studying shelf geology and the use of low frequency sound sources for the study of the sub-bottom echoes, enabled more far-reaching, rapid and accurate plotting than ever before. The various oceanographic projects which formed part of the International Geophysical Year researches also resulted in more definitive knowledge of selected areas.

The present state of oceanographic mapping is that some areas, such as the English Channel, have been extensively surveyed by soundings; other larger areas of ocean require very much more research, but, as equipment improves and expeditions are mounted, piece by piece the map is being filled in. The International Indian Ocean expedition during the past few years has added greatly to knowledge of the relief of the ocean floor. It has been exciting also to witness the gradual uncovering of the vast mountain ridge running through the north and south Atlantic oceans; the *Meteor atlas* published the results of the lines of stations made by the *Meteor* in the Atlantic between 1925 and 1927 and research has been as continuous as possible since then. The *Füglister atlas—Atlantic Ocean atlas of temperature and salinity profiles and data from the International Geophysical Year of 1957-1958,* by F C Füglister (Woods Hole Oceanographic Institution atlas series, volume 1) incorporates the work of the Woods Hole Oceanographic Institution and of the British *Discovery II* of the British National Institute of Oceanography. Many unpublished contour charts exist of parts of the Pacific and Atlantic oceans, used mainly in research on other related studies.

This outline of the gradual creation of the conditions necessary for international co-operation in charting and mapping oceanic conditions has been necessarily brief; mention should be made, however, of the formation of the National Oceanographic Data Center in Washington DC, supported by the Bureau of Commercial Fisheries, the National Science Foundation, the Coast and Geodetic Survey, the Department of the Navy, the Atomic Energy Commission, the United States Weather Bureau, the Coast Guard, the Geological Survey and the Department of Health, Education and Welfare of the United States. Oceanographers have made great progress in the use of computers for the analysis of data; the NODC has been accumulating data at the rate of ten thousand readings a month from bathythermograph observations taken by monitoring stations and research vessels throughout the world. The information on the data cards was transferred to magnetic tapes, the typed instructions including the exact locality and the degree of enlargement. M Piver, E Fredkin and H Stommel produced a computer compiled oceanographic atlas in 1963 as a working tool and basis for further experiment.

One of the major Russian post-war atlases has been the *Morskoy atlas,* edited by I S Isakou in three volumes, 1950-58. The Soviet Admiralty and Naval General Staff are among the major contributors, also the Administration of the North Sea route and the Soviet Arctic Institute. Credits are frequently given to British Admiralty charts and to those of the United States Hydrographic Survey. The paper was specially made for the purpose and the printing is of exquisite quality, by the Naval Cartographic and Hydrographic Staff and the Chief Administration for Geodesy and Cartography. General maps of the oceans are at 1 : 50M, with smaller areas at from 1 : 2M to 1 : 100,000. Representation of coast and shore features is particularly successful. The North Sea route is shown in detail. The work, which was begun in 1923, was conceived as a fundamental manual on the geography of the seas and oceans for both scientists and seamen and it is unique in its attempt to portray the relationships between the elements of the lithosphere, the hydrosphere and the biosphere. The cartography is quite beautiful. The first volume contains maps of the oceans, seas, bays and major ports, excluding Russian ports, at scales of 1 : 3M to 1 : 30M; a detailed picture of the relief of the sea bed is shown by bathymetric contours and land relief is indicated by hypsometric tints. The second volume deals with historical data and oceanography. The history of geographical discoveries and exploration of the oceans of the world is traced; detailed information is given on the sea and ocean beds, data for temperature, density and salinity, currents and tides and the vegetable and animal life in the seas and oceans; also all aspects of climate and a comprehensive section on terrestrial magnetism, cartography and astronomy. Volume III maps the history of naval warfare.

The Institut Géographique National, Paris, is preparing a ' Carte générale bathymétrique des oceans ', at 1 : 10M, sponsored by the Bureau Hydrographique International. Also in production is a most detailed work in many parts, by the American Geographical Society, a *Serial atlas of the marine environment,* edited by W Webster. As many correlations of variables as possible are being presented, the use of electronic computers greatly assisting the sorting of the many thousand items of information. Much careful thought has gone into the cartographic designs and format of the atlas, which aims to present the results of American research on the physical oceanographic environment and the distribution of the species of animal life, chiefly in the continental shelf areas of the North Atlantic and Arctic basins. Base maps have been prepared on a conformal, oblique stereographic projection devised by O M Miller, centred at 54 degrees north and 38 degrees west, so that all sheets are comparable without distortion. Fourteen sectional maps cover the water and adjacent continental shelf, plotted at 1 : 2,500,000 for reduction to 1 : 4M. The Arctic and North Atlantic basins are

14*

being mapped at 1:5M and 1:10M respectively; also, a single-sheet base map covers the whole area at 1:20M. Printing is done in four or five colours on loose double sheets of transparent material to facilitate comparisons. Every folio has its own specific interest; no 5 is particularly significant, as it forms the first part of a ' North Sea synopsis ' in preparation for FAO by a committee of the International Council for the Exploration of the Sea, published under joint agreement between FAO, the American Geographical Society and the Advisory Group of the *Serial atlas* . . . Each folio is issued in a special binding, on guards, so that the maps can be opened out flat, and each is accompanied by text and a bibliography.

Die weltmeere: taschenatlas mit den wichtigsten tatsachen aus meteorologie und nautik (Gotha, Haack, fifteenth edition 1956; previously named *Seeatlas*) consists of twenty three maps of the oceans and seas, showing fathom depths and currents. The text discusses the interaction of meteorology and oceanography, navigation and ocean traffic regulations, with illustrations of navigational aids, lights and types of vessels.

Other atlases of the oceans include the *Oceanographic atlas of the polar seas* by the United States Hydrographic Office, 1957; *Ice atlas of the northern hemisphere*, 1946; *World atlas of sea surface temperature charts*, 1944 and *Atlas of surface currents in various oceans*, 1944-47. In addition, there are the thousands of charts produced annually by relevant bodies in most countries, which, as stated in the preface, are beyond the scope of this work.

Of the vast literature available on this subject, the following are a few works in which maps or information concerning cartographical projects are to be found : the periodical *Limnology and oceanography*, American Society of Limnology and Oceanography, Lawrence, Kansas, quarterly from 1956, prepared by the University of Michigan, Ann Arbor, is one of the most comprehensive of the journals dealing with the chemical, geological and biological phenomena operating in open water; *Progress in oceanography*, edited by Mary Sears (Pergamon Press, 1963-), a forum for reports on work completed or in progress; also *Oceans: an atlas-history of man's exploration of the deep*, edited by G E R Deacon (Hamlyn, 1962); Günter Dietrich and Kurt Kalle: *Allgemeine meereskunde: eine einführung in die ozeanographie* (Berlin, Gebrüder Borntraeger, 1957); translated as *General oceanography: an introduction*, by Feodor Ostapoff, which has maps and charts in a separate pocket; M J Dunbar, editor : *Marine distributions* (Royal Society of Canada, 1963), based on a symposium held in June 1962 to outline the work being done in Canadian oceanography to map the features of the waters around Canada and in the Northern Atlantic across to

Europe; M N Hill and others, editors: *The sea: ideas and observations on progress in the study of the sea* in three volumes (New York, Wiley, 1962-63). *Oceanography,* edited by Mary Sears (Washington, DC, American Association for the Advancement of Science, 1961), is a collection of invited lectures presented at the International Oceanographic Congress, New York, 1959; and, on a special aspect, J L Mero: *The mineral resources of the sea* (Amsterdam, Elsevier, 1965) includes useful maps.

THE MAPPING OF VEGETATION, SOIL, AGRICULTURE AND ECONOMIC RESOURCES

Carl Ritter was one of the first geographers to experiment with systematic conventional symbols for forests, agricultural land and agricultural products. His ' Product charts ' for European countries were published between 1804 and 1806; distribution of plants and animals, however, was shown not by symbols but by their names. In the *Geographical atlas according to the principles of Carl Ritter,* 1851, is a chart indicating the distribution of camels, also date palms and other products. These experimental efforts encouraged the development of thematic maps, especially in the atlases published by the Cartographic Institute Justus Perthes at Gotha. A hundred years or so later, the preparation of thematic maps became urgent as a basis for planning. 'Accurate maps are a prerequisite to the proper development of the world resources, which, in many cases, lie in relatively unexplored regions ' (Resolution 131 (VI) of February 19 1948, of the Economic and Social Council of the United Nations). For such maps, accurate and correctly interpreted statistics must complement the work of surveyors.

John Morton was one of the pioneers in working out a classification of soils in the first half of the nineteenth century, also G C L Krause with his *Bodenkunde und klassification des bodens* (Gotha, 1832) and from this time dates the modern concept of applied soil science. K D Glinka has been a leading specialist in this field and has published, among other works, *The great soil groups of the world and their development,* 1908; translated from the Russian by C F Marbut (Ann Arbor, Edwards, 1927; Allen and Unwin, 1928, second edition 1937). Russian scientists have continued to take the lead in this aspect of thematic mapping, both in the practical preparation of maps and in the production of technical manuals on the subject. Most of them remain in Russian; one important monograph which has been translated into English is M S Simakova: *Soil mapping by color aerial photography* (Israel Program for Scientific Translations; Oldbourne Press, 1965).

The following classification system of world soils, worked out by Soviet soil scientists, was used as the basis for the new soil map of the world and of separate continents included in the *Fiziko-geografičeskij atlas mira*: *1* Polar soil formations, arctic and tundra soils; *2* Boreal soil formation; *3* Sub-boreal soil formation; *4* Sub-tropical soil formation; *5* Tropical soil formation. This system permitted the compilation of soil maps of the world and of different continents in accord with the other geographical maps of the atlas—climatic, botanical, geomorphological, geological. A *Soil survey manual* was published by the United States Department of Agriculture in 1951 (Agricultural handbook no 18); this work, which sets out the official principles of soil survey as practised in the United States, is intended primarily for those engaged in field classification and mapping of soils. All these ideas were discussed at a symposium on the methods of study in soil ecology held under the co-sponsorship of UNESCO, Natural Resources Research Division, and the International Biological Programme in Paris, November 1967, in which more than a hundred scientists from thirty countries took part. At the same time, plans were being discussed for a ' Soil map of the world' project, sponsored by FAO and UNESCO; the first draft of the map was presented at the Ninth International Congress of Soil Science, held in Adelaide, 1968.

In many parts of the world soil surveys have been carried out as part of irrigation projects, to obtain improved land use, for erosion control or for the study of special soil conditions. In addition to the examples mentioned in Chapter 3, there is the *World map of salt-affected soils,* for example, prepared by UNESCO and the Board of the Sub-commission of Salt-affected Soils, of the International Society of Soil Science, presented to the first meeting, held at the Research Institute of Soil Science and Agricultural Chemistry of the Hungarian Academy of Sciences, Budapest. The *Atlas of soil profiles,* compiled by W L Kubiena, was published by CSIC in Madrid and made available in Britain by Murby in 1954. The *Atlas zur bodenkunde,* the first volume to appear of *Meyers grosser physischen weltatlas,* has been mentioned above; there are soil maps also in *Meyers Duden-weltatlas,* available in Britain through Harrap; and George Philip has published a generalised world soils map, based on the work of Louis A Wolfganger.

Most countries now have their own growing literature on this subject, much of it frequently containing maps or reporting on surveys; for example, the *Canadian journal of soil science,* 1921-; *Soil science,* 1949-, the journal of the British Society of Soil Science;

the *Zeitschrift für landeskultur . . .,* 1960- and the *Australian journal of soil research,* 1963-. Monographs are numerous, many of them reporting on the results of symposia, conferences or special projects. FAO has published *Papers* and *Documents* on this subject for the past twenty years, nearly all of them containing at least one map; *Soil surveys for land development,* 1953, is a particularly useful text on the techniques of soil classification and mapping, accompanied by maps and a bibliography. As already mentioned, many of the monographs are in Russian, by such scientists as D G Vilenski, S S Neustruyev, I P Gerassimov and M A Glazovskaya, as well as by K D Glinka above mentioned. Specific areas of the world have particular problems, explained in such works as P A Yeomans: *The challenge of landscape: the development and practice of Keyline* (Sydney, Keyline Publications, 1958) and L S Berg: *Loess as a product of weathering and soil formation,* translated from the Russian in 1964, both of which contain maps. The National Research Council published *Soil exploration and mapping* in 1950 (Washington, DC, Highway Research Board, bulletin no 56) and J W Fox: *Land use survey: general principles and a New Zealand example* (Auckland University College, 1956) should also be noted.

The German Federal Institute for Topographical Research, at Hanover, has carried out a number of important studies in the problem of agricultural productivity in various parts of the world. Results of research were exchanged at a UNESCO symposium in 1968 on salt deposits and their effect on soils; among other reports, Dr Omer B Raup, of the Geological Survey of Denver, Colorado, described a new mineral which had recently been discovered in a salt deposit in south-east Utah.

An ' International vegetation map ', at 1 : 1M, under the direction of Henri Gaussen, at the Institut de la Carte Internationale de Tapis Végétal at the University of Toulouse, is in preparation. The first sheets to be issued, in 1958, provided a new regional map of Tunisia, by Gaussen and A Vernet, and four sheets, 1961-, of areas in India, published jointly by the India Council of Agricultural Research and the French Institute of Pondicherry. Gaussen's system of vegetation mapping was first shown in the vegetation map of the *Atlas de France.* Much discussion has revolved for some time on the question of the representation of vegetation types on topographic maps, on land use maps and on specialised vegetation maps. Information concerning vegetation cover and type is of obvious importance in the interpretation of or development of any area, from the point of view of land use, planned farming and forestry or improvement of communications, and yet, in many areas the changes in vegetation cover make it difficult to give accurate impressions in map form.

Much more research must be done on the analysis of vegetation to be mapped, as well as on a universal agreement on the meaning of terms and symbols depicting the varieties of vegetation. On the crop distribution map in the *Atlas of Great Britain and Northern Ireland,* for example, broad types of cultivation such as 'market gardening' were shown over a wide area, using the simple device of printing 'rhubarb', 'cauliflowers', etc to indicate predominance in certain areas, in much the same way as Carl Ritter did.

In 1960, at Toulouse, a colloquium was organised by the Centre National de la Recherche Scientifique on the methods applicable to the cartography of vegetation. Twenty specialists discussed the facts to be mapped and the compilation, reproduction and circulation of vegetation maps. Resolutions were adopted, with the object of securing international agreement on terminology, choice of scales and colours for vegetation maps. Agreement was reached on the preparation of an *International bibliography of vegetation maps,* noted below.

The appearance of monographs and articles underlines the importance of this aspect of the physical environment in the understanding of individual areas and as a basis for planning; *Vegetation mapping,* for example, by A W Küchler (New York, Ronald Press, 1967), begins with an historical introduction, then continues to examine the basic considerations in the classification of vegetation and the various approaches to mapping; part 3 covers the technical aspects of vegetation mapping, with a final summary of the practical use of vegetation maps. A W Küchler has also been responsible for initiating the *International bibliography of vegetation maps,* published in the University of Kansas Library series, beginning in 1965. The first volume, 1965, by Küchler and Jack McCormick, covers North America, Greenland, Mexico, Central America and the West Indies. Only published maps are included of natural and semi-natural vegetation. The data noted were title of map, date of preparation, colour and scale legend, in the original, if a western European language, author, place of publication and publisher. Arrangement is geographical, then chronologically by date of preparation, followed, if necessary, by division according to scale and local area. 'Europe' is the subject of the second volume.

In *The geographical review,* XLVI 1956, is included an article by A W Küchler, 'Classification and purpose in vegetation maps', and in the issue for 1964, no 2 appears 'An international vegetation map on the million scale'. An article by D Brown: 'Methods of surveying and measuring vegetation (Farnham, Royal Commonwealth Agricultural Bureau, 1954, Commonwealth Bureau of Pastures and Field Crops, bulletin no 54) is illustrated and includes a bibliography. Also informative are G McGrath: 'Further thoughts on the representation of vegetation on topographic and planimetric maps', *The cartographic journal,* December 1966, including a bibliography; 'Quality and quantity in the representation of vegetation on topographical maps', by the same author, International Geographical Congress, London, 1964, section IX Cartography; and by Devis Riley and Anthony Young: *World vegetation* (CUP, 1966), including a map.

Many organisations throughout the world issue publications on the interrelated topics of soil, land use and agriculture of all kinds; the publications of the International Society of Soil Science, for example, at the Royal Tropical Institute, Amsterdam, and the *Annals* of the Royal Agricultural College of Sweden. Information concerning any of these publications may be had from the Commonwealth Bureau of Soil Science, which has its headquarters at Rothamsted Experimental Station, Harpenden. The International Geographical Union Commission for Agricultural Typology was established at the Twentieth International Congress in 1964, with the aim of establishing uniform criteria, methods and techniques of distinguishing types of agriculture, of furthering on this basis, the typology of world agriculture and promoting more detailed studies of agricultural types. Two questionnaires, distributed in 1965 and 1966 to people likely to be interested, elicited much valuable material, helpful in the basic mapping. *Soil surveys and land use planning,* edited by L J Bartelli (Wisconsin, American Society of Agronomy, 1966), contains papers given originally at a symposium; these papers, which are illustrated by maps and diagrams, are particularly interesting on the application of such surveys to urban land use planning, the siting of housing, the planning of roads and other such modern problems.

One of the most pressing problems facing modern man is the exploitation of his environment to meet complex demands. In addition to the numerous surveys and reports already indicated, the importance of the subject promoted the full-scale report which was called for by the United Nations Economic and Social Council at its thirty ninth session; UNESCO, FAO, WHO and a number of individual specialists prepared a draft report presented to the UNESCO Advisory Committee on Natural Resources Research in June 1967. The report will serve as a basis for the work of the Intergovernmental Conference of Experts on the Scientific Basis for Rational Use and Conservation of the Resources of the Biosphere, organised by UNESCO, held in Paris, September 4-13 1968. The new element in modern environmental studies is the recognition of the interrelation between the changing factors of natural and artificial environments.

The new term ' ecosystem ' is used to define the unity of study of the natural environment; this is of extreme variety in size and in character, but always possessing a unity deriving from the complex interrelationships of its living communities between each other and with their environment. The study must be dynamic, not static, and no factor must be neglected, otherwise faulty deductions may be made.

The first volume of the *Atlas of the world's resources: the agricultural resources of the world,* by William Van Royen (New York, Prentice-Hall, for the University of Maryland, 1954), is one of the most useful atlases yet produced on this subject; compiled by the Department of Geography, University of Maryland and the College of Business and Public Administration, in co-operation with the United States Department of Agriculture and the United States Department of the Interior, it deals first with agricultural regions and land use, then with the distribution of individual crops, farm animals and products, with annotations, explanatory text, arranged by political units, and bibliographies. In the same year was published the *Atlas de géographie alimentaire* (La Documentation Française, 1954, recueils et monographies, no 28), in which maps analyse the world situation qualitatively and quantitatively, and from various aspects, such as that of health, with accompanying text.

Many of the publications of agricultural research bodies contain maps intermittently. One series, *Farm studies and the teaching of geography,* by I V Young, published by the Association of Agriculture in collaboration with the Standing Sub-committee in Geography of the University of London Institute of Education, based on the association's farm study scheme, is perhaps not so widely known as it might be. The studies of farms from all parts of Britain and other countries are published in the form of looseleaf folders, containing maps, including a detailed survey map of the farm, photographs and printed sheets giving details of the soil, land formation, agricultural history and markets, water supply, cropping and general farm practice.

Research into individual problems, such as, for example, those that come within the jurisdiction of the Anti-Locust Research Centre, has been greatly facilitated by systematic mapping and co-ordination with the mapping of other physiological factors. Journals most frequently including maps of the various aspects of the subject are *World crops: the international journal of agriculture; Agricultural situation,* issued by the United States Agricultural Marketing Service, with a supplement of original documents; the *Revue internationale des industries agricoles* from the Centre de Documentation, Commission des Industries Agricoles, Paris; the *Zeitschrift für Acker-*

und Pflanzebau, a long-standing journal, 1853-, including English summaries; *The agricultural history review,* published by the British Agricultural History Society, Department of Agricultural Economics, University of Manchester; *Études rurales: revue trimestrielle d'histoire, géographie, sociologie et économie des campagnes,* from the École Pratique des Hautes Études, Sorbonne, Sixième Section: Sciences Économiques et Sociales avec le Concours du Centre National de la Recherche Scientifique; and *Acta agronomica,* semi-annual 1951- from the Publishing House of the Hungarian Academy of Sciences, Budapest. The last mentioned, in particular, often contains excellent maps; the text is in English, French, German and Russian, with summaries in the language not used in the article. Specific maps of great signifiance may be found in individual issues, for example, the diagrammatic map of 'Current FAC projects' in the July-August number of *Ceres.*

The themes discussed at the International Geographical Union symposium on 'Agricultural geography', of which the papers were edited by E S Simpson, are effectively demonstrated by maps and diagrams (University of Liverpool, Department of Geography, 1965, research paper, no 3). Maps are included in *Climatic factors and productivity,* a synopsis of discussions held in March 1963 at the University College of Wales, Aberystwyth (memorandum no 6, 1963); in René Dumont: *Types of rural economy: studies in world agriculture,* translated into English by Douglas Lagnin from *Économie agricole dans le monde* (Paris, Librairie Dalloz, 1954; Methuen, 1957, reprinted 1964); P P Courtenay: *Plantation agriculture* (Bell, 1965); OECD: *Regional rural development programmes with special emphasis on depressed agricultural areas including mountain regions,* 1964, which carries maps in a separate pocket and forms the report summarising the conclusions of a seminar held in 1963, including special country studies for France, Italy, the Netherlands, Sweden and the USA; E Otremba: *Allgemeine agrar-und industriegeographie,* edited by R Lütgens (Stuttgart, Franckh'sche Verlagshandlung, second revised edition 1960), a scholarly analysis of industrial and agricultural landscapes and rural population, well illustrated with photographs, maps and figures; C O Sauer: *Agricultural origins and dispersals* (American Geographical Society of New York, 1952), a reprint from the second series of Bowman Memorial Lectures; Guy Harold Smith, editor: *Conservation of natural resources* (New York, Wiley; Chapman and Hall, second revised edition 1958); and W S and E S Woytinsky: *World population and production: trends and outlook* (New York, Twentieth Century Fund, 1953).

Monographs on a specific aspect of cultivation usually include maps. Several sketch-maps aid in the understanding of J W F Rowe: *The world's coffee: a study of the economics and politics of the coffee industries of certain countries and of the international problem* (HMSO, 1963); the 'World crop' books series issued by Hill/Wiley Interscience are also well illustrated with photographs and maps, for example, F L Wellman: *Coffee: botany, cultivation and utilization,* 1961, and R F Peterson: *Wheat: botany, cultivation and utilization,* 1965.

The *World forestry atlas* has been in process of compilation for some thirty years at the Bundesforschungsanstalt für Furst-und Holzwirtschaft, at Reinbek, near Hamburg. This is the centre's most important single project, completed in 1967. Over sixty maps show the distribution of forests of different kinds throughout the world. The publications of other leading research organisations in forestry sciences occasionally include maps, such as those of the various sections of the International Union of Forest Research Organizations; the International Centre for co-ordination of scientific and technical information in agriculture and forestry, at Prague; the relevant divisions of FAO and the Commonwealth Forestry Institute, Department of Forestry of the University of Oxford; *Unasylva,* published bi-monthly by FAO since 1947, then quarterly from 1950, includes systematic surveys of forestry and forest products and articles on world conditions, development and research. From FAO come also the *Forestry and forest products studies,* 1950-, *Forestry development papers,* 1954- and *Forest economics and statistics,* 1948-. *The Commonwealth forestry review* is the official journal of the Commonwealth Forestry Association, the medium for the exchange of information on all aspects of forestry, including the marketing and utilisation of timber and forest products; the journal occasionally includes maps in demonstration of an article and also the reviews and technical notes provide a source of information on mapping, as well as current forestry projects.

The quarterly journal of forestry, 1906-, journal of the Royal Forestry Society of England and Wales, gives a similar service for the area covered, as does also *Forestry,* published by the Society of Foresters of Great Britain, 1925-. Of the numerous activities of the Timber Research and Development Association, London, the series of wall charts issued as part of the teaching aids division are the relevant publications here. The *Revue forestière française,* an excellent journal edited by the Ecole Nationale des Eaux et Forêts, is illustrated, but only occasionally includes maps. A new *International review of forestry research* was begun by the Academic Press of New York in 1965. *The forest area of the world and its potential productivity,* by S S Paterson, issued in 1956 by the Royal University of Göteborg, Department of Geography, contains maps throughout the text and a world map, 1 : 30M, in a separate cover. As an example of an analysis of a specific species, *Cork and the cork tree,* by G B Cooke (Oxford, Pergamon Press, 1961), includes a distribution map.

The Arid Zone Programme of UNESCO, begun in 1951, to promote research in the various scientific disciplines concerned with the problems of the arid regions of the world, has developed into an outstanding example of international scientific co-operation. To make available the results of such research, UNESCO publishes the 'Arid Zone research series', which includes the proceedings of symposia, reports and maps; similarly, the work edited by E S Hills: *Arid lands: a geographical appraisal* (UNESCO; Methuen, 1966), which constitutes a 'reader' in the subject, containing contributions by seventeen authors from eight countries. The Arid Zone Programme continues within the Natural Resources Research Programme of UNESCO; an allied publication, *Nature and resources: a newsletter about scientific research on environment, resources and conservation of nature,* the bulletin of the International Hydrological Decade, does not itself contain maps, but is a valuable source of information on maps in progress or completed in this field of research.

Tropical science, 1959-, which replaced *Colonial plant and animal products,* 1950-, the quarterly journal of the Tropical Products Institute, is an excellent production, usually containing a number of maps in the text. A section on the mapping of tropical vegetation is included in *Study of tropical vegetation: proceedings of the Kandy Symposium, 1958,* and several maps are integrated with the text; this is one of the fundamental publications which have emerged from the UNESCO Humid Tropics Research Programme. Another, also containing maps, is *Problems of humid tropical regions, 1958,* which reports on individual tropical regions around the world. Maps form an essential part of the *Journal of tropical geography,* published by the Geography Department of the University of Malaya, 1953-, of *Agronomica tropical,* 1951- by the Centro de Investigaciones Agronomicas and of the *Rivista di agricoltura subtropicale e tropicale,* 1907- from the Istituto Agronomico per l'Ottremere, Florence. Maps are also to be found in:

D L Linton: 'The tropical world'; an inaugural lecture, University of Birmingham, May 1961.

Sir Harold Tempany and D H Grist: *An introduction to tropical agriculture* (Longmans, 1958).

Gordon Wrigley: *Tropical agriculture: the development of production* (Batsford, 1961).

At the 1949 International Geographical Congress at Lisbon in

1949, a project was brought forward by Professor Samuel Van Valkenburg to form a commission to look into the possibilities of a Land Use Survey on a world scale. UNESCO supported the venture and a commission was established which met at Worcester, Massachusetts, in the following year. As a result, the World Land Use Survey was set up, with two main aims: first, the examination of methods of work, especially of the use of air photographs, and secondly, to co-ordinate work throughout the world on lines which would enable surveys carried out nationally to be readily comparable with one another. A uniform basic framework would be used, but sub-divisions within the broad structure would be allowed to meet particular needs. The commission reported to the Washington congress in 1952 and the full classification agreed on was set out in the *Report of the Commission on World Land Use Survey for the period 1949-1952* (IGU and UNESCO, 1952). A series of maps with accompanying memoirs was planned to cover the whole world, using the 1:1M scale maps, in conjunction, whenever possible, with soil surveys, demographic studies, etc. Also in the report were given the findings of Professor Dudley Stamp's direction of a preliminary survey of work in the Old World Division, as a pilot scheme to test the validity of the proposals. In August 1951, during the annual meeting of the British Association for the Advancement of Science, a discussion on the surveys of under-developed areas reviewed existing work in many countries. A bibliography compiled of land use maps and reports revealed inadequate coverage in many parts of the world. A further *Report* was made to the Rio congress in 1956 and by then there were ' many and measurable results '. Some of the pioneer studies were carried out by the Directorate of Overseas Surveys, especially in Africa; eastern Pakistan began an aerial survey of the Indus basin; the Canadian government initiated surveys of forest land in particular; land use in Cyprus was mapped for the first time and Japan gave priority to land use study.

The object of the Land Use Survey was to show by examples what could be done by various methods to encourage countries to carry out land use surveys for themselves; especially if these surveys could be made concurrently and using the proposed classification of land use, valid comparisons could be made. The classification agreed ran as follows:

1 Settlements and associated non-agricultural lands (dark and light red).

2 Horticulture (deep purple).

3 Trees and other perennial crops (light purple).

4 Crop land, continual and rotation cropping (dark brown), land rotation (light brown).

5 Improved permanent pasture (light green).

6 Unimproved grazing land, used (orange), not used (yellow).

7 Woodland, dense (dark green), open (medium green), scrub (olive-green), swamp forest (blue-green), cut over or burnt over forest areas (green stipple), forest with subsidiary cultivation (green with brown dots).

8 Swamps and marshes (blue).

9 Unproductive land (grey).

The commission recommended that survey should be carried out at the scale most suitable for the country, but that uniform publication should be 1 : 1M for the whole world. Uniformity of treatment was considered of paramount importance, but it was understood that individual surveys would make further sub-divisions of the major categories as required and the descriptive *Memoirs* would be regarded as integral parts of the whole work. Especially in developing countries, there exists a growing interest in land use surveying and, in a number of cases, these countries have turned to the commission for advice and guidance. *Occasional papers* have been produced irregularly from 1956- and regional monographs since 1958, published by Geographical Publications, Bude, Cornwall. The *First report* covered 1949-52, the *Second*, 1952-56, both of these being out of print. The *Third* and *Fourth reports*, 1956-60 and 1960-64, were included in the International Geographical Union *Newsletter* for 1960 and 1964 respectively. There were no documentary files for the period 1965-66, but the *Report* for the period Summer 1966 to Summer 1968 appeared in the *Newsletter* XIX 2 1968.

One of the achievements of Sir Dudley Stamp was the creation of a series of ' Publications of the World Land Use Survey ', consisting of monographs and occasional papers, obtainable from Geographical Publications, Bude. The first monograph concerned land use in Hong Kong, with a coloured map, 1 : 80,000, 1958, by T R Tregear, now under revision; ' The evolution of the rural land use pattern in Cyprus ', with a map at 1 : 250,000, by D Christodoulou, followed in 1959; ' Land use and population in Tobago, with a map at 1 : 63,360 ' by D L Niddrie, appeared in 1961 and ' Land use in the Sudan ' by J H G Lebon, in 1965. The first few occasional papers were M M Cole: *The Transvaal Lowveld*, 1956; D N McMaster: *The agricultural geography of Uganda*, 1962; L D Stamp: *Land use statistics of European countries*, 1965; L E Virone: *Borgo a Mozzano: technical assistance in a rural community in Italy*, 1963; L E J Brouwer: *The North Sea* and *The law of the continental shelf*, by Lord Shawcross, 1964; 'A resort map of Europe' by G W S

Robinson, was produced at about this time by the University of Southampton.

After the death of Sir Dudley Stamp, in 1966, Professor Hans Boesch was appointed interim chairman of the Commission on World Land Use Survey. Interesting and ever more specific developments are now continually in progress; aerial photographic surveying has proved invaluable, especially in those areas difficult of access. Two informative articles by Alice Coleman are relevant at this point: ' Cartographic methods for use in land use memoirs ', *Transactions* of the Institute of British Geographers, XXXVI 1965; and ' Some technical and economic limitations of cartographic colour representation on land use maps ', *The cartographic journal*, December 1965.

A Regional Sub-commission for East-Central Europe Land Use was formed and conferences have been held; the second and third were particularly fruitful. At the second conference in 1964 in Budapest, reports were presented on the progress of research into various types of land use in Yugoslavia, Czechoslovakia, Bulgaria, Rumania, Hungary, East Germany, Poland and the Soviet Union. The *Proceedings* of the conference were published in 1967. During the period 1964-68, co-operation between these countries increased and numerous surveys were undertaken by groups of geographers; a number of scholars from various west European, American and Asiatic countries also studied the methods and carried out field work in Poland and the other east European countries. One of the first major studies to emerge was *Land utilization in East-Central Europe: case studies* edited by Jerzy Kostrowicki, *Geographica Polonica*, 5 1965. The third conference is taking place at the time of writing, in Yugoslavia. The publication, *Land utilization in Eastern Europe: studies in geography in Hungary*, 1967 is a collection of papers of central importance, including much information on the progress of mapping in the countries concerned. Symposia on applied and agricultural geography were held in Rumania in 1966 and 1967.

The International Geographical Union Commission on Interpretation of Aerial Photographs has issued a comprehensive *Index to the use of aerial photography for rural land use studies* (Bad Godesberg, Selbstverlag der Bundesanstalt für Landeskunde und Raumforschung); the documents listed comprise reports on specific land use projects, articles and books dealing with various aspects of photo-interpretation and papers which have used air photographs to illustrate the text and it is the first in a new series ' Geographical interpretation of aerial photographs ' issued by the periodical *Landeskundliche Luftbildauswertung im Mitteleuropäischen Raum*. The bibliography is in English, but the preface is given in English, French and German. An *Air photo atlas of rural land use* is also in progress.

The study of the spatial distribution of the various aspects of economic activity is continuing to increase in importance; economic factors play a large part in the geography of production and trade, and geographers and cartographers can evoke concepts from the presentation of data in visual form. Experiment in the effectiveness of different methods is urgent and speed of publication is required in this area of mapping if cartographical work is to be helpful in policy making. Most countries now compile reliable and regular statistics of all major aspects both of domestic economy and of external trade. Increasingly during the past decade, groups of countries have combined to issue comparable statistics and the work of the United Nations and of the specialised agencies has been a central factor in co-ordinating information on a world scale. In addition to all the current continuing services, publications such as *The growth of world industry, 1938-1961 : international analyses and tables,* in English and French, illustrated by charts, have an obvious value in assisting an understanding of the interplay of economic forces in all parts of the world.

Many maps of industries still do little more than indicate their general location. In this context, Olsson's *Economic map of Europe,* 1953, was an interesting advance, for, using employment figures, town types based on industry were indicated, as, for example, ' textile towns '. This map demonstrated the value of more exact data and local maps have since appeared using material obtained from individual plants to show the importance of individual industries in relation to their regional setting. Maps for research and planning are frequently never published at all; the large-scale plans for the preparation of economic atlas plates are fundamental working source material. Many other, frequently pioneer, maps are to be found within the files of periodicals, reports and monographs.

The chief difficulty in devising economic maps is to do justice to the complexity of the data and the relationships between them; in attempting to achieve clarity of interpretation, oversimplification is not always easy to avoid. Complex symbols are inevitable; the *Atlas of Britain and Northern Ireland,* for example, showed some interesting experiments in the use of many-faceted star-shaped symbols, each point denoting different data. Economic maps cannot usually be made to show any but the broadest generalisations at a glance, in the way that topographical maps are able to do. Careful study must be applied, with a close understanding of the mapping policy; at the same time, it must be borne in mind that the economic factors depicted can be fully appreciated only in conjunction with the physical features—frequently with the social characteristics also —of the area. Some of the data mapped can be seen in the field, notably in arable or cattle country, in mining or quarrying areas, whereas factory buildings tell little or nothing of the character and value of products and by-products. The use of colours adds greatly to the clarity of interpretation and it would be helpful if a universal colour scale could be agreed upon, as has been achieved to a large extent in geological mapping. Computers are making an increasing volume and diversity of data quickly available and a beginning has been made in the production of statistical maps by automated methods.

All these problems and considerations are magnified in the case of the production of an economic atlas, which may be consulted for several different purposes and a higher degree of evocative visual effect may be desirable than in individual maps prepared for skilled readers. The *Oxford economic atlas of the world,* 1954, which went into a second edition in 1959, with partial revisions in 1960 and 1961 and a third revision edition in 1965, was the product of the Cartographic Department of the Clarendon Press, Oxford, in collaboration with the Economist Intelligence Unit. The atlas is in two main parts, consisting of commodity maps, grouped in ten sections, and a country by country statistical index. In the latest edition, comprising 286 pages of maps and text, all distribution symbols were checked, the industrial section was enlarged and the statistical index was extended, containing more detailed information on the newly developed countries. The index includes also information on production and comparative figures for imports and exports, taken from United Nations and other official sources. Emphasis is on agricultural and mineral products; data on lithium, antimony and nuclear fuels have been added in the revisions, also a world soil map and a map of climatic regions, according to the Köppen classification. Relief, population and communications maps are included for reference. In series with the general economic atlas are regional atlases, the complete set comprising *The Middle East and North Africa,* 1960; *The USSR and eastern Europe,* 1963; *Africa,* 1965; *Canada and the USA,* 1967; *Europe; Latin America; India, China and Japan; South East Asia; Australia and New Zealand.* In these regional atlases, general reference maps are followed by topical maps, notes and statistics, bibliographies and gazetteers. The *Shorter Oxford economic atlas of the world,* in a third edition, with revisions 1965, contains the plates of coloured maps of the parent atlas, in ninety seven pages, but omits the statistical index.

A new and useful reference atlas produced in 1952 was *Weltatlas: die Staaten der Erde und ihre Wirtschaft,* edited by Professor Edgar Lahrmann in collaboration with Professor Heinz Sanke (Leipzig, VEB Bibliographische Institut, 1952), revised and enlarged in 1964 (Gotha, VEB Hermann Haack). Each economic map faces a corresponding location map, at the same scale, for comparison of the various factors involved. The basis of the economic maps is land utilisation, which is shown by solid colouring. A uniform system of conventional signs and symbols, divided into twelve categories, is superimposed to indicate more detailed information. Mineral resources are shown by symbols and letters, mostly in black, industries in red or blue. It might be kept in mind that in Germany economic geography is not usually separated from politics. In this atlas, Germany is naturally treated in the greatest detail; general maps are followed by five plates at 1:1·4M. Other European countries are at scales ranging from 1:1·65M to 1:7M. Outside Europe and parts of America and Asia, scales are much smaller and the information given is frequently very generalised.

The *Atlas of economic development,* prepared by Norton Ginsburg (University of Chicago P, 1961), comprises a series of global maps with commentary and statistics amounting to 135 pages. The main theme of the atlas, which is to isolate the variable factors affecting economic development in each country, is explained in the commentary by B J L Berry. Part II of the *Atlas of the world's resources* is devoted to *Mineral resources of the world,* compiled by William Van Royen, O Bowles and E W Pearson (New York, Prentice-Hall, 1952). The accompanying text reviews the most important technical factors affecting the exploitation of each of the minerals, the basic processing methods involved and the principal uses of each mineral in the modern economy. Another *Mineral resources of the world* atlas was published in 1952 by the University of Maryland.

A firmly established economic atlas is the 'Atlas of economic geography', compiled by Johannes Humlum, which appeared first in 1936, with revised editions in 1944 and 1947, published by Gyldendal of Copenhagen. The atlas covers the geography of population, agriculture, livestock and animal products, fishing, forestry and associated manufactures, minerals, industries and water power, the majority of the major maps being drawn on the equal-area Olaf Kayser Projection, at 1:225,000,000. The statistical information given on the maps, together with the descriptive text, makes this a useful work of reference with notes and list of sources, but no index. A fourth edition of the atlas volume was brought out under the title *Kulturgeografisk atlas* (Copenhagen, Gyldendal; Meiklejohn, 1955).

441

This time, text and maps were published separately, both being revised and enlarged. Legends and notes are given in Danish, German, French and English. All the maps, in red and black, are striking in their use of symbols. The text volume, in Danish, lavishly illustrated with photographs and diagrams, allows full understanding of the mass of material dealt with, pointing out the interrelationships and current trends in world production and trade.

A second major work by Scandinavian geographers, the *Atlas of the world commodities, production, trade and consumption*: *an economic-geographical survey*, by Olaf Jonasson and Bo Carlsund (Göteborg, Akademiforlaget/Gumperts . . . Scandinavian University Books, 1961), was a combined effort from the Göteborg Graduate School of Economics, under the direction of Professor Jonasson. Coverage includes the staple commodities of world trade, with the major emphasis on animal and vegetable raw materials, but dealing also with minerals, coal and petroleum, with a section on international sea-borne shipping. The maps, mostly diagrammatic, are skilfully contrived in black and white, accompanied by brief annotations and statistics.

Most familiar of all commercial atlases in English is the unique *Mercantile marine atlas,* first published by George Philip in 1904 and constantly revised since. A complete revision was carried out by Harold Fullard for the sixteenth edition of 1959. Large-scale charts on the Mercator Projection present full information on the ocean highways of the world from the point of view of trade and commerce; it is not intended for navigational purposes and the land areas shown are of secondary importance. Tables of distances between ports are given and much other port information, together with 135 port plans; alternative spellings of place names and lists of British Consulates, Trade Commissions, British Chambers of Commerce abroad, Lloyd's Agencies and United States Consulates and Chambers of Commerce abroad increase the reference value of the atlas. The 1959 edition contained a new map of the Persian Gulf, showing developments in the oilfields and ports, refineries and pipelines; a world oil map shows load-line zones and seasonal areas, oilfields, refineries and pipelines. Details of the improved St Lawrence seaway are also included in this edition. Shipping routes, canals, navigable rivers, docks and air routes are all clearly indicated and the index comprises about twenty five thousand entries. *Philip's mercantile map of the world,* 1:37M, containing maps coloured according to political divisions, is especially suitable for general use in business concerns or shipping offices; *Philip's commercial map of the world,* 1:22,500,000, is another map in international use. *Lloyd's maritime atlas: including a comprehensive list of ports and shipping places of the world* was first produced in 1951, under the direction of the shipping editor at Lloyd's, R C E Lander. A sixth major revision was issued in 1966. The contents include a world map showing bad weather areas and seasons, regional maps, a list of ports and shipping places of the world, arranged geographically, with reference to the maps, and an index.

The Institute for Shipping Research at Bremen published *The world shipping scene*, a new international atlas for marine navigation (Weststadt-Verlag [IRO-Verlag], Munich, 1963), under the direction of Dr Gustav Adolf Theel. Five years' work in planning and production resulted in an atlas which aimed to bring together information on all the harbours of the world, shipbuilding yards and shipping routes. 11,084 ports in all are included. In addition to the principal maps, there are particularly useful maps of small sections of coastline and minor harbours throughout the world. Information is given on the capacity of the ports and the depths of their harbours, analyses of the building capacities of 959 shipyards, mentioning the types of ships built, the sites of refineries and oil wells and the shipping routes associated with them. Another section traces the history of marine navigation and trade with regard to the leading maritime nations.

The *Atlas international Larousse: politique et économique* is another work valuable both for use in further research and for practical purposes. 108 pages of maps, many of them double-page spreads, and forty five pages of source statistics, were edited by I du Jonchay and Sándor Radó, with explanatory notes in French, English and Spanish. Four introductory physical maps of the northern and southern hemispheres, the old and the new worlds, show relief by layer colouring; some climatic data are included in this section. In the main part of the work, hill-shading is shown on some maps and, where suitable, relief is indicated by hachuring. On the economic maps, a variety of symbols have been used, including pictorial symbols.

A third atlas by Philip, the *Commercial concise atlas,* also edited by Harold Fullard, should be mentioned. This is a straightforward, factual atlas, containing, in the 1965 edition, eighty eight coloured plates of maps and diagrams and a twenty three page index. Claiming to be the oldest annually published reference atlas in existence is the Rand McNally *Commercial atlas and marketing guide,* 1876-, which covers the world, with chief emphasis on the development of America. Successive editions comprise in effect an economic history, showing, for example, the effect of the expansion of communications. The 'Map of world crude oil production and refining capacity' is a fine example of the mapping of a specific commodity.

443

Published by the Petroleum Information Bureau, London, at a scale of approximately 1:40M, the map shows estimated quantities of crude oil and natural gas liquids produced in each country in 1963, and the total annual intake capacity of the refineries processing crude oil. Maps of individual commodities are to be found in such journals as the *International nickel magazine,* published by International Nickel, previously as the *Inco-Mond magazine* and *Mond magazine* from 1956 and under the new title since 1965.

A bibliography of statistical cartography, by P W Porter, was published by the Department of Geography, University of Minnesota, 1964. *New viewpoints in economic geography: case studies from Australia, New Zealand, Malaysia, North America,* by J Rutherlord, M I Logan and G J Missen (Sydney, Martindale Press, 1966), presents some evocative ideas and factual commentaries, supported by maps. A vast periodical literature has grown up on this subject, but those having an emphasis on the methodology and principles of economics rely on diagrams and tables of statistics, as a general rule, to illustrate the textual matter. Many maps are found in *Economic geography,* quarterly from 1925 from the Graduate School of Geography, Clark University, Worcester, Mass; *Geographical papers,* a new journal begun in 1964 by the London School of Economics and Political Science, Department of Geography, is relevant, also *The journal of development studies: a quarterly journal devoted to economic, political and social development,* edited by the School of Oriental and African Studies, University of London, in co-operation with the Department of International Economic Studies, University of Glasgow and the Department of Economics, University of Manchester; *Geography and industries,* semi-annual 1956-, in Chinese and English, by the Fu-Min Geographical Institute of Economic Development, Taipei and *Reports on economic geography and anthropology,* a new series published by the Institute of Geography, University of Naples.

Much more bibliographical work needs to be done in recording the existence and location of thematic maps. Individual monographs or papers in this field, illustrated by maps or discussing methodology are legion; a few relevant examples include:

J W Alexander: *Economic geography* (New York, Prentice-Hall, 1963). Particularly well documented by maps, charts, statistics and a bibliography; valuable sections on the location of industry and statistical techniques are included.

Roberto Amalgià: *Elementi di geografia economica generale* (Milan, Giuffrè, 1947). With maps and a bibliography: a classic work.

Herbert Alnwick: *A geography of commodities* (Harrap, fifth revised edition 1965).

Hans H Boesch: *A geography of world economy* (Van Nostrand, 1964). Deals with economic geography within a broad framework of physical, historical and cultural factors. Well illustrated with maps, plans and charts.

Jean Chardonnet: *Géographie industrielle* (Paris, Editions Sirey, 1962).

G C Dickinson: *Statistical mapping and the presentation of statistics* (Arnold, 1963, reprint 1967). An excellent and eminently practical text, using maps in illustration, and appending a useful 'List of Census Reports for Great Britain 1801-1931', but, unfortunately, with a totally inadequate index.

Gustav Fochler-Hauke: *Verkehrsgeographie* (Braunschweig, Westermann, second edition 1963).

Rolt Hammond: *Air survey in economic development* (Muller, 1967). The emphasis is on equipment and methods, with summaries of economic schemes carried out in many parts of the world.

Y Lacoste: *Géography du sous-développement* (Paris, PUF, in the 'Magellan' series: la géographie et ses problèmes).

P W Porter: *A bibliography of statistical cartography* (University of Minnesota, Department of Geography, 1964). Nearly one thousand titles, mainly in English, are arranged alphabetically.

Lech Ratajski: 'Maps of industry: their methodological and cartometrical attributes' (Prace G [Warsaw] LVI 1966). In Polish, with an English summary.

— and Winid Bogodar: *Kartografia Ekonomiczna: metody upracowania map gospodarczych* (Warsaw, Panstwowe Przedsiebiorstwo Wydanictw Kartograficznych, 1960).

W W Ristow:'A half century of oil company road maps', *Surveying and mapping,* December 1964.

R S Thoman: *The geography of economic activity: an introductory world survey* (McGraw-Hill, 1962). With maps and diagrams effectively integrated with the text.

R S Thoman and D J Patton: *Focus on geographic activity* (McGraw-Hill, 1964). Thirty original case studies of economic units, such as mines, farms, factories; illustrated by maps.

W Van Royen et al: *Fundamentals of economic geography: an introduction to the study of resources* (New York, Prentice-Hall; Constable, fifth edition 1964).

W A Willox: 'Maps for natural resource studies', *Report of proceedings,* Conference of Commonwealth Survey Officers, 1963 (1964).

MAPPING BIOGEOGRAPHICAL FACTORS

Since biologists and ecologists have co-operated with geographers, especially in the mapping of natural factors, a number of works of central importance have become available. Marion I Newbigin, a pioneer in the study of plant and animal geography, stressed repeatedly the value of well-constructed sketch-maps to demonstrate the facts under discussion. Later experts in this field include Alexander von Humboldt, C Troll, Henri Gaussen, A W Küchler, V N Sukachev and L S Berg, their methods proceeding for some time individually or in small groups before scientific mapping became an accepted part of co-operative projects. In Bartholomew's *Physical atlas of zoogeography,* 1911, more than two hundred maps of the world show the distribution of mammals, birds, reptiles and amphibians, together with ninety two pages of explanatory text and index. The International Society for Plant Geography and Ecology was founded in 1937; the society holds an annual symposium and it was at the 1950 symposium that the plan for a vegetation map of Europe was brought into being. *The chromosome atlas of cultivated plants,* by C D Darlington and A P Wylie (Allen and Unwin, second edition 1955) contains very specialised maps and a useful bibliography. Distribution maps of plant diseases have been issued from Kew Observatory, London, since 1942, at the rate of some twenty four maps a year, including revisions, under the auspices of the Commonwealth Agricultural Bureau. The unique *Herring atlas* should also be mentioned; edited by W C Hodgson, it was published by the International Council for the Exploration of the Sea, Copenhagen, 1951. Since the introduction of punched card systems to analyse the data on flora and fauna, work in this field has been able to progress more rapidly. M E D Poore and V C Robertson wrote on this topic in *An approach to the rapid description and mapping of biological habitats,* published by the Nature Conservancy in 1964.

Blumea, tijdschrift voor de systematick en de geographie der planten, 1954- from the Rijks-Herbarium, does not often carry maps, but it is an international source for information concerning current research and mapping. Monographs including particularly effective maps include:

P J Darlington, jr: *Zoogeography: the geographical distribution of animals* (New York, Wiley; Chapman and Hall, 1957).

Sven Ekman: *Zoogeography of the sea* (Sidgwick and Johnson, 1953). Translated from the Swedish by E Palmer.

James Fisher and R T Peterson: *The world of birds: a comprehensive guide to general ornithology* (Macdonald, 1964). Contains beautiful illustrations and 199 clear maps showing the distribution of families of birds.

Henri Gaussen: *Géographie des plantes* (Paris, Colin, 1954).

H A Gleason and Arthur Cronqvist: *The natural geography of plants* (Columbia UP, 1964). Mainly American examples are included.

Nicholas Polunin: *Introduction to plant geography and some related sciences* (Longmans, 1960).

F A Schilder: *Lehrbuch der allgemeinen Zoogeographie* (Jena, Gustav Fischer, 1956).

Heinrich Walter: *Die vegetation der Erde in ökologischer Betrachtung* (Jena, Gustav Fischer, 1962).

MAPPING POPULATION AND SETTLEMENT

It is now axiomatic that aspects of population mapping should be included in national and regional atlases, yet until well into the eighteenth century, little account was taken of the role of man vis-à-vis his physical environment, and not until 1933 that Sten de Geer put forward the idea of plotting the distribution of population on a uniform scale throughout the world. The suggestion was then received with acclamation, but the existing information on population in different countries was still uneven. At the International Geographical Union conference in Lisbon, it was agreed that experiments should be made with the mapping of population density and a report presented to the Washington congress, 1952; an atlas of world population was an ideal to be kept in mind. Further and more positive measures were agreed at the 1956 Rio de Janeiro congress; a special commission was appointed ' to formulate some guiding principles on mapping the results of the population censuses of 1960 '. Through the national committees, contacts were made with population geographers throughout the world; the International Statistical Institute and the Statistical Department of the United Nations were also approached. Population maps from thirty one countries were exhibited at the 1960 congress at Stockholm and much discussion ensued on the problem of collecting precise statistical data for the smallest possible geographical areas, to have maps in all statistical publications, showing limits and names of all the administrative units mentioned and to maintain direct contact with census departments in order to agree on standards of efficiency and comparability, especially with regard to symbols and scales.

The Laboratory of Photography at the Royal Institute of Technology in Stockholm engaged in research on three-dimensional symbols for population maps, that is, mainly spheres and cubes. Following many meetings of eminent geographers until June 1958, the ideal of a uniform world population map was considered unlikely of

447

achievement, but universal agreement was reached on the necessity for a general map of world population at 1 : 1M, or at 1 : 4M in some areas. Dot symbols were chosen as the most suitable, in association with proportional circles and cubes. Full details were included in the *Report* of the International Geographical Union Commission on a World Population Map, prepared by W William-Olsson, which was published in *Geografiska annaler,* 1963, and also made available later as a separate reprint. In the *Report* is also a short article, ' Population census results as basic material for the World Population Map ', by Kurt Horstmann, with a diagrammatic map showing the periods of census returns throughout the world, accompanied by a chronological table of census returns, quoting area and population figures. A further report is included : on the comparative study of the different cartographic symbols used in the experiments on population mapping carried out by Gösta Ekman, Ralf Lindman and W William-Olsson, sponsored by the Air Force Office of Scientific Research of the Air Research and Development Command, United States Air Force, through its European Office; it was first mimeographed and sent to members of the commission, then published in *Perceptual and motor skills,* December 1961. The symbols are amply illustrated and further references are given.

At the International Geographical Union Congress, London, 1964, the Commission on the Geography and Cartography of World Population was established, to encourage and advance geographical studies of population, with particular reference to the relevance and application of these studies to practical problems; to continue and extend the work of the commission on a world population map, by encouraging further the production of maps of the distribution of population and by advising on the development of techniques for mapping all kinds of population data. The commission organised an exhibition of population maps at the congress, of which a catalogue was prepared and issued. The programme of work was formulated at a meeting in Belgrade in 1965; two bibliographical projects were agreed, a bibliography of population maps and a glossary of the terminology used in population geography, on which Professor Radó and his colleagues are working at the Institute of Cartography, Budapest. The bibliography is nearing completion; the glossary is to be prepared in several languages. Liaison has been established with the Institute of Ethnography, Moscow, and proposals for the *Soviet atlas of world population* were circulated for comment to members and corresponding members of the commission. Comments received were collated and transmitted to Moscow and it is hoped to continue this co-operation as work on this major atlas progresses.

A grant was made available by the National Science Foundation of the United States to finance a commission in the first phase of a 'Co-operative study of population pressures on physical and social resources in developing countries'. The commission held a symposium in September 1967, at Pennsylvania State University; papers were circulated in advance, so that all the available time could be given to discussion. A *Report* was prepared for the International Geographical Union Congress in 1968 and full publication of all papers and discussions is planned.

The *Atlas narodov mira* (Atlas of the peoples of the world), edited by S I Bruk and U S Apenchenko is unique; published by the Akademiya Nauk SSSR in 1964 with the co-operation of the Chief Administration of Geodesy and Cartography and the N N Miklukho-Maklai Institute of Ethnography. World and regional colour plates show the location and population densities of some nine hundred distinct ethnic and racial groups. The average scale used is approximately 1 : 10M, with larger scales and a greater variety of maps given for areas of the greatest ethnic variety. Such a task as this is monumental in its scope and complexity; in the published maps some omissions may be discerned, some over-simplification or controversial presentation, but all credit is due to this work as a pioneer effort in depicting one of the fundamental preoccupations of the geographer—the contemporary settlement patterns produced by centuries of man's activities and wanderings on the face of the earth. Maps, legends and contemporary text are in Russian, but, in the majority of cases, the index gives English equivalents of names. In general, the basis of grouping is by language; but in several cases, minority groups have been ignored; for example, checking a current issue, African races are not adequately represented in American cities. A subsidiary section of maps presents continental and regional maps of population density and distribution; tables give population figures and twenty eight pages of text discuss population, race, language, religious and ethnic groups.

The uneven distribution of population within a country—the extremes within the British Isles, for example—make difficulties in the effective mapping of distribution. Aerial survey has considerably aided this aspect of mapping; the shapes of communities are immediately apparent on aerial photographs. Much research on the various problems has been carried out by the Population Studies Group of the Institute of British Geographers. Decisions must be made on the actual data to be mapped, whether actual figures, percentages or deviations from a mean, etc; great new possibilities have been opened up with the use of data processing and computer methods.

A useful atlas in a spiral binding is the *Atlas for anthropology*, compiled by Robert F Spencer and Eldon Johnson (Iowa, Dubuque,

William C Brown, 1960). Folded maps in black and white outline cultural areas and tribal groups; language families of the world and the racial distribution of mankind, with sections on the ' Old world ' and the ' New world '. An earlier ' World atlas of population ' was edited by F Burgdörfer and published in Hamburg in 1954.

Original maps or sketch-maps are to be found from time to time in most of the journals devoted to aspects of human geography. Among the most relevant to our purposes are *Homme: revue française d'anthropologie*, 1961-; *Homo*, 1948-; *Landscape: magazine of human geography*, 1951-; *Human geography*, 1947- from Kyoto University, in Japanese with English summaries; *Mankind*, 1931- from the Anthropological Society of New South Wales; *Population*, 1946- from the Institut National d'Etudes Démographiques; *Demography*, 1963- by the Population Association of America, University of Chicago; the United Nations *Population Studies; Kulturgeografi: tidsskrift for befolkningsgeografi . . .*, 1949-, edited by Professor Humlum at Arhus, Denmark, and *Man* published by the Royal Anthropological Institute from 1965, taking the place of the *Journal*, which had been in existence since 1901.

The major monographs containing original maps or information about new projects in population mapping include:

W G V Balchin, editor: *Geography and man: a practical survey of the life and work of man in relation to his natural environment* (New Era Publishing Company, second edition 1955). Three volumes.

Jacqueline Beaujeu-Garnier: *Géographie de la population* (Paris, two volumes, Librairie de Médicis, 1956, 1958).

Jean Brunhes: *La géographie humaine* (Paris, Librairie Félix Alcan, 1934, fourth edition revised by M Jean-Brunhes Delamarre and Pierre Deffontaines). In two volumes, with a third volume of plates and maps.

G F Carter: *Man and the land: a cultural geography* (New York, Holt, Rinehart and Winston, 1964).

Peter Haggett: *Locational analysis in human geography* (Arnold, 1965).

Ellsworth Huntington: *Principles of human geography* (New York, Wiley, sixth edition, revised with E B Shaw 1951).

M A Lefevre: *Principes et problèmes de géographie humaine* (Brussels, Editorial Office, 1945).

A V Perpillou: *Human geography*, translated by E D Laborde (Longmans, 1966).

Max Sorre: *Les fondements de la géographie humaine* (Paris, Colin, three volumes, 1947-52).

Griffith Taylor: *Environment and race: a study of the evolution, migration, settlement and status of the races of man* (OUP, 1927).

Griffith Taylor: *Environment, race and migration: fundamentals of human distribution, with special sections on racial classification and settlement in Canada and Australia* (Toronto UP, third edition, enlarged 1949).

W L Thomas et al, editors: *Man's rôle in changing the face of the earth,* an international symposium under the co-chairmanship of C O Sauer, Marston Bates and Lewis Mumford (University of Chicago P for the Wenner-Gren Foundation for Anthropological Research and the Natural Science Foundation, 1956).

Paul Vidal de la Blache: *Principes de géographie humaine* (Paris, Colin, fifth edition, 1955).

P L Wagner and M W Mikesell, editors: *Readings in cultural geography* (University of Chicago P; University of Toronto P, 1962).

Kurt Witthauer: *Die Bevölkerung der Erde: Verteilung und Dynamik* (Gotha: Haack, 1958).

One of the pioneers of urban geography as a special academic study was Lewis Mumford, especially with his *The culture of cities,* an exhaustive survey and source book for this aspect of social geography (Secker and Warburg, 1938), which includes a group of city plans at various stages of development, with notes on the reasons underlying the development. In his introduction, he wrote ' If Germany is perhaps foremost in urban historic scholarship, France has to its credit the best work on the geographic foundations of cities,' quoting the work of Le Play, Comte, Reclus, Vidal de la Blache, and Brunhes, Lavedan and Poëte. British geographers have not entered this field until recently. Sir Patrick Geddes was a pioneer thinker in planning and sociology, whose ideas have exerted considerable influence on geographers interested in urban geography. He worked closely with French scholars, especially Frédéric Le Play.

The Town and County Planning Act of 1947 required every county and county borough in England, Wales and Scotland to present a development plan to the Minister of Housing and Local Government, recording its history, topography and geology, population, communications, etc. *English county,* published by Faber for the West Midland Group on Post-War Reconstruction and Planning in 1946, including illustrations and maps, was the first industrial area survey and a good example of geographical analysis of an area. The *South-east study* was of great significance in Britain, in that, irrespective of individual proposals, the need for a comprehensive plan was universally recognised. The need for cartographical analysis immediately gives the geographer a definite role in any investigational team: once correct, large-scale maps have been prepared, correlation and interpretation become possible, the need for constant study of the interplay of geographical, economic and sociological factors becomes obvious and the radius of urban spheres of influence can be gauged. One of the themes of the British Association for the Advancement of Science, section E, Geography, meeting at Southampton in 1964 was 'Planning and geography in Great Britain'; a symposium was held on regional planning problems especially on the growth of 'megalopolis' with reference to Greater London and the Midlands.

The regional volumes prepared for the annual meetings of the British Association for the Advancement of Science, by a team led, traditionally, by a geographer, present most useful studies of cities within their regions. Similarly, the works prepared for any geographical congress, for example, the *Guide to London excursions,* edited by K M Clayton for the Twentieth International Geographical Congress, London, 1964, with its excellent maps, sketch-maps and diagrammatic maps. The current interest in urban geography was highlighted in December 1967, when the annual cartographic exhibition and conference held in Budapest took as its theme 'Cities of the world today and tomorrow'. About five hundred maps from fifty six countries were on display, including maps of traffic in cities, development ideas of all kinds and land utilisation maps. Great Britain submitted the greatest number of town planning maps; development plans of Liverpool and Coventry were displayed, together with the planning studies of projected new cities. The Town Development Bureau of Montreal exhibited maps and slides on the Development Plan of Montreal, which is planned for execution through the next thirty years. Urban mapping has reached a high standard, being characteristic in each case of the cartographic qualities of the nation. For example, the Swedish maps are in bold, strong colours; the Malmö sheets, at 1:5,000, are typical, showing deep-blue water areas, blue-grey built-up areas and bright red public buildings. Streets are left white and the black lettering shows up well. The Netherlands produces excellent town plans; the sheets for 's-Gravenhage at 1:5,000 and 1:10,000, for instance, present fine specimens of cartography in greens, greys and light shades of orange and brown; streets and paths are left white. Again, there are clear colours and fine line work, shown to advantage in the 1:5,000 city survey of Berne. In each of these examples, the respective municipal survey organisations have been responsible for their own surveys.

Orell Füssli's maps, such as that of Lausanne at 1:10,000 and of Zürich at 1:20,000, show good design and use of blocks of colour, black being reserved for lettering. Kümmerly and Frey have experimented with the use of black for main road casing lines as well as for lettering; and public buildings have been shown partly in 'perspective', combining blue and white shading with red. Attempts at 'bird's-eye' view mapping have been made by several publishers during the past few years, some recalling the portrayal of buildings in early maps. The 'Plan de Paris à vol d'oiseau', drawn by Peltier (Blondel la Rougery, 1964) is an obvious example.

The particular training of geographers enables them to co-ordinate and analyse the factors involved in, for example, the planning of transport, population units or the location of industries in a way that accountants, economists, sociologists or experts in individual industries are not equipped to do. In the post-war Six year Plan of Poland, for example, a special organisation, 'Geoprojekt', employed many geographers; maps and memoirs, presenting detailed studies of environment and physiography formed the basis of new planning for urban networks, the development of conurbations and of special purpose towns, while climatic studies were co-ordinated and taken into account in the expansion of agriculture, industrial/urban growth and the establishment of health resorts.

The trend towards computer aided road design has greatly increased during recent years, especially in conjunction with photogrammetry to make an accurate survey at a suitable scale, with contours and, perhaps, a terrain model. New low level photography is required, with a fairly dense network of ground control. The soil survey procedure normally applied in Britain consists of an evaluation of existing information prepared by the Soil Survey or Institute of Geological Sciences, site reconnaissance and detailed site investigation. Air-photo interpretation usually reveals drainage patterns, vegetation changes resulting from soil changes and other factors, more quickly, clearly and over a wider area than can be done by ground survey. Geophysical methods are now normally used in detailed survey, for example, some method of seismic and electrical resistivity, in conjunction with carefully considered borings.

From air photographs a mathematical digital ground model can be made of the topography, taking the form of a square grid of spot heights automatically recorded directly on to paper tape. Large models are divided into blocks, each of which may have different and virtually unrestricted grid spacing appropriate to the terrain. Blocks may be square or rectangular and of any size up to 4,000 points. Local topographic anomalies are defined by small close-gridded blocks superimposed on the main data. Two computer programmes are available. The first organises the photogrammetric data into ordered blocks. The second is used to extract ground long-section and cross-section data from the digital ground model, for any route at any width. Levels are interpolated at every station and tangent point and at any interval left or right from the centre line. By this method, terrain data for alternative routes are quickly and easily prepared.

Huntings Surveys and Consultants have developed a service which employs photogrammetry, geophysics, electronic computing and automatic drawing, to assist highway engineers. Overlapping aerial photographs are taken with a six inch focal length wide angle lens camera to the extent of the area under construction. Stereoscopic plotting apparatus is then used to make a contoured strip map from these photographs. Experienced geologists, soil engineers and geophysicists investigate the site to locate bed-rock, to classify the overlying strata or carry out any other process that may be necessary; the contoured map and geological reports enable the engineers to select precise locations for their routes. Broadly speaking, two special methods are available for the automatic preparation of drawings: by the use of an off-line flat-bed tape or card controlled co-ordinatograph, or by an on-line Calcomp graph-plotter.

Some of the outstanding monographs in this context are:

Jacqueline Beaujeu-Garnier and Georges Chabot: *Traité de géographie urbaine* (Paris, Colin, 1963).

K A Boesler: *Die städtischen funktionen* (University of Berlin, *Abh geogr*, 1960).

F Bolt: 'The case for urban cartography', *Cartography*, VI 1 1966.

R E Dickinson: *City and region: a geographical interpretation* (Routledge and Kegan Paul, 1964).

Pierre George: *Le ville: le fait urbain à travers le monde* (Paris, PUF, 1952).

Peter Hall: *The world cities* (Weidenfeld and Nicolson, 1966).

A J Hunt: 'Problems of population mapping', Institute of British Geographers, *Transactions,* 1965.

A J Hunt and H A Moisley: 'Population mapping in urban areas', *Geography,* 1960 79-89.

J N Jackson: *Surveys for town and country planning* (Hutchinson University Library, 1966).

Mary Jelliman: 'A cartographic analysis of the Glasgow 1961 Census', *The cartographic journal,* June 1967.

W L G Joerg: 'Geography and national land planning', *The geographical review,* April 1935.

Knut Norborg, editor: *Proceedings* of the International Geographical Union Symposium in Urban Geography, Lund, 1960 (Royal University of Lund, Department of Geography, 1962).

A G Ogilvie: 'The mapping of population, especially on a scale 1:1M', International Geographical Union Commission for the Study of Population Problems *Report,* 1952.

A E Smailes: *The geography of towns* (Hutchinson, 1953, eighth impression 1964).

M C Storrie and C I Jackson: 'A comparison of some methods of mapping census data of the British Isles', *The cartographic journal,* June 1967.

J Staszewski: *Vertical distribution of world population* (Polish Scientific Publishers, 1957). In English.

Griffith Taylor: *Urban geography: a study of site, evolution pattern and classification in villages and towns and cities* (Methuen, second edition revised 1951; reprinted 1958).

Articles in architectural journals are sometimes illustrated by maps and plans, especially *The architectural review;* also the new journal *Regional studies,* May 1967- from the Regional Studies Association.

The chief cartographical use of transport data has been the straightforward mapping of existing roads, railways, etc. An innovation was the insertion of agreed road classification followed by experiments in showing the flow lines of goods and of commuter traffic patterns, which helped in revealing potential demands. Linear symbols are those mostly used to date; the Ministry of Transport maps of the trunk roads system showing road capacity compared with actual movement of traffic have proved useful in the planning of road improvements. Analysis of ferry transport figures has provided the basis for improvement in design and time-tabling or the provision of a second ferry and bridge, as at Torpoint and Saltash. The current indications of renewed interest in canals is a welcome sign. Further research, however, is necessary into more effective methods of mapping all communication systems, together with capacity, gradients and other influential features, before maps can be used effectively in the total planning of local or national areas. It must also be borne in mind that volume of traffic varies from day to day and from hour to hour, also that different kinds of traffic make different demands on road widths, surfaces and gradients.

The Michelin 'Carte routière' of France and other European countries at 1:200,000 is a good example of the presentation of much data while retaining clarity; the 'Turist karta' of Generalstabens Litografiska Anstalt, Sweden is another. The Automobile Club of Missouri has produced an outstanding 'Street map of St Louis', compiled by Gousha; triple black lines with red in-fill show controlled access dual highways distinct from the other varieties of roads. This and many other firms have given much thought to the problem of the speedy publication of up-to-date information at low cost; Rand McNally, also, have made many experiments in communication mapping, especially urban. The complex nature of American cities presents a great challenge to cartographers. From road maps, the following information should emerge: where roads exist, what types of traffic may use them, which is the best route and of what quality it is, also information which will help in following the chosen route. Many classifications and devices have been used to portray this information, for example, in Great Britain, the Ministry of Transport numbers may be prefixed by M, motorway, A or B. The one inch Ordnance map used separate colours for A roads, B roads and other roads.

The report *Traffic in towns: a study of the long term problems of traffic in urban areas,* one of the *Reports* of the steering group and working group appointed by the Ministry of Transport (HMSO, 1963) is illustrated by maps and figures; A W Gatrell contributed 'The design and production of road maps' paper to the Fifth Meeting of the Symposium of the British Cartographic Society, Swansea, 1950, on 'Current development in cartographic practice and photomechanical techniques' (*The cartographic journal,* June 1966) and the problem is also dealt with clearly in A Morrison's article 'Principles of road classification for road maps of western Europe and North America', in *The cartographic journal,* June 1966, with a bibliography appended.

Road international, quarterly from 1950, published by the International Road Federation, London, is a superb periodical, illustrated with photographs, diagrams and maps, especially of new road projects; according to the editors, 'its controlled circulation takes it to the desks of men involved in the roads and road transport field throughout the world'. André Michelin, foreseeing the rapid development of car travel, issued his first guide-book designed chiefly for the motorist; in about 1913, the first of the Michelin separate sheets at varying scales were produced, flat or folded. At present, some seven million copies of maps of Europe and various parts of Africa are issued and they are probably among the most widely used road maps in the world. For details, the article 'Les services de tourisme du Pneu Michelin: histoire et évolution des publications cartographiques Michelin' should be consulted, in *The yearbook of cartography,* 1963. Every year now, it seems, new publishers of documents for tourists appear; Hallwag motoring maps are long established, however, covering the European countries at various scales, and available in this country through Philip; also Nagel's *Travel guides,* available through Muller, provide encyclopedic information and specially prepared maps. *The Sunday Times*

RAC road atlas (Nelson, 1968), provides, in handy form, comprehensive, reliable and up-to-date information. 104 pages of four-colour maps, at five miles to the inch, and 100 town plans in two colours are printed with great clarity. A new road classification is used and there are 144 pages of information.

World railways, edited by Henry Sampson and published by Sampson Low, gives ' a world-wide survey of equipment and operation of the railways of the world '. Illustrated reports are included on major railways throughout the world and of proposed developments, with many maps; a separate section covers underground railways. A pioneer map of world air routes, by F V Botley, appeared in *The aeroplane* of November 27 1953, drawn on a new version of an octahedral gnomonic projection, devised to give the best possible representation of inter-continental routes and great circles. Since that time, much progress has been made in mapping air routes at international level. The International Civil Aviation Organisation developed in 1947 from the provisional ICAO founded in San Francisco in 1944. The Sixth World Map Conference of the ICAO was particularly important, when proposals were made for new chart requirements. In 1960, at the Stockholm International Geographical Congress, proposals were discussed for the production of a common base map from which the 1:1M International Map of the World and the ICAO charts might be derived. The standard air charts and maps are produced at scales 1:1M, 1:500,000 and 1:250,000. An *Aeronautical chart catalogue* was begun in 1951, presenting a world list of charts conforming to the ICAO standards, with details of price and how they may be obtained; supplements keep the catalogue up to date.

The comprehensive *Géographie universelle des transports* is in four volumes, published by Chaix of Paris. Generously illustrated with photographs, maps and sketch-maps, the volumes deal with French railways, including North Africa and other French colonies; European railways, except those of France and the USSR; railways in America, Australia, New Zealand, Africa and Asia; and, fourthly, an analysis of road, sea and air transport. The annual *Exposé des travaux de l'Institut Géographique National: texte et planches* (Paris; International Geographical Union, Ministère des Travaux Publics et des Transports) includes maps, diagrams and bibliographies.

Monographs on transport geography have increased during recent years; those containing significant maps include:

J H Appleton: *A morphological approach to the geography of transport* (University of Hull, occasional papers in Geography, 1965).

Stanislaw Berezowski: *Geografia transportu* (Warsaw, Polish Scientific Publishers, 1962).

A C Hardy: *Seaways and sea trade: being a maritime geography of routes, ports, rivers, canals and cargoes* (Routledge, 1927). A sound approach to its date.

K J Kansky: *Structure of transportation networks* (Chicago University, Department of Geography, research paper no 84, 1963).

A C O'Dell: *Railways and geography* (Hutchinson University Library, 1956).

Eugène Pépin: *Géographie de la circulation aérienne* (Paris, Gallimard, 1956).

M Perpillou: *Géographie de la circulation* (Paris, Centre de Documentation Universitaire, 1954).

K R Sealy: *The geography of air transport* (Hutchinson University Library, 1957, second revised edition 1965).

THE MAPPING OF HEALTH AND DISEASE
Books and memoirs on various aspects of ' medical topography ' appeared from the early years of the nineteenth century and very interesting they are, but the majority of them contained no maps. It was the severe epidemics of cholera from 1831 that caused the earliest known mapping of a disease in Britain. Dr Robert Baker included a ' Cholera plan ' of Leeds in his ' Report of the Leeds Board of Health ' in 1933; Dr Thomas Shapter compiled a map of the deaths from cholera in Exeter, 1832-34, to accompany his *History of the cholera in Exeter in 1832;* the great German geographer, Augustus Petermann, who was at that time living in England, brought geographical reasoning to bear on the incidence of the disease and produced a cholera map of the British Isles, 1852, showing areas affected in 1831-33, comparing them with the density of population, clearly brought out on the population map which he compiled for the 1851 Census of Great Britain.

The *Physikalischer atlas of Heinrich Berghaus* contained maps of the geographical distribution of disease, this section being published in 1847; the 1848-49 edition of A K Johnston's *The physical atlas of natural phenomena* did not include any such maps, but in the second edition, 1856, a world map shows the geographical distribution of health and disease, with text by A K Johnston. In 1855, Dr John Snow plotted the deaths from cholera in the Broad Street area of London in September 1854 for inclusion in the second edition of his *On the mode of communication of cholera.* Reproductions of all these maps illustrate Professor E W Gilbert's paper, prepared for the Washington meeting of the International Geographical Congress, Commission on Medical Geography, 1952, under the title ' Some early English maps of the geography of health and disease ' and included in the June 1958 issue of *The geographical journal* with the title ' Pioneer maps of health and disease in England '. Also reproduced in this article are Petermann's map of deaths from cholera in London in 1832 (1852) and a series of maps by Dr H W Acland illustrating the incidence of the disease in Oxford. A classic pioneer work was A Hirsch: *Handbuch der historisch-geographischen pathologie*, 1860-64.

Four levels of study can be distinguished from this time—international studies such as the American Geographical Society maps directed by Dr Jacques May, mentioned below; mapping at national level, such as those in the Indo-Pakistan sub-continent; regional studies, as in Africa, and local studies, of which many examples were carried out in England and Wales.

The International Geographical Union Commission on Medical Geography was set up at the Lisbon congress in 1949, to extend previous work, particularly in relation to diets and their consequences, cancer distribution and the *Atlas of disease mortality*, the mapping of blood groups and blood factors. The *Reports* of this commission, 1952-, are documents of central importance. The *Report* to the Rio de Janeiro congress, 1956, concentrated on the theme ' The study of geographical factors concerned with cause and effect in health and disease '. A *Newsletter* is issued three times a year, in which are notes of current work and selected bibliographical information.

At the Washington congress, 1952, the Commission on Medical Geography suggested that the term ' ecology of health and disease ' be substituted for ' medical geography ', thus stressing both the environmental aspect and the emphasis on physiology and pathology. The first stage in the study of the environmental influences on disease patterns is the accurate collection, interpretation and mapping of data. ' Geogens ' is the term given to the relevant geographical factors; these include physical factors, climate, relief, soils, hydrography and terrestrial magnetism; human influences on the natural pattern, distinguishing between natural and artificial densities and distributions, considered in conjunction with the variants in standards of living, communications and any social customs or habits that may affect mental or physical health. Any local flora and fauna that may present hazards directly or indirectly to the health of the inhabitants must be taken into account. The presence of some geogens produces obvious results; in the majority of areas, however, interactions are complex, and painstaking and specific mapping is one of the most evocative methods of bringing new correlations to light. Studies at local or regional level are the most rewarding; they

459

may be finally linked together to form comparison on continental or world scale. The long, historical introduction to *The Atlas of disease mortality in the United Kingdom* should be remembered at this point; Dr Melvyn Howe, of the Department of Geography and Anthropology, University College of Wales, an expert, especially in cancer research, states, ' We must try to associate the death with the place where the life was lived if these maps are to have any meaning at all.' For many decades, the General Register Office in Britain has published an annual analysis of mortality data (Registrar-General's Statistical Review of England and Wales, part III particularly, the commentary), but the method of recording statistics has changed since the compilation of this atlas; current practice is to record death at a hospital only after a six month residence period. Even so, an exaggerated death rate may be given and the essential relevant factor may be disguised, having exerted its effect some years before.

An interim atlas concerning the mapping of disease mortality has been published in Australia; a new series of *Memoirs* on the subject is in progress in Germany. Distribution of cancers is being studied, in addition to Dr Melvyn Howe, by Dr Indra Pal, professor of geography, Balwant Rajput College, Agra, India. Dr May is investigating the mapping of blood groups in their correlation with disease distributions. Dr N C Mitchel's work on multiple sclerosis and goitre in Northern Ireland has brought to light much useful information; Dr Ortiz is studying skin diseases in Mexico; Dr Latapi's speciality is leprosy and he has been among the foremost in the campaign against the disease. Professor A T A Learmonth has for many years studied malaria, cholera and typhoid, especially in India; maps are included in the printed record of his talk to the Royal Geographical Society on ' Medical geography in India and Pakistan ', *The geographical journal,* March 1961.

The *Atlas of diseases,* produced in sheets, 1950- by the American Geographical Society, under the direction of Dr Jacques M May, was issued with the quarterly numbers of *The geographical review,* in which appeared descriptions supplementary to the annotations on the map sheets themselves. The maps include world distribution of major diseases such as cholera, leprosy, and yellow fever; these and the poliomyelitis map all demonstrate the different fields of research in which progress has been encouraged by the application of these cartographical techniques. Two studies of human starvation are included: ' Sources of selected foods ' and ' Diets and deficiency diseases '. With each map are bibliographies and references; in addition, Dr May's three volume work, *Studies in medical geography,* 1958, 1961, should be used to obtain a full appreciation of the maps. Bibliographical references are included in the work.

The German 'Atlas of epidemic diseases,' edited by H Zeiss (Gotha: 1941-5), was enlarged after the war and published as the *Welt-Seuchen-Atlas,* world atlas of epidemic diseases, edited by Professor Ernst Rodenwaldt and Professor Heimut Jusatz in three volumes (Hamburg, 1952, 1956, 1961) sponsored by the Bureau of Medicine and Surgery, Navy Department, Washington. This is an outstanding study, emphasising the relationships between physical environments, men and diseases, and showing the historic and contemporary patterns throughout the world, with particular emphasis on Europe. The editors believed from the outset that climatological maps must form an integral part of the work, hence their inclusion of global maps of temperature, precipitation, etc. There was still, however, no cartographic representation of the various climates of the earth on a relevant classification. The 'World maps of climatology' by H E Landsberg and others, edited by Professor Rodenwaldt and Professor Jusatz, fill this gap; on an equal area projection, at 1:45M, they can be compared with one another and with the maps of epidemic diseases in the world atlas. The five most relevant maps are 'Mean January sunshine', 'Mean July sunshine' and 'Total hours of sunshine' by H Lippmann; 'Generalised isolines of global radiation' by H E Landsberg, and 'Seasonal climates of the earth' by C Troll and K H Paffen.

Further to this work, a series of monographs are in preparation. The first to appear was the study on Libya by Dr Helmuth Kanter, geomedical monograph series no 1, edited by H J Jusatz (Heidelberg Academy of Sciences, Springer Verlag, 1967); the work includes 163 pages of maps and plates. Other regional volumes will appear as they are finished, including those on Ethiopia and Afghanistan, which are nearly complete. 'Libya', published in German, with an English translation, considers the physical geography of the country, illustrated by eleven coloured maps at 1:3M and 1:7·5M and ground photographs, covering geology, landscape, water conditions, climate, vegetation and animals. Some analysis of the population, nomadic and semi-nomadic, is made and an appendix quotes population statistics. Diagrammatic maps depict health services and environmental sanitation, while further maps plot the diseases mainly suffered by the Libyan people, with three maps of disease distribution. With the increasing use of electronic computers and storage of full data on magnetic tape, it will become more possible to isolate individual data or to combine certain factors for cartographic representation.

The leading learned society periodicals including maps and information on specific mapping projects include the *Transactions* of the International Society of Geographical Pathology, Basel; the publications of the Commonwealth Mycological Institute, Kew, information centre for plant diseases and applied mycology, and of the Bureau of Hygiene and Tropical Diseases, London, the documentation centres for this aspect of the subject. *Studies in medical geography* has been published irregularly since 1958 by the American Geographical Society. Among specific monographs, the following are valuable :

A Leslie Banks : ' The study of the geography of disease ', *The geographical journal*, June 1958. Includes bibliographical references.

P A Buxton : *The natural history of tsetse flies* (School of Hygiene and Tropical Medicine, 1955).

R Doll, editor: *Methods of geographical pathology* (UNESCO, 1959).

E W Gilbert : ' Pioneer maps of health and disease in England ', *The geographical journal,* June 1958.

P Manson : *Tropical diseases,* fourteenth edition, 1954.

J M May : *The ecology of malnutrition in five countries of eastern and central Europe* (New York, Hafner, 1963, studies in medical geography series). Concerns East Germany, Poland, Yugoslavia, Albania, Greece.

Malcolm Murray : ' The geography of death in England and Wales ', *Annals of the Association of American Geographers,* 1962, in which maps are used as the basis of the argument.

L D Stamp : *Some aspects of medical geography* (OUP, 1964, University of London, Heath Clark lectures, 1962).

THE GEOGRAPHICAL MAPPING OF HISTORY AND CURRENT AFFAIRS
In the modern mapping of historical and archeological data, two alternatives are currently in use—the depiction of features from the past which have survived, at least in part, into the present, or the compilation of a series of maps, each attempting to show the characteristics predominating at certain past times. An obvious difficulty in the compilation of the second type of historical map is the lack of consistent and comparable records. For example, between the Domesday survey and the introduction of systematic census reports in Britain, there is no comprehensive record. Early local maps abound in most library and archive collections, but even these and the intermittent records that do exist have not been exploited to the full. Two outstanding projects, however, show how useful and generally interesting such interpretations can be, namely the Ordnance Survey series of historical maps and the volumes of the *Domesday geography of England,* compiled by H C Darby and others, 1952-. If well executed, historical maps are able to present

the complex conditions of any period in a more forceful and effective manner than can be done in words; ' The atlas of Czechoslovakian history' is a good example of a carefully conceived and scholarly national venture of this kind.

In 1959, Soviet cartographers compiled a fine 'Atlas of geographic discoveries and explorations', intended as a companion to academic studies of the history of geographic discovery in relation to the development of the history of geographic and scientific knowledge generally, edited by K A Salichtchev and others (Moscow, Glavnoe Upravlenie Geodezii i Kartografii, 1959). Eighty maps cover the period from antiquity to the present, showing the natural environment, distribution of ancient tribes and communication between them, and political divisions at key periods of time. In the main section, the routes followed by explorers and expeditions are shown by colours indicating nationality, the selection of explorations having been chosen with a view to their scientific and practical results. Successive theories regarding the earth are indicated and facsimiles of maps shown at different periods illustrate the development of scientific cartography to the mid 1950s. A really imaginative production throughout, it is perhaps particularly good on the Far East and the polar regions.

Muir's series of historical atlases, published by Philip, are probably the best known in English-speaking countries. *Muir's atlas of ancient and classical history,* in its fifth edition, is edited by George Goodall and R F Treharne; twenty plates, containing twenty seven coloured maps and diagrams, with an illustrated introduction, cover the period from the fifteenth century BC to AD 395. *Muir's historical atlas: mediaeval and modern,* in its ninth edition, revised by Harold Fullard and R F Treharne, includes new maps showing post-war developments in Europe, Africa, India and the Near and Middle East. The emphasis is on political detail, relief features being included only when they have been of particular historical significance. *Muir's historical atlas: ancient, mediaeval and modern,* edited by Harold Fullard and R F Treharne, consists of *Muir's historical atlas: mediaeval and modern* combined with sixteen pages of maps, introductory matter and index from the *Atlas of ancient and classical history,* covering, in its ninth edition, the period from the fifteenth century BC to 1961.

A new publication, Philip's *Atlas of modern history,* edited by Harold Fullard and prepared by the Atlas Sub-committee of the Historical Association under the chairmanship of Professor R F Treharne, was designed to provide an accurate and up-to-date teaching atlas. It is an attractively produced reference atlas of forty eight pages, with index, covering the period 1740 to 1961, with a special interest in the integration of the contemporary world. There is also the *Intermediate historical atlas* prepared under the direction of the Historical Association, which was issued by Philip in a nineteenth edition in 1960.

W R Shepherd, compiler of *Shepherd's historical atlas,* has also been the promoter of a long line of historical atlases. Published by Philip, the eighth edition 1956, reprinted in 1959 and 1962, contains a new supplement of historical maps for the period 1929 to 1961, prepared for the atlas by C S Hammond and Company. Over two hundred pages of maps illustrate ancient, classical, medieval and modern history from 2100 BC to the 1960s. There is a 115 page index. The *Rand NcNally atlas of world history,* or, as it is also called, the *Historical atlas of the world,* edited by R R Palmer, was published in 1957. It comprises forty pages of maps for quick reference on the course of world history and, to its date, is one of the most reliable works.

Westermanns atlas zur weltgeschichte, 1956, covers ancient to modern history in 160 plates, but there is no index of place names. The *Grosser historischer weltatlas,* completed in 1962 by the Bayerischer Schulbuch-verlag of Munich, is the work of many collaborators. Cartography is clear and the whole production has been carefully designed, with a useful annotated text and index. The *Nouvel atlas historique,* by P Serryn and R Blasselle, distributed in Britain by Harrap, includes over a hundred maps and forty pages of textual commentary, tables and an index of names. Sir Gavin de Beer's *Atlas of evolution* is an interesting conception (Nelson, 1964); a concise text includes many line diagrams, photographs and drawings, in which the bold use of colour makes an immediate visual impact. The quality of the map content could be improved, but the printing, done in Holland, is very fine; there is a bibliography and an adequate index.

Everyman's atlas of ancient and classical geography, first published in 1907, was revised in 1952 and again in 1961, published by Dent. In the latest edition, there are eighty maps and a particularly valuable essay by J Oliver Thomson on the development of ancient geographical knowledge and theory. The Elsevier *Atlas of the classical world,* edited by A A M van der Heyden and H H Scullard, was made available for the English-speaking market by Nelson in 1959. It is notable for its linking of history with geography, from prehistoric Greece to the decline of the Roman Empire before the barbarian migrations of the fourth century. Seventy three maps are included, each concerned with one particular theme. The standard of cartography is uniformly high; good topographical maps give the background of the Hellenic civilisation, but not for the

Roman. 475 fine photogravure illustrations cover all aspects of the two civilisations and aerial photographs show sites in Greece and Italy. The maps are accompanied by an informative text. The *Atlas of the early Christian world*, compiled by F van der Meer and Christine Mohrmann, was translated by Mary F Hedlund and H H Rowley (Nelson, 1958). The maps end with the late seventh century; early maps bring out the concentration of the Christian churches in the eastern Mediterranean, central Italy, north-west Africa (modern Tunisia) and southern Spain and plans are included of Alexandria, Christian Rome, Jerusalem and Ravenna. The life and art of the early Christians are illustrated and the text contains extracts from early Christian and other writings.

The *Penguin atlas of medieval history*, edited by Colin McEvely, in a second edition 1964, presents the history of Europe and western Asia from the late Roman Empire under Julian to the decline of the Duchy of Burgundy in 1478. There are some generalised economic maps, but the main emphasis is on political units. *An atlas of world affairs*, by Andrew Boyd (Methuen, 1957-, fourth edition, 1963), contains maps by W H Bromage. It is intended as a guide to current affairs, explained with maps and with text giving the basic facts of some of the complex problems and developments of recent years. It is also available as a University paperback. The ' Serial map service' provided a monthly review of world affairs between 1939 and 1948, published by Phoenix House, after which an annual volume of geographical progress was issued, including plans and sheet maps—the *Serial map service atlas*, with commentaries on aspects of economy and trade.

The historical atlas and gazetteer supplementing Arnold Toynbee's *Study of history* may stand as a work in its own right, prepared with Edward Myers (OUP, 1959, issued under the auspices of the Royal Institute of International Affairs). It is in three parts: a gazetteer of all place names mentioned in the ten volumes of the text, an atlas section of seventy three maps and an index of all names occurring on the maps. The maps themselves are mainly in black and white, with brown sometimes added. Emphasis is naturally on the rise and development of civilisations and cultures, their physical environment and their relations to other countries, especially in Europe and Asia, stressing the dynamic situations which evolved, for example, in the Nile valley and in Mesopotamia. The one map of North America concerns the Indians and their successors, the cattle men. If carefully scanned, the maps yield a vast amount of information and there is a valuable appendix by Professor Toynbee on Hittite sites and locations. Similarly useful are the atlas volumes of the Cambridge Histories series and the

465

Thematic maps & atlases: History

Recent history atlas, 1870 to the present day, by Martin Gilbert (Weidenfeld and Nicolson, 1966), which contains 121 pages of maps by J R Flower.

466

5
MAP
LIBRARIANSHIP

IT IS SURPRISING that, although the value of maps has so long been recognised in Britain and in many British libraries, the fundamentals of map librarianship are only beginning to be considered on any but a local basis. In Britain, there is no professional training in map librarianship and no substantial literature on the subject, as there is for other specialities, such as music, medicine or government publications. The elements of such training are inherent in the c203 option in part II of the Library Association qualifying examination for Associateship, ' Bibliography and librarianship of geography ', but in the time allowed even keen students can obtain no more than an introduction to the two distinct aspects of the subject—knowledge of the content and treatment of early and of modern maps and atlases. Individual instances are not infrequent of students acquiring permanent inspiration from this course to devote some of their professional expertise to the improvement of map collections in public, special or academic libraries, but these are pioneers. Some map curators, in their turn, have during the past few years admitted that their collections could be improved in acquisition policy, arrangement or maintenance and have expressed the wish for some authoritative training.

The third meeting of the British Cartographic Society, held at Swansea in 1965, discussed as a matter of priority, ' Problems of maintenance, storage, classification and procurement for map libraries '. The *Papers* presented, together with the ensuing discussion, were printed in *The cartographic journal* for June 1966 and are included in the list of references at the end of this section.

A Map Curators' Group of the British Cartographic Society, convened by Dr Helen Wallis, held its first meeting at the Royal Geographical Society in December 1966. Meetings are to be held twice a year and a *Newsletter* is being published. This exchange of information is already proving valuable and it will be interesting to see what practical results ensue.

The major map collections have grown and have been organised empirically; they are in the charge of scholars, many of whom are among the leading authorities in their subject. But in the public libraries, there is rarely such knowledge and, even in school, college and special libraries, acquisition, arrangement, storage and maintenance frequently leave much to be desired.

ACQUISITION

The increased use of maps for various purposes has led to a corresponding increase in the literature concerning them, but it is still far from easy to find definitive information about current output on specific regions or subjects. Librarians in the great national geographical libraries and in the libraries of university departments must be knowledgeable on the state of world mapping. Of great value here is the *Bulletin* of the International Cartographic Association, 1962-. The association was formed to encourage the science and techniques of cartography and to co-ordinate cartographical work. Membership is open to all member countries of the International Geographical Union who are willing to co-operate actively in the exchange of ideas. National committees, mentioned in the course of the third chapter of this book, have been formed to act in liaison between national cartographic programmes and the ICA. The first General Assembly was held in Paris in 1961, thereafter in conjunction with the International Geographical Congress, beginning with the London congress in 1964. Technical conferences are also held, the first at the Institut für Angewandte Geodäsie in Frankfurt-am-Main in 1962, when some of the themes discussed were automation in practical cartography, the training of cartographers, and map copyright. The Second Technical Conference was held in 1964 in Edinburgh, as one of the symposia of the congress. The published *Papers* and discussions of all these meetings of experts and the reports of research in the multitude of publications which are always inspired by such occasions, form the central informative documents. The *Bulletin* itself is published usually twice a year; the main language is German, although verbatim reports appear in the language in which they were originally presented.

World cartography, issued irregularly since 1951 by the Cartographic Office, Department of Social Affairs of the United Nations, New York, provides another mine of information on activities, progress and plans in the field of cartography throughout the world. In English and French, the fascicules include longer or shorter articles, reports, notes and valuable bibliographies. Also, for example, index maps of the state of publication of the IMW 1:1M and other co-operative undertakings are to be found from time

468

to time. Fascicule VII 1962 reports on activities, plans and progress in the various branches of cartography, taking into account present needs in all countries. This issue reiterates one of the main purposes of this journal, the stimulation of the dissemination of information among scientists, technicians and cartographic agencies, on an international basis. After a longer interval than usual, Fascicule VIII 1967 appeared, containing the *Reports* on cartographic activities submitted by the participants in the United Nations Inter-regional Seminar on the Application of Cartography for Economic Development, held at Elsinore in October 1965.

The United Nations Regional Cartographic Conferences for Asia and the Far East, of which the first was held in 1955 at Mussoorie, are also reported in *Proceedings* and *Technical papers,* which are invaluable for the informed discussion on current topics and for reviews of achievement in individual areas, not entirely confined to the Far East.

Numerous organisations throughout the world issue publications on the subject, the importance of which to any individual library depends on the extent and depth of its work; among organisations making regular contributions to the literature of mapping, at research level, are the International Association of Geodesy, the International Society for Photogrammetry, the Centre de Documentation Cartographique et Géographique, the Institute of Geodesy, Photogrammetry and Cartography of Ohio State University, the Ontario Institute of Chartered Cartographers, the Commission on Cartography of the Pan-American Institute of Geography and History and the British Cartographic Society. In 1941, the American Congress on Surveying and Mapping was founded, with headquarters in Washington, DC, having sub-divisions for cartography; control surveys; property surveys; topography; instruments and education. The publications arising from these congresses, as from all other similar meetings, are of essential importance. Librarians in university departments specialising in particular parts of the world need to scan specific publications, such as *Canadian cartography* and the *Acta geodetica et cartographica Sinica,* Peking.

The International Association of Geodesy co-ordinates documentation in the field of geodesy and organises international geodetic studies and investigations. The Central Bureau in Paris is responsible for an index containing analyses of all published articles on geodesy; this appears in the *International geodetic bibliography* (Butterworths Scientific Publications, 1928-30-). *Papers* and *Reports* are published in the Association's *Bulletin géodésique,* quarterly from 1924, with a new series beginning in 1946.

All librarians in charge of map libraries will find the *International yearbook of cartography,* annual since 1961, both informative and interesting. This has been a truly imaginative venture, under the general editorship of Professor Eduard Imhof of Zurich until 1967, in collaboration with experts from Austria, Great Britain, Italy, Germany, Sweden, the Netherlands, France and the USA. The *Yearbook* is published by C Bertelsmann Verlag of Gütersloh, together with George Philip in London; Armand Colin, Paris; the Esselte Map Service, Sweden; Freytag-Berndt und Artaria, Austria; the Istituto Geografico de Agostini, Italy; Art Institut Orell Füssli AG, Switzerland; and Rand McNally of the USA. The languages used are German, French and English, with summaries in the two languages not used. So far, there has been no index and enquiries have produced no evidence of a plan for indexing, but, beginning with the 1963 issue, the titles of the articles in preceding yearbooks have been cited in each volume. The first volume dealt mainly with thematic maps, the second, enlarged issue stressed topographic maps, with sections also on the United Nations Bangkok conference and new atlases, illustrated by maps and diagrams. The third, 1963, featured a number of articles on the compilation and design of topographic and thematic maps and editorial policy widened to include pictorial and plastic representations of the earth's surface. The fourth issue took atlas production as its specific theme. The fifth contained *Papers* read at the 1964 Technical Conference of the International Cartographic Association. A variety of topics were treated in the sixth volume, which contained also the statutes of the International Cartographic Association, in English, French and German, with the aim of making its work more widely known to organisations not yet members. In 1967, Professor Dr Konrad Frenzel took over the chief editorship; the volume again increased in size and a great variety of articles was included, especially on thematic maps and the mapping of statistics; this seventh volume contains also all the lectures and discussions presented at the Third International Conference for Cartography, held at Amsterdam. Brief sections giving current information and reports are included at the end of each volume and a feature of the greatest value is the inclusion of specimen plates of various map services which might not otherwise be easily seen in Britain.

The first systematic attempt to list French maps was made by the Cartographic Department of the Bibliothèque Nationale in 1936. Following this venture, the establishment of an annual international bibliography was proposed in 1938. Impeded by the war, however, the first issue of the *Bibliographie cartographique internationale* did not appear until 1949, published by Armand Colin for the Comité National Français de Géographie and the International Geograph-

470

ical Union, with the support of UNESCO and the Centre National de la Recherche Scientifique. This volume listed maps published during 1946 and 1947 in eight countries. Since that date, the *Bibliography* has grown steadily. Editing is done at the Bibliothèque National from lists submitted by participating countries. Minor or unreliable maps are excluded and the ' secret document ' classification, which is still frequently applied to maps, also limits its completeness; each entry is accompanied by a brief annotation, and it is no doubt the best ' check list ' yet available, bearing in mind the time lag in production. All kinds of thematic maps, plans and surveys are included and, at the end of the volume, a separate section is devoted to recent catalogues of maps issued by each country. The whole work is extensively indexed.

The *Bibliographie géographique internationale* also includes maps, but much more selectively.

The *Bibliotheca cartographica* was planned by a working group, the Arbeitskreis Bibliographie des Kartographischen Schrifttums, set up in 1956 by the Deutsche Gesellschaft für Kartographie, as an international bibliography to cover all new publications relating to all aspects of the theory and practice of cartography. It is published twice a year, more than 520 publications being examined for each issue by more than fifty experts and international organisations throughout the world. Editing is done at the Institut für Landeskunde within the Bundesanstalt für Landeskunde und Raumforschung, together with the Deutsche Gesellschaft für Kartographie, and publication is based at the Institute for Cartography and Topography, University of Bonn. Two issues have usually appeared each year since 1957, comprising some twenty thousand references; publication languages are German, English, French and Russian. References appear under the following headings : bibliography, map collections, documentation, general publications, history of cartography, institutes and organisations, topographic and landscape cartography, thematic maps and cartograms, atlas cartography, use and application of maps, special purpose maps, relief and relief maps, block diagrams, globes. A *General catalogue of cartographic literature,* the first volume covering the period from 1900, is also in preparation.

A basic research tool and acquisition check list is the *Annotated index of aerial photographic coverage and mapping of topography and natural resources undertaken in the Latin American member countries of the OAS,* prepared by the Pan-American Union, Department of Economic Affairs, in nineteen volumes (Washington, 1964-65). This work is an exhaustive country by country inventory of all known photographic, topographic and planimetric mapping

471

and maps of geology, soils and land capability, vegetation, ecology, land use and forests. The maps are located on index diagrams and full bibliographic citation is given for each item.

Limited in extent but more quickly available are the accessions lists of the great map collections, the reviews in relevant journals and the (usually) annual catalogues compiled by map publishers.

New geographical literature and maps, the classified list of accessions to the library and map room of the Royal Geographical Society, was begun by the society in 1951, and is now issued twice a year, in June and December. All new maps and atlases acquired are included and, from 1958, the scope was widened to include all articles contained in twenty of the most important geographical periodicals in English, French and German; a selection is made from 150 others. An annual list of completed theses has been included since 1960. From 1967, the map information provided in the issues has been extended to include details of individual map sheets for all except the large-scale series and, for the first time, details of maps published in *The geographical journal* are also being included.

Maps and atlases are also included in *Current geographical publications,* the record of additions to the Research Catalogue of the American Geographical Society of New York, begun in 1938 by Elizabeth Platt and still the responsibility of the librarian and library staff. Ten issues a year are published, in which references to periodical articles as well as books are arranged according to the classification used in the Research Catalogue, regional placing taking preference over thematic. The scale, issuing body and the subject of the maps are cited in each entry. An index to maps in books and periodicals is maintained in the Map Department of the society; this was reproduced in ten volumes in 1967 by G K Hall, forming an invaluable reference work.

All map librarians need to scan these two last mentioned lists for new items relevant to their collections. Essential also is the section ' Cartographical survey ', previously ' Cartographical progress ', in *The geographical journal,* in which attention is drawn to new or completed mapping projects in all parts of the world; this is in addition to the full reviews included in the ' Review section ', on the appearance of a new atlas or major map. At the end of the year, the four issues are issued as a separate publication. *The geographical review,* main organ of the American Geographical Society, also includes reviews of new maps and atlases and, from time to time, a major article surveys recent production of new atlases throughout the world. Reviews of major cartographical works of particular interest in Britain are included in *Geography,* journal of the Geographical Association, and in *The Scottish geographical magazine* section ' Reviews of maps '.

The British Museum *Accession lists* for maps is printed as a supplement to the *Catalogue of printed maps* . . .; an essential reference and acquisitions source lately available is *The photo-lithographic edition of the catalogue of printed maps, charts and plans,* which records the holdings of the British Museum map room to the end of 1964, in fifteen volumes (1967). The *Accessions lists* published by the Bodleian map room and of other university departments are also essential tools for the librarians of map collections in Britain. The librarians of research collections need also to scan some of the other notable *Accessions lists* and indexes circulated by the great map collections throughout the world, e.g. those of the Département des Cartes et Plans, Bibliothèque Nationale and the *Catalogue des cartes* of the Institut Géographique National; the *Geographical bulletin,* bibliographical series and the Accessions lists of the Canada, Department of Mines and Technical Surveys, Geographical Branch; the *Catalogo delle Pubblicazioni,* Istituto Geografico Militare, Florence; the *Katalog över landkart,* Norges Geografiske Oppmåling, Oslo; the *Catalogo de cartas e publicaçoes Instituto Geografico é Cadastral,* Lisbon; the *Kartor, atlaser, böcher* published by the Generalstabens Litografiska Anstalt, Stockholm; the *Karttalvettelo* of the Maanmittaushallitus, Helsinki; the *Index to printed maps* of the Bancroft Library, University of California, Berkeley; the *Catalogue* of maps of the Geodaetisk Institut, Copenhagen, and many others, some of which have already been mentioned in chapter 3 in the context of the particular country. The *Generalkatalog der Deutschen landeskunde,* Institut für Landeskunde, Bad Godesberg, is one of the most complete individual catalogues. Russian mapping output is particularly well documented within the country; in addition to the central journals, each state is responsible for its own catalogues. To give one example, the national output of maps and atlases of the Byelorussian Soviet Socialist Republic has been listed since 1925 in the *Annals* of the publications of the BSSR, issued monthly by the Chamber of Literature of the BSSR, attached to the V I Lenin State Library of the BSSR.

The Map Division of the Library of Congress aims to keep at least one copy of each edition of every map, atlas, globe and other form of cartographic publication having any reference value. The bibliographical work which is constantly being undertaken by the division is therefore particularly comprehensive. A *Bibliography of cartography* is maintained, of which microfilm copies are available. The library has issued printed cards for maps and atlases from the beginning of the card printing programme; a more comprehensive

473

cataloguing of maps began in 1946, though still on a selective basis. Cards for maps and atlases printed before 1953 were included in the *Library of Congress author catalog;* from 1950, entries were included also in the *Subject catalog.* With the reorganisation of the catalogue in 1953, a separate part, *Maps and atlases,* was issued and this policy continued until 1955 (1956), when entries were again included in the main sequences *Books: authors* and *Books: subjects.* Bibliographies of central importance are frequently published, among them the *List of geographical atlases in the Library of Congress,* which is the most extensive bibliography of atlases available and an indispensable reference work. The basic work was in four volumes, 1909-20, by Philip Lee Phillips; a fifth volume, with bibliographical notes, was completed by Clara Egli Le Gear in 1958, containing references to over two thousand world atlases acquired by the library between 1920 and 1955; a further supplement was completed by the same compiler in 1963, on the same plan. A seventh volume comprises a consolidated *Author list* and the eighth volume is the index. *Author* and *Topographical lists* are also included in the first four volumes. Naturally, these volumes contain a vast amount of information on early atlases as well as on the modern.

The Special Libraries Association of New York has devoted much thought to geographical collections and maps. The Geography and Maps Division was organised as a unit of the Washington Chapter in 1941 and the first number of the *Bulletin* appeared in November 1947, thereafter twice a year until 1953, when it became quarterly. Its articles, notes and news items, lists of new maps, books and bibliographies are of the greatest practical use to librarians in charge of map collections; from time to time the findings of surveys and recommendations are included. One of the most important of these was the *Final report* on the cataloguing and classification practices of the larger American map libraries, in the issue for April 1956. *Surveying and mapping,* quarterly from 1941, is another journal that should be scanned by the map librarian, especially in an academic institution requiring to keep up to date with current technical developments. The organ of the Congress on Survey and Mapping, Washington, it was created ' to advance the sciences of surveying and mapping and to contribute to public interest in the use of maps '. It covers all aspects of mapping, both military and civil. *The cartographer,* issued twice a year by the Ontario Institute of Chartered Cartographers, and *Cartography,* from the Australian Institute of Cartographers, Melbourne, are both essential for their notices and reviews and, in the latter, the section ' Cartographic abstracts '. *The cartographer* stresses particularly cartography as a graphic art and aims to provide a forum for the exchange of information and ideas.

Kartographische nachrichten, in six issues a year 1951-, by the Deutsche Gesellschaft für Kartographie, is an erudite and essential source of information, published by Bertelsmann Verlag, Gütersloh. There is a cumulative index, 1951-55. Another journal of the greatest scholarship, the *Referativnyi zhurnal—geografiya,* bi-monthly from 1951, monthly since 1956, by the Akademiya Nauk sssr, Moscow, includes abstracts and references to articles on cartography, maps and atlases.

The national current bibliographies containing maps have been mentioned in the relevant places in this text. Unfortunately, the *British national bibliography* does not include them, so that information covering new British maps must be culled from many sources. *British books* includes the most notable publications. The Ordnance Survey issues a monthly *Publications report* and a frequently revised sales list; for a 2s 6d annual subscription these regular issues, together with all other advance notices and handouts, will be sent. Information on other maps covering Britain is to be found in the *Publications lists* of the Geological Survey, Second Land Use Survey, Soil Survey, etc, and the relevant sections of the catalogues of private firms. The co-ordination of British cartographical activity was strengthened in 1960 when the Council of the Royal Society approved the formation of a Sub-committee for Cartography which could devote itself entirely to cartographic matters, previously discussed by the National Committee for Geography, appointed in 1961. The Sub-committee advises the National Committee for Geography on all cartographic matters, with the particular purpose of formulating British policy in international proceedings, encouraging the exchange of ideas and experiences and the advancement of cartographic knowledge in this country. It is also charged with the co-ordination of the work of cartographic bodies in Britain, when this might be advantageous. The sub-committee is representative of the major governmental, academic and private organisations concerned with the production of maps and, in turn, the committee nominates national representatives to attend international cartographic meetings. Working parties have been set up to study problems connected with the training of cartographers, the standardisation of cartographic terms, the preparation of a glossary of terms used in Great Britain, and automation in cartography, in order to assist the special commissions in these subjects.

In Britain, the Royal Geographical Society has always included cartography in its work and maintains its own cartographic staff; in the *Bulletin* of the Society of University Cartographers for December 1966 is an article, by G S Holland, which outlines the chief cartographic productions of the society during the past thirty years. The Royal Scottish Geographical Society, during the 1960-61 session, set up a Cartographic Committee to establish a separate Cartographic Section of the society. The first independent cartographic body to come into existence was, however, the British Cartographic Society, which was created as the result of discussions at the University of Edinburgh in September 1962 and at the University of Leicester in 1963. A *List* of holdings of the society's library, which is housed in Edinburgh Public Library, is being issued and will become progressively valuable as the library grows. Most important for acquisition purposes is the bi-annual *Journal*, June 1964-; in the first issue were included an account of cartographic activity in Britain and five papers from the Cartographic Symposium held at the University of Leicester. In addition to the articles and the notes and news sections, which enable the librarian to be knowledgeable concerning contemporary developments in the subject, the ' Recent maps ' section, sub-divided by country, and the ' Recent literature ' section will be of permanent usefulness to all concerned with the maintenance of a cartographic collection. New sheets of the IMW 1 : 1M and of the ICAO charts, for example, are announced and important new atlases and other major works are extensively reviewed. A cumulative index to the first four volumes is to be issued with the fifth volume.

National and research map collections are usually as comprehensive as possible in coverage of the home country and representative of the map output of other countries. Legal deposit and exchange account for much of this stock. University collections must be determined by any special regional or thematic commitments undertaken by the departments concerned. Political difficulties, either temporary or more permanent, may limit acquisition from some parts of the world; for example, the export of Survey of India maps is restricted and it is usually difficult to obtain maps from South American countries or from Russia. Collet's Russian Bookshop can assist in obtaining those cartographical works from the USSR which are published for general export and their periodic *Newsletters* are invaluable for obtaining information concerning Russian publications which would otherwise be missed.

All map librarians need also to consult the issues of volume B of *Geographical abstracts*, which covers cartography, edited by Dr E M Yates. A considerable amount of information of cartographic interest is included also in the annual volumes of *The geographical digest*, edited by Harold Fullard (Philip, 1964-). For any map librarian wishing to contact the national cartographical agency of any country a list, with addresses, is given in *Orbis Geographicus*, volume I. Relevant bibliographical information for individual countries has been indicated throughout chapter 3 above; to these must be added the series of *Mémoires et documents*, 1949- of the Centre de Documentation Cartographique et Géographique, Centre Nationale de la Recherche Scientifique, Paris, which concentrates in each issue on a special area. Section A comprises high level articles, followed by section B, ' Documentation cartographique ' and section C, ' Documentation bibliographique '. Volume I, 1949, was devoted to Canada; volume II, 1951, to Belgium; III, 1952, to the British Isles, and so on.

Die Kartographie, 1943-1954: eine bibliographische übersicht, compiled by H G Kosack and K-H Meine (Lahr, Schwarzwald, Astra Verlag, 1955) is a general bibliography of references relating to cartography for the period stated, with a section on historical cartography.

A useful check list to its date is *Map collections in the United States and Canada: a directory*, edited chiefly by Marie C Goodman, a project of the Special Libraries Association, Geography and Maps Division, carried out between 1946 and 1954. 527 collections are included, excepting the following, because of their comprehensiveness—the American Geographical Society, the University of California at Los Angeles; Canada, Department of Mines and Technical Surveys, Geographical Branch; the University of Chicago; Harvard College Library; New York Public Library; the Free Library of Philadelphia; Stanford University Library; the United States Government Departments and Yale University. Another American handy list is *Whyte's atlas guide*, compiled by F H Whyte (New York, Scarecrow Press; Bailey Bros and Swinfen, 1962), which classifies by region and subject the maps in twenty atlases most likely to be found in American libraries. Its principal use is as a guide to specialist maps covering such features as airways, bridges, dams, minerals, etc. *Maps*, by P M de Paris, was the first in the series ' Library resources in the Greater London area ' (Library Association, Reference and Special Libraries Section, South Eastern Group, 1953).

The many excellent catalogues issued by private cartographical organisations usually include a mixture of maps of the home country and other world or regional series, according to their publishing policy. The Clarendon Press, Oxford, Bartholomew, and Philip in Britain, Esselte of Stockholm, Westermann of Braunschweig, all issue annually revised, well-produced catalogues, annotated and illustrated. Edward Stanford, London, now acts as a clearing house for maps and maintains an excellent bibliographical service. The

International map bulletin: a list of maps and atlas publications by many official overseas surveys contains entries for over one thousand map series; individual maps and catalogues are entered from more than 180 countries.

The *RV katalog,* issued by the Reise-und Verkehrsverlag, Stuttgart, is a looseleaf sales catalogue kept up to date by monthly supplements, the *Kartenbrief.* A handy paper-backed volume, *Der kleine katalog: landkarten, reiseführer, globen, atlanten aus aller welt,* has a natural emphasis on German works, but maintains a balanced coverage also for the rest of the world amounting usually to some nine hundred numbered items. Useful index maps and sheet line diagrams of series are included in its pages. These, with the *Zumstein katalog mit register,* first published in 1964 by Zumsteins Landkartenhaus, Munich, are probably the most comprehensive map catalogues available. The latter, which is intended as an annual publication, is proving one of the most important sources of information on world maps, atlases and globes; the inclusion of specimen maps increases its value.

Opportunities for actually seeing maps and atlases on or even before publication have increased in recent years. The host country of the International Geographical Congress every four years stages exhibitions of materials of interest to geographers, many of them specially published for this event, and any exhibitions involving the interest of publishers and various academics, such as the International Printing Machinery and Allied Trades Exhibition and other familiar Book Fairs provide the opportunity of handling maps and atlases not hitherto seen outside the home country. International cartographic exhibitions have taken place in many countries, as, for example, in Hungary annually since 1962, sponsored by the State Office of Geodesy and Cartography (National Office of Lands and Mapping, 1967-), specialising each year in one particular category; national atlases, automobile maps, tourist maps, wall maps for schools, geographical atlases and globes for schools and, in December 1967, city maps under the title ' Cities of the world today and tomorrow'. Hundreds of maps and other cartographical materials from many countries are gathered together on these occasions. National exhibitions, similar in character on a more restricted scale, are frequent. In Britain, for example, the annual conference of the Geographical Association is the occasion for a publishers' exhibition; here again, emphasis is usually on one special theme, but there is also great variety, including displays of Ordnance Survey maps, land use survey maps, relevant government publications, sometimes FAO and other United Nations productions, representative maps, atlases and monographs from all the map-making firms

478

and publishers interested in geographical subjects. Frequently, pre-publication copies are on display and it is the occasion for special catalogues and other publicity material.

ARRANGEMENT
It is convenient for geography students at all levels if the map collections can be stored in or near the geographical library, but there may be some conflict of interests here, as students of other disciplines also require to use maps of various types. Location of materials naturally depends on the subject priorities of individual institutions and the accommodation available; experience has shown, however, how frustrating for the worker it can be to have books, periodicals and atlases, perhaps, in one place and the collections of sheet maps in another department, even in a separate building.

In the case of closed stacks, arrangement according to acquisition order or by size, using a location number, has proved more practical in many libraries, because better use is made of the stack space available. Whatever the system adopted for shelving books and maps, however, the catalogue entries must organise all holdings in an arrangement that satisfies the requirements of users. The system adopted has depended in most cases upon the time when the library collections were started and recorded.

The published general bibliographical schemes of classification have taken Geography into consideration inadequately or not at all. In most general libraries using the Dewey Decimal Classification, 'Geography' is represented by regional descriptions, frequently mixed with light travel literature and history; while material concerned with climate, vegetation, industries, transport, etc is not recognised as having any geographical interest at all. The basic eccentricities of the Dewey Decimal Classification with regard to Geography are only too well known to geographers who try to use a library thus classified, the position being often aggravated by misconceptions on the part of the classifiers themselves. To quote Dr Meynen on this point: 'Unless he has done some work in Geography himself, the librarian will be unable to recognise at once the specific geographic character of a book or an article'. This applies, of course, to the application of all general bibliographical schemes. Further, the use of the DDC in the arrangement of entries in a bibliographical reference work limits the value of such works for geographers.

In the seventeenth edition of the Dewey Decimal Classification, greater thought has been given to the placings for geography and some options are allowed. The basic division 'General geography' is still allocated the 910-919 placing:

911 Historical geography
912 Graphic representations of the earth's surface: atlases, maps, charts, etc (but ' map projections ' are still at 526·8)
913 Geography of the ancient world
914 Geography of modern Europe
915 Geography of modern Asia
916 Geography of modern Africa
917 Geography of North America
918 Geography of South America
919 Geography of the rest of the world.

No geographically knowledgeable classifier will get far with that! ' General geography of specific continents, countries, localities is placed by editors' preference in 913-919, but, optionally may be placed in 930-999, with " general history of these areas ",' says the Preface; and, again, ' Traditionally, DDC has placed the various branches of geography under the specific topics that are areally considered, for example, Economic geography, 330·9; Phytogeography, 581·9; Medical geography, 614·42. However, with the growing academic and lay interest in geographical matters, some general libraries have come to prefer an arrangement that brings all Geography together. This is supplied optionally, at 910·1, Topical geography, which is divided by subject, for example, Economic geography, 910·133; Phytogeography, 910·158; Medical geography at 910·161·4. Earth sciences are still, however, at 550. But, under 910 is 'Areal differentiation and traveler's observation of the earth (Physical geography, formerly 551·4) and man's civilisation upon it. Use 910·001-910·008 for standard subdivisions; but class charts and plans in 912.

910·02 The earth (Physical geography)
                                 Add notations 14-16 to 910·02
  ·03 Man and his civilisation
                                 Add area notation 1-910·03
  ·09 Historical and regional treatment
  ·091 Geography of regions limited by continent, country, locality   Add area notation 1-910·09
  ·093-009 Discovery, exploration, growth of geographic knowledge   Add area notations
       Class geography of specific continents, countries, localities in 913-919
  ·1 Topical geography
  ·2-·3 Miscellany and dictionaries, encyclopedias, concordances of travel
  ·4 Accounts of travel
  ·5-·9 Other standard subdivisions of travel '.

480

H E Bliss was the great theoretician, but he did not take into account the definition, especially during the past two decades, of the subject discipline Geography. Bliss attempted close analysis of the inter-relation of geographical subjects, but it must be remembered that his thinking was basically that of the early years of the century, although it was perpetuated in the second edition of *The bibliographic classification,* 1952. In the introduction to the latter are to be found some curiously involuted concepts: ' Geography, especially Physical geography, or Physiography, as mainly the descriptive science of the Earth's *surface,* is more special than Geology, merging into Physical geology and more particularly into division, Physiographical geology. But Geography, as comprising Biogeography and Human geography, extends beyond Geology, and in this aspect it may be regarded as co-ordinate with Geology and with Biology. Meteorology is regarded as subordinate to Geography and co-ordinate with Physiography '. Human geography is separated from Physical geography and Biogeography. DQ, Geography in general, is only part of the whole subject; regional geography is included under Physical geography, following this reasoning: ' Regional geography, the study of regions and zones with regard to ecological-geographical conditions and distribution, may be regarded either as subordinate to Physical geography or as co-ordinate with it, and therefore with Biogeography. It is distinct from this last and from Political and Historical geography, in that it does not describe habitats, countries and peoples, or races ' (Volume I, 1940, pp 90-91). Bliss quite failed to understand the purposes and methods of regional geography; further, he allocated ' Human geography to Ethnography and Social-political history, Political and Historical geography being ancillary to the latter '. Further again, Bliss follows through a contorted argument—' Economic geography is to be distinguished from Commercial geography, from Human geography and from Historical geography. Briefly, Economic geography comprises more of the physical, mineral and biological, while Commercial geography is less physiographic than the Economic and less theoretical than the Anthropogeographic, which is more biologic, ecologic and ethnologic '. But Economic *geology* is with *Geology*!

The Library of Congress prepared new schedules for geographical works: ' to meet the needs of special collections the library is now proposing an alternative schedule which will collocate all of these works in class G ', to quote the Head of the Subject Cataloging Division. This class embraced Geography, anthropology, folklore, manners and customs, recreation (third edition, Washington, Superintendent of Documents 1954). In the Colon classification, Geography is placed in chapter U, having the following main divisions:

16

mathematical geography; physical; geomorphology; oceanography; meteorology; biogeography; anthropogeography; political geography; economic geography; travel, expeditions and voyages.

In Geography, to a greater extent than in other subjects, any scheme must cater for all aspects of the subject and be capable of satisfactory revision as required. It is the wide scope of this dynamic subject that has necessitated special treatment; previous schemes have been devised theoretically, in isolation from the subject.

The hospitality of the Universal Decimal Classification in coping with special subjects has made it the most used of the general bibliographic schemes in those libraries which have not devised their own schemes. It was therefore to the UDC that geographers looked when considering a scheme that would be internationally acceptable. The Commission on Classification of Geographical Books and Maps in Libraries was constituted during the International Geographical Congress, Washington, 1952; a group of specialists under the chairmanship of Professor André Libault, with a number of correspondent members, including Mr G R Crone, then librarian and map curator of the Royal Geographical Society, conducted discussions on this vital problem, revealing similar difficulties and similar empirical methods of dealing with them throughout the world. The commission's brief was to promote interest in the classification of geographical literature, to stimulate the formation of geographical sections in general libraries and to co-ordinate the work of organisation concerned with the indexing of geographical literature.

An interim explanatory brochure was published for the 1956 Congress at Rio de Janeiro which soon became o p. The greatest difficulties were inherent in the age of the great collections, whose curators began to organise them not only before the systematic approach to the ' organisation of knowledge ', but before the concept of ' Geography ' was fully developed.

In his introduction to the *Final report on the classification of geographical books and maps,* 1964, Professor Libault pays tribute to the skill and devotion of Professor Dr E Meynen, on whose work the final proposals largely rest. He has managed to create an internationally acceptable scheme, embodying such combinations as should make possible the application of the scheme, in general or particular, to any geographical library. The scheme has been recognised by the International Federation for Documentation, which has since issued papers incorporating further points of improvement and updating. The British Standards Institution is at present engaged upon the preparation of an English full edition of the extended UDC for publication by the end of 1969.

Since experience has shown that the 1956 *Report* and the *Final Report* . . . are not as widely known as they should be among either British documentalists or British geographers, a brief analysis of their content may be helpful here, quoting, in particular, some of the basic tenets of Professor Meynen's concept of Geography. The *Final report* . . . was presented to the Twentieth International Geographical Congress, London, 1964, by the International Geographical Union Commission on the Classification of Geographical Books and Maps in Libraries (Institut für Landeskunde, Bad Godesberg, 1964). Part I is a most lucid account by Dr Meynen ' On the classification of geographical books and maps and the application of the Universal Decimal Classification (UDC) in the field of geography'; part II is a ' Draft of a regional classification according to physio-geographical areas of the earth for application in the Universal Decimal Classification (UDC)' by E Meynen, B Winid and M Bürgener; and part III is a ' Draft of a regional classification of the oceans and seas for application in the Universal Decimal Classification (UDC)' by Dr Meynen. Four appendices are included: ' Regional classification on physio-geographical areas of the earth for application in the UDC—common auxiliaries of the main units ' (Proposal); ' Regional classification on physio-geographical areas of the earth for application in the UDC—common auxiliaries of the main units and further subdivision ' (Proposal); ' Regional classification of the oceans and seas for application in the UDC—common auxiliaries of the main units and further subdivision ' (Proposal). The whole text is presented in English, French and German and fixed inside the back cover are a series of maps, showing the proposed notations for classification of the world, Europe, Asia, Africa, North and Central America, South America, Oceania, Arctic (north polar regions), Antarctica (south polar regions), oceans and seas.

As Dr Meynen points out in his ' On the classification of geographical books and maps . . .', 'A carefully constructed, clear-cut classification and a systematization uniformly used for references and bibliography would mean enormous progress for our method of work.' The general system outlined is based upon the concept of geography as leading ' from the natural elements of a land to biological life, and finally to the creation of its cultural landscape ', thus:

Physical geography
    Geological structure and relief forms
    Weather and climate
    Water and water management

Biological life
   Vegetation
   Fauna
   Human race
Human geography
   Economy
   Settlement
   Government and Administration

It determines not only the work of the individual geographer, but also the organization of many bibliographies.' In addition, further concepts must be considered. ' Geography as a scientific doctrine presupposes a theory of the doctrine of landscape. It comprises problems of terms, problems of logic and methodology. Literature on the history of geography, the biographies of great scholars and teachers, the literature on the development of research institutes or organizations form a group of their own . . . Formerly, literature on journeys very often was the primary source of geographical knowledge; therefore, in older bibliographies, they form part of regional literature. Today, we are of the opinion that this literature forms a special group of systematic geography that precedes the regional geography of our times. It must be mentioned here that present-day reports of journeys and study tours are generally no longer a means of promoting geographical knowledge. They are rather reports on the techniques and organization of journeys and research . . . The work of the geographer presupposes some knowledge of research techniques and presentation as well as of the results of other disciplines . . . mathematical knowledge of the earth, mathematical geography (today only partly carried on by geographers), cartography, technique of taking and evaluating photographs, geographical nomenclature, statistics, etc . . . Last not least, the topic " geography and teaching (didactics of geography) " forms a large group of literature of its own '. A comparative schedule is included in this first part of the *Final report* . . . ' Comparative schedule of classification systems used by geographical bibliographies and reference periodicals ', that is, the UDC 91 Geography, *Bibliographie géographique internationale, Current geographical publications, Generalkatalog der deutschen landeskunde* and the *Referativnyi zhurnal—geografiya.*

Professor Meynen's analysis continues with detailed comments on the history of the UDC, the place of Geography in the UDC and the organisation of the UDC, with an accompanying table ' Example of subdivision of the UDC main tables 91—Geography ' which lead directly to the tabulation of the revised UDC 91—Geography, based on the proposals of the International Geographical Union Commission on Classification of Geographical Books and Maps in Libraries, cumulated in 1962. The *Final report* . . . ' sets out proposals for a regional classification as an auxiliary '. A working group is continuing the commission's work with the aim of establishing an international centre of geographical and cartographical intelligence, in co-operation with national centres, using a uniform system of classification based on the commission's proposals. The commission envisages the storage of all such information in a form compatible with modern developments in the use of electronic computers, using perhaps magnetic tape, and foresees the possibility of using these techniques for the reproduction of maps for scanning by the reader, so that he can choose which he needs to study. With the advent of machine readable catalogues and the expansion of the MARC service, this forecast seems by no means fanciful.

An information retrieval system for maps is being operated already in the University of California, Los Angeles, map library, described in the UNESCO *Bulletin for libraries,* January-February 1966. The whole collection of some two hundred thousand maps at this time is classified by the Library of Congress classification and useful comments are made in this article on the application of the scheme to map classification.

The classification used for the arrangement of the *Bibliographie géographique internationale* dates back to the system established by Raveneau in 1891, revised to a certain extent in recent years by Professor Libault; while the other most familiar individual classification, adopted by the American Geographical Society of New York for its *Research catalogue,* was devised by S W Boggs for the Council of the Association of American Geographers, published in 1907 as *Library classification and cataloging of geographical materials.* This is a very detailed classification, able to cope with the classification of articles and pamphlets as well as larger documents.

The classification compiled by G R Crone, when he was librarian and map curator of the Royal Geographical Society, is interesting in that it has a philosophical, truly ' geographical ' approach. In the foreword to 'A classification for Geography ', published by the society in 1961 (Library series no 6), Mr Crone explains that ' it is the result in part of the experiences of the library staff and of an attempt to produce a scheme related to the requirements of Geography today. The ruling idea has been to move from the separate subjects in which the geographer may be interested through their interrelations and on to the final synthesis.' In this scheme, the main classes are ' Exploration of the environment '; ' the Physical environment '; ' the Biological environment '; ' Integration and relationships '; ' Geographical synthesis '; ' Geographical studies: the end-product'; 'Aids to geographical studies'. Each of these classes contains at least two sub-divisions, each again sub-divided into several sections. The individual nature of the collection for which the classification was devised is revealed in the sub-division 'Organisation of expeditions', with sections 'Hints to travellers', 'Equipment', etc, and in the detailed treatment of the sub-division 'Geodesy, survey, cartography'. The level of practice and use is also implicit. In the hands of less knowledgeable classifiers, the scheme would allow of some cross-classification between, for example, the 'Geographical synthesis' class, which includes Human geography, the social environment, city regions, regional planning, etc, and 'Geographical studies: the end-product', where regional studies proper are found, including regional description and cultural landscape (Landeschaftskunde). The class 'Aids to geographical studies' includes almanacs, ephemerides, bibliographies, biography, catalogues, dictionaries, directories, encyclopedias, gazetteers, place names, glossaries, guides, maps and atlases, periodicals, readers' guides, statistical methods, world history and yearbooks.

In accord with the Oxford approach to geographical studies, the emphasis in the classification adopted by the department library is regional, with the systematic aspect subsidiary. The classification of the Association of American Geographers is used for a classified index of periodical articles, for which a really detailed scheme is needed. In the Oxford scheme itself, numerals are used for the regional schedule, for example, 200—British Isles; 220—England and Wales; 230—Wales; 300—Europe; 340—France . . . 800—Australia. Letters, A-Z, with some omissions, are used for systematic sub-divisions; A—Educational geography; B—Mathematical geography . . . H—Climatology and meteorology . . . T—Terminology; Y—Atlases. Precise shelf arrangement is achieved by adding a running number to the class number, so that a book on the climate of France might bear the number 340·H·23. The number of pamphlets is preceded by the letter P, which is not used in the schedules.

Another detailed scheme devised for a particular purpose and working admirably in that context, is the War Office classification, which demonstrates most of the points to be taken into consideration in the organisation of a large map collection. 'Provision is made for all maps available and for those which may become available in the future'. Letters of the alphabet are allocated to the main divisions:

A The Universe (astronomic charts, etc)
B The world
C Europe
D Asia

E   Africa
F   North America
G   Central America
H   South America
I   Australasia
J   Pacific Ocean (and islands not adjacent to continents)
K   Atlantic Ocean (and islands . . .)
L   Indian Ocean (and islands . . .)
M   Arctic regions
N   Antarctic regions.

Each division is then sub-divided, each being given an arabic number. In the continental divisions, the sub-divisions are allocated to countries which are arranged in alphabetical order and numbered consecutively, for example: c1—Europe general; c2—Europe, seas, gulfs, channels, etc; c4—Austria; c5—Balkans . . . A mnemonic feature is included, since the sub-division ' 1 ' is always ' seas, gulfs, channels, etc '. Further sub-division is achieved by means of a colon, for example, c1 : 1—northern Europe. Town plans are placed within the relevant classification, with the name of the town added. The scheme is intended to be operated always in conjunction with a standard atlas and gazetteer for definition. Other general rules govern niceties of placing. A sequence number in brackets is added to the class number to complete the reference; for example, the first and second maps to be acquired of a province of Norway would be c36 : 13(1) and c36 : 13(2), so that each sheet has a unique reference number. The staff record is the *Handlist,* in which the entry is abbreviated title, scale, date and number of sheets, for identification. The complete *Handlist* forms an inventory of the whole collection, arranged in the same order as maps are housed in the map presses. ' It should be borne in mind that the handlist is the most valuable of all library records and great care should be exercised in its maintenance and use.' (Parsons: *see* reference at end of section.) Finally, sequence numbers are allotted to map series within the classification divisions, for example, 1-10 denotes series larger than 1 : 250,000; 11-20, series of 1 : 250,000 and larger than 1 : 500,000, etc. Small letters, (a) (b), etc are used to indicate reproductions and variants. The handlisting of atlases and guide-books follows a similar pattern.

A M Ferrar, of the Bodleian Library, advocated the use of symbols consisting of a capital and lower case letter directly suggesting the region, for example, Eu for Europe. Logically developed, this system gives exact delineation plus mnemonic quality.

The Geographical Association has given much thought to the whole problem of the arrangement of geographical materials. At the annual conference in 1962, an exhibition was mounted to display the application of a library classification system to materials of geography teaching. A classification scheme for the association's library was evolved, having taken into account the individual systems used in other comparable collections. All material is classified by region or by subject, using a numerical notation 100 to 900 for the former and a literal one, a-p, for the latter, both notations being capable of detailed sub-division. In the subject hierarchy, for example, the letter ' g ' identifies works on economic geography, ' gc ' distinguishes material relating to agricultural products and manufactures based on agriculture, and ' gct ' refers specifically to documents of information concerning textile manufactures. Within each final sub-division, thus distinguished by a combination of numerals or letters or both, items are arranged alphabetically according to the names of the authors. The regional factor is given preference in placing, with added entries in the catalogue for related regions or subjects as necessary.

In the scheme drawn up for school use by the Cheltenham Ladies' College (second edition 1958), a separate class P has been allocated to Geology and geography. It is a numerical scheme, in three main divisions, ' Geology ', ' Geography ' and ' Geography by countries '. All are sub-divided. The Geology section includes meteorology and climate. ' Geography ' includes Map-making and reading, Surveying, Travel and exploration, as well as the other usual aspects. Sub-divisions, such as English counties or American states, are intended to be arranged alphabetically, but geographers are invited to make their own decisions for the sub-divisions, possibly using a decimal point after the main number.

A simplified version of the extension of the UDC 91 classification, described above, was made available for school use in 1965.

Much of all this theorising on the potentials of various classification schemes is, however, rapidly assuming only historical interest, as increasing numbers of the world's great library and map collections move towards the use of computerised catalogues. Full cataloguing remains the essential process. The cataloguing of early maps involves very specialised knowledge and, even in the case of modern maps, considerable experience is desirable. The majority of modern maps do bear a title, but, if no title is given on the sheet, it may be necessary to provide one, as definite as possible. Precise regional entry is the practice of the Royal Geographical Society; the British Museum rules state ' the main entry of every atlas, map, etc, is placed under the generally accepted name of the geographical or topographical area which the work delineates '. The rules of the Library of Congress, however, state that the regional heading of the entry is to be referred to the widest possible geographical unit in which it occurs. Effective use of mechanical aids such as computers does, however, require very great precision in the use of the terms in which concepts are expressed, otherwise much of the vast potential economy of the retrieval system will be vitiated. At national and international level geographers are making great efforts to secure agreement on terminology in the interests of universal understanding of research, and the published glossaries embodying the results of their discussions will be valuable also if required for computer programming.

For modern maps, the government department, survey office or private publisher is, in the majority of cases, stated on the map sheet. Even now, however, the date is not always given and this should be established as nearly as possible. Indication of scale is essential; this should be given as stated on the map, but it is helpful to reduce it also to a uniform fraction or proportion. All other relevant information should be noted; most important are the projection used, series status, including revision dates, names of cartographer and printer, method of relief representation and/or of subject data shown, inset maps and plans. It is important to state the exact limits of the map and its size and physical format. It may be monochrome or in several colours. If the map concerned has been produced in connection with some special research or has any other particular qualities, the annotation should include this information; also reference to any other maps or documents which may have special bearing on the map or assist in its full interpretation. Index diagrams to map series should be available for use in conjunction with the catalogue and added entries need to be made for all significant features. With regard to place names, the principles adopted by the Permanent Committee on Geographical Names or other recognised lists should be used.

Frequently, separate sections of the catalogue or individual indexes are made for ' Series ', ' Place ', ' Scale ', etc. The map collection of the American Geographical Society of New York is particularly fully catalogued; items are filed under detailed subject headings, based on the principles of Samuel Boggs. A completely revised edition of *Cataloging and filing rules for maps and atlases in the Society's collection* was issued in 1964. The most useful manual on the subject is S W Boggs and Dorothy Cornwell Lewis: *The classification and cataloging of maps and atlases*, first published in 1932, with a second edition, published by the Special Libraries Association 1945. In the introduction, the main features and problems of map classification and cataloguing are stressed, beginning with the general assertion that, ' The differences between maps and

489

books are numerous and fundamental. In approaching the problem of map cataloging, it is therefore not in the interest either of specialists in library science or of the map users who are to be the catalog's beneficiaries to begin with the assumption that maps are to be regarded as if they were simply books in another format '. Much of the poor cataloguing of maps and atlases in general libraries arises from just this assumption.

A committee of the Geography and Map Division of the Special Libraries Association also made a special study of the cataloguing of maps, atlases, globes and relief models, of which the first report was published in the *Bulletin* of the division, no 3, December 1948. The committee continued to investigate practice in the large American libraries and a second report appeared in *Bulletin* no 13, October 1953. The committee's final report, with comments by Charles W Buffum, Map Division, Library of Congress, and B M Woods, was issued in *Bulletin* no 24, April 1956.

Many map curators have compiled their own rules for guidance, some of which have been published; notable examples include *Rules for descriptive cataloging in the Library of Congress* (relevant section), 1949, and *Règles adoptées par le Département des Cartes et Plans pour la conservation des collections et la rédaction des catalogues*, 1951, Bibliothèque Nationale.

STORAGE AND MAINTENANCE

The basic document to date for English readers is *The storage and conservation of maps: a report prepared by a committee of the Royal Geographical Society*, 1954, which was published separately and in the June 1955 issue of *The geographical journal*. The committee concerned with the drafting of the *Report* was the Library and Map Committee of the RGS, composed of members representing the RGS, the British Museum map room, the Ordnance Survey, the Directorate of Military Survey, the Library Association, the British Records Association, the Society of Local Archivists, the British Standards Institute, the Hydrographic Department of the Admiralty, the Bodleian map room, the University of Cambridge Department of Geography and the University College, Swansea, Geography Department. Other brief, but useful, sources are included in the list of references at the end of this section.

The method of storing sheet maps depends on the size and purpose of the collection. The aim should be to preserve them in the state in which they came from the publisher; equally important is it to protect them from the effects of atmosphere, dirt and handling. If they are to be handled constantly, as in a public library local collection, for example, one set may be kept freely accessible in the library for general use; the most used are usually the Ordnance Survey six inch series in British local collections and these may be bound around the edges with suitable material and kept in suitably sized boxes. The most constant demand for maps for loan is usually for the one inch series; here again, a duplicate set of dissected and folded linen-backed maps may be acquired which may be inserted in manilla pockets made to size, along the spine of each being marked the location number used and the area covered by the map.

The most used method in academic and research libraries of storing sheet maps is flat in large-sized shallow drawers; too many should not be piled on top of one another, otherwise it will be difficult to pull any one sheet out. The whole contents of each drawer may be kept in a manilla portfolio, or they may be sub-divided into a suitable number of folders; the main point being that no attempt should be made to pull out individual map sheets. It is axiomatic that both the outside of the drawer and each individual folder should be clearly labelled. The drawers must not be overloaded. In some cases, an additional fold of cloth or other suitable material has been inserted in each drawer, to protect the contents from dust. The front of the drawers should be hinged so that it falls out of the way when opened. The number of drawers in a unit depends on use; if storage only is the objective, then the cabinets may be built to any required height, but if the room is to be used for study purposes also, then the top of the cabinets should be at a convenient height for consultation of the maps required. In any case, some flat space must be allowed for staff use. The choice between wood or steel cabinets is largely a matter of taste and cost. From experience, it has been proved that well-constructed wooden cabinets wear just as well as steel, which tends to chip and the locking devices tend to be less co-operative than in wooden ones. The argument previously put forward in favour of steel cabinets in case of fire was disproved during the war, when it was found that paper in wooden cabinets was only charred, whereas the steel had generated so great a heat inside the drawers that a much greater proportion of the contents was destroyed. Steel drawers lined with asbestos may provide the perfect solution. In tropical countries, of course, steel would be the obvious choice. Steel was the choice of the Royal Scottish Geographical Society, when the map room was reorganised and refurnished in 1958 and experiments were made in design; map cases in a darker shade of blue against pale blue walls were installed. Two large tables were provided for readers, one including a tracing panel. No windows were made, but an oval lay-light almost fills the ceiling area, above which is the fluorescent lighting. The map room of the National

Library of Wales at Aberystwyth is equipped with rows of wooden cabinets so placed that their combined tops make for ample working surface.

Vertical filing methods take various forms. Generally speaking, this method is preferable for smaller maps; it is also convenient in the case of modern map series having the index number at the top of the sheet. According to contents, suitable forms of folder should be used to contain maps in short series, maps on a particular subject, or, in a long series, a sufficient number of sheets that can be handled easily. One or two permanent divisions should be placed in the drawers, otherwise the folders of maps become unmanageable when any number of them have been withdrawn for use. At the same time, the total contents of each drawer should not be packed too tightly or the folders at the back tend to become damaged. Maps should not be stored vertically loose, without folders; even with the protection of the folder, the lower edge may become crumpled if care is not taken. There are various forms of vertical or horizontal suspension filing. If this method is adopted, the side by which each map is suspended must be reinforced. The Roneo Flushline Vertical Planfile has been designed to store plans, maps or charts freely without the necessity for suspension bars or folders. The principle lies in the form of the half-dozen or so vertical 'divides', which are slightly curving in formation instead of being straight; the slight friction caused by the curve holds the sheets firm and avoids the danger of crumpling the bottom edge. Closed, the cabinet is free-standing and neat, with the top providing a working surface. It opens by a hinge at the front, so that the desired partition can be pulled forward by hand and the sheets lifted out. In the closed position, the 'divides' close up to hold all the sheets steady. The standard model holds fifteen hundred sheets, but takes up less room in depth than a cabinet in which the drawers must frequently be pulled out to their fullest extent. The Stanford Planfile, designed specially for Ordnance Surveys maps and plans, can accommodate up to two hundred sheets by a method of suspension filing and an index to the contents can be mounted on the inside of the lid.

Wall maps and all maps on rollers should be handled carefully; when rolled, they should be tied firmly by the tapes provided and either hung vertically from hooks or, preferably, be kept horizontally in racks or troughs. The Olympia map stand and Rigid Map Hook have been designed by Westermann for the compact storage and display of wall maps in a small space. George Philip and Son offer a wide range of apparatus for holding and suspending wall maps. Aluminium is being widely used for a variety of fittings; a lightweight rail of aluminium and plastic, for example, is available for holding sheet maps firmly for display without damaging the edge, at the same time giving edge to edge vision of the map sheet.

Small- to medium-sized public libraries holding perhaps a few authoritative ' handy ' atlases, *The times atlas of the world* and *The atlas of Britain and Northern Ireland,* can easily shelve these works in the normal oversize sequence or, preferably, in one atlas stand. Holdings of sheet maps may comprise the one inch Ordnance Survey, seventh series, which can be housed in a stout looseleaf binder made for the purpose; the local collection should contain at least the six inch and the twenty five inch maps for the area, requiring a cabinet, folder or box made to size for the former and a cabinet for the latter. The larger public libraries are in the same position as the academic and special libraries, except in the sheer bulk of map stock. The New York Public Library map collection, for example, is among the largest in the world. The time is long overdue for British public libraries to take as much care in the acquisition and exploitation of their map collections as they do of their music and other special materials. Complete sets of current Ordnance Survey editions should be acquired, plus a representative selection of the most interesting of foreign map production; local collections should be much more systematic than they frequently are in building up and making known their map collections concerning the area. Such a policy requires, of course, knowledgeable staff, proper equipment and full documentation, but the resulting public service would be well worth the effort.

Small atlases may often be kept on normal quarto or folio shelves. The larger world or national atlases should be stored on special atlas stands, of which there are various designs, single or double sided, and usually having a slightly sloping top, with a ridged edge to prevent the volume slipping when it is in use. One or at most two atlases should be kept on each shelf. Some of the largest atlases, especially the looseleaf variety in boxes, for example, the *Atlas van Nederland,* present special problems and usually require specially constructed shelving. If large atlases are stored vertically, care must be taken that they do not slant sideways, thus damaging their pages, even when they are specially guarded. Plenty of flat working space is essential when maps and atlases are being used, for each reader will need three or four times the space normally allotted in a library.

When maps taken from monographs, periodicals, directories, guidebooks and the town guides issued by local authorities are kept for information purposes, either vertical filing or a series of suitably sized pamphlet boxes may be most serviceable, according to use. Most of these categories of maps may have only one year of active life and therefore require no other protection. If such maps are to

493

be used for a longer time, it may be necessary to dissect them and mount them on linen; the paper surfaces will crack if they are not dissected. In all cases, ease of access and protection from dust, damp or undue warmth, as from radiators, must be the chief considerations.

Opinions differ widely on the desirability of using lamination processes to protect the surfaces of maps; as these processes are improving all the time, the original objections are to a large extent met, but in any case only a minority of maps in a collection are suitable for this kind of protection.

Repair of modern maps largely consists of mending tears or damage due to friction at the edges of folds. Cellophane should not be used, unless it can effectively mend the tear from the back; it is usually better to mount the whole map, taking extra care to press the torn edges exactly together.

Microfilming of duplicate copies of maps is a useful way of saving space in a working library. Modern colour film exposed through a 35 mm camera can produce reasonably good copies of coloured maps, though the difficulties are greater than in letterpress reproduction. Careful selection must be made of the microfilm reader provided, for not all of them are suitable for map scansion. One successful viewer is the Recordak Archival Viewer, manufactured by Williamsons and marketed by Kodak; it provides two scales of viewing, projecting downwards to an inclined surface on which tracing can be made. Further research is still needed in this kind of equipment.

The map library should contain a tracing table and should be equipped to provide photostat copies of maps, subject to the usual copyright limitations.

REFERENCES

S W Boggs and D C Lewis: *The classification and cataloging of maps and atlases* (Special Libraries Association, second edition 1945).

J Burkett: *Special library and information services in the United Kingdom;* section 8 on map libraries by G R Crone (Library Association, second edition 1965).

G R Crone: ' The cataloguing and arrangement of maps ', *Library Association record,* 1936, 98-104.

G R Crone: *Existing classification systems.* Report of the Commission for the Classification of Books and Maps in Libraries (IGU Eighteenth International Geographical Congress, Rio de Janeiro, 1956).

G R Crone: 'Notes on the classification, arrangement and cataloguing of a large map collection', *The Indian archives,* VII, 1, 1953, 8-13.

G R Crone and E T Day: 'The Map Room of the Royal Geographical Society', *The geographical journal,* March 1960.

Romañ Drazniowsky: 'Bibliographies as tools for map acquisition and map compilation', *Cartographer,* III, 2, 1966, bibliog.

Romañ Drazniowsky: *Cataloguing and filing rules for maps and atlases in the Society's collection* (American Geographical Society of New York, 1964, mimeographed and offset publication no 4).

L Dufresnov: *Catalogue des milleures cartes géographiques générales et particulières* (Amsterdam, Meridian, 1965).

A M Ferrar: 'The management of map collections and libraries in university geography departments', *Library Association record,* May 1962, 161-165.

M E Fink: 'A comparison of map cataloging systems', *Bulletin,* Special Libraries Association, Geography and Map Division, December 1962.

M Foncin: 'Some observations on the organisation of a large map library', *World cartography,* III, 1953, 33-40.

G H Fowler: 'Maps', *Bulletin* of the Technical Section, British Records Association, no 16, 1946.

A C Gerlach: 'Geography and map cataloging and classification in libraries', *Bulletin,* Special Libraries Association, Geography and Map Division, May-June 1961.

C B Hagen: 'Map libraries and automation', *Bulletin,* Special Libraries Association, Geography and Map Division, December 1966.

Clara E Le Gear: *Maps, their care, repair and preservation in libraries* (Library of Congress, Reference Department, Division of Maps, Washington DC, 1949; revised edition 1956). Includes an extensive bibliography of predominantly American works.

R J Lee: *English county maps: the identification, cataloguing and physical care of a collection* (Library Association, 1955). Mainly concerning earlier maps, but some useful comments applicable to modern ones.

D C Lewis: 'Maps: problem children in libraries', *Special libraries,* March 1944.

D H Maling: 'Some thoughts about miniaturisation of map library contents', *The cartographic journal,* June 1966.

Emil Meynen: 'A geographic classification of geography material as based upon the Dewey decimal classification', *Annals* of the Association of American Geographers, XXXVI, 1947, 209-222.

Emil Meynen: *UDC 91—Geography*. Proposal for a revision of the group UDC 91—Geography. (International Geographical Union Commission on the Classification of Books and Maps in Libraries; Ninth General Assembly and Eighteenth International Geographical Congress, Rio de Janeiro, 1956.)

Emil Meynen: 'Die wissenschaftliche kartensammlung grundsätzliches und hinweise', *International yearbook of cartography*, III, 1963. Includes some sketches of equipment and a bibliography.

I Mumford: ' What is a map library? ' *The cartographic journal*, June 1966.

E J S Parsons: *Manual of map classification and cataloguing prepared for use in the Directorate of Military Survey* (War Office, 1946).

P L Phillips: *Notes on the cataloging, care and classification of maps and atlases, including a list of publications compiled in the Division of Maps* (Washington DC, Government Printing Office, revised edition 1921).

R T Porter: ' The library classification of geography ', *The geographical journal*, March 1964.

R A Skelton: ' The conservation of maps ', *Bulletin of the Society of Local Archivists*, no 14, October 1954. A discussion of map storage and use problems in general and the British Museum plans and practices in particular.

Special Libraries Association, Geography and Map Division: *Cartographic research guide, Part IV: The catalogue library and the map librarian* (New York, SLA, 1957).

United States Department of the Interior, Geological Survey: Federal Map Users' Conference, October 1964, *Proceedings.*

Helen Wallis: ' The rôle of a national map library ', *The cartographic journal*, June 1966.

Ernst Winkler: ' Das system der geographie und die dezimalklassifikation ', *Geogr Helv*, 1946, 1, 337-349.

M L Wise: ' Geography in the public library ', *Library Association record*, December 1954.

B M Woods: ' Map cataloging ', *Library resources and technical services*, Autumn 1959.

INDEX

The 'word-by-word' method of alphabetisation has been used; alternative headings or cross-headings are given as necessary. Capitals, eg FAO, are treated as if printed in full; the first part of double-barrelled names forms the main entry.

Admiralty charts 'Geographical handbook series' *contd*)
  Syria 247
  Tunisia 355
Hydrographic Department: *Antarctic pilot* 400
representation on RGS Library and Map Committee 490
Adolphe, H: *Bibliography of Mauritius, 1502-1954,* Auguste Toussaint and H A 268
Adonias, Isa: *La cartografia da região Amazonica: catalogo descritivo (1500-1961)* 335,
Adriatic coasts, maps 205-239
Adriatic Institute 238
Advance Revision Information Service, OS 133
Advance atlas of modern geography (Bartholomew) 90
Ady, P H: *The Oxford regional economic atlas—Africa* 346
A E G, Germany: instruments for automated cartography 37
Aerial Photographic Unit, Cambridge University 56-57
Aerial Photography Institute, Central Geodetic Department 259
Aerial Photography and Photogrammetry, Congresses 24
Aerial Survey Methods and Equipment, UN Seminar 24
'Aerial surveys and intergrated studies' conference, Toulouse 17
Aero Service Corporation, Philadelphia 18
Aerodist 13
Aerofilms Library 22
Aero Service, work in Arabia 255
Aerograph air-brush 78
Aerology 417
Aeronautical Chart and Information Centre, automated management system 304
  Catalogue 457
mapping services 301
The aeroplane 457
Afdeling Hydrografie, Ministerie van Marine, The Hague: *Meteorologie Nederlands Niewe Guinea . . .* 384
Afforestation, Wales 166
Afghan Tourist Association: map at 1 : 2M 257

Afghanistan: Geological and Mineral Survey 257
geomedical monograph 461
German Geological Mission: 'Geological man of Africa 1 : 1M' 257
maps 74, 257-58, 269
geological 257
'Survey of land and water resources', FAO 257
Africa: aerial surveys 17
'An agroclimatic mapping of Africa' 349
An atlas of Africa, J F Horrabin 346
An atlas of African history, J. D Fage 347
atlas of climate 349
An atlas of foreign affairs, Andrew Boyd and Patrick van Rensburg 346-47
The changing map of Africa 347
'The climate of Africa', B W Thompson 349
climatological atlases 348-49
Climatological Unit 348
Commission for Technical Co-operation in Africa 'Soils map of Africa' 350-51
control surveys 353
disease distribution, regional studies 459
East European World Map, sheets 70
Economic Commission for Africa, UN 347
economic research 347-48
Economic transition . . . , M J Herskovits and Mitchell Harwitz 348
edition, Philips' educational atlases 104
The ethnographic survey of Africa, Daryll Forde 345-46
fauna and flora, conservation 347
Geological Surveys Association, *Geological map of Africa* 349-350
Géologie de l'Afrique, Raymond Furon 350
International Seminar on African Primary Products and International Trade, Edinburgh 348
IMW 1 : 1M sheets 64

Agassiz, Alexander 423
Agency for International Development 297
Agraratlas über das gebeit der Deutschen Demokratischen Republik, Hermann Haack 210
The agrarian history of England and Wales, ed: H P R Finberg 155
Agricultural Afghanistan, I N Vavilov and DD Bukinich 257-58
An agricultural atlas of England and Wales 156-57
Agricultural atlas of Missouri 307
An agricultural atlas of Scotland, 1931 169
Agricultural atlas of Sweden 184
The agricultural atlas of the Ukrainian SSR 262
Agricultural Economics Research Institute, Oxford 157
The agricultural geography of Uganda, D N McMaster 437
The agricultural history review 433
Agricultural Institute, Soils Division, Wexford 173
Agricultural origins and dispersals, C O Sauer 433
Agricultural Research Council, London 154
Agricultural situation, US Agricultural Marketing Service 432
Agricultural typology, IGU Commission 131
Agriculture, advisory work 154
 Association, Great Britain 432
 Atlas of Western Australian agriculture bibliographical references 432-34
 Board, surveys of Scotland 169
 co-ordinated planning 453
 irrigation 276
 Malta: *Agriculture in Malta: a survey of land use,* H Bowen-Jones 233
 maps 91, 106, 164, 169, 173, 180, 209-10, 273-74, 308, 338, 429-35
 Ministry, Great Britain: Bulletins 155
 control of OS department 124
 Research monographs 155
 Statistics Branch, assistance with agricultural atlases 157
 atlases 157

Agriculture (*contd*)
 Survey of Grasslands of England and Wales 150
 Agriculture, 1954, graphic summary: land utilization, farm machinery and facilities, farm tenure, US Department of Agriculture 306
 research 429-30
 Agriculture en Roumanie: album statistique 238
 Royal Highland and Agricultural Society of Scotland 171
 'Agriculture in Russia' 262
 Soil Commission on Land and Water Use of the European Commission on Agriculture 117
 Surveys 152-53
 USSR studies 262-63
 US Department: *Soil Survey Manual* 428
'An agroclimatic mapping of Africa', M M Bennett 349
Agronomica tropical, Centro de Investigaciones Agronomicas 435
Ahaggar Highlands, mapping 77
Ahlmann, Hans W *et al,* eds: *Sverige Nu: Atlas over folk land och naring* 184
Ahmad, K S: *A geography of Pakistan* 277
Ahwaz, map 256
Air brush, use on astrafoil 40
Air conditioning, requirement in use of plastics 34
Air Force: use of maps 150
 Missile Test Center Library, Florida 17
Air photo atlas of rural land use 17, 439
Air Photo Interpretation, Sub-Commission 16-17
'Air photography and geography', symposium 25
Air routes (the most significant) 93, 457
Air Survey Company of India Limited, Dum Dum 72
Air Survey Company of Pakistan Limited, Karachi 72
'Air survey of the New Zealand Alps', N E O'Dell 395
Airborne Profile Recorder 23
Aircraft Operating Company (Aerial Surveys), Johannesburg 380

504

Artesian basins, Australia, maps 92, 389
well, location 412
Ascension Island, map 112
Asia: bibliographical references 267
climate 417
Economic survey of Asia and the Far East, UN 266
' Formulating industrial development programmes, with special reference to Asia and the Far East ', UN 266
Magazine 282
maps 64, 72, 74, 77, 82, 98, 104, 116,
geological 266
historical 465
oil 266
' Mineral resources development ' series, UN 266
' Programming techniques for economic development . . .' UN 266
railways 457
seminar on development of basic chemical . . . industries, Bangkok 267
Proceedings 267
Small industry bulletin for Asia and the Far East, UN 266
surveys 412
Towards economic co-operation in Asia . . . , David Wightman 266
training of cartographers 57
University of Cambridge Group for Afro-Asia Social Studies 352
Asia and the Far East, UN *Regional Cartographic Conferences* 65, 386
Asia Minor: maps 114, 121
Asia, Soviet: mapping 265
Asian Population Conference, Delhi 266-67
Assam: mapping of frontier with Tibet 281
Association of African Geological Surveys 350, 351
Association of Agriculture: *Farm studies and the teaching of agricultural* series 432
Association of American Geographers: *Annals* 280, 331, 374
cartography in the USA, report 312
geographical materials, classification and cataloguing 485, 486

Association of American Geographers (*contd*)
Offshore geography of northwestern Europe: the political and economic problems of delimitation and control 122
publications 312
sub-committee for population map of Anglo-America 309-10
Association of Young Farmers' Clubs 152
Association pour l'Atlas de la France de l'Est 208
Association pour l'Atlas de Normandie 202
Association pour l'Etude Taxonomique de la Flore d'Afrique Tropicale 364
Associazione Italiana di Cartografia 229
Associazione Italiana degli Insegmanti di Geografia 233
Astrafoil 34, 38, 40, 136
Astralon 38, 78
Astrascribe 248
Astronomy: mapping 94, 425
Aswan High Dam: planning maps 359
l'Atelier Méchanograph que de l'Institute Géographique National 200
Athens Centre of Planning and Economic Development: ' Research monograph series ' 241
Athens: map 241
Meridian 240, 241
Atlante del Colonie Italiane 7
Atlante fisico-economico d'Italia 232
Atlante internationale della consociazoine Turistica 82, 84
Atlante mondiale, Istituto Geografico de Agostini 86
Atlantic Ocean: *Atlas of temperature and salinity profiles and data from the International Geophysical Year of 1957-58* 111
central ridge 111, 424
see also Great Meteor Bank
continental shelf 425
contour charts 424
maps 98, 111-12, 373, 421
geological 115
research 423, 424, 426-27
vessels for acquisition of weather data 418

505

Atlases (*contd*)
physical 408-411
planning 48
purpose 81
revision 12
stands 493
storage 493
UN Commission on National Atlases 109-110
world 80-107
bibliographical references 108
Atlasi, Geographical Society of Belgrade
Atmosphere: effect on sheet maps 490
mapping 409
Atter, Willi: *Pfalzatlas* 210
Atwood, W W: *The physiographic provinces of North America* 300
Aubreville, A *et al*: 'A vegetation map of Africa south of the Tropic of Cancer' 364
Auckland: maps 74, 392, 394
Aurada, Fritz: 'Entwicklung und Methodik der Freytag-Berndt Schulwandkarten' 78
'Aus der Arbeit des Kartographischen Institutes Bertelsmann' 87
Australasia: Royal Geographical Society, publications 385
Australia: aerial surveys 17
Antarctica research 399, 400
Atlas of Australian resources 389
Atlas of Australian rural industries 390
Atlas of Australian soils 389-90
An atlas of population and production for New South Wales 390
BANZARE Antarctic expedition, co-operation 397-98
Bureau of Mineral Resources, Geology and Geophysics 387-88
Charting a continent, G C Ingleton 386
Commonwealth . . . Bureau of Meteorology 388
Department of National Development 386, 387, 388
Department of Natural Resources, Canberra: *Classified index to resources maps of Australia* 388
government in South Pacific 384
Indian Ocean, expeditions 267
'The large-scale cadastral map

Australia (*contd*)
coverage of Australia and the parish maps of New South Wales ' 390
Linnean Society of New South Wales 387
A map book of Australia, A Ferriday 390
maps 72, 74, 82, 88, 92, 100, 104, 385, 387-88, 389, 394, 411, 444, 460
maps of Antarctic 398
meteorology 388
National Mapping Council: *Indexes* 388
Map catalogue 388
National Mapping Office 398
Philip's educational atlases 104
railways 457
Regional atlas of New South Wales 389
relationship with New Zealand, maps 389
State Geological Survey 387
Memoirs 388
state mapping agencies 386
state Royal Societies 387
topographical series 386-87
'The use of maps in Australia ', G C Irving 390
Australia, West: *Atlas of Western Australian Agriculture* 390
see also CSIRO and under the names of individual Provinces
Australia and New Zealand Association for the Advancement of Science 285, 395
Australia and New Zealand: Oxford regional economic atlas 440
Australian Committee on Antarctic Names 400
The Australian geographer 389, 390
Australian geographical studies 385
Australian Institute of Cartographers 18
Australian Institute of Agricultural Science 385
Australian journal of soil research 429
Australian meteorological magazine 388
Australian National Research Council 384
Austria: maps 204, 212-15

509

Belgian Lambert Projection 195
Belgian Ministry Geographical Institute, Congo mapping 371
Belgium: Antarctic research 399
Atlas du Survey National 196
biographical references 196
'Carte géologique de la Belgique' 196
Comité National de Géographique ...195-96
maps 119, 194-96, 204
geological 116
Mémoires et documents, Vol II, on Belgium 477
Queen's University, David Keir Library 18
'rush' editions, national topographical maps 195
society publications including cartographical information 196
'Belgique carte administrative' 195
Belgique et Pays Bas: documentation bibliographique 196, 477
Belgrade: Geographical Society, publications 239
Geological Institute 238-39
Bellagio, Italy: Conference on Economic Research in Africa 348
Bellamy, C V: Geological map of Cyprus 253
Ben Nevis: os tourist map 129
Benares Hindu University, National Geographical Society, publications 274
Bench marks lists, os 125
'Benelux road map', Shell 196
Bengal in maps ..., S P Chatterjee 275
Bengal, West, land use maps 275
Bennett, H A: on Survey Production Centre 68-69
— *et al*: 'The land map and air chart at 1 : 250,000' 80
Bennett, M M: 'An agroclimatic mapping of Africa' 349
Benson-Lehner Data Reduction Systems 36
Berg, L S: interest in biogeographic factors 446
Loess as a product of weathering and soil formation 429
Berge der welt 414
Bergen, maps 179

Berghaus, Heinrich: *Physical atlas* 408-409, 458
theories on mapping 408
Berichte zür Deutschen Landeskunde 207, 211
Berkey, C P and F K Morris: *Geology of Mongolia* 283
Berkshire, geological map 147
Berlin: *Atlas* 162, 208
atlas production 71
Department of Geodesy and Cartography 72
Free University, research station 356-57
maps 74, 103, 113, 211
Bermuda: *Fodor's guide* ... 329
map 112
Bernard, Augustin and R de Flotte de Roqueraire: *Atlas d'Algérie et de Tunisie* 354
Berne: city plans 452
Berry, B J L: commentary in *Atlas of economic development,* Norton Ginsburg 441
Berry, L: *The development of Hong Kong and Kowloon as told in maps,* T R Tregear and L B 284
Bertelsmann Cartographic Institute, Gütersloh 16
Atlas of Central Europe 204
Hausatlas 87
international atlas 86-87, 119
Kartographischer Nachrichten 455
Yearbook of cartography 470
Besairie, Henri: 'Carte géologique' (Madagascar) 382
Beyond the Urals, Violet Connolly 265
Beyrouth, Service Technique du Cadastre 247
Bhariya oasis, topographical mapping 358
Bhat, L S: *An atlas of resources of Mysore State:* A T A Learmonth and L S B 273
Bhutan, maps 74, 280
Bible lands, maps 101
Biblio 204
Bibliografia aborigen de Costa Rica 328
Bibliografia boliviane 338
Bibliografia cartográfia do Brasil 335
Bibliografia cartográfia do Nordeste (Brazil) 335

Bouju, P A *et al*: *Atlas historique de la France contemporaine 1800-1965* 203
The boundaries of towns and urban areas (Fiji), G T Bloomfield 383
Boussard, Jacques: *Atlas historique et culturel de la France* 203
Bowen, E G: Patagonia, maps 341
Wales ... 166
' The Welsh colony in Patagonia 1865-1885 ...' 341
Bowen-Jones, H: *Agriculture in Malta* ... 233
— *et al*: *Malta: background for development* 233
— ed: *An atlas of Durham City* 433
Bowles, O *et al*: *Mineral resources of the world* atlas 441
Bowman Memorial Lectures 433
Bown, Lalage and Michael Crowder, eds: *Proceedings, Congress of Africanists, Accra, 1962* 352
Boyd, Andrew: *An atlas of world affairs* 465
— and Patrick van Rensburg: *An atlas of African affairs* 346-47
Boyle, R, automated cartography 34-35
Boxes: method of storing maps 493
Bracing 126
Bracknell, British Meteorological Office 418
Automatic Picture Transmission Systems 418
Bradford Pothole Club, original surveys 149
Brazil: *Atlas pluviométrico do Brasil* 334-35
bibliographies containing maps 335
Centre of Applied Hydrology 334
Conselho Nacional de Geográfia 333-34
Departmento Nacional do Café: *Atlas corografico de la cultura cafeeria* 334
development projects 335
Geological Survey 334
Instituto Geográfico de Estado Minas Gerais 333
maps 333-34
geological 334
1 : 1M 68
national atlas 334

Brazil (*contd*)
photogrammetric surveys 333
Prefeítura do Distrito Federal, maps 333
Serviço de Meteorologico do Brasil: *Atlas climatologico do Brasil* 335
Servico Geográfiço do Exécito, maps 333
Sociedad Brasileira de Cartografia 333
statistical atlas 334
Brazilian Chamber of Commerce and Economic Affairs in Great Britain 342
Brazilian Embassy Commercial and Information Service 342
Brazilian Engineering Club, ed: IMW 1 : 1M map of Brazil 334
Brazzaville, maps 358-70
Breasted, J H: *European history atlas* 120
Breasted-Huth-Harding, historical maps 76
Bremen, plan 211
Brentwood, plan 162
'A brief summary of national activity in Italy during the past few years ', C P de Divelec 231
Brill, of Leiden, publications 352
Brisbane, Queensland University Library 18
Bristol: *Gloucester and Bristol atlas* 163
Bristol: National Agricultural Advisory regional service 154
Philip's educational atlases, Bristol edition 102
maps 165
training of cartographers 60
Bristol and Gloucestershire Archaeological Society 163
' Britain in the Dark Ages ', Ordnance Survey 132
British Admiralty, Hydrographic Department 224
British Airports Authority 123
British Antarctic Survey (formerly The Falkland Islands Dependencies Survey) 399, 403
' British Antarctic territory, Alexander Island and Charcot Island ', 1 : 1M map 400
British Association for the Advancement of Science, regional volumes

British Museum (*contd*)
representation on Library and Map Committee of the Royal Geographical Society 490
British national bibliography, exclusion of maps 475
British National Committee for Geography 58, 122-23
British National Institute of Oceanography 424
British Permanent Committee on Geographical Names 43, 83, 90, 91, 99-100, 489
British Petroleum: road maps 158, 191-92, 201, 213, 228, 235, 239, 241, 245, 256, 318, 356, 381, 389
British Records Association 490
British Society of Soil Science, *Soil Science* 428
British Standards Institute 490
British Trans-Arctic Expedition, 1968-69 396-97
British Virgin Islands, maps 329
The Broads, Ordnance Survey district map 130
Brocks, Karl and Otto Meyer, *Meteor* research 111
Brodie, J W: 'Coastal chart series 1 : 200,000' 393
Bromage, W H: maps in *An atlas of world affairs* 465
Brouillette, B and P Dagenais: 'La province de Québec' 316
Brouwer, L E J: *The North Sea* 112, 437
Brown, D: 'Methods of surveying and measuring vegetation' 430-31
Brown, E L: *Deciduous forests of eastern North America* 306
Brown, Glen: Saudi Arabia field surveys 360
Brown, H: *The relief and drainage of Wales* ... 166
Browne, S R: new edition of Sir T W Adgeworth David: *Geology of the Commonwealth of Australia* 388
Bruges, College of Europe 117-18
Brugmans, R *et al*: *Sciences humaines et intégration Européenne* 122
Bruk, S I and U S Apenchenko, eds: *Atlas narodov mira* 449
Brunei: *Colonial reports* 298

Brunhes, Jean 451
Brüning, Kurt: *Deutscher Planungasatlas* 208
Brussels: Société Belge de Photogrammétrie, *Bulletin* 24
Commission Belge de Bibliographie: *Bibliographie géographique de la Belgique* 196
Conference on international marine weather system 423
Buch, J L: *Land utilization in China* 282
Buchan, A, ed: *Atlas of meteorology* ... 420
Buchanan, Keith: 'Principal systems of farming in Great Britain', *Sir Dudley Stamp and K B* 152
Bucher, W H, comp: 'Geologico-tectonic map of Venezuala' 332
Budapest: Eötvös Loránd University, Cartographic Department 57, 234
international map exhibitions 234, 452
guides and maps 234
Institute of Cartography 448
State Office of Geodesy and Cartography 69
Buffum, C W: on cataloguing of maps 490
Buhler, Paul: *Schriftformen und Schrifterstellung* ... (*Switzerland*) 216
Buildings, public: maps 146
shown in 'perspective' 453
Bukinich, D D: *Agricultural Afghanistan*, I N Vavilov and D D B 257-58
Bulandshabar District: population map 272
Bulgaria: Academy of Sciences, publications 240
Geodetic and Cartographic Services 69
maps 116, 239-240
land use 438
road 237
tourist 239
national bibliography 240
Office of Geodesy and Cartography, publications 239
Scientific and Technical Union of Bulgarian Geodesists 240
Bulgarian Bibliographical Institute, national bibliography 240

517

518

Canterbury University (New Zealand), Department of Geography 395
Cape Breton Island, map 313
Cape Town: maps 73, 74, 381
University, Department of Land Surveying 18
Cappelen: 'Greater Oslo Plan' 179
1 : 1M map, Northern Scandinavia 179
Norge 178
Norway, maps 179
Caracas, map 332
Cardiff: National Agricultural Advisory regional service 154
Oxford plastic relief maps 139
Care of maps: references 494-96
Caribbean: bibliographical references 330
charts of Alexander Agassiz 423
Fodor's Guide ... 329
maps 329-330
Skinner's West Indian and Caribbean yearbook 329
Carl Zeiss, Jena 48
Carl Zeiss Precision Co-ordinographs 46
Carlsberg Foundation 191
Carlsund, Bo and Olaf Jonasson: 'Atlas of the world commodities, production, trade and consumption ...' 442
Carnegie Institution of Washington and the American Geographical Society: *Atlas of the historical geography of the United States* 310
Carolina: *Atlas* ... 308
University: *Officials' views on the uses of and need for topographic maps in North Carolina* 308
Carpathians: geological maps 116
influence on weather 236
'Carta corográfico de Portugal' 228
'Carta de Moçambique' 379
'Carta del Uruguay' 342
'Carta della densita della populazione in Italia' 232
'Carta general de las Republic Mexicana' 322
'Carta geológicó de Portugal' 228
'Carta itineraria de Portugal' 228
'Carta militar de la Republica de Cuba ...' 330
'Carta mineira de Portugal' 228

'Carta stradale d'Europa' 121
Carta topografica d'Italia 230
Cartactual 79, 234
'Carte administrative de la Tunisie' 355
'Carte del'Afrique central' 373
'Carte d'Algérie' 354
'Carte de Belgique' 195
'Carte de chemins de fer français' 201
'Carte de l'Etat Major', France 197
'Carte de France' 198-99
'Carte de Madagascar' 382
'Carte de reconnaissance des sols d' Algérie' 354
'Le carte des precipitations de l'Afrique du Nord' 420
'*Carte du monde au millionième* ... 65
'Carte forestière de la France' 200
'Carte générale bathymétrique des Océans 425
Carte général du monde 1 : 1M 72
'Carte géologique de l'Afrique Equatoriale françaises et du Cameroun' 371
'Carte géologique de l'Afrique Occidentale française' 371
'Carte géologique de la Belgique' 196
'Carte géologique de la Côte d'Ivoire', 1 : 1M 366
'Carte géologique de la Syrie et du Liban', 1 : 1M 247
'Carte géologique de la Tunisie' 355
'Carte géologique de Congo' 372
'Carte géologique générale du Grand-Duché de Luxembourg' 204
'Carte phytogéographique de l'Algérie et de la Tunisie' 354
'Carte routiere', Michelin 455
Carter, Harold: *The towns of Wales: a study in urban geography* 166
'Cartimat', data processing system 48
La cartografia da região Amazonica ... 335
La cartografia Mallorquina 234
Cartograms 226, 260, 471
The cartographer 304, 474, 475
Cartographers, training 15, 16, 18, 43, 55-62, 78, 122, 125, 135, 234, 380, 475
Cartographia, Budapest 69, 117

Chronique 13

Ciráo, Aristides de Amorin, ed: *Atlas of Portugal* 228-229

Circles, proportional; use in population maps 448

Cities: 'Atlas of all cities in Japan' 288

in Australia 390

co-ordinated planning 473

maps 163-64, 451-55

pioneers in urban research 451

'Cities of the world today and tomorrow', Budapest conference 152, 478

see also under Town plans and bibliographical references 454-55

The city-port of Plymouth: an essay in geographical interpretation 164

Civil engineering: use of data processing machines 37

Civil Engineering Department, University of Leeds 18

Clarendon Press, Oxford; Cartographic Department 8, 14, 16, 34-35, 44, 45, 47, 57, 73, 75-77, 87-88, 104, 110, 139, 143-45, 165, 242-43, 263, 276, 310, 346, 440, 477

Clarion atlas 104

Clark, A C: maps in *Resorts of Western Europe* 120-21

Clark University, Graduate School of Geography: *Economic geography* 444

Clarke, G R: *Soil Survey handbook* 154

— *The study of the soil in the field* 155

Clarke, John I: *Sierra Leone in maps* 366

Clarke, P J H, ed: *Kenya junior atlas* 104

Classification: maps 474, 479-90
bibliographical references 494-96
soils 428

The classification and cataloging of maps and atlases, Samuel Boggs and D C Lewis 489

'Classification and purpose in vegetation maps', A W Küchler 430

Classification of Geographical Books and Maps, Final Report 482-85

Clayton, K M, ed: *Guide to London excursions* 71, 163-64, 452

— 'Map of the drift geology of

Clayton, K M, ed (*contd*)
Great Britain and Northern Ireland' 148

Clevinger, W R: text in *Atlas of Washington Agriculture* 308

Cliffs, mapping 411

Climate: bibliographical references 422

classification 262

factor in vegetation control 166

mapping 14, 76, 77, 91, 93, 99, 100, 142, 144-45, 166, 207, 210, 222, 245, 262, 274, 404, 417-22

automatic 419

in disease studies 461

effect of electricity 112

microclimate studies 417

regions 421

(the most significant references only; *see also* under Meteorology; Weather)

The climate of Africa, B W Thompson 349

Climate of Japan, T Okado 288

'Climatic atlas of Czechoslovakia' 219

'Climatic atlas of Hungary' 235

Climatic factors and productivity 433

The climatic map of the United States 305

Climatological and oceanographic atlas for mariners 421

Climatological atlas for Africa south of the Sahara 380

Climatological atlases of Africa 348

Climatological atlas of the British Isles 151

Climatological atlas of Canada 320

Climatological atlas of India 274

'Aclimatological atlas of Japan' 286

'The climatological atlas of the USSR' 262

Climatological research series, McGill University 422

Climatology; Atlas mira 84

local variations 417

micro-climatology 417

Cloud and weather atlas 421

Clouds: *International cloud atlas* 417

maps 421

photographs 418, 421

shadow on aerial photographs 21

Clownes, E R and G S Holland: mapping in west Nepal 280

de Paris, P M: *Maps* 477
de Payerimhoff, P: 'Carte forestière de l'Algérie et de la Tunisie', R Maire and P de P 354
de Roqueraire, R de Flotte: 'Atlas d'Algérie et de Tunisie' 354
de Smet, L: *Bibliographique géographique de la Belgique* 196
de Smet, R E: *Cartes de la densité et de la localisation de la population de l'Ancienne Province de Léopoldville* 372
de Terán, Manuel, ed: *Geografía de España y Portugal* 227
De Vere Horizontal Camera 60
Deacon, G E R, ed: *Oceans: an atlas-history of man's exploration of the deep* 426
Dead Sea Scrolls 252
Deaths: reliability of hospital records 156
Debenham, Frank: *Antarctica . . .* 402
  British (Terra Nova) Antarctic Expedition . . . *Report* 402
  The Reader's Digest Great World Atlas 97-98
  The world is round 104
Debrecen, branch of the Hungarian Geographical Society 236
Deciduous forests of eastern North America, E L Brown 306
Deetz, C H: *United States Coast and Geodetic Survey* 304
Defant, A: *Atlas zur Schichtung und Zirkulation des Atlantischen Ozeans*, G Wüst and A D 111
Definition of 'cartography' 9-10
Definitions of surveying, mapping and related terms 10
Definitions of terms used in geodetic and other surveys 10
Defontaines, P: wall maps of Spain and Portugal 227
Delft: International Training Centre for Aerial Survey 17
  Topografische Dienst 192-93
  Atlas van Nederland 193-94
Delhi: maps 271, 273
  University, 'Continuous village surveys' 272
Deltas: Russian mapping 19-20, 42
d-Mac Limited, Scotland: cartographic digitisers 36

Demography 450
Demography, mapping 14, 145
Denmark: artificial satellites, co-operation in use in surveying 20
Atlas of Denmark 191
economic geography 177
engineering surveys 188
Geological Institute, Copenhagen 191
Guide prepared for International Geographical Congress, Stockholm 192
List of official maps 192
maps 74, 103, 176, 189-92
  geological 115, 116, 191
  Hallwag motoring map 191
  road maps 191-92
world atlas, co-operation in 86
Dennert and Pape, Hamburg: surveying and mapping equipment 36
Denoyer-Geppert Company, Chicago 76, 91, 331
Department of Agriculture and Fisheries, Scotland: data for current mapping 169
Types of farming in Scotland 169
Department of Education and Science: interest in training of cartographers 59
Department of Forestry Library, Ottawa 18
Department of Geodesy and Cartography, Berlin: world maps 72
Department of Geodesy and Cartography, SSSR: 409-10
Department of Mines and Technical Surveys, Surveys and Mapping Branch: *Cartographic manual* 25
Surveying manual 25
DSIR: Ross Dependency Research Committee, Wellington 396
  sponsorship of Bickmore Boyle system of Automatic Cartography 35
Departmental atlas of the Republic (Guatamala) 326
Dépôt de la Guerre et de Topographie (later Institut de la Guerre Militaire, Brussels 194, 198
Dépôt des Cartes et Plans 198
Derby, National Agricultural Advisory regional service 154
A description of the Ordnance Maps of Northern Ireland 172

Display screens 20
Dissection of maps 494
Distance measurement, equipment 21, 22
' Distribution of land improvement districts by size and work in Japan as of March 1906 ' 287
' Distribution of population in Europe ' 119
Division des Travaux Topographiques, France 197
Dixey, F: *Colonial geological surveys* . . . 71
Dixon, O M: 'The selection of towns and other features on atlas maps of Nigeria ' 369
Dobbie McInnes, Glasgow: automatic cartography 34, 47
Documentation, maps and atlases 7, 8
La Documentation Française: *Atlas industriel de la France* 202
Documentation of Asia 267
Dodecanese, Geographical handbook series 241
Dodge, J V *et al*, eds: *Wilhelm Goldmanns grosser Weltatlas* 97
Dollfus, Jean: *Atlas of western Europe* 119
Domesday geography of England, comp: H C Darby 462
Domesday survey of Britain 462
Dominian, Leon: *The frontiers of language and nationality in Europe* 122
Dominica Island, West Indies, map 329
Dominican Republic: index atlas 343
Doppler Navigator 23
Doris, J L, ed: *Regional planning atlas of Tasmania* 390
Dorset, geological mapping 147
Dot symbols 260, 309, 310, 448
Douai, maps 200
Douala, map 370
' Draft of a regional classification according to physiographical areas of the earth for application in the Universal Decimal Classification (UDC)' 483
Drainage, factor in vegetation control 166
Drakensburg, map 381

Drawers, method of storage of map sheets 491
Dresch, J: *Atlas et commentaire des cartes sur les genres de vie de montagnes dans le massif de Grand Atlas Marocain* 354
Drift map, *Atlas of Great Britain and Northern Ireland* 144
' Droning Maud Land ', map 398
Drumlin 25
Dublin: Geological Survey 173
Institute for Advanced Studies: *Linguistic atlas and survey of Irish dialects* 174
maps, Ordnance Survey 172
population 174
National Library, index of maps 175
National Survey of soils 173
Stationery Office, annual list of government publications 175
Trinity College Library: ' Ireland in maps exhibition ' 172
Dudal, R *et al*: ' Soil map of Europe 1 : 2,500,000 ' 117
Dudek, Arnost and Felix Ronner: *Report on geology in Iraq* 246
Duigman, M V: *Shell guide* (Ireland), Lord Killanin and M V D 175
Dumont, M-E and L de Smet, comp: *Bibliographique géographique* 196
Dumont, P and J S Baltus: ' cultural map of Europe ' 118
Dumont, René: Types of rural economy . . . 433
Dunbar, C O: introduction to *Atlas of paleographic maps of North America* 300
Dunbar, M J, ed: *Marine distributions* 426-27
Dunedin, map 74
Dunkle, J R: text in *Atlas of Florida* 308
Dunsyre, maps 168
Duplex map, 168
Duplex map, ' London ' 163
du Jonchay, I and Sándor Radó, eds: *Atlas international Larousse* . . . 443
du Toit, A L: ' Geological map of South Africa ' 380
— *The geology of South Africa* 380
Durban, maps 381

Europe (*contd*)
International Tectonic map 114-15
maps 70, 72, 74, 77, 78, 79, 82, 84, 88, 92, 98, 99, 100, 104, 107, 114, 117, 118, 119-22, 411, 430, 446, 456, 457, 463, 465
monographs including useful maps 122
Oxford regional economic atlas 440
palaeographic atlas 155
plastic relief model maps 76
' Product charts ', Carl Ritter 427
The resorts of western Europe 120-21
social and economic integration 118-19
L'Europe centrale, Pierre George 205
' Europe: how far?', W H Parker 113
Europe in maps . . . 117
' L'Europe moins la France ' 114
see also Council of Europe; and under names of individual countries
European Commission on Agriculture 117
European Economic Community 120
European Fisheries Convention 122
European history atlas 120
European Organisation for Experimental Photogrammetry Research 12
Everest, George: survey of India 269
Everest, Mount, exploration and mapping 279
Everyman's atlas of ancient and classical geography 464
' Evolution of interrupted map projections ', R E Dahlberg 34
' The evolution of the map of Alexander and Charcot Islands, Antarctic ', D J H Searle 400
' The evolution of the rural land use pattern in Cyprus ', D Christodoulou 437
Exchange and Gift Division, Library of Congress 18
Exercises, in educational atlases 104
Exeter, cholera maps 458
University, Department of Geology: survey of South Devon 148
Exhibitions of maps 405, 478-79
map 129, 130
' Exmoor ', Ordnance Survey Tourist map 129, 130

Experimental cartography, Oxford symposium 14, 24-25, 37, 153
Experimental Husbandry Farms, detailed soil maps 154
Explicacion del mapa geológico de España 225
Exploration, mapping 91, 440
Exposé des travaux de l'Institut Géographique National . . . 457
Eyre, J D: ' Key to the *Economic atlas of Japan* ' 288
Eyre, S R and G R J Jones: *Geography and human ecology* . . . 161

Faber school atlas 102
Fabre, Jean: explanatory notes to the ' Coal map of Africa ' 351
Fact book on manpower, Institute of Manpower Research, India 272
Faeroes, maps 191
Fage, J D, comp: *An atlas of Africa history* 347
Fairchild Camera and Instrument Corporation, New York 18
Fairey Air Surveys Limited 8, 16, 22-23, 45, 57, 59, 72-73, 254
Nigeria 7
Rhodesia 72
Zambia 72
Fairey Plotterscope 23
Falcon world atlas (Philip) 93
The Falkland Islands, surveys 403
Falk-Verlag, Hamburg: maps of Austria 213
Far East, bibliographical references 267
maps 265-67, 463
see also under individual country headings
Farafra oasis, irrigation map 358
Farijûm area, topographical mapping 358
Farm studies and the teaching of geography series, I V Young 432
Farrell, B H: *Power in New Zealand* . . . 395
Fauna *see* Plant geography
Faure-Muret, Anne: commentary on ' Geological map of Africa ' 350
Federal Government of Germany, expeditions to Indian Ocean 267
Federal Institute for Topographical Research 412

Great Britain (*contd*)
soil 154-55
Stanford's maps 147
town planning 452
marine research 112
Ministry of Transport road classification 456
oceanography research 423-24
regional studies 160-64
Soil Survey 153-54, 453
South Pacific, government 384
Sunday Times Royal Automobile Club road atlas of Great Britain and Ireland 159
Tabula Imperii Romani, contribution 68
weather charts 151
see also under British Isles and United Kingdom
' Great map of western Palestine ' 250
Great Meteor Bank 112
The great soil groups of the world and their development 427
Greece: aerial photographs of historical sites 465
Athens Centre of Planning and Economic Development 241
Institute for Geology and Subsurface Research 241, 242
maps 240-242
geological 241, 242
Ministry of Public Works, Photographic Section 240
National Statistical Service 241
Social Sciences Centre: *Economic and social atlas of Greece* 241
Topographic Service, Athens 240
see also under Ellas and Hellenic
Greek Army Geographical Service, Athens 240
Green, F W H: ' Urban hinterlands —15 years on ' 157
Green *see* Kirke-Green, A H
Greenford, map 163
Greenland, maps 73, 74, 115, 116, 191, 299, 320, 430
Greenwich Meridian 123, 192
Grid networks 36, 45, 97
see also under National grid
Die griechischen landschaften, Frankfurt 241
Griffiths, I L: 'A linguistic map of Wales ' 167

Grindley, G W *et al*: geological map of New Zealand 392
Grivas, G C: ' Reconnaissance soil map ' (Cyprus) 254
Grollenberg, L H, ed: 'Atlas of the Bible ' 251-52
Der grosse Bertelsmann weltatlas 86-87
Der grosse Brockhaus, atlas section 107
Grosse Elsevier atlas 97
Grosser historischer weltatlas 464
Grosser Shell—Atlas of Germany and Europe 114
Grosser Deutsche Kolonial atlas 71
The growth of world industry, 1938-1961 . . . , UN 439
Guatamala: Anales de la Sociedad de Geografia e Historia . . . 326
'Area map of the Republic ' 326
Departmental atlas . . . 326
Instituto Geográfico Nacional 325
' Mapa de la Republica de Guatamala ' 326
Maps, geological 226
topographical 325-26
Preliminary atlas . . . and text 326
Revista de la Economica Nacional 326
trade journals 326
Guernsey, ' British landscape through maps ' series 175-76
maps 145-46, 175
Guide to London excursions 71, 163-64, 452
A guide to the compilation and revision of maps, US Department of the Army 25
Guide to Scotland (Chambers) 171
Guides 78, 79, 120, 160, 185, 456
Guildford, plan 162
Guinea, maps 363, 364
Gulf of Akaba, tectonic sketch-map 360
Gulf of Bothnia, map 176
Gulf of Suez, tectonic sketch-map 360
Gulf Stream, studies by Alexander Agassiz 423
Gunning, H C ed: *Canadian journal of earth sciences* 413
Guyana, maps 332-33
see also under British Guiana

Gyldendal, Copenhagen: *Danmark-atlas med Faerøerne of Grønland* 191
economic atlases 441-42
educational atlases 103
Gyldendals Verdens-atlas 84

Haack, Hermann: Geographische-Kartographische Anstalt, Gotha 78-79, 82, 96-97, 206, 210, 281, 323, 345
Grosser weltatlas 96
' Sowjetunion, bergbau und industrie ' 264
Weltatlas: die Staaten der Erde und ihre Wirtschaft 441
Die Weltmeere . . . 421, 426
Habel, Rudolf, ed: *Haack grosser Weltatlas* 96
Hachures, method of relief representation 40, 51, 82, 84, 97, 128, 134, 212, 215, 219, 443
Hadego photosetting machine 215
Hadramaut, maps 254, 255-56
Hadrian's Wall, Ordnance Survey map 132, 133
Die häfen der nördlichen Adria und ihre Beziehungen zur Österreichischen Aussenwirtschaft 205
Hägerstrand, T: thematic mapping 182
Hagerup, Copenhagen: *Atlas of Denmark* 191
The Hague: Cartografisch Instituut Bootsma 193
Geological Society 193
Haile Selassi I Prize Trust 352
Hailey, Malcolm: *An African survey —revised 1956 . . .* 364
Haiti: index-atlas, Pan American Union 343
Halifax, map 314
Hall, D G E: text in *Atlas of south east Asia* 291
Hall, G K: reproduction of Index of maps in books and periodicals, American Geographical Society 472
Hall, R B and Toshio Noh: *Japanese geography . . .* 289
Hallam, A and L A Stevens, trans: Raymond Furon: *Géologie de l'Afrique* 350
Hallissey, T: *Handbook of the geology of Ireland* 173

Hallwag, Berne: 79, 113, 114, 121, 122, 191, 211, 224, 245, 456
Hamburg University 111
Hamlyn's new relief world atlas 99-100
Hammond, C S and Company: new supplement to *Shepherd's historical atlas* 464
Hammond's advanced reference atlas 94
Hammond's ambassador world atlas 94
Hammond's diplomat world atlas 95
Hammond's historical atlas 94
Hammond's library world atlas 94
Hammond's pictorial travel atlas of scenic America 311
Hammond's world atlases 94-95
Hampshire edition, Philip's educational atlases 102
A handbook for travellers in India, Pakistan, Burma and Ceylon 272
Handbook of the geology of Ireland 173
Handbook of Korea 290
Handbook of the Leeward Islands 329
Handbuch der historisch-geographischen pathologie 459
Handling, effect on sheet maps 490
Handy reference atlas of the world (Bartholomew) 93
Hargreaves, R P: *Maps in New Zealand provincial council papers* 393
Maps of New Zealand appearing in the British Parliamentary Papers 393
Harley, J B, ed: David and Charles' facsimile reprint of first edition Ordnance Survey one-inch map 129
Harlow, plan 162
Harnickell, Elizabeth *et al*: *Klimadiagramm weltatlas* 421
Harper Admas Agricultural College, Soil Survey Centre 154
Harrap: *3-D junior atlas* 104
Harrington, H J: ' Recent exploration of Victoria Land north of Terra Nova Bay ' 400
Harris, J F: *Summary of the geology of Tanganyika* 377
Harrison, R E: *Fortune atlas* 34
— *Ginn world atlas* 100

Harvard College Library, map collection 477

Harvard University: *Atlas de Cuba* 329-30

Harwitz, Mitchell: *Economic transition in Africa* . . . 348

Hassinger, Hugo: *Burgenland atlas* 214

Hatching, use as symbol in mapping 42, 126, 139

Hatfield, plan 162

Hatherton, Trevor, ed: *Antarctica* 400

Haughton, S H: revision of A L du Toit: 'Geological map of South Africa' 380

Hauptverwaltung der Deutschen Bundesbahn: 'Eisenbahnen in der Bundesrepublik Deutschland' 211

Haut Commissaire de la République française au Caméroun: *Atlas* . . . 370

Hawaiian Islands, maps 300, 304

Hazard, H W: *Atlas of Islamic history* 243

Hazlewood, L K: 'Planning the second series of *The Atlas of Australian resources*' 389

Health, mapping 458-62

Health resorts, planning 453

Heaney, G F: 'Rennell and the surveyors of India' 269
— 'The Survey of India since the second world war' 270

Hebrides, maps 116, 159

Hedlund, Mary trans: *Atlas of the Bible* (Grollenberg, shorter version) 252
— and H H Rowley: *Atlas of the early Christian world* 465

Heidelberg Academy of Sciences, Geomedical Research Unit: maps of climate 420

Height, automatic measurement 37
use of colours for relief features 40-41

Helicopters, use by Ordnance Survey 125

Hellenic Geographic Society, national committee 240-41

Hellner, W F, ed: Gyldendal's educational atlases 103

Helsinki, maps 103, 186, 188-89
University Institute of Geography:

Helsinki (*contd*)
Atlas of Finland 185
Institute of Geology, maps 188

Hemel Hempstead, plan 162

Henry, Prince, the Navigator: fifth centenary national atlas 229

HMS Challenger 423

HMSO: *The projection for Ordnance Survey maps and plans and the national reference system* 134
— Sectional Lists 148, 151, 419

Herbert, E: *Spanish and English maps of Hispanic America* 331

Herbert, W W: 'In Amundsen's tracks on the Axel Heiberg glacier' 401

Herbertson, A J: *Atlas of meteorology* . . . 420

Hereford: Caving Club: original maps 149
edition of Philip's educational atlases 102
Ordnance Survey map 165

Herm, Bartholomew maps 155

Hermann Haack Verlag, Gotha *see* under Haack

Hermannsson, Halldor: *The cartography of Iceland* 189

Herring atlas 446

Herrmann, Adolph: *A historical and commercial atlas of China* 282

Herskovits, M J and Mitchell Harwitz, ed: *Economic transition in Africa* 348

Hertfordshire Natural History Society and Field Club publications 149

Hickey, G C: *Village in Vietnam* 294

Highsmith, M R: *Atlas of Oregon agriculture* 307
— *Atlas of the Pacific northwest* . . . 306

'Highways of Canada and the northern United States', Canadian Government Travel Bureau 316

Hill, C A and Associates 17

Hill, D A: *The land of Ulster* . . . 173

Hill, M N: on research in the Indian Ocean 268
— *et al*: *The sea* . . . 427

Hills, E S: *Arid lands* . . . (Unesco) 435

547

Hills, features, 'sugar-loaf' 39
  shading 40, 77, 78, 79, 84, 85, 88,
    91, 92, 94, 96, 101, 102, 103, 107,
    113, 114, 121, 129, 130, 131, 134-
    135, 144, 159, 161, 165, 168, 176,
    179, 208, 210, 211, 213, 225, 257,
    259, 280, 288, 294, 299, 300, 316,
    329, 332, 337, 346, 377, 382, 390,
    401, 443
Hilton, T E: *Ghana population atlas
  ...* 367-68
  — *Practical geography in Africa*
    345
Himalaya: bibliographical references
  275
  mapping 21, 92, 279
Hirsch, A: *Handbuch der historisch-
  geographischen pathologie* 459
Hispanic Council, London 342
Hispanic America *see* America, South
Hissar District, population mapping
  272
*A historical and commercial atlas of
  China* 282
Historical Association: Atlas Sub-
  Committee 463
  Intermediate historical atlas 464
Historical Association of Canada,
  Historical atlas ... 319
The historical atlas and gazetteer,
  supplement to Arnold Toynbee:
  Study of history 465
A historical atlas of Canada 320
A historical atlas of Canada, ed.
  D G G Kerr 319
Historical atlas of Cheshire 163
An historical atlas of China (formerly
  *A historical and commercial atlas
  of China*) 282
The historical atlas of the Holy Land
  252
Historical atlas of India, Charles
  Joppen 274
*An historical atlas of the Indian
  Peninsula* 274
Historical atlas of the Muslim peoples
  242
*Historical atlas of town plans for
  western Europe* 120
Historical atlas of the United States
  311
A historical atlas of Wales 165
*An historical atlas of Wales from
  early to modern times* 165, 166

'Historical map of Ireland' 174-75
'Historical maps of Scotland' 168
'Historical map of Wales and Mon-
  mouth' 166
*Historical records of the Survey of
  India* 270
*Historisches-geographisches Karten-
  werk* 96, 119
History, mapping 14, 89, 94, 119, 220,
  319-20, 462
*The history of the retriangulation of
  Great Britain, 1935-1962* 138
Hockyer, A A: paper on the work
  of the Survey Production Centre
  68-69
Hodgkiss, A G: maps of Harrogate
  163
Hodgson, R D and E A Stoneham:
  The changing map of Africa 347
Hodgson, E C, ed: *Herring atlas* 446
Hogg, L A: 'The land map and air
  chart at 1 : 250,000' 80
  — on the work of the Survey
  Production Centre 68-69
Hokkaido: maps 286
  University, Faculty of Science
  Journal 287
Holland, G S: on the cartographic
  production of the Society of Uni-
  versity Cartographers 475-76
  — and E R Clownes: mapping in
  West Nepal 280
Hollingsworth, T H: 'People on the
  map' 146
Hollwey, J R 59
Holmes, J M: *An atlas of popula-
  tion and production for New South
  Wales* 390
'Holy Land', cartographical publica-
  tions 251-53
  map in plastic relief 76
Home Counties, motoring maps 140,
  163
  see also under London and the
  names of individual boroughs
  and towns
*Homme: revue française d'anthropo-
  logie* 450
Hommes et terres du Nord 204
Homo 450
Hong Kong: *Colonial annual reports*
  284
  Crown Lands and Survey Office of
  the Public Works Department,

India (*contd*)
  maps 74, 92, 266, 269-76, 463
  health and disease 459, 460
  geological 270-71
  IMW 1 : 1M 64
  land use 271
  metals and minerals 270-71
  metric system, introduction 270
  political 271
  population 271-72
  tectonic 270
  transliteration of names 270
  tribal 272
  United Kingdom mapping 70
  vegetation 429
  Mining, Geological and Metallurgical Institute 271, 274
  Ministry of Cultural Affairs, New Delhi, *State gazetteers* 272
  Ministry of Food and Agriculture: *The Indian agricultural atlas* 273
  Ministry of Information and Broadcasting . . . 272
  National Atlas Organisation 271-73
  Naval Hydrographic Branch 269
  Rural planning 272
  State gazetteers 272
  Survey 8, 279, 476
   see also under Survey of India
  UN Cartographic Conference for Asia and the Far East, *Proceedings* . . . 276
   see also under names of individual States
India, China and Japan, Oxford regional economic atlas 440
India: a reference annual . . . 272
   The Indian agricultural atlas 273
Indian Geographical Society 274
Indian Meteorological Department, *Memoirs* 274
Indian Ocean 267-69
  bibliographical references 268-69
  charts by Alexander Agassiz 423
  international expeditions 424
  maps 98, 373, 421
' The Indian Ocean routes ' 268
Indian Statistical Institute, regional survey of Mysore State 273
Indian weather review 274
Indiana University, Centre of African Studies 352

Indien Entwicklung seiner Wirtschaft und Kultur 96
Indien historisch-geographisches Kartenwerk . . . 274
Indo-china, Admiralty handbook series 293
Indochina: Association nationale pour l'Indochine française 293
   Atlas météorologique . . . 292
  maps 291, 293
  Service Géographique 292, 293
  Service Géologique 293
Indo-China: a geographic appreciation 294
Indonesia: bibliographical references 297-98
  Bureau of Mines and Geological Survey 297
  Geographical Institute 296-97
  maps 291, 296, 298
  land use 297
  national atlas 297
  relationship with Australia 389
Indonesian journal of geography 297
Indus Valley: maps 102
  pioneer studies of land use 436
  Treaty between East and West Pakistan 276
   see also Lower Indus Project
' Industrial atlas of the Soviet Union ' 263
Industrial Census Report, 1951 104
Industrial Meteorological Association, Tokyo 288
Die industrie der Schweiz 217
Industries: mapping 14, 91, 96, 145, 158, 439
  rural, *Atlas of Australian rural industries* 390
Information, recording 15
' Information sheets ', Hunting Surveys Limited 48
Informe semestral, Costa Rica 438
Ingleton, G C: *Charting a continent* (Australia) 386
Inglis, H R G: *The contour road book of Ireland* . . . 175
Inks, choice in map preparation 39
' Inland cruising map of England ' (Stanford) 160
Insects, mapping 146
Institut de la Carte Internationale de Tapis Végétal 197, 429

551

Institut de Géographie, Brussels 196
Institut de Géographie: *Cahiers de géographie de Québec* 320
Institut de Géographie de l'Université de Lille 202
Institut de Géologie et de Géographie Physique, Université de Liège 196
Institut de Recherches Sahariennes, *Travaux* 356
Institut Français d'Afrique Noire 351, 358
Institut für Angewandte Geodäsie 206, 208, 211, 468
Institut für Erdmessung (later Institut für Angewandte Geodäsie) 206
Institut für Landeskunde 206, 210, 471, 483
Institut Géographique Militaire, Brussels 184-95, 196
Institut Géographique National 10, 56, 72, 104-105, 114, 197-99, 200, 201, 204, 293, 332, 354, 355, 356-57, 365, 366, 370, 373, 383-84, 410, 425
Catalogue des cartes 473
Institut National d'Etudes Démographiques: *Population* 450
Institut National de la Statistique et des Etudes Economiques 201
Institut National pour l'Etude Agronomique de Congo Belge 351
Institute for Shipping Research Bremen: *The world shipping scene* atlas 443
Institute of Australian Geographers: *Australian geographical studies* 385
Institute of British Geographers 157, 164, 449
Institute of Chartered Cartographers, Ontario 16
Institute of Geodesy, Photogrammetry and Cartography, Ohio State University 16, 469
Institute of Geological Sciences 148, 412, 453
Reports 414-15
Institute of Oceanography, Kiel 112
Institute of Pacific Relations 283, 384, 386
Instituto Brasiliera de Geografía e Estatistica: *Atlas do Brasil,* 334
Instituto de Investigacíones Geologicas 'Lucas Mallada', Madrid 225

Instituto Geografico e Cadastral, Lisbon 227, 473
Instituto de Geografía, Santiago 340
Instituto Geografico 'Agustin Cadazzi', map of Colombia 336
Instruments: use in automatic cartography 35-36
electronic, use by Ordnance Survey 125
electronic distance measuring 12
meteorological 417
Sub-Division of American Congress 16
surveying 23
Inter-African Bureau for Soils and Rural Economy, Paris 351
'Soils map of Africa' 350-51
Inter-African Pedological Service 351
Inter Africa Research Fund 363
Inter-America Committee for Agricultural Development 342
Inter-American Geodetic Survey of the USA 301, 325, 326-27, 339
Inter-Departmental Committee on Aerial Surveys 25
Intérêt des stéréogrammes aériens dans l'enseignement de la géographie physique 415
Intermediate historical atlas (Philip) 464
International African Conference on Hydrology 351
International African Institute 345-46, 351
International African Seminar . . . 348, 349
International Association of Friends of the Arab World, Lausanne 255
International Association of Geodesy 16, 22, 469
International Association of Scientific Hydrology 413
International Association of Sedimentology, Wageningen 412
The international atlas of West Africa 364-65
International Bank for Reconstruction and Development 252, 330, 343
International bibliography of photogrammetry 17
International bibliography of vegetation maps, ed A W Küchler 430

International Biological Programme ... 428

International Cartographic Association 13-15, 47, 81, 122, 204, 206, 212, 215, 218, 221, 24, 227, 229, 231, 234, 240, 248, 256, 258, 269, 286, 295, 300, 302, 313, 321, 333, 380, 386, 391, 405, 468, 470
  Bulletin 15, 468
  Multilingual dictionary ... 11
  Newsletter 15
  Statutes 470
  training of cartographers 58

International Cartographic Union 77

International Centre for Co-ordination of Scientific and Technical Information in Agriculture and Forestry, Prague 434

International Civil Aviation Organisation 12, 43, 65, 67, 457

International cloud atlas, WMO 517

International Commission for the Scientific Exploration of the Mediterranean Sea, publications 223

International Committee on Geographical Names, UN 44

International Conference for Cartography 470

International Conference of American States 289

International Conference on the Organisation of Research and Training in Africa ... (Lagos Plan) 347

International Congress of Africanists, *Proceedings* 352

International Congress of Photogrammetry 19

International Congress of Soil Science, Adelaide 389-90, 428

International Council for the Exploration of the Sea 63, 426
  Herring atlas 446

International Economic Studies Department 444

International Federation of Surveyors 10-11

International geodetic bibliography 469

International Geographical Commission: Population map 1 : 1M 322

International Geographical Union 12, 14, 15, 47, 63, 64, 65, 68, 70, 71, 81, 110, 131, 155, 180, 189, 192,

International Geographical Union (*contd*) 226, 236, 239, 271, 272, 279, 297, 317, 323, 345, 360, 411, 415, 431, 433, 435-36, 438-39, 447, 448, 449, 457, 458-59, 478, 482-85
  Biographie cartographique internationale 470-71
  Land use in semi-arid Mediterranean climates 223-24

International Geological Congress 114, 256, 266, 322, 354, 406
  Papers 411

' International geological map of Europe ' (Unesco) 115-16

International Geophysical Year 83, 86, 111, 215-16, 315, 339, 397, 399, 412, 419, 423-24
  Annals 397, 412, 419

International Hydrographic Bureau 12, 43

International Hydrological Decade 348, 413
  Nature and resources 412, 435

International Indian Ocean Expedicion, Unesco 268, 424

International journal of bioclimatology and biometeorology 422

' International map of the Roman Empire ', Ordnance Survey Report 68

International Map of the World, 1 : 1M 41, 43, 63-64, 65, 66-68, 69, 88, 131, 259, 281, 303, 310, 331, 332, 334, 360, 457, 468
  bibliographical references 80, 108
  reports 64, 65, 66, 67

International Meteorological Organisation 417
  see also World Meteorological Organisation

International nickel magazine 444

International Oceanographic Congress 427

International Oceanographical Research Programme 348

International Printing Machinery and Allied Trades Exhibition: display of overseas maps 478

' International Quaternary map of Europe ' 116-17

International review of forestry research 434

International Road Federation, London: *Road international* 456

Ireland (*contd*)
Soil Survey 173
Sunday Times RAC road atlas of Great Britain and Ireland 159
Ireland: a general and regional geography, T W Freeman 174
Ireland in maps . . . 172
Irish geography 174
Irrawaddy valley, maps 92
Irrigation surveys 22, 23, 73, 276
Irving, G C: 'The use of maps in Australia' 390
Irwin, B st G and W A Symons, on Ordnance Survey revision policy 133
Isaac Wolfson Foundation 156
Isakou, I S ed: *Morskoy atlas* 424-25
Islam: *Atlas of Islamic history* 243
'The spread of Islam' map 242
'The world of Islam in the Middle Ages' 242
Islands kortlaegning, N E Nørland 189
Isle of Man: government department reports 175
maps 116, 145-46
land utilisation 152
Ordnance Survey 128-29, 175
notes in os map catalogue 137
Natural History and Antiquarian Society, publications 175
The Isle of Man: a study in economic development 175
Isle of Thanet Geographical Association: *Second Land Use Survey of Britain* 152-53
Isograms 11
Israel: bibliographical references 250-51
Exploration Society 250
Geological Society, *Bulletin* 248
Geological Survey, maps 248
maps 74, 243, 265
land use 250
soil 248
'Touring map . . .' 249
'Physical master plan' 249-50
Soil Conservation Service 249
Survey of Palestine, later Survey of Israel 248
The Israel physical master plan, Jacob Dash and Elisha Efrat 250

'Israel's territorial gains as a result of the June war' 251
Istanbul: maps 74, 113
University, economic map of Turkey 245
Geographical Institute, *Review* 245
Istituto Agronomico per l'Ottremere, Florence 435
Istituto 'De Agostini', Novara 56. 71, 78, 86, 97, 229, 470
Istituto Geografico Militare, Florence 229, 341
Bollettino 233
cartography courses 56
Catalogue 233, 473
Istituto Geografico Polare 396
Istituto Italiano d'Arti Grafiche, tourist map of Italy 232
l'Italia, Roberto Almagià 233
Italy: Consiglio Nazionale della Richerche 231
Direzione Generale del Catasto 231
historical sites, aerial photographs 465
Istituto Geologica: 'Carta geologica d'Italia' 231
Litografia Artistica Cartografica 232
maps 74, 78, 84, 113, 229-33
early maps of Christian churches 465
geological 116, 231-32
land use 231, 437
overseas mapping, contribution 71
road 232-33
soil 231
Tabula Imperii Romani 1 : 1M, contribution 68
Ministero dell'Industria e del Commercio 231
National Central Library, Florence 233
rural development programmes 433
Servicio Geologico d'Italia 232
training courses for cartographers 56, 229
l'Italia nell'economica delle sue regioni 233
Italian Cartographic Association, training courses 56
Itenberg, I M *et al*: small *Atlas mira* 85
Ivory Coast, maps 363, 364, 366
Iwo Jima, map 74

555

Johnson, James H: 'Population changes in Ireland, 1951-1961' 174

Johnson, John, ed: *Atlas of Canada, 1906* 318

Johnston, A K: *The physical atlas* ... 409, 458

Joint Advisory Board: experiments on lettering and type faces 45

Joint Operations Graphic Land Map and Air Chart 68-69, 79-80

Joint United States-Mexican Defense Commission, mapping of border territory 301

Jonasson, Olaf and Bo Carlsund: 'Atlas of the world commodities, production, trade and consumption ... 442

Jonasson, Olaf *et al*: 'Agricultural atlas of Sweden' 184

Jones, Emrys: 'The London atlas' 162

— and Ieuan L Griffiths: 'A linguistic map of Wales, 1961' 167

Jones, G R J: *Geography and human ecology* ... 161

Jones, J I: *A geographical atlas of Wales* 165

— *A historical atlas of Wales* 165

Jonglei Investigation Team, *Report* 361

Joppen, Charles: *Historical atlas of India* 274

Jordan, E L, ed: *Hammond's pictorial travel atlas of scenic America* 311

Jordan: bibliographical references 252-53

  economic survey 252

  'Europa' survey 244

  land use survey 252

  water resources programme 252

The journal of development studies ... 444

The journal of the Geological Society of Japan 287

Journal of the Geological Society of Tokyo (later *Journal of the Geological Society of Japan*) 287

Journal of Nigerian studies 369

Journal of physics of the earth 286

Journal of tropical geography 435

Journal officiel du Cambodge 294

Juan Fernandez Archipelago, map 340

Jugoslavia, Admiralty geographical handbook series 239

see also under Yugoslavia

Jungles: aerial surveying 21

Junior atlas (Collins-Longmans) 104

Junior atlas (Nelson) 94

Jusatz, H J, ed: geomedical monograph series 461

— *Welt-Seuchen-Atlas* 461

— 'World map of climatology' 461

Jutland Peninsula, map 117

Kabylie region, maps 354

Kaduna, Ministry of Land and Survey 18

Kahas, J: 'Climatic atlas of Hungary' 235

Kaigaeon, triangulation point 280

Kajamaa, M: 'Report on International Cartographic Association' 189

Kalle, Kurt: *Allgemeine meereskunde* ... 426

Kampala, maps 373

K'ang Hsi, Emperor: 'Atlas of the empire' (China) 283

'Kanjiroba Himal and adjacent areas in the Karnali region of western Nepal', map 278-80

Kansas in maps, R W Baughman 308

Kansas State Historical Society 308

Kansas University: *International bibliography of vegetation maps* 430

Kanter, Helmuth: geomedical monograph on Libya 461

Karachi, map 271

Karan, P P: *Bhutan* ... 280

Karakoram, research 233

Kares, J F, ed: 'Soviet and East European agriculture', seminar, papers and maps 262-63

Kariba Dam, surveying 23

Karnali region, western Nepal: Kanjiroba Himal expedition 279

Karte, Geographical Society of Belgrade 239

Kartenbrief 478

Kartograficii, Vállalata: road maps of Hungary 235

Die Kartographie, 1943-1954: eine bibliographische übersicht 477

Kartographische gelandedarstellung, Eduard Imhof 415

557

Logan, M I *et al*, eds: *New viewpoints in economic geography* 444
Logan, W E: *Geology of Canada* 315
Lomb Multiplex equipment 60
Lomonsov Ridge, map 396
London atlas 157, 162, 208
London: cartographers, training 59
  Climatic Society 162
  Duplex map 163
  East European world map, sheets 69
  Geographical Institute (Bartholomew) 75
  Geological Society 412
  Greater: Philip's educational atlases 102
    symposium on regional planning 452
    see also Home counties
  Guide . . . 452
  ICA Conference, 1964 122
  map office, national atlas of Britain 141
  Ordnance Survey, in the Tower 123
  maps 74, 83, 103, 113, 124, 126, 128, 129, 161-64
  disease 458
  geology 147, 148
  history 162
  redevelopment 164
  meteorological conference 417
  Royal Meteorological Society, publications 151
  School of Economics and Political Science, Department of Geography, 'Cyprus land use map' 253
  Geographical papers 444
  Topographical Society 162
  Underground Railway 163
  University College, Department of Photogrammetry and Surveying 18
  Institute of Education, *Farm studies* . . . 432
  School of Oriental and African Studies 444
    see also under Home counties and under names of individual organisations, *eg* Royal Geographical Society
'The London map directory . . .' (Geographia) 163
'London, Westminster to the City', map 163

Long Ashton, soil survey centre 154
Longmans, Green and Company Limited: Cartographic Department 139
Collins-Longmans educational atlases series 103-104
— *The Reader's Digest complete atlas of the British Isles* 145-46
textbooks 382
Lossdale, R E: *Atlas of North Carolina* 308
Loose-leaf format 48, 79, 84, 89, 96, 97, 98, 118, 131, 183-84, 193-94, 196, 202, 205, 212, 235, 274, 307, 308, 348, 389, 392
Lord, C L and E H, eds: *Historical atlas of the United States* 311
Los Angeles, Munger Oil Information Service 351
Lough Foyle, base line for Ordnance Survey triangulation 124
Louis, V E St J M: *A motorist's guide to the Soviet Union* 264
Loukotka, Čestmir: 'Ethno-linguistic distribution of South American Indians' 331
Low Countries, map 102
  see also under Belgium, Netherlands
Lower Indus project 22
Luke, *Sir* Harry: *Cyprus* . . . 253
Lumitype 215
Lund: Geographical Society 185
  South Swedish Geographical Society 185
  University, Geography Department 185
  work on modern geographic data systems 182
Lund studies in geography 185
Lundqvist, Gösta: Director, Esselte Map Service 180
  population maps 183
  'Some trends in Swedish map reproduction' 183
Luso-Brazilian Council, London 342
Lutgens, R, ed: E Otremba: *Allgemeine agrar-und industrie-geographie* 433
Luton Technical College: interest in training of cartographers 59
Luxembourg: Administration du Cadastre et de la Topographie 204
  Carte géologique . . .' 204
maps 119, 204
Service Géologique 204

Lynch, J: maps of Harrogate 163
— 'Preparing and printing soil maps in Ireland' 173
Lyon, map 201
Lysted, Robert A: *The African world* ...352

Maanmittaushallitus, Helsinki, *Karttalvettelo* 189, 473
Macaulay Institute of Soil Research 154
McCagg W O *et al*: *An atlas of Russian and east European history* 264-65
McCormick, Jack: contribution to *International bibliography of vegetation maps* 430
Macculloch, John: geological survey of Scotland 169
— *Description of the Western Islands of Scotland* 169
McDonald, E A: on mapping Antarctica 400
McEvely, Colin, ed: *Penguin atlas of medieval history* 465
McEwan, Peter J M and R B Sutcliffe, eds: *The study of Africa* 352
MacFadden, C H: *A bibliography of Pacific area maps* 386
McGill University, Geography Department: *Climatological research series* 422
Sub-Arctic Research Laboratory 397
McGrath, G: 'Further thoughts on the representation of vegetation on topographic and planimetric maps' 431
— 'Quality and quantity in the representation of vegetation on topographic maps' 431
McGraw-Hill international atlas 87
Machines, photo-setting 215
see also Automated cartography
McHugo, M B: 'Mapping the Falkland Islands Dependencies and British Antarctic territory, 1956-64' **403**
Machwe, V, ed: *Documentation on Asia* 267
Mackinder, *Sir* Halford 75
McLintock, A H, ed: *A descriptive atlas of New Zealand* 393-94

McMaster, D N: *The agricultural geography of Uganda* 437
— 'A subsistence crop geography of Uganda' 375
McMurdo Sound, map 400
Madagascar: 'Carte géologique' 382
Geological Service, *Catalogue des archives et périodiques* 383
maps 74, 345, 363, 381
Unesco review of natural resources 348
'Madagascar: cartes de densité et de localisation de la population' 382
see also Malagasy Republic
Madras: map 271
University, Indian Geographical Society, publications 274
Madrid: maps 227
by Department of Geodesy and Cartography, Berlin 226
East European world map, sheets 69
Maggs, K R A: *Land use survey handbook*, Alice Coleman and K R A M 152
Magnetic tape, use in automated Cartography 35, 36
Magnetographs, nuclear 111
Magyar nemzeti bibliográfia 236
Maintenance of maps 494-96
Mair, Volkmar: 'Strassenkarten aus Mairs Geographischen Verlag' 210-11
Maire, R and P de Peyerinhoff: 'Carte forestière de l'Algérie et de la Tunisie' 354
Mairs Geographischen Verlag 79, 210-11
'Benelux road map' 196
Europe, maps 114
guides 120
Majorca, maps 234
Makerere, University College, Geography Department: *National atlas of Uganda* 375
population maps 373-74
Malacca, map 295
Malagasy Republic: Geological Service, *Catalogue des archives et périodiques* 383
Institut Géographique National 382
population maps 382
see also Madagascar

Malaria studies 460
Malawi: bibliographical references 378
Clarion atlas, Malawi edition 104
Geological Survey 378
  Annual reports 378
  Bulletins 378
  maps 363, 378-79, 391
Malaya, ed Norton Ginsburg and C F Roberts, jr. 296
Malaya: British Association of Malaysia, publications 295
  engineering surveys 188
  Federation of Malaya Survey Department 295
  Gazette 296
  Geological Survey, publications 295
  The geology of Malaya, J B Scrivenor 295
  land use survey 295
  maps 290, 291
  University, Department of Geography, *Journal of tropical geography* 435
    Singapore Improvement Trust and Department of Geography, land use maps 295
Malaysia: bibliographical references 296
  Directorate of Overseas (Geodetic and Topographical) Survey maps 295
  economic map 444
  Junior atlas 104
  maps 294-96
  Philip's educational atlas, Malaysia edition 104
  Primary school atlas . . . (Nelson) 296
Malaysia in brief, Department of Information, Kuala Lumpur 295
Mali: Institut National de Topographie de Mali 357
  ICA charts 357
  levelling network 357
  maps 363
    Directorate of Overseas Surveys 357
  IGN, Dakar Branch 357
  Ten-year plan of topographic surveying 357
Malinin, S N *et al,* ed: 'Atlas of Byelorussian SSR' 260

Malivanék, R: *Mapping for economic and technical purposes in Czechoslovakia* 220
Malmø, mapping 184, 452
Malmstrøm, V H: *Norden . . .* 180
Malta: *Clarion atlas* edition 104
  maps 233
Malta: background for development, H Bowen-Jones 233
Malta: an economic survey, Barclay's Bank 233
Man, Royal Anthropological Institute 450
Man and Africa . . . 352
Manchester: Geological Association 150
  training of cartographers 59
  University, British Agricultural History Society: *The agricultural history review* 433
    Department of Economics, co-operation with *The journal of development studies . . .* 444
Manchuria, maps 290
Manhattan, inset plan 77
Manitoba: Department of Industry and Commerce, *Economic atlas . . .* 319
  geological map 316
Mankind 450
Mannerfelt, C M and Gösta Lundqvist: 'Some trends in Swedish map reproduction' 183
The manual for the compilation and preparation for the publication of the State map of the Soviet Union on the millionth scale 259
A map book of Australia, A Ferriday 390
A map book of France, comp: A J B Uussler and H H Allen 203
Map Curators' Group, British Cartographic Society 16, 467
  Newsletter 467
Map of Brazil, Club de Engenharia 68
'Map of China', Hungarian cartographic work 234
'Map of Far East oil', B O Lisle 266
'Map of forests of the Soviet Union' 263
Map of Hispanic America, American Geographical Society 65, 68, 331

567

Maps (*contd*)
  bibliographical references 48-55
  draughtsmanship 9, 10
  lettering 45
  office-compiled 9
  pictorial representation 39
  printing 9, 38-39
  production firms 72-78
  relief model 40, 75-76
  three-dimensional 14, 146
  repair 494
  reproduction 10, 34-35, 38-39
  resources development 9, 15, 70-71,
    81, 91, 98, 110, 145, 243, 254, 316,
    323, 356, 387, 388, 390, 404
  soil 99, 106, 117, 154-55, 166, 169,
    188, 231, 243, 246, 254, 257, 259,
    261-62, 319, 327, 350-51, 354, 389-
    90, 393, 405, 411, 428
  statistics 43, 76, 144, 444
  storage 490-496
  bibliographical references 494-96
  tourist 40, 72, 77, 79, 120-22, 129,
    174-75, 176, 184-85, 192, 201, 213,
    224, 234, 239, 311, 323, 455, 456-
    57
  transport 14, 91, 142, 160, 453-58
  air routes 457
  canoeing 160
  inland waterways 160
  railways 173, 201, 457
  roads 74, 79, 120-22, 140-41, 142,
    159-60, 174-75, 187, 191-92, 196,
    201, 210-11, 225, 228, 234, 239,
    245, 248-49, 257, 264, 311-12,
    368, 370, 375, 381, 389, 405, 453-
    457
  tribal 373
  urban 22, 452-53
  vegetation 14, 77, 92, 99, 102, 166,
    200, 217, 224, 243, 262, 306, 322,
    364, 375, 378, 405, 446, 472
  vineyards 203, 225
  wall 33, 75, 76, 77, 78
    see also under the names of indivi-
      dual places and organisations
Maps and atlases, Library of Congress 474
*Maps and politics: a review of the
  ethnographic cartography of Macedonia*, H R Wilkinson 239
*Maps in New Zealand provincial
  council papers*, R P Hargreaves
  393

*Maps of Canterbury and the west
  coast: a selected bibliography*
  (New Zealand) 393
Maps of Costa Rica . . . 328
*Maps of New Zealand appearing in
  the British Parliamentary Papers* 393
*Maps of Soviet Central Asia and
  Kazakhstan* 265
Maps, topographical and statistical
  109
Marbut, C F: trans *The great soil
  groups of the world* . . . 427
Marie Byrd Land, map 400
Marine climatic atlas of the world,
  us Navy 421
Marine distributions, ed: M J
  Dunbar 426-27
Mariner IV, shots of Mars 99
MARC Service 485
Marpiti, maps 384
Market gardening, mapping 430
Marketing maps of the USA . . .
  comp: M C Goodman and W W
  Ristor 312
' Markets and media survey of Great
  Britain ' (Geographia) 158
Mars, *Mariner IV* shots 99
Marseille, map 201
Martin, Ruth *et al*, eds: *Wilhelm
  Goldmanns grosser Weltatlas* 77
Maryland, College of Business and
  Public Administration: co-opera-
  tion with *Atlas of the world's re-
  sources* 432
University, Department of Geo-
  graphy: *Atlas of the world's
  resources* 432
*Mineral resources of the world
  atlas* 441
MIT Department of Civil and Sanitary
  Engineering, Photogrammetry La-
  boratories 17-18
Matley, I M *et al*: *An atlas of
  Russian and east European history*
  264-65
Matthew, *Sir* Robert: *Belfast regional
  survey and plan* 174
*Matthew Fontaine Maury, scientist
  of the sea*, F L Williams 423
Maurel, J B: *Geografía urbana de
  Granada* 227
Mauritania: maps 363, 364
  Directorate of Overseas Surveys
  357

Meteorology (contd)
Institut de Méteologie et Physique du Globe de l'Algérie 355
IGY programme 419
maps 186, 317, 393
observing stations 417
synoptic 417
Meteorology for glider pilots, C E Wallington 419
see also under Climate, Weather
' Methods of surveying and measuring vegetation ' 430-31
Methuen and Company Limited, textbooks 382
Metric system 34, 93, 98, 135
Mexico: *Anuaria bibliográficos* 324
atlases 323
Boletin de la Biblioteca Nacional 324
City, maps 322
Coleccion Geografica Patria, ' Carta general de la Republica Mexicana ' 322
Comité Coordinador del Levantamiento de la Carta Geografica de la Republica Mexicana 322
Dirección General de Geografía 321
Geological Society, *Boletin* 323-24
Geological Survey 322
Guide to Mexico 323
Instituto Nacional para la Investigacion de Recursus Minerales, *Boletines* 323
maps 321-24
border with United States 301
Rand McNally 299
tectonic 322
tourist 323
vegetation 430
Military Cartographic Department 322
Office of Naval Research, land form map 322
' 1 : 1M population map ' 322
skin diseases, studies 460
University, Department of Geography 321-32
' 1 : 1M *population map of* Mexico ' 322
' Mexiko: Land der Olympiade 1968 ' map 323
Mexican Association of Professional Geographers, publications 323

Mexican Society of Geography and Statistics, publications 323
Mayer, Otto: director of *Meteor* research 111
Meyers Duden-Weltatlas 105, 428
Meyers grosser physischen weltatlas 428
Meyers neuer handatlas 105
Meynen, E, Director, Institut für Landeskunde 206
The Atlas östliches Mitteleuropa 205
concept of geography 483
geographical books and maps, classification 479, 483-84
Michelin, Service de Tourisme, Paris: guidebooks 120, 456
maps 201, 225-26, 227, 360, 381, 455, 456
publications 79
Michigan University: Center for Japanese Studies 289
Center of African studies 352
University Press, *Economic atlas of the Soviet Union* 264
Microfilm, of maps 494
Microfilm viewers, for maps 494
Micronesia, maps 385
Micropalaeontology, surveys 22
Middle East: *Atlas of Islamic history* 243
bibliographical references 243-44
Development Division 252
Institute, Washington: *Middle East Journal* 243
maps 74, 88, 116, 242-44, 463
relief representation 41
The Middle East and North Africa, Oxford regional economic atlas 440
Middle East Journal 243
Middle Eastern Studies 243
Midlands: Oxford plastic relief map 139
New Town Society, ' West Midlands Green Belt map ' 161
symposium on regional planning 452
West, Group on Post-war Reconstruction and Planning 451
Midlands and East Anglia, road maps 140
Midwest Federation of Mineralogical Societies, Chicago: *Earth Science* 414

Montevideo, plan 342
Monthly abstract bulletin, Eastman Kodak Company 24
Montpélia, maps 384
Montreal: development map, *Atlas of Canada* 318
maps 313
Town Development Bureau: development plan 317
' Montreal and vicinity ', map 318
Montserrat Island, West Indies, maps 329
Moolman, J H: text for ' Population distribution of the Union of South Africa ' 381
Moon, maps 88, 99
Mooréa maps 384
Morgan, D L: text in *Pioneer atlas of the American West* 306
Morgan, W T W: *Nairobi: city and region* 376
— and N M Schaffer: *Population of Kenya . . .* 375-76
Morphology, maps 100, 205
Morrison, A: ' Principles of road classification for road maps of Western Europe and North America ' 456
— ' Space photography and its geographical application ' 419
Morocco: atlases 354
Institute Scientifique Chérifien, Rabat 353
Service Géographique (later Section Géographique du Service Topographique) 353
Service Géologique, *Notes et mémoires* 353
Spanish: map 225
Station Météorologique de Montagne, Ifrane 353
Station de Recherches Présahariennes 353
thematic surveying and mapping 353
topographical maps 353
Morris, F K: *Geology of Mongolia* 283
Morskoy atlas 8, 260, 424-25
Mortality, maps 460
Morton, John, classification of soils 427
Moscow: Academy of Sciences. ' Ethnographic map of the Middle East ' 242

Moscow (*contd*)
Chief Directorate of Geodesy and Cartography 85
Ground Surveying Institute (later Moscow Institute of Geodetic, Aerial Photographic and Cartographic Engineers) 258
Institute of Ethnography 448
Institute of Geodetic, Aerial Photographic and Cartographic Engineers, *Transactions* 24, 258
maps 103, 263
World Weather Watch, major world centre 418
' Motoring map of the eastern Mediterranean ' 224
A motorist's guide to the Soviet Union ' 264
Motoring maps *see* under Maps
Mott, O G: paper on mapping of Antarctica 403
Mount Kenya, relief map 376
' Mount Revelstok National Park . . .' 316
Mountain ranges, aerial surveying 21
Mountain Rescue Posts, on Ordnance Survey ' Cairngorms ' tourist map 129
Mountain world 337
Mouratov, M V *et al*: Tectonics of Europe . . . 114-15
Mozambique: *Atlas . . .* 379
Geographical Mission 227
maps 378, 379
geological 379
road 381
Servicos de Geologia e Minas: *Boletin* 379
Mugahya, Yunia, ed: *Uganda junior atlas* 375
Muir, A: new soil maps 243
Muir's historical atlases 155, 463
Multilingual dictionary, International Federation of Surveyors 10-11
Multilingual dictionary of cartographic terms 11-12
Multiple sclerosis, research in Ireland 460
Multiplex 60, 286
Mumford, Lewis: *The culture of cities* 451
Munby, L M, ed: *East Anglian studies* 161
Munger map book 351

Munsell system of colour standardisation 41-42

Muret *see* Faure-Muret, Anne

Murray, John: *Norwegian cirque glaciers* 180

Musée Royal de l'Afrique Centrale 346

Museo Argentino de Ciencies Naturales, Buenos Aires, publications 341

Museum of Practical Geology 412

Musil, Alois: *Oriental explorations and studies* 255

Muslim University Geographical Society, publications 274

Mussio, Giovanni: 'Practical lessons of cartography in geography study' 231

Mutton, Alice: *Central Europe* . . 205

Mycology, applied 462

Mylar, use in map preparation 39

Mysore State: *Atlas of resources* 273
Geological Department, publications 271
population study, UN 272
A regional synthesis (second volume of *Atlas of resources*) 273

Nadar *see* G F Tournachon

Nagata, Takesi, ed: papers of Pacific-Antarctic Sciences Symposium 402

Nagel, travel guides 456

Nairobi: maps 373, 375, 376
Mines and Geological Department 268
The Royal College 18
Survey of Kenya 72
University, Department of Geography: *Population of Kenya* . . . 375-76

Nairobi: city and region, ed: W T W Morgan 376

Name Placement Projector 35

Names, geographical *see* under Nomenclature; Placenames; Terminology

Nammacher, Fred: maps in *Atlas of Washington agriculture* 308

Nanking: Institute of Geography, research 284
University, Department of Agricultural Economics 282

Naples: tourist map 232
University, Institute of Geography: *Reports on economic geography and anthropology* 444

Napoleon, comment on maps 10

Natal: road map 381
University, Department of Land Surveying 18

National Academy of Sciences, National Research Council 312

National Aeronautics and Space Administration 20

National Agricultural Advisory Council 151

National Atlas Committee, British Association 141

'National atlas of China' 283

'National atlas of Czechoslovakia' 220

The national atlas of disease mortality in the United Kingdom (RGS) 155-56

'National atlas of Hungary' 235

National atlas of India 273

National atlas of Kenya . . . 376

National atlas of Tanganyika 377

National atlas of Turkey 245

National atlas of Uganda 375

'The national atlas of the United States of America', A C Gerlach 304

National Book League *et al*: *A select history of British scientific periodicals, geology* 414

National Bureau of Standards, Washington DC 18

National Cartographical Centre, Tehran, training courses 56

National Coal Board, statistics 145

National Committee for Cartography, Great Britain 122, 475

National Council of the Cartographers of the USSR 11

'National Diplomas in Cartography, Surveying and Photogrammetry' 39

National Forest Survey, Sweden 184

National Geographic atlas of the fifty United States 304

National Geographic atlas of the world 89-90

National Geographic magazine, map supplements 89

National Geographic Society: maps 72, 294, 299, 304, 311, 314
publications 312
National Geographical Society of India, publications 274
National Geological Survey of China 283
National Grid: 32, 124, 127, 128, 133, 136, 139, 143, 147-48, 160, 170
National Institute of Economic Agriculture, Milan 231
National Library of Scotland 170
National Library of Wales, Map Room 491-92
National Oceanographic Council 423
National Oceanographic Data Center, Washington DC 424
National Office of Lands and Mapping, Budapest 234
National Parks 130, 139, 142, 166
'The National Plans' Ordnance Survey, Professional Papers series 136
National Ports Council 160
National Research Centre for Cartographic Sciences 16
National Research Council, Ottawa 18
National Research Council, USA: *Soil exploration and mapping* 429
National Science Foundation, support of National Oceanographic Data Center 424
'The national survey map of New Hampshire', National Survey, Chester, Vermont 307
The National Trust atlas 158
Natural Environment Research Council Canada 37, 61, 148, 412, 413, 415
Natural Environmental Research group 16
Natural resources, mapping 388
research 431-32
Research Division, Unesco 428
Research Programme, Unesco 235
Research series 17
Nature aid resources . . . 117, 350, 412, 435
Nature Conservancy 153, 155, 446
Nature reserves, on Ordnance Survey maps 129
Wales 166
Navigation, marine history 443
Navy Department, Washington, *Welt-Seuchen-atlas* 461

Near East, *Atlas of Islamic history* 243
mapping 113, 463
Negev, map 74
Negretti, Henri, balloon photography experiments 20
Nelson and Company, Limited: atlases 76, 94, 369
Atlas of the world . . . 241
Cartographic Department 76, 139
Kenya junior atlas 104
Primary school atlas for Malaysia and Singapore 296
Shorter atlas of the classical world 241
Nelson's Canadian atlas 320
Nelson's Canadian junior atlas 320
Nelson's Canadian school atlas 320
Nelson's historical atlas of Canada 320
Nelson's school atlas 94
'Neolithic Wessex', Ordnance Survey map 132
Nepal: bibliographic references 279
maps 74, 279-80
surveying expeditions 279-80
Netherlands: *Atlas of cultural history* 194
atlases 19, 42
Geological Survey, Haarlem 192, 193
Government Office for Water Supply maps 106-107, 192-94, 204
geological 116
Ministry of Agriculture, Fisheries and Food Supply 194
Ministry of Education, co-operation in *Atlas van Nederland* 193
'The Netherlands and surrounding countries', in *Atlas van Nederland* 194
overseas mapping, contribution to 71
Polderauthorities 414
reclamation programmes 414
Royal Netherlands Geographical Society 193
rural development programmes 473
Society for Economic and Social Geography 193
Soil Survey Institute 173
Topografische Dienst, Delft 192-93
town plans 452
world atlases, co-operation in 86

Netherlands Antilles Cadastral Survey Department 329
Netherlands East Indies, Admiralty geographical handbook 297
Neundörfer, Ludwig, ed: *Atlas of social and economic regions of Europe* 118
Neur Welt-atlas, Editions Rencontre, Lausanne 97
Neustruyev, S S, soil scientist 429
A new atlas of China . . . 282
New atlas of the Chinese Republic 282
New atlas of the world (Cassell) 93
'New ANARE landings in Australian Antarctic territory', 1960, Phillip Law 400
New Brunswick University, Department of Surveying and Engineering 18
The new comparative atlas (Bartholomew) 91
New Forest, Ordnance Survey tourist map 129, 130
New France, map 320
New geographical literature and maps 472
New Georgia, maps 383
New Guinea : encyclopaedia 384
maps 74, 290, 297, 384, 387
annual record 386
glacier mapping 413
New Hampshire ' The national survey map . . .' 307
New Hebrides, maps 383
New international world atlas (Hammond) 94
The new Israel atlas 249
'A new landform map of Mexico', Erwin Raisz 322
New maps of the Chinese provinces 282
New secondary atlas (Collins-Longmans) 104
New South Wales, State mapping agency 386
see also under Australia
' New techniques used in producing the national map of Spain, 1 : 50,000 ' 224
New viewpoints in economic geography . . . 444
New York: Centre of African studies 352

New York (*contd*)
Eastman Kodak Research Library 18
Lockwood, Kesslen and Bartlett Inc 18
maps 103
plan 77
Public Library, map collection 477, 493
State: forestry atlas 308
Richards atlas . . . 308
New Zealand: Alps, aerial surveying 391
Antarctic Research Programme 398-99
Antarctic Society, *Antarctic: a news bulletin* 402
Atlas of New Zealand geography 394
Australian and New Zealand Association for the Advancement of Science 395
bibliography of maps 393
BANZARE Expedition, co-operation 397-98
' Christchurch and environs ' map 392
' Coastal chart series ' 393
Department of Lands and Survey, Wellington 391, 392, 398
maps of Antarctica 401
DSIR, co-ordination of thematic mapping 393
Geographical Society 395
Canterbury Branch 393
Geological Society, Antarctic expeditions 398
maps and memoirs 392
hydrographic charts 393
' Map of Bay of Islands ' 392
' Map of Blenheim and environs ' 392
' Map of Dunedin and environs ' 392
maps 74, 88-89, 92, 104, 385, 391-95
continental shelf 393
economic 444
glacier 413
land utilisation 393
mosaic 391
relationship with Australia 389
' strip map ' series 391, 392
thematic 393
Meteorological Service 393
National Advisory Corporation 393

New Zealand (*contd*)
Oceanographical Institute 393
The Oxford atlas . . . 394-95
The Oxford regional economic atlas
440
railways 457
Royal Society, *Transactions* 395
Soil Bureau 393
' Soil map of the North and South
Islands of New Zealand ' 393
South Pacific, government 384
' New Zealand's field work in the
Antarctic ' 398
The New Zealand geographer 395
Newbigin, M I 446
Newcastle: National Agricultural
Advisory regional Service 154
training of cartographers and sur-
veyors 80
University, Department of Geogra-
phy, *Planning reports* 161
*The regional atlas of north-east
England* 161
Newfoundland: maps 299
' Terra Nova National Park . . .' 316
Newlyn Tidal Observatory 124
Newnes' pictorial knowledge atlas 100
Niagara Province, Ontario, land use
maps 317
' Nicaragua ' map 327
bibliographical references 328
General Directorate of Cartography
326
Inter-American Geodetic Survey,
assistance from 326-27
maps 326-27
soil 327
National Geological Service surveys
327
us Air Force, assistance from 327
Nickles, M: ' Carte géologique de
l'Afrique Equatoriale française et
du Cameroun ' 371
Nicobar Islands, maps 74, 271, 297
Nicosia, map 254
Niddrie, D L: ' Land use and popu-
lation in Tobago . . .' 437
Nielsen, Niels, ed: *Atlas of Denmark*
191
Danmark, gazetteer 192
Niger River, mapping 77, 364
Niger Territory: mapping 357-58,
363
Topographical Service 358

Niger Valley, irrigation programmes
358
Nigeria, Overseas Economic Survey
369
*Directory of the Federation of
Nigeria* 369
' Experience with a new mapping
system . . .' 369-70
Federal Survey of Nigeria, Lagos
368
Geological Survey, publications 369
Journal of Nigerian Studies 369
maps 363, 368-70
planning 188
Northern: Department of Sur-
veying 18
A sketch map . . . 369
Timber Association, reports 369
Nigerian geographic journal 369
Nigerian publications 370
Nijhoff, Martinus: *Atlas van Neder-
land* 71
Nile Delta: inset maps 242
Nile River, mapping 77
historical 465
' strip ' 243
topographical 358
*Nine glacial maps, northwestern
North America* 315
Noh, Toshio: *Japanese geography*
. . . 289
Nohon shuho, weekly national bib-
liography 289
Nomenclature 87
in atlases 82
in maps 43
geographical 63
*Report of the group of experts on
geographical names* 45
standardisation 289
see also Place names; Termino-
logy
Norden, John 123
*Norden: crossroads of destiny and
progress* 180
Nordenskiöld Foundation 188
Nordisk Världs atlas 33
Norfolk and Norwich Naturalists
Society, *Transactions* 149
Norfolk: Ordnance Survey district
map 130
studies of river system 161
Norfolk Broads, canoeing maps 160
Norge 178

Oboli, H O N: *A sketch-map of Nigeria* 369
Occupations, maps 91
Oceania, maps 144
Océanographie géologique et géophysique de la Méditerranean occidentale ... 224
Oceanography, ed: Mary Sears 427
Oceanography: 13
computers, use in mapping 48
international congresses 327
international co-operation 423
maps 14, 92, 93, 104, 144, 404, 421, 423-27
relationship with meteorological features 421
techniques 423
The oceanographic atlas of the polar seas, US Navy Hydrographic Office 396, 426
Oceanographic Institute of the USSR, congresses 384
Oceans ... G E R Deacon 426
O'Connor, S M: 'East African topographic mapping' 374
— *An economic geography of East Africa* 374
O'Connor, Maeve: *Man and Africa* ... , ed: Gordon Wolstenholme and M O 352
O'Dell, A C: *The Scandinavian world* 177
O'Dell, N E: on aerial surveying of the New Zealand Alps 391
— 'Air survey of the New Zealand Alps' 395
Odense, map 103
O'Donovan, John 131
Office de la Recherche Scientifique Technique Outre-Mer, Paris 197, 351
Office Française de la Recherche Scientifique et Technique d'Outre-Mer 358
l'Office National Météorologique (previously Bureau Central Météorologique) 200
'The official map of Sweden, modern series' 182
Official Records of the Economic and Social Council 44
Officials' views on the uses of and need for topographic maps in North Carolina 308

Offshore geography of northwestern Europe ... 122
Ogilby, John 123
Ohio State University 18
Institute of Geodesy, Photogrammetry and Cartography 16, 469
Oil: maps 254, 266, 356, 444
production 443
shale 306
world map in *Mercantile marine atlas* 442
surveys 22
'Oil and natural gas map of Asia and the Far East' 266
Okado, T: *Climate of Japan* 288
Okinawa: maps 74, 288
Olaf Kayser Projection 441
O'Leary, T J: *The ethnographic bibliography of South Carolina* 331
Olson, Don: text in *Atlas of Washington agriculture* 308
Olsson *see* W William-Olsson
Olympia map stand 492
'On the classification of geographical books and maps ...' 483
'On the highland frontier of North Bhutan' 280
On the mode of communication of cholera, John Snow 458
Ontario: Department of Mines and Technical Surveys, Ottawa 18
'Ontario geology and principal minerals' map 316
'Ontario: province of opportunity' 316-17
Department of Planning and Development, Surveys 317
Institute of Chartered Cartographers 16
publications 469, 474, 475
'Ontario geology and printed materials map' 316
'Ontario: province of opportunity' 316-317
Oran, maps 354
Orbis geographicus 15, 476-77
'Orbis terrarum Europae' 117-18
Ord, H W, ed: International Seminar on African Primary Products and International Trade, papers 348
Ordnance, Board of 123
Ordnance Survey of Ireland 122
Belfast 172

Peak District, Ordnance Survey Tourist map 129
Pearson, E W *et al*, comps: *Mineral resources of the world atlas* 441
Peat, studies in the USSR 262
Pecka, K: *Organisation of work on the National Atlas of Czechoslovakia and The atlas of Czechoslovack history* 220
Pécs branch, Hungarian Geographical Society 236
Pésci, Marton *et al*: *Ten years of physico-geographical research in Hungary* 236
Peiping National Academy, Institute of Geology, support of soil studies 283
Peking-Tsinan district, map 281
Pemba, population map 373-74
Penck, Albrecht 63, 64, 66
Penguin atlas 100
Penguin atlas of medieval history 465
Penrose *et al*: *Graphic arts technicians handbook* 48
' Pentland Hills map ' 168
' People on the map ', diagrammatic map 146
Percheron, Maurice: *L'Indochine moderne* ... 293
Perejda, A and V Washburne: translation of legends, *Bolshoy Sovetskiy atlas* 85
Pergamon Press, Cartographic Department: 76, 139
  East European world map, UK agent 70
  The London atlas 162
  The Pergamon general world atlas 98-99
  PWN atlas of the world 100
Periodicals, containing pioneer maps 439
Permafrost 306, 413
' The permafrost map of Alaska ' 306
Permatrace 136
Perret, Robert: ' Essai d'une carte structurale de la France ' 200
Perring, F H and S M Walters, eds: *Atlas of British flora* 155
Persia, maps 243
Persian Gulf, maps 113, 243, 442

Perthes, Justus: 55
Stieler's Atlas of modern geography 82
  thematic maps 427
Perthes Geographische Anstalt, Gotha (later, Darmstadt) 206
Peru: *Anuario bibliográfico peruano* 337
  Asociación de Geógrafos del Peru, publications 337
  Geological Society, *Boletin* 337
  Hydrographic and Lighthouse Service, maps and charts 336-37
  index-atlas, Pan-American Union 343
  Instituto Nacional de Investigacion y Fomento Mineros, *Boletin* 337
  maps 336-37
  Military Geographic Institute 336
  National Aerial Photography Service, maps 336
  National Land Reform Office 337
  National Office of Natural Resources Evaluation 337
  National Town Planning and Town Development Office 337
  National Planning Institute 337
  Sociedad Geográfico de Lima, publications 337
Petermann, Augustus: cholera maps of the British Isles 458, 459
Peterson, R F: *Wheat: botany cultivation and utilization* 434
Petrie, G and J S Keates: ' Topographic science at the University of Glasgow ' 60
Petroleum: developments in Africa 351
  Information Bureau, London 444
  maps 243
' Petroleum and natural gas map of Canada ' 318
Petrology, surveys 22
Pettijohn, F J and P E Potter: *Atlas and glossary of primary sedimentary structures* 410
Pfalzatlas 210
Philadelphia: Aero Service Corporation 18
Philbrick, A K: *The pattern of Asia* 284
Philby, H St J: map of northwest Arabia 254

Philip, George and Son, Limited: 75
Africa, general maps 345
apparatus, for storage and display 492-93
Atlas of modern history 463-64
Australia, educational editions 104
general map 388-89
catalogues 477
Chambers of Commerce atlas 33
China, regional wall map 281
China in maps 281
Commercial concise atlas 443
Comparative wall atlases 75
Cosmopolitan world atlas, revised edition 95
educational publications 101-102
Europe, commercial map 113-14
The geographical digest 476
graphic relief wall maps 75
Hallwag Europa touring series 121
Hallwag political maps of Europe 113
Intermediate historical atlas 464
' International world map ' 75
International yearbook of cartography 470
Malaysia, educational atlases 104
general map 296
maps 91-92, 93-94, 139
Muir's series of historical atlases 463
Our United States— its history in maps 311
plastic relief maps 76
Secondary school atlas 102
Shepherd's historical atlases 464
Soviet Union in maps . . . 264
Spain and Portugal, road maps 225
three-dimensional relief representation 104
soils, world maps 428
wall maps 75
Yugoslavia, road map 239
Philip, S: maps of Finland 188
Philip's commercial map of the world 442
Philip's mercantile map of the world 442
Philip's modern school atlas 101-102
Philip's new school atlas 101-102
Phillimore, R H: *Historical records of the Survey of India* 270
Philippine Islands: Bureau of Coast and Geodetic Survey, Manilla 298
Census atlas . . . 298

Philippine Islands (*contd*)
Geographical Society 299
maps 291, 297, 298-99, 385
geological 298
mineral distribution 298
' Urban and rural population distribution map . . .' 298
Philippine geographical journal 299
Phillips, C W: Ordnance Survey map, ' Hadrian's Wall ' 133
Phillips, P L: *List of geographical atlases in the Library of Congress* 474
— *A list of maps of America in the Library of Congress . . .* 299
Photogeology 24
' Photogeological map of western Aden Protectorate ' 254
Photogrammetria 16, 19
Photogrammetric engineering 24
Photogrammetric record 18
Photogrammetric Society 18
Photogrammetry: 17-19, 24-32, 41, 60, 453
training courses 56-57
UN Technical Committee 12
Photogrammetry Inc, Rockville 17
Photogrammetry-Survey Department, University College, London 18
Photographs, in atlases 81, 93, 96 97, 99, 100, 146
library collections 70
Photography 13
aerial 16-17, 19, 20-21, 37, 67, 68, 70-71, 104-105, 109, 187, 188, 191, 194, 195, 407, 410, 411, 436, 438, 454
in *Atlas of Denmark* 191
balloon experiments 20
bibliographical references 25-32
deep sea 423
interpretation 453
space rocket 99
Photo-interpretation 24
international symposia 19, 25
UN Technical Committee 12
The photo-lithographic edition of the catalogue of printed maps, charts and plans, BM 473
Photolithography 46, 125, 128, 134, 151
Photo-maps 23, 129, 386-87
Photomosaics 23, 126, 127, 186-87, 362

Photonimographs 81
Photoscribe Head drafting machine 36
Photosetter 215
Photostat, copying of maps 494
Photo-theodolite 21
Photozincography 125, 128
The physical atlas . . . A K Johnston 409
The physical atlas of natural phenomena 458
Physical atlas of zoogeography 446
' The physical geography of Poland ' 223
' The physical master plan of the Israel coastal strip ' 249-50
' Physical master plan of Jerusalem ' 250
' Physical master plan of the northern Negev ' 250
Physical-statistical atlas of Austria-Hungary 213
Physikalische atlas . . . , Heinrich Berghaus 408-409, 458
Physiographic diagram of Europe 114
The physiographic provinces of North America 300
Physiography of Burma, H L Chibber 278
' The physique of Wales ' 166
Pictomaps 23
Piggott, Mary, ed: *Government information and the research worker,* Ronald Staveley and M P 141
A pilot survey of fourteen villages in Uttar Pradesh and Punjab 272
Pinpoints, aerial surveys 21
Pioneer atlas of the American West (Rand McNally) 306
' Pioneer maps of health and disease in England ' 459
Piver, M *et al:* computer compiled oceanographical atlas 424
Placenames 67, 76, 80-81, 82, 83-84, 88, 90, 98, 99, 107, 119, 171
catalogue entries 489
photographic setting 78
transliteration 85, 90
Placid, E: *Munger map book* 351
' Plan of Paris à vol d'oiseau ' 453
Planimeters 47

Planning: of atlases 82
bibliographical references 454-455
regional 81
' Planning and geography in Great Britain ', British Association for the Advancement of Science 452
' Planning maps of Britain ', 1 : 625,000 141-42
see also under Great Britain, national atlas sheets
Planning proposals for the Belfast area 174
Planning reports, University of Newcastle-upon-Tyne, Department of Geography 161
' Planning the second series of *The Atlas of Australian resources* ' 389
Plantation agriculture, P P Courtenay 433
Plants: bibliographical references 446-47
diseases 462
distribution 427
International Society for Plant Geography and Ecology 446
mapping 14, 77, 100, 146, 155
in micro-climates 417
Plastic relief maps 14, 78, 139, 146
' Plastic relief model map of Europe ' (Philip) 76
' Plastic relief model map of the world ' (Philip) 76
Plastics: use in map production 14, 34, 36, 38, 77, 134-35
apparatus for display of map sheets 492-93
Platt, Elizabeth, ed: *Current geographical literature and maps* 472
Plotters 36
Plumb, T W: ' Progress on Australia's national atlas ' 389
— and L K Hazlewood: ' Planning the second series of *The Atlas of Australian resources* ' 389
Plymouth, original maps 164
The pocket world atlas (Bartholomew) 93
Pointe-Noire, map 370
Poland: Academy of Sciences, Geographical Institute 222
bibliographical references 223
Cartographic Service, sheets of East European world map 70

585

'The population structure of Australian cities ', Peter Scott 390
Population studies, UN 450
Population Studies Group, Institute of British Geographers 157, 449
Port of London Authority: contribution to *The London atlas* 162
publications 160
Porter, P W: *A bibliography of statistical cartography* 444
— 'East Africa: population distribution as at August 1962 ' 374
Portfolios, manilla, method of storing sheet maps 491
Porto: maps 228
University, publications 229
Ports: plans 160, 421, 425, 442, 443
Port Harcourt 368
Portugal: Army Cartographical Service 227
bibliographical references 229
British bulletin of publications . . . 342
Geological Survey, Lisbon 228
' Grandes Carreteras de España y Portugal ', Michelin 227
mapping 71, 112, 225, 227-29
geological 228
land use 228
road 225, 228
marine research 112
national bibliography 229
Observatório Central Meteorológico, publications 22
Overseas Astronomic and Gravimetric Mission 228
Overseas Geography Centre 227-28
Spain and Portugal, Geographical handbook series 227
Post Office, ship-shore service 145
Postal atlas of China, Nanking 282
' Post-war progress in cartography in the Federal Republic of Germany ' 207
Post-war Reconstruction and Planning, West Midland Group: *English county* 451
Potato Marketing Board, *British atlas of potato varieties* 157
Potchefstroon University, Ferdinand Postma Biblioteek 18
' Potential natural vegetation of the conterminous United States ' 306
Pothole societies, surveys 149

Potter, P_ E: *Atlas and glossary of primary sedimentary structures* 410
Potteries, Oxford plastic relief map 139
Pottinger, Don: ' Map of Scotland of old ' 168
Pounds, N T G: text in *An atlas of European affairs* 120
— and K C Kingsbury: *An atlas of Middle Eastern affairs* 243
Power in New Zealand: a geography of energy resouces 395
' Powiat ', Poland 222
Practical atlas (Philip) 93
Practical geography in Africa 345
' Practical lessons of cartography in geography study ' 231
Prague: Geological Society 246
State Publishing House 86
Pre-Cambrian, Gulf of Suez 360
The Pre-Cambrian along the Gulf of Suez and the northern part of the Red Sea 359-60
Precipitation, maps 421
' Predominant types of commercial farms in Canada, 1961 ' 316
Preliminary atlas of the Republic of Guatamala 326
' Preliminary draft plan for the economic and social development of the country in twenty one years ' (Guatamala) 326
' Preliminary Plot ' method 23
The Prentice-Hall world atlas 95-96
' Preparing and printing soil maps in Ireland ' 173
' Present status of cartography and related surveys in Japan ' 287
Pretoria Publicity and Travel department 381
Priestley, *Sir* Raymond *et al,* ed: Antarctic research . . . 399
Primary atlas (Collins-Longmans) 104
Primary school atlas for Malaysia and Singapore (Nelson) 296
Prince Edward Island, land use maps 317
Princeton UP: *The southeast in early maps* 311
' Principal mineral areas of Canada ' map 316

Queen Maud range, maps 401
Queen's University, Belfast, David Keir Library 18
Queensland, state mapping agency 386
University Library, Brisbane 18
' The question of adopting a standard method of writing geographical names on maps ' 45
Quito University, map of Ecuador 336

Races of man, maps 93, 95
Radar altimeter 23
Radar equipment 112
Radar signals: conversion to map symbols 22
Radcliffe Observatory, publications 151
Radio probes 112
Radó, Sándor: *Atlas international Larousse* . . . 443
— bibliography and glossary of population maps 448
— Cartactual 79
— world map 69, 70
Raiatea-Tahaa, maps 384
Railways: surveys 22, 112, 173
maps 201, 381
underground 457
Rainbows, photographs of 421
Rainfall, maps 225, 228, 355, 420
Raisz, Erwin 55
— *Atlas de Cuba* 329-30
— *Atlas of Florida* 308
— land forms photographs 94
— 'A new landform map of Mexico ' 322
Rajchman, Marthe: *An Atlas of Far Eastern politics* 282
— *A new atlas of China* . . . 282
Rand McNally and Company, Chicago: *Atlas of western Europe* 119
cartographers, meeting 13
Commercial atlas and marketing guide 443
communications maps 456
Cosmopolitan world atlas 97
Guide to Mexico 323
The historical atlas of the Holy Land 252
International yearbook of cartography 470

Rand McNally and Company, Chicago (*contd*)
Pioneer atlas of the American west 306
Standard world atlas 93-94
United States, road maps 311
world atlases 95
Rand McNally atlas of world history 464
' Rand McNally Imperial map of Mexico ' 323
Rand McNally road atlas 299
Rand McNally's business atlas of the Great Mississippi valley and Pacific slope, 1876 306
Rank Cintel Organisation, prototype automatic reading planimeter 47
Rapports et procès verbaux, International Commission for the Scientific Exploration of the Mediterranean Sea 223
Rasht, map 256
' Ratio maps of China's farms and crops ', G T Trewartha 282
Raup, O B: exploration of salt deposits in South Utah 429
Ravenna: early maps of Christian churches 465
Rawson, R R and K R Sealy: ' Cyprus land use map ' 253
Ray Society *et al: A select list of British scientific periodicals: geology* 414
The Reader's Digest AA book of the road 159
The Reader's Digest complete atlas of the British Isles 145-46
The Reader's Digest great world atlas 97-98
The Reader's Digest world atlas 34, 95
Reading: National Agricultural Advisory regional service 154
Soils Survey Centre 154
Readings in the geography of North America . . . 300
' Read-out ' equipment 36
Real Estate Office, maps of Helsinki 188-89
' Recent exploration of Victoria Land north of Terra Nova Bay ' 400
Recent history atlas, 1870 to the present day, Martin Gilbert 466

589

Riley, D and A Young: *World vegetation* 131

Rimington, G R C: *Report* of the Conference (UN Cartographic Conference for Asia and the Far East, Bangkok) 13

Rimli, Eugène Th 97

Rio Grande, Federal University, Institute of Hydraulic Research 334

Ripa, L C: *Surveyor's manual* 25

Ristow, W W: *Marketing maps of the USA* ... 312

Ritter, Carl 430

— 'Product charts' 427

Ritter, Wigand: *Fremdenverkehr in Europa* 114

Rivers, mapping 77, 141-42, 150, 160

Riviera, maps 74, 113,

Road and travel atlas of Europe (Kümmerly and Frey) 121

Road atlas of Great Britain (Bartholomew) 159

Road atlas of Norway 179

Road classification 34-37, 179, 405, 457

Road map of Europe, ed: Alliance International de Tourisme 121

'Road map of Israel' 248-49

'Road map of Scotland' (Bartholomew) 171

Road Research Laboratory, Harmondsworth 18

Roadmaster motoring atlas of Great Britain 159

Roads: design, use of geophysical methods 453

maps 88, 120-22, 147, 160, 174, 191-92, 196, 201, 210-11, 228, 235, 239, 245, 248-49, 257, 264, 311-12, 368, 370, 375, 381, 389, 453-57

exhibition in Budapest 234

planning methods 453-54

Road international 456

surveys 22

Wales, improvement 166

see also under Maps

Roberts, C F, jr, ed: *Malaya* 296

Robertson, I M L: 'Population enumeration on a grid square basis ...' 170

Robertson, V C: *An approach to the rapid description and mapping of biological habitats* 446

Roberty, Guy: 'Carte de la vegetation de l'Afrique Tropicale Occidentale, 1 : 1м' 351

— vegetation maps 351

Robinson, A H: on type faces for maps 44

Robinson, G W: soil mapping in North Wales 166

— Soil Survey of England and Wales 154

Robinson, G W S: *The law of the continental shelf* 437-38

— and M G Webb: *The resorts of western Europe* 120-21

Robinson, J L 313

Robinson, Ronald, ed: African development planning 348

Rock forms, representation 41, 188

Rockets, photography experiments 20

Rodenwaldt, Ernst and Heimut Jusatz, eds: *Welt-Seuchen-atlas* 461

— 'World maps of climatology' 461

Rohtak District, population mapping 272

'Roman Britain', Ordnance Survey map 132

Roman Empire, international map 68

Romania: maps 205

geological 116

see also under Rumania

Rome: early maps of Christian churches 465

East European world map sheet 69

Meridian 230

Romer, Professor: 'Geographical statistical atlas of Poland' 222

Romney, D H, ed: *Land in British Honduras* ... 324-25

Roneo Flushline Vertical Planfile 492

Ronne Antarctic research expedition, 1947-48 401

Ronner, Felix: *Report on geology in Iraq* 246

Roolvink, Roelof: *Historical atlas of the Muslim peoples* 242

Ross Dependency area: New Zealand explorations 398

Rothamsted: Experimental Station 131

Soil Survey of England and Wales 154

Royal Society: Cartographic Sub-Committee 45
Indian Ocean, research 268
modern surveying, symposium 21-22
National Committee for Geography 14, 122
Natural Environment Research Council, sponsorship of 16
Sub-Committee for Geography 475
Triangulation of Great Britain 123
Royal Society of Arts, *Journal* 398
Royal Society of Canada: *Marine distributions* 426-27
Royal Thai Survey Department, Bangkok 292
Royal Tropical Institute, Amsterdam 131
Royal Tunbridge Wells, plan 162
Rugg, D S: 'Post-war progress in cartography in the Federal Republic of Germany' 207
'Ruhr coal mining district atlas' 209
Rules for descriptive cataloging in the Library of Congress 490
Rumania: *Agriculture en Roumanie* . . . 238
'The economic map of the Rumanian People's Republic' 237
Geodetic and Cartographic Services 69
geographers, appled research 236-37
Geographical Research Institute 236-37
Geo-Karta Institute, Belgrade 238
Geological Institute 236
maps 236-38
administrative 237
land use 237, 438
road 237
microclimatic studies 237
Monografia geographica a republicii populare Romîne 237
national bibliography 238
Revue de géologie et de géographie 414
Topografica Militara 236
Rumanian Natural Sciences and Geographical Society 237
Rumanian Society of Geography, *Bulletin* 236
Rural Bank of South Wales: *Atlas of Australian rural industries* 390
'Rush' editions, Belgian topographic maps 195

Russell *see* Tullis Russell and Co Ltd
'Russia in the thirteenth century' 260
Russian history atlas 264
Russian Trade Exhibition, 1961; exhibition of tectonic map 114
Rutherlord, J *et al,* eds: *New viewpoints in economic geography* . . . 444
Ruysch, map of 1508 91
RV der kleine katalog . . . 212
RV Kalalog 212
Kartenbrief 478

Sabarís, L and P Defontaines: wall-map of Spain and Portugal 227
Sahara 356
Sahara: bibliographical references 356
'Carte géologique du Sahara . . .' 356
Institut de Recherches Sahariennes 355, 356
mapping 77, 356-57
Saigon: Vietnamese-American Association Library, *Catalogue* 294
St Anthony's College, Oxford: Soviet Affairs Study Group 265
St John, maps 314
St Lawrence Seaway, maps 98, 314, 319-20, 442
Saint Lucia Island, West Indies, map 329
St Louis, Missouri, street map 455
Sakhalin region, atlas 261
'Sales manager's and population maps of Great Britain' (Geographia) 158
Salichtchev, K A *et al,* ed: 'Atlas of geographic discoveries and explorations' 463
Salinity profiles 111
Salisbury Plain: base line for Ordnance Survey triangulation 124
Salt deposits, effect on soils 246, 429
Saltash: planning of bridges 455
Salzburg atlas 214
Sampson, Henry, ed: *World railways* 457
San Francisco and vicinity; road map 312
Sanceau, *Major* 64
Sanke, Heinz, ed: *Weltatlas: die staaten der Erde und ihre Wirtschaft* 441

Sinclair, D J, ed: Faber *School atlas* 102
Singapore: *Gazette* 296
Improvement Trust and Department of Geography, University of Malaya: land use maps 295
maps 291, 296
Primary school atlas (Nelson) 296
State Development Plan 296
Sinkiang, maps 283
Sinusoidal Projection 32, 33
Sirius dyeline copier 60
Site reconnaissance 454
Sites, historical 145
'The siting and development of British airports', K R Sealy 166
Skelton, R A: official history of the Ordnance Survey, in preparation 138
A sketch-map of Nigeria 369
Skerry, surveys 188
Skin diseases, studies in Mexico 460
Skinner's West Indies and Caribbean yearbook 329
Sloane, C S: *1925 Census atlas of USA* 309
Slope, factor in vegetation control 166
Sloss, L L *et al,* eds: *Lithofacies maps . . .* 411
Slovak Academy of Sciences 218
Smailes, P J: 'The large-scale cadastral map coverage of Australia . . .' 390
Small industry bulletin for Asia and the Far East, UN 266
Smiley, J McA, comp: maps in *Atlas of Islamic history* 243
Smith, C G: 'Israel's territorial gains as a result of the June war' 251
Smith, Gordon: new water balance map 243
Smith, Guy Harold, ed: *Conservation of natural resources* 433
Smith, W D: *Geology and mineral resources of the Philippine Islands* 298
Smith, William: geology maps 147
Smith and Son's globe-making business 75
Snow, John: mapping of deaths from cholera 155, 458
— *On the mode of communication of cholera* 458

Snow, mapping 395-96, 413
Snowdonia National Park: Ordnance Survey district map 130
Social studies atlas (Tokyo) 288
Sociedade Brasileira de Cartografia, Rio de Janeiro 333
Sociedade Brasileira de Fotogrametria 24
Sociedad Geográfica de Colombia, *Boletín* 336
Società Geografica Italiana, geographical research 233
Société Africaine de Travaux et d'Etudes Topographiques 370, 371
Société Belge de Cartographie, Belgische Vereniging voor Kartografie 194
Société Belge de Géologie, de Paléontologie et d'Hydrologie, Brussels, publications 196
Société d'Editions Géographes Maritimes et Coloniales 71
Société des Etudes Océaniennes, publications 384
Société du Géographie d'Alger et de l'Afrique du Nord, *Bulletin* 355
Société Européenne d'Etudes et l'Information 119
Société Française de Photogrammétrie, Section Laussedat de la Société Française de Photographie et de Cinématographie 24
Société Géologique de France, *Bulletin* 414
Société Géologique de Belgique, Liège, *publications* 196
La Société Météorologique de France 201, 422
Societies, local geological 149, 412
Society for International Cultural Relations, Tokyo 288
Society of Foresters of Great Britain, *Forestry* 434
Society of Local Archivists, representation on Library and Map Committee, RGS 490
Society of Mining Geologists of Japan, *Mining geology* 287
Society of University Cartographers 58-59
Bulletin 475-76
see also under individual names
Söderlund, Alfred: 'Distribution of population in Europe' map 119

598

Stamp, *Sir* Dudley (*contd*)
  Land use statistics of European countries, 1965 437
  Land Utilisation Survey of Great Britain 152-53
  'Publications of the World Land Use Survey' 437
  World Land Use Survey, first Chairman 438
  pilot scheme 436
  — Keith Buchanan: 'Principal systems of farming in Great Britain' 152
Standard atlas of Japan, Tokyo 288
The standard atlas of Korea 290
The standard reference atlas (Philip) 93
'Standard symbols for use on topographic maps of Antarctica' 402
Standard world atlas (Philip; Rand McNally) 93-94
Stanford, Edward, Limited: maps 75, 113, 147, 152, 153, 160, 176, 257, 296, 299, 360, 388-89, 477-78
  Planfile 492
  Stanford's geological atlas of Great Britain and Ireland 150
  Stanford's Whitehall atlas 93
Stanford Food Research Institute studies 349
Stanford Research Institute of California 22
Stanford University: centre of African studies 352
map collection 477
Stanley, W F: prototype automatic reading planimeter 47
Stanley Precision Planimeter 47
Stansbury, M J: 'The cartographic course at the Oxford College of Technology' 61
Stapledon, *Sir* R G and William Davies: *Survey of grasslands of England and Wales* 150
Starcross: National agricultural advisory regional service 154
Starvation mapping 460-61
State Publishing Office, Moscow 10
'States and districts of India' 270
'States and trends of geography in the United States, 1957-1960' report 312
Statistical map of Bombay State 273

A statistical atlas of Brazil 334
Statistical atlas of the United States 309
'Statistical cartography at the US Bureau of the Census' 309
Statistical cartography, bibliography 444
Statistics: mapping 43, 76, 81, 93, 94, 95, 97, 99, 100, 105, 107, 118, 144, 145, 156-57, 158, 166, 202, 226, 427, 439, 460
data processing 157
'Status of topographic, geologic and soil mapping and vertical aerial photography . . .' (Pan-American Union) 324
Stavanger, map 179
Staveley, Ronald and Mary Piggott: *Government information and the research worker* 141
Steer, K A: 'Scotland, south-east', map 168
Stefansson, V: on the Arctic 396, 397
Steier-Märkischen Landesregierung: *Atlas der Steirmärk* 214
Steinberg, D J *et al,* eds: *Cambodia . . .* 294
Steiner Verlag, Weisbaden: *Atlas der Deutschen agrarlandschaft* 209-10
Stephen, K H: 'Cartographic training at Survey Production Centre' 61
Stephen F Austin State College, Texas 18
'State geobotanic map of the Soviet Union' 259
'State geologic map of the Soviet Union' 259
'State soil map of the Soviet Union' 259
Stembridge, J H: *An atlas of the USSR* 264
Stereoplanigraph 286
Stereoplotting 21, 188
Stevens, L S, trans: Raymond Furon: *Géologie de l'Afrique* 350
Stewart, I G and H W Ord, ed: *Papers,* International Seminar on African Primary Products and International Trade 348
'Sticking up' method 45, 224
Stieler atlas of modern geography 82, 105

Times Roman, type used in *The Times atlas of the world* 82
Timmers, J J M: *Atlas van de Nederlandse beschaving* 194
Timor: Geographical Mission 227
   maps 287, 290
Ting, N K: reference atlas of China 281
Ting, V K: *New atlas of the Chinese Republic* 282
Tinker, Hugh: *The Union of Burma* ... 278-79
Tints, importance in maps 82
   hypsometric 41, 42, 78, 107
   layer 130, 141, 168
Tiros satellites 418
Titles, of maps, on catalogue entries 488
Tobago: land use surveys 437
   maps 329
Togo, maps 363
Tohoku University, *Science reports* 287
Tokyo: Geographical Society 290
   maps 74, 288
   Meteorological Research Institute 422
   Ministry of Information, *Antarctic record* 402
   University, Faculty of Science *Journal* 287
   Institute of Human Geography 287
Tools, in map-production 38, 39, 40, 77
'Topocart', stereocartographic instrument 37
Topographic instructions, US Geological Survey 25
'Topographic map of the United Kingdom of Libya', US Geological Survey 356
'Topographic Science and the University of Glasgow' 60
Topographical Research Centre, Hanover 412
Topography, standards of presentation 63
   Sub-Division of American Congress 16
   terms 10
   UN Technical Committee 12
Toronto: Convention and Tourist Bureau, map 317
   development map, in *Atlas of Canada* 318

Toronto (*contd*)
   Dominion Bank, Calgary: 'Petroleum and natural gas map of Canada' 318
   University, Geophysics Laboratory: 'Glacial map of Canada' 315
Torpoint: planning of ferries 455
Toulouse: 'Aerial surveys and integrated studies' conference 17
   Colloquium on cartography of vegetation 430
   University, Institut de la Carte Internationale de Tapis Végétal 429
   sponsorship of 'Aerial surveys ...' conference 17
Touring atlas 'Cartographia', Scandinavia 176
Touring Club Italiano 56, 78, 121, 229, 231, 232
'Touring map of Ireland' 174-175
'Touring map of Israel' 249
Tourists, special maps for 40, 120-22, 158-59, 166, 176, 192, 201, 213, 311, 323, 324, 455, 456-57
Tournachon, G F (Nadar) 20
Tornadoes, photographs 421
Toussaint, Auguste and H Adolphe: *Bibliography of Mauritius, 1502-1954* 268
'Towards a national atlas', in *Government information and the research worker* 141
Towards economic co-operation in Asia ... 266
Tower of London, former map office of the Ordnance Survey 123
Town and Country Planning Act, 1947 451
Town Development Bureau, Montreal 452
Town planning: bibliographical references 454-55
   Institute, Council for national survey 141
   use of data processing machines 34-37
Town plans (including city plans) 22, 73, 77, 86, 89, 91, 103, 114, 120, 121, 127, 159, 161, 162, 168, 171, 175, 176, 178, 187, 193, 201, 202, 211, 217, 219, 227, 235, 238, 239, 291, 299, 300, 304, 309, 318, 320, 332, 340, 342, 343, 368, 370, 373, 377, 389, 392, 451, 457

607

Town stamps 79
Towns: co-ordinated planning 453
symbols 103, 142
The towns of Wales . . . 166
Toynbee, Arnold: on Hittite sites and locations, appendix to *The historical atlas and gazetteer* 465
— *Study of history, with The historical atlas and gazetteer* 465
Tracing table, provision in map libraries 494
Trade: Commissions list in *Mercantile marine atlas* 442
maps 439
Report, British Honduras 325
Trade and Travel Publications: *South American handbook* 343
Traffic: patterns 455
surveys 188
Traffic in towns . . . 456
Training, of cartographers 55-62, 234, 258
see also under Education
' The training and education of cartographers and cartographic draughtsmen ', Symposium 61
Transbaikalia, regional atlas 261
Transliteration 44, 90
Transport: bibliographical references 457-58
data 455
maps 14, 91, 160, 453-58
see also under Ministry of Transport
The Transvaal Lowveld, M M Cole 437
Transverse Mercator Projection 32, 133, 193, 277, 290, 292, 303, 313, 314, 327, 329, 332, 357, 367, 386, 387, 390, 391
see also under Mercator
Trap, J P: *Danmark,* gazetteer 192
Travel guides, Nagel 456
Travels and discoveries in north and central Africa 345
Traversing 19
Treasure, P J: maps of Harrogate 163
Tregear, T R: land use monograph on Hong Kong 437
— and L Berry: *The development of Hong Kong and Kowloon as told in maps* 284

Treharne, R F, ed: Muir's historical atlases 463
' The Trent Basin ', Ordnance Survey map 132
Trewartha, G T: ratio maps of China's farms and crops 282
Triangulation 19, 123, 186
' Tribal map of India ' 272
' Tribal map of Negro Africa ' 345
' Tribal map of east Africa and Zanzibar ' 373
Tricart, Jean: *L'Europe centrale* 205
Trigonometrical and Topographical Surveys Department, Rhodesia 18
Trinidad and Tobago: Barclay's Bank economic survey 329
Colonial Office, *Annual reports* 329
maps 329
Survey Department 329
Troll, Carl 15
climate maps 420, 461
interest in biogeographic factors 446
Tromso, map 116
Trondheim, maps 116, 179
Tropical Plant Research Foundation, Washington 330
Tropical Products Institute, *Tropical science* 435
Tropical science, (formerly *Colonial plant and animal products*) 435
Tropical soils and vegetation, Proceedings of Abidjan Symposium 364
Tropics, soil surveys 428
Trossachs, Ordnance Survey Tourist map 129
Trudel, Marcel: *Atlas historique du Canada français des débuts à 1867* 320
Tsao-Mo on ' Program of cartographic work in China ' 281
' Tschechoslowakei strassenkarte ' 220
Tseng, S T: reference atlas of China 281
' Tsetse control and livestock development . . .' 375
Tubuai, maps 384
Tufescu, Victor, ed: *The atlas geografic Republic Socialista România* 237-38
Tullis Russell and Company, Limited, printing papers 38
Tulupnikov, A I *et al*: 'Atlas of agriculture in the USSR ' 262

609

20

Union of South Africa (*contd*)
  Trigonometrical Survey Office 379-80
  USSR: Academy of Sciences, Congress of Oceanographic Institutes 384
  'Fiziko-geografičeskij atlas mira' 409-10
  Soil geographical zoning of the USSR 262
  Admiralty and Naval General Staff: *Morskoy atlas* 8, 260, 424-25
  agricultural studies 262
  Antarctic research 399
  Arctic expeditions 396
  Bureau of Agro-Meteorology 262
  Central Geodetic Department 258
  Central Hydrological and Meteorological Offices 395
  Central Scientific Research Institute of Geodesy, Aerial Photography and Cartography 258
  Chief Administration of Geodesy and Cartography, *Atlas narodov mira* 449
    Morskoy atlas 425
  Climate classifications 262
  Corps of Military Engineers 258
  current national bibliographies 265
  Department of Navigation, 'Arctic studies' 395
  forest studies 263
  Geographical Society, Congress 396
  Indian Ocean expedition 267
  Institute of Ethnology 449
  maps 20, 42, 82, 84-86, 102, 106, 116, 258-65
    agricultural 262
    availability 476
    catalogues 473
    'commodity flow' 263
    economic 106, 263-64
    educational 106
    First Five-Year Plan, mapping 263
    'forecast' maps 261
    geodetic network 258-59
    geological 115, 116, 259, 261
    historical 260, 264
    hydrogeology 261
    land use 438
    IMW 1 : 1M 259
    peat areas 262
    regional atlases 260-61
    road 264

USSR (*contd*)
  soil 259, 427-29
  soil studies 261-262
    methods 154
    technical manuals 427-29
    statistical 263
    tectonic 114
    topographical, availability 8
  bibliographical references 265
  Meteorological Office, Moscow 395
  Ministry for Geology and Natural Resources 85
  Naval Cartographic and Hydrographic Staff: *Morskoy atlas* 425
  Norway, communication links with 179
  Peat Reserves, Central Department 262
  railways, influence on agriculture 263
  soil scientists 428
  State Oceanographical Institute, *Arctic studies* 395
  State Publishing House for Geodesy and Cartography 85, 86
  The Times atlas of the world, treatment of USSR 82-83
  training of cartographers 57, 258
  The USSR and eastern Europe, Oxford regional economic atlas 440
  see also under Soviet
Union Syndicale des Industries Aéronautiques 18
Upper Volta, maps 363, 364
United Arab Republic (Egypt Region): Egyptian Geographical Society *Journal* 360
  'Europa' survey 244
  Geological Survey, Cairo 359-60
  maps 243, 358-60
    colours, in topographical mapping 358-59
    geological 360
    'Northern Sinai' series 358
    United Kingdom 70
  The Pre-Cambrian along the Gulf of Suez . . . 359-60
  The surface features of Egypt . . . 359
  Survey of Egypt 358-59
  Tabula Imperii Romani 1 : 1M, contribution to 68
United Kingdom: aerial mapping 21
  Antarctic Placenames Committee 400

610

United Kingdom (*contd*)
Antarctic research 399
automated cartography 34-36
disease mortality mapping 155-56
— Ghana Technical Assistance Scheme 367
training in cartography 58
world mapping, contribution 70
support for IMW 1 : 1M 64
surveying in Iraq 246
see also Great Britain
United Nations: Aerial Survey Methods Seminar 24
African resources, assistance in planning 347
Bangkok Conference 470
Cartographic Office, aims 12
' cartography ', definition 9
Commission for Asia and the Far East, geological and other maps 266
Commission on National Atlases 109-10
bibliographical references 109
co-ordination of information 439
Department of Social Affairs, Cartographic Office: *World cartography* 468
Economic and Social Council, New York 11, 63, 427
Economic bulletins (Asia and Far East) 266
Economic development in the Middle East . . . 234
economic statistics 440
Economic survey of Asia and the Far East 266
'Formulating industrial development programs, with special reference to Asia and the Far East ' 266
Geological Survey, geological map of Indonesia 297
Group of Experts on Geological Names 11
The growth of world industry 1938-61 . . . 439
Indian Ocean expedition, sponsorship of 268
International Committee on Geographical Names 44
IMW 1 : 1M, Central Bureau 65, 66
Inter-regional Seminar, ' The application of cartography for economic development ', 1965 469

United Nations (*contd*)
' Mineral resources development ' series 266
The Mysore population study 272
Population studies 450
Regional Cartographical Conferences for Africa 351, 362
Regional Cartographical Conferences for Asia and the Far East 12-13, 44, 200, 266, 278, 281, 287, 292, 312, 386
Proceedings 12-13, 267, 469
Reports 80, 200, 278
Technical papers 469
Secretariat 44
Small industry bulletin for Asia and the Far East 266
standardisation of geographical names 45
Statistical Department 447
Survey of Indian mapping 276
Technical Conference of the IMW 1 : 1M, 1961 122
Toward the economic development of the Republic of Viet-Nam 294
United Kingdom of Libya, Annual reports 356
Unesco: 63, 65
Advisory Committee on Natural Resources Research 431
Aerial surveys conference 17
Arid Zone Research Programme 223, 224, 246, 435
Bibliographie cartographique internationale, support 470-71
' Change in climate ' symposium 420
Fluctuations of glaciers, 1959-1965 . . . 413
geological surveys 246, 349-50
Humid Tropics Research Programme 435
Lagos Plan, assistance with 347
Problems of humid tropical regions 435
surveys: ' Bioclimatic map of the Mediterranean zone ', with FAO 223-24
' International Quaternary map of Europe ' 116-17
' Metallogenetic map ' 117
Symposium on Photo-Interpretation, Ottawa 25
Symposium on salt deposits, effect on soils 429

611

United States of America (*contd*)
mapping techniques 77
maps 84, 91, 93-94, 95, 107, 299, 300-12
facsimile 306
forests 306
geology 305
land forms 410
IMW 1 : 1M, support of 64
' Permafrost map of Alaska ' 306
relief 305
road 311-12
bibliographical references 312
Military Academy, teaching on landforms 411
National Academy of Sciences 303, 399
National Automobile Club, road map 311
National cartographic report, 1964 302
National Science Foundation 448, 449
Naval Oceanographic Office 36
Navy: Hydrographic Office: *Ice atlas of the northern hemisphere* 413
Oceanographic atlas of the polar seas 396
strip charts 400
support of National Oceanographic Data Center 424
trimetrogon aerial photographs 401
oceanographical research 423-24
Oxford regional economic atlas 440
photogrammetric organisations 17-18
population studies 449
soil survey methods 154
Soil Survey, *annual reports* 312
South Pacific, government 384
Tiros Operational Satellite system 418
' Vacation lands of the United States and Canada ' 299
War Department, Washington 330
Water Information Center: *Water atlas of the United States* . . . 305
Weather Bureau 305, 419, 421, 424, 418
The United States and Canada: Oxford regional economic atlas 300
see also under names of individual States

Universal atlas (Cassell) 93
Universal Decimal Classification: UDC 91 adaptation for classification of geographical books and maps 482, 484, 488
Universal Postal Union world map 43
Université Libre de Bruxelles en Afrique Centrale, Centre Scientifique et Médical 372
Universities, introduction of cartography courses 55-56
The university atlas (Philip) 91-92
University cartographers, Society 58-59
' Up the Poles ', The *Sunday times* review of *Pergamon world atlas* 99
Upper Mantle Project 348
Upper Volta, levelling network 357
Uppsala University; Geographical Department 185
Geological Institution 183
Ural Mountains, mapping 113
' Urban and rural population distribution map of the Philippines ' 298
' Urban hinterlands—15 years on ' 157
Urban Studies, Council: contribution to *The London atlas* 162
Uruguay: Hydrographic Bureau 342
index atlas, Pan-American Union 343
Instituto Geologico del Uruguay, geological map 342
maps 341-42
national series 342
thematic 342
National Bibliographical Group 342
' Use of the International Map of the World series in China ' 281
' The use of maps in Australia ' 390
Ussher Society, geological and geographical studies of south-west England 150
Utrecht: Conferences 417
University, Geography Institute 227
Uttar Pradesh, maps of population 272

Valencia, map 332
Van Bemmelen, R W: *The geology of Indonesia* 297
Van der Gragt, F: *Europe's greatest tramway network* . . . 211

613

Washington (*contd*)
National Bureau of Standards 18
Organization of American States: *Annotated index of aerial photographic coverage and mapping of topographical and natural resources* . . . 322-23
State Department of Agriculture: *Atlas of Washington agriculture* 308
World Weather Watch 418
see also under United States and under individual names, *eg* Coast and Geodetic Survey, US Geological Survey
Wassermann, Basle, town plan 217
Water atlas of the United States . . . 305
Water: conservation programmes 150, 252
Water Resources Board: ' Water supplies in south-east England ' 161
Water Resources Research 414
Water spouts, photographs 421
Water supplies, geological surveys 408, 413-14
Waterbalance, map 243
' Waterstaatkart ', Netherlands 414
Waterways, inland 160
Waterways atlas of the British Isles 160
Watkins, F H: *Handbook of the Leeward Islands* 329
Watson, J Wreford, ed: *Atlas of Edinburgh* 170
— *Concise world atlas* 94
— *Nelson's Canadian atlas* 320
Watts Radial Line Plotter 60
Weather: automatic plotting of data 22
factor in aerial surveying 24
forecasting: Wiesbaden conference 419
maps 43, 151-52, 164, 360, 417-22
construction from forecasts 419
world data 418
Weather in the Black Sea 420
Weather in Home Fleet waters 420
Weather map 151, 419-20
Weather Reports—no 9—stations, codes, transmissions, WMO 417
see also under Climate; Meteorology
Webb, D A, ed: *A view of Ireland* . . . 174

Webb, M G: *The resorts of western Europe* 120-21
Webster, W, ed: *Serial atlas of the marine environment* 425
Weddell Sea, coast maps 400
Weir, T R, ed: *Economic atlas of Manitoba* 319
Weisse, Hildgaard: *Indien historisch-geographisches Kartenwerk* . . . 274
Wellington, J H: *Southern Africa* . . . 382
Wellington, maps 74
Wellman, F L: *Coffee* . . . 434
Welsh, D trans: *Izvestiya* review on the *Oxford economic atlas of the USSR and eastern Europe* 263
Welsh, D R trans: *Atlas of Mesopotamia* . . . 246
' The Welsh colony in Patagonia . . .' 341
Weltatlas: die Staaten der Erde und ihre Wirtschaft 441
Die Weltmeer: taschenatlas mit den wichtigsten tatsachen aus meteorologie und nautik 421, 426
Welt-Seuchen-atlas 461
Welwyn Garden City, plan 162
Wen-hao, Wong: reference atlas of China 281
Werner, interrupted projection 33
Wesley's historical atlas of the United States 311
Wessex Cave Club, original surveys 149
West Africa: the French-speaking nations, yesterday and today 364
West Country, Bartholomew six-inch map series 140
West Highland Survey: an essay in human ecology 171
West Indies: *British bulletin of publications* . . . 342
maps 73, 74, 299
Directorate of Overseas Surveys 329
vegetation 430
West Indies: a catalogue of books, maps, etc 330
West Midlands, marketing map 158
' West Midlands Green Belt map ' 161
West of Scotland College, soil survey centre 154

Woods Hole Oceanographic Institution 111, 424

Woodward, M B, ed: *Stanford's geological atlas of Great Britain and Ireland* 150-51

Working Party, British National Committee for Geography: report on training of cartographers 58

Works, Ministry of: previous control of Ordnance Survey 124

World Aeronautical Chart 1 : 1M 67, 322

World atlas of population ' 450

World atlas of epidemic diseases 420

World atlas of sea surface temperature charts 426

World cartography 11, 44, 45, 66, 80, 182, 186, 259, 299, 326, 332, 335, 336, 355, 372, 468-69

World climate from 8,000 to 0 BC ... 420-21

World complex atlas for the officer 85-86

' World crop ' books series 434

World crops: the international journal of agriculture 432

World forestry atlas 434

World Forestry Congress 225

The world geographic atlas: a composite of man's environment 100

The world is round, Frank Debenham 104

World Land Use Survey, Commission 225, 317

World land use surveys 287, 295, 318, 362, 375, 436-37

Memoirs 437

Newsletters 437

Occasional papers 112, 437

Reports 436, 437

Regional monographs 437

' The world maps of climatology ' 420, 461

' World map, scale of 1 : 2,500,000 ', Sándor Radó 70

World Meteorological Office 417, 418, 419, 420

publications 417, 418, 419

World Population Map of the IGU 237, 317-18, 448

World population and production ... 433

World railways 457

' World route chart ' (Bartholomew) 74

The world shipping scene atlas 443

World vegetation, D Riley and A Young 131

World War I, effect on Ordnance Survey mapping 124

World Weather Watch 418

Worldmaster atlas (Rand McNally) 95

The world's coffee ... 434

Worthington, E B: *Science in Africa* ... 352

— *Science in the development of Africa* 352, 364

Woytinsky, W S and E S: *World population and production: trends and outlook* 433

Wright, G E and F Filson: *Westminster historical atlas of the Bible* 252

Wright, J K, ed: *Atlas of the historical geography of the United States* 310

Wrobel, A: *The Voyevodship of Warsaw* ... 223

Wüst, G and A Defant, comp: *Atlas zur Schichtung und Zirkulation des Atlantishen Ozeans* 111

Wye: National Agricultural Advisory regional service 154

soil survey centre 154

Wylie, A P: *The chromosome atlas of cultivated plants* 446

Yale University: Centre of African Studies 352

map collection 477

Forestry library 17

Yarnall, H E: text to *A new atlas of China* ... 282

Yates, E M, ed: *Geographical abstracts*, vol B 476

Yates, R A: *British weather in maps* 151, 164

Yellow fever, mapping 460

Yemen, ' Europa ' survey 244

Yenching Institute, Harvard: *A historical and commercial atlas of China* 282

Yeomans, P A: *The challenge of landscapes* ... 429

Ymer 185

Yorkshire: maps 102, 140, 161